*Advances in*

# BOTANICAL RESEARCH

## VOLUME 62

*Advances in*

# BOTANICAL RESEARCH

## Series Editors

# RECENT TRENDS IN MEDICINAL PLANTS RESEARCH

*Volume Editors*

**LIE-FEN SHYUR**
*Agricultural Biotechnology Research Center,*
*Academia Sinica,*
*Taipei, Taiwan*

**ALLAN S. Y. LAU**
*Department of Paediatrics and Adolescent Medicine,*
*Li Ka Shing Faculty of Medicine,*
*University of Hong Kong,*
*Hong Kong, China*

**VOLUME 62**

ELSEVIER

AMSTERDAM • BOSTON • HEIDELBERG • LONDON
NEW YORK • OXFORD • PARIS • SAN DIEGO
SAN FRANCISCO • SINGAPORE • SYDNEY • TOKYO
Academic Press is an imprint of Elsevier

Academic Press is an imprint of Elsevier
32 Jamestown Road, London NW17BY, UK
The Boulevard, Langford Lane, Kidlington, Oxford, OX51GB, UK
225 Wyman Street, Waltham, MA 02451, USA
525 B Street, Suite 1900, San Diego, CA 92101-4495, USA
Radarweg 29, PO Box 211, 1000 AE Amsterdam, The Netherlands

First edition 2012



ISBN: 978-0-12-394591-4
ISSN: 0065-2296

For information on all Academic Press publications
visit our Web site at store.elsevier.com

Printed and bound in USA
12 13 14 15 11 10 9 8 7 6 5 4 3 2 1

# CONTENTS

## Taxol, Camptothecin and Beyond for Cancer Therapy

WAI-LENG LEE, JENG-YUAN SHIAU AND LIE-FEN SHYUR



## Effects of Beta-Glucans on Different Immune Cell Populations and Cancers

DANIEL MAN-YUEN SZE AND GODFREY CHI-FUNG CHAN



## Developing Phytocompounds from Medicinal Plants as Immunomodulators

CHIH-CHUN WEN, HUI-MING CHEN AND NING-SUN YANG

## Chinese Herbal Medicines for Rheumatoid Arthritis

CHI ZHANG, MIAO JIANG, HONGTAO GUO, GUANG ZHENG, JING YANG AND AIPING LU



## *Ligusticum chuanxiong* and Its Decoctions: Effects and Underlying Mechanisms of Action on Stroke

CINDY L. H. YANG, TERRY C. T. OR, JONATHAN S. H. LAU AND ALLAN S. Y. LAU



## Potential of 'Omics' Technologies for Implementation in Research on Phytotherapeutical Toxicology

THOMAS EFFERTH AND HENRY JOHANNES GRETEN

## Herbal Medicine: Toxicity and Recent Trends in Assessing Their Potential Toxic Effects

CHIT SHING JACKSON WOO, JONATHAN SEE HAN LAU AND HANI EL-NEZAMI



## Development of Drugs from Plants: Regulation and Evaluation

MICHAEL RIEDER AND JOHN R. BEND

# CONTRIBUTORS TO VOLUME 62


BHARAT B. AGGARWAL *Cytokine Research Laboratory, Department of Experimental Therapeutics, The University of Texas MD Anderson Cancer Center, Houston, Texas, USA*

JOHN R. BEND *Department of Paediatrics, Division of Clinical Pharmacology; Department of Physiology & Pharmacology; Department of Pathology, Schulich School of Medicine & Dentistry, University of Western Ontario, London, ON, Canada*

PAUL PUI-HAY BUT *School of Life Sciences, The Chinese University of Hong Kong, Shatin, New Territories, Hong Kong, PR China*

GODFREY CHI-FUNG CHAN *Department of Paediatrics & Adolescent Medicine, Li Ka Shing Faculty of Medicine, The University of Hong Kong, Pokfulam, Hong Kong*

HSIAO-SUNG CHAN *Institute of Biochemical Sciences and Technology, Chaoyang University of Technology, Taichung, Taiwan*

HUNG-CHI CHANG *Institute of Biochemical Sciences and Technology, Chaoyang University of Technology, Taichung, Taiwan*

EMILY CHIN-FUN CHEN *Institute of Biochemical Sciences and Technology, Chaoyang University of Technology, Taichung, Taiwan*

HUI-MING CHEN *Agricultural Biotechnology Research Center, Academia Sinica; Department and Institute of Pharmacology, National Yang-Ming University, Taipei, Taiwan*

FU-SHIN CHUEH *Department of Health and Nutrition Biotechnology, Asia University, Taichung, Taiwan*

THOMAS EFFERTH *Department of Pharmaceutical Biology, Institute of Pharmacy and Biochemistry, Johannes Gutenberg University, Mainz, Germany*

HANI EL-NEZAMI *School of Biological Sciences, Faculty of Science, The University of Hong Kong, Hong Kong SAR, China*

HENRY JOHANNES GRETEN *Heidelberg School of Chinese Medicine, Heidelberg, Germany; Biomedical Sciences Institute Abel Salazar, University of Porto, Porto, Portugal*

HONGTAO GUO *Institute of Basic Research in Clinical Medicine, China Academy of Chinese Medical Sciences, Beijing; Shanghai University of Traditional Chinese Medicine, Shanghai, China*

SUBASH C. GUPTA *Cytokine Research Laboratory, Department of Experimental Therapeutics, The University of Texas MD Anderson Cancer Center, Houston, Texas, USA*

SUSHIM KUMAR GUPTA *Institute of Biochemical Sciences and Technology, Chaoyang University of Technology, Taichung, Taiwan*

MIAO JIANG *Institute of Basic Research in Clinical Medicine, China Academy of Chinese Medical Sciences, Beijing, China*

REN-WANG JIANG *School of Life Sciences, The Chinese University of Hong Kong, Shatin, New Territories, Hong Kong; Guangdong Province Key Laboratory of Pharmacodynamic Constituents of Chinese Medicine and New Drugs Research, Institute of Traditional Chinese Medicine and Natural Products, Jinan University, Guangzhou, PR China*

CHAO-LIN KUO *School of Chinese Pharmaceutical Sciences and Chinese Medicine Resources, China Medical University, Taichung, Taiwan*

JONATHAN S. H. LAU *School of Biological Sciences, The University of Hong Kong, Pokfulam, Hong Kong Special Administrative Region, China*

ALLAN S. Y. LAU *Molecular Chinese Medicine Laboratory, Li Ka Shing Faculty of Medicine; Cytokine Biology Group, Department of Paediatrics and Adolescent Medicine, The University of Hong Kong, Pokfulam, Hong Kong Special Administrative Region, China*

WAI-LENG LEE *Agricultural Biotechnology Research Center, Academia Sinica, Taipei, Taiwan*

GE LIN *School of Biomedical Sciences, The Chinese University of Hong Kong, Shatin, New Territories, Hong Kong, PR China*

AIPING LU *Institute of Basic Research in Clinical Medicine, China Academy of Chinese Medical Sciences, Beijing, China*

TERRY C. T. OR *Molecular Chinese Medicine Laboratory, Li Ka Shing Faculty of Medicine; Cytokine Biology Group, Department of Paediatrics and Adolescent Medicine, The University of Hong Kong, Pokfulam, Hong Kong Special Administrative Region, China*

SRIDEVI PATCHVA *Cytokine Research Laboratory, Department of Experimental Therapeutics, The University of Texas MD Anderson Cancer Center, Houston, Texas, USA*

SAHDEO PRASAD *Cytokine Research Laboratory, Department of Experimental Therapeutics, The University of Texas MD Anderson Cancer Center, Houston, Texas, USA*

MICHAEL RIEDER *Department of Paediatrics, Division of Clinical Pharmacology; Department of Physiology & Pharmacology; Department of Medicine, Schulich School of Medicine & Dentistry, University of Western Ontario, London, ON, Canada*

PANG-CHUI SHAW *School of Life Sciences; Institute of Chinese Medicine, The Chinese University of Hong Kong, Shatin, New Territories, Hong Kong, PR China*

JENG-YUAN SHIAU *Agricultural Biotechnology Research Center, Academia Sinica, Taipei, Taiwan*

LIE-FEN SHYUR *Agricultural Biotechnology Research Center, Academia Sinica, Taipei, Taiwan*

BOKYUNG SUNG *Cytokine Research Laboratory, Department of Experimental Therapeutics, The University of Texas MD Anderson Cancer Center, Houston, Texas, USA*

DANIEL MAN-YUEN SZE *Department of Health Technology and Informatics, The Hong Kong Polytechnic University, Kowloon, Hong Kong*

HSIN-SHENG TSAY *Institute of Biochemical Sciences and Technology, Chaoyang University of Technology; Department of Agronomy, National Chung-Hsing University, Taichung, Taiwan*

CHIH-CHUN WEN *Agricultural Biotechnology Research Center, Academia Sinica, Taipei, Taiwan*

CHIT SHING JACKSON WOO *School of Biological Sciences, Faculty of Science, The University of Hong Kong, Hong Kong SAR, China*

YAN-TONG XU *School of Life Sciences, The Chinese University of Hong Kong, Shatin, New Territories, Hong Kong; Tianjin State Key Laboratory of Modern Chinese Medicine and Institute of Traditional Chinese Medicine Research, Tianjin University of Traditional Chinese Medicine, Tianjin, PR China*

CINDY L. H. YANG *Molecular Chinese Medicine Laboratory, Li Ka Shing Faculty of Medicine, The University of Hong Kong, Pokfulam, Hong Kong Special Administrative Region, China*

JING YANG *Institute of Basic Research in Clinical Medicine, China Academy of Chinese Medical Sciences, Beijing, China*

NING-SUN YANG *Agricultural Biotechnology Research Center, Academia Sinica, Taipei, Taiwan*

CHI ZHANG *Institute of Basic Research in Clinical Medicine, China Academy of Chinese Medical Sciences, Beijing, China*

GUANG ZHENG *Institute of Basic Research in Clinical Medicine, China Academy of Chinese Medical Sciences, Beijing; Information Science and Engineering School, Lanzhou University, Lanzhou, China*

# PREFACE

Medicinal plants have been used for many millennia in mankind for preventing or curing diseases; however, only about 10% of higher plants on earth have been investigated for their bioactive compounds, but in most of these plants, a small part of the phytocompounds have been studied in detail. Drug discovery from medicinal plants in the nineteenth century led to the identification of several plant secondary metabolites as early drugs, including morphine, codeine, digitoxin, and quinine, which are still widely used today for different medications. In the twenty-first century, much of the world's population still relies on plants as a primary source of traditional or alternative medication. With the steadily growing use of herbal supplements, public concerns on safety in addition to efficacy are raised. Burgeoning interest in medicinal plants has increased scientific scrutiny of their therapeutic potential, toxicity, safety, and herb–drug interactions, thereby providing clinicians with important information to adequately treat or supervise their patients. In the past decade, a great deal of interest and effort from scientists worldwide has contributed to the identification of active ingredients from traditional or folk herbal medicines as therapeutic or chemopreventive agents and that has become an important area of nutritional and pharmaceutical sciences. Nevertheless, the bioactive chemicals and their relative concentration and distribution present in herbs may vary considerably depending on genetics, growing conditions, plant parts used, time of harvesting, postharvest treatment and preparation, etc. Collectively, these have spurred the researches on agronomy and phytochemistry of medicinal plants. An array of scientific studies has also been undertaken in the area of plant, cell, tissue, and organ cultures to produce valuable metabolites from traditional medicinal plants. Further, emerging 'omics' approaches have been adopted in drug lead discovery and have revealed novel mechanisms of action from phytomedicines and phytocompounds.

This thematic book series summarizes the current and most updated medicinal plant research regarding the molecular authentication and quality assessment of medicinal plants, *in vitro* propagation of medicinal plants for bioactive metabolites production, systems biology approaches in addressing the efficacy, and potential toxicity of phytomedicines using 'omics' technological platforms. Metabolomics approach can be used in medicinal plant research on metabolite fingerprinting, bioactive compounds identification, plant species classification, quality control, and toxicology. Comparative genomics and proteomics, on the other hand, provide important information of biological indexes or fingerprints which are indicative of responsiveness to bioactive phytocompound treatment in different cell lines or disease models. Although toxicogenomics of medicinal plants and herbal products is still in its infancy, in this book series, the potential of 'omics' technologies for implementation in research on phytotherapeutical toxicology has been discussed.

A major volume of reviews on the current research of elucidating bioefficacy and deciphering the mechanisms of action and molecular targets of specific bioactive phytocompounds or traditional herbal medicines for cancers, rheumatoid arthritis, cerebrovascular, stroke, and other inflammation-mediated chronic diseases, which are the main concerns of human health, are reported in this thematic book. More than half of the current anticancer drugs in clinical use are plant natural products or their derivatives. The successful experiences with the discovery and development of paclitaxel and camptothecin as anticancer drugs have been significant milestones for the design of alternative strategies and remedies from plant-derived agents in cancer management. The most updated researches on both representative plant-based drugs are summarized. On the other hand, application of traditional or new medicinal plants for use as immunomodulators in treating immune-related diseases is still relatively limited. New experimental approaches for evaluating the immunomodulatory activities of candidate phytomedicines are discussed. The drug development and regulatory process is facing a number of challenges, many of which are not unique to plant-based drugs but rather reflect the changing nature of biomedical science and drug regulation. Some reforms, which may make the process of developing drugs from plants both more swift and straightforward, are reported in this book series. The aim of this thematic book is to expose researchers, academics, and students to the multidisciplinary bioactive natural products and phytomedicines research arena.

Lie-Fen Shyur and Allan S.Y. Lau

# CONTENTS OF VOLUMES 35–61

*Series Editor (Volumes 35–44)*

*J.A. CALLOW*
*School of Biosciences, University of Birmingham,*
*Birmingham, United Kingdom*

## Contents of Volume 35

## Contents of Volume 36

## Contents of Volume 37

### ANTHOCYANINS IN LEAVES

*Edited by* **K. S. Gould and D. W. Lee**

## Contents of Volume 38



## Contents of Volume 39



## Contents of Volume 40

## Contents of Volume 41

## Contents of Volume 42



## Contents of Volume 43

## Contents of Volume 44

*Series Editors (Volume 45–60)*

*JEAN-CLAUDE KADER*
*Laboratoire Physiologie Cellulaire et Moléculaire des Plantes, CNRS, Université de Paris, Paris, France*
*MICHEL DELSENY*
*Laboratoire Génome et Développement des Plantes, CNRS IRD UP, Université de Perpignan, Perpignan, France*

## Contents of Volume 45

### RAPESEED BREEDING

## Contents of Volume 46

### INCORPORATING ADVANCES IN PLANT PATHOLOGY

## Contents of Volume 47

## Contents of Volume 48



## Contents of Volume 49

## Contents of Volume 50

## Contents of Volume 51

## Contents of Volume 52

## Contents of Volume 53

## Contents of Volume 54



## Contents of Volume 55

## Contents of Volume 56



## Contents of Volume 57

## Contents of Volume 58

## Contents of Volume 59



## Contents of Volume 60

*Series Editors (Volume 61–)*

*JEAN-PIERRE JACQUOT*
*Professeur, Membre de L'Institut Universitaire de France, Unité Mixte de Recherche INRA, UHP 1136 "Interaction Arbres Microorganismes", Université de Lorraine, Faculté des Sciences, Vandoeuvre, France*
*PIERRE GADAL*
*Professor honoraire, Université Paris-Sud XI, Institut Biologie des Plantes, Orsay, France*

## Contents of Volume 61

# Authentication and Quality Assessment of the Antitussive Herb Baibu (Radix Stemonae)

PAUL PUI-HAY BUT,[*,1] PANG-CHUI SHAW,[*,†] GE LIN,[‡] REN-WANG JIANG[*,§] AND YAN-TONG XU[*,¶]

**School of Life Sciences, The Chinese University of Hong Kong, Shatin, New Territories, Hong Kong, PR China*
[†]*Institute of Chinese Medicine, The Chinese University of Hong Kong, Shatin, New Territories, Hong Kong, PR China*
[‡]*School of Biomedical Sciences, The Chinese University of Hong Kong, Shatin, New Territories, Hong Kong, PR China*
[§]*Guangdong Province Key Laboratory of Pharmacodynamic Constituents of Chinese Medicine and New Drugs Research, Institute of Traditional Chinese Medicine and Natural Products, Jinan University, Guangzhou, PR China*
[¶]*Tianjin State Key Laboratory of Modern Chinese Medicine and Institute of Traditional Chinese Medicine Research, Tianjin University of Traditional Chinese Medicine, Tianjin, PR China*



[1]Corresponding author: E-mail: paulbut@hotmail.com

Advances in Botanical Research, Vol. 62

0065-2296/12 $35.00
DOI: 10.1016/B978-0-12-394591-4.00001-5

# ABSTRACT

The Chinese herb Radix Stemonae, or Baibu in Chinese, has been used for treating coughing for over 15 centuries. According to the Pharmacopoeia of the People's Republic of China, it comes from the tuberous roots of three species of *Stemona*. The properties of multiple origins of source plants and existence of multiple antitussive components made the authentication and quality assessment extremely difficult. This review provides an up-to-date coverage of scientific advances in species identification and authentication among the three genuine species and substitutes derived from the genus *Asparagus*, characterization of the *Stemona* alkaloids including chemical diversities among the three genuine species and among the major commodity derived from *Stemona tuberosa* from different locations, identification of the antitussive components and mechanisms of action, clarification of adverse reactions, and, based on all these information, development of quality assessment methods.

# I. INTRODUCTION

The empirical knowledge accumulated in many cultures in incorporating natural products for relief from diseases and maintenance of health is a cultural heritage. Tapping into such knowledge has provided leads for the development of many pharmaceuticals including the antimalarial artemisinin and antileukemic arsenic trioxide as well as nutraceutical products such as ginkgo leaf and ginseng preparations. In China, traditional Chinese medicines have continued to flourish and play an important role in primary healthcare. The demand for them is increasing in many developed countries in the West, where orthodox medicine has a dominating advantage. For further advance in the globalization and modernization of Chinese medicines, one of the necessary improvements is the satisfactory characterization and standardization of Chinese herbs. Two of the more obvious issues are (a) species specification to allow proper identification and authentication and (b) characterization of both bioactive and toxic components. Radix Stemonae is chosen as an example to illustrate the relevance and complications in these two issues.

Radix Stemonae is called Baibu in Chinese. It was first recorded as an antitussive and noted with mild toxicity in the herbal Mingyi Bielu (Supplementary Records of Famous Physicians). This herbal was compiled around

500 A.D. by Tao Hongjing. It started and established for subsequent Chinese herbals the practice of adding new knowledge of the compilers next to verbatim descriptions and comments of previous herbals (But, 1995). This practice has made it possible for tracing the identity and uses of an herb down the ages. Radix Stemonae is still frequently used today in combination with other herbs for the treatment of cough (Tie, 2010). A recent randomized double-blind clinical study found it helpful to cough patients (Wong *et al.*, 2006). There are, however, several problems in the use of this herb, including (a) multiple origins of source plants, (b) substitutes found in the market, (c) multiple antitussive components, (d) chemical races in the same species and (e) adverse effects. In order to make best use of this herb, proper authentication, identification of the active and toxic components and quality control is necessary.

# II. SPECIES SPECIFICATION AND AUTHENTICATION

## A. SOURCE PLANTS OF RADIX STEMONAE

According to the Pharmacopoeia of the People's Republic of China (2010 edition, hereafter as China Pharmacopoeia), Radix Stemonae comes from the tuberous roots of three species of *Stemona*, namely, *S. japonica* (Blume) Miquel, *S. sessilifolia* (Miquel) Miquel, and *S. tuberosa* Loureiro. These three species are here collectively called pharmacopoeial species for convenience. *S. japonica* and *S. sessilifolia* grow on hillsides in eastern China, whereas *S. tuberosa* has wider distribution and is found in similar habitats in eastern China as well as in south and southwest China (Ji and Duyfjes, 2000). The Flora of China (Ji and Duyfjes, 2000) has recognized four more species of *Stemona* in China, namely, *S. kerrii* Craib, *S. mairei* (H. Léveillé) K. Krause, *S. parviflora* C. H. Wright, and *S. shandongensis* D. K. Zang. The last species has been confirmed as conspecific with *S. sessilifolia* (Li and Fu, 2007).

All the six *Stemona* species in China are herbs or herbaceous vines with leaves bearing parallel veins, typical of monocotyledon plants. Their flowers have four perianth segments, four stamens and a superior ovary. The stamens are erect, inserted at the base of the perianth segments; the anthers are basifixed, each often bearing an appendage on the adaxial surface and an extended connective beyond the linear anther lobes. The superior ovary has one chamber with basally attached ovules (Rudall *et al.*, 2005). The fruit is a capsule; the seeds are oblong, longitudinally ribbed and bearing beard-like elaiosomes (Fig. 1). Their roots are present in form of a bunch of fusiform tuberous roots. The remaining three species of *Stemona* in China are rather small and rare. In market surveys conducted between 1964 and 1989,

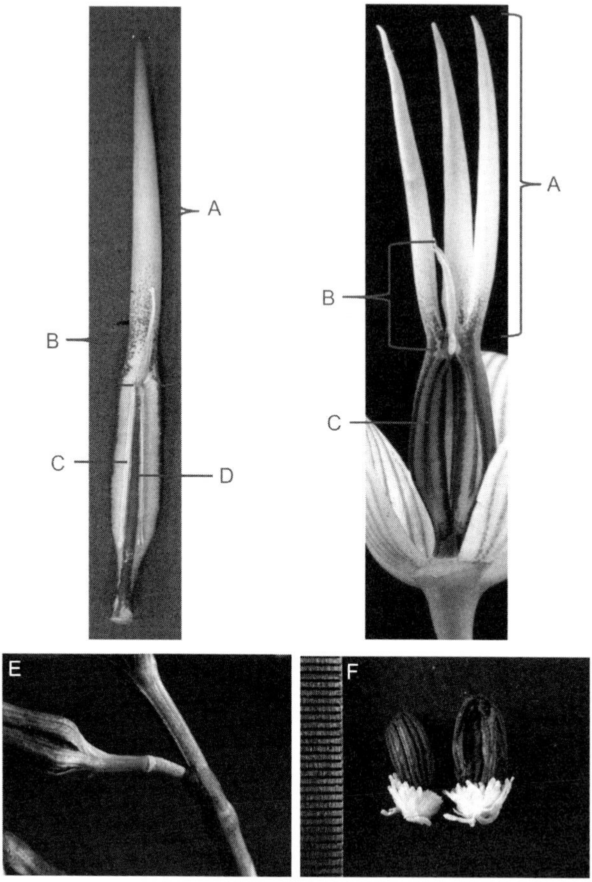


Fig. 1. Morphological characters shared by *Stemona*. (A) connective extension, (B) sterile apical appendage of anther, (C) anther, (D) filament, (E) articulate pedicel in *S. japonica*, and (F) seeds with white elaiosome. (See colour plate section at the back of the book.)

*S. mairei* and *S. parviflora* were spotted in samples of Radix Stemonae from only three places (Cong and Xu, 1997). Recent surveys did not find them in samples collected (Fan *et al.* 2009b).

When in flowers, the three species recognized in China Pharmacopoeia are very distinct from one another.

*1.* Stemona japonica

*S. japonica* is an herbaceous vine with erect stems. The leaves are whorled, with petioles of 1–4 cm long. The leaf blades are ovate, ovate–lanceolate, or ovate–oblong, 4–9 cm long and 1.5–4.5 cm wide; the base is subtruncate to rounded, rarely rounded-cordate or cuneate, the margins are slightly undulate and the apex is acuminate. The inflorescences have one to several flowers borne on the mid-vein of leaf blades; the pedicel or peduncle is slender and 0.5–4 cm long. The perianth segments are green, lanceolate, 10–15 mm long and 2–3 mm wide. The stamens and protruded connective are purple red and slightly shorter than the perianth; the filaments are stout and about 1 mm long. The anthers are 2–2.5 mm long; the appendages are adherent, with a silky hair on either side. The capsules are oblong, 10–14 mm long, and bearing 2–3 seeds.

*2.* Stemona sessilifolia

*S. sessilifolia* is somewhat shrubby with erect stems of 30–70 cm tall. Its leaves grow in whorls of 2–5 at each node, with short or no petioles. The leaf blades are obovate– or ovate–elliptic to ovate–lanceolate, 3.5–6 cm long, 1.5–4 cm wide, with cuneate base and shortly acute apex. Individual flowers are borne in leaf axils at lower parts of the stem with pedicels that are 1–1.5 cm long and articulate at around mid-length (Fig. 1E). It is subtended by a scale like bract of about 8 mm long. Each flower has four pale green, ovate–lanceolate perianth segments of 10–15 mm long and 2–4 mm wide. The four stamens and the prolonged connective are purple red and slightly shorter than the perianth, with stout filaments of 2–4 mm long and two anthers of about 3 mm long on the adaxial surface of the stamen. An appendage of 5–7 mm long grows on the connective surface between the anther lobes. The capsules are ovoid, 7–9 mm long, with one or two seeds.

*3.* Stemona tuberosa

*S. tuberosa* is an herbaceous vine. Its stems are often branched and can cover large areas. Its leaves are opposite or whorled, rarely alternate, with petioles of 3–10 cm long. The leaf blades are ovate to ovate–lanceolate, 6–24 cm long and 5–17 cm wide, membranous, with 7–13 veins; the base is cordate, the margins slightly undulate and the apex acuminate. Inflorescences are racemes, with 1–3 flowers; the peduncle or pedicel are axillary or rarely borne on petioles, 2–5 or rarely up to 12 cm long. The perianth segments are greenish with purplish veins, 3–8 cm long, and 0.7–1 cm wide, with an acuminate apex. The stamens and protruded connective purple red, slightly shorter than the perianth; the filaments are stout, 2–5 mm long. The anthers are linear, about 10 mm long; the appendages are partially adherent of about 5 mm long. The capsules are ovoid–oblong, 2.5–6 cm long and 1–3 cm wide, bearing several seeds.

## B. PHARMACOGNOSTIC DIFFERENTIATION OF RADIX STEMONAE

The tuberous roots of *Stemona* are the parts for therapeutic uses. Those of *S. tuberosa*, up to 2 cm in diameter and 30 cm in length, are evidently bigger than those of *S. japonica* and *S. sessilifolia* (about 0.5–1.2 cm in diameter and 8–14 cm in length). Moreover, plants of *S. tuberosa* generally grow bigger and each bears larger numbers of tuberous roots. These advantages are reflected in the fact that majority of the market samples of Radix Stemonae (Fan *et al.*, 2009b) comes from this species.

Transverse sections of the tuberous root of these three *Stemona* species showed similar anatomical arrangements. The velamen located at the outermost periphery contains several layers of cells with suberized cell walls. Below the velamen is the exodermis composed of a layer of thin-walled cells. The cortex is occupied by parenchyma cells. The stele located in the central part of the tuberous roots is represented by the endodermis composed of rectangular cells with distinct Casparian strips, vascular bundles with phloem alternating with xylem, and a pith of parenchyma cells (Kongkiatpaiboon *et al.*, 2010; Wen, 2002; Xu *et al.*, 2006; Zhao, 2005). However, the tuberous roots of *S. tuberosa* differ from those of the other two species by the presence of scattered fibres in the cortex and pith and also by the absence of thickened stripes or striations on the velamen cell walls (Kongkiatpaiboon *et al.*, 2010; Li *et al.*, 2011; Wang, 2011; Wen, 2002; Xu *et al.*, 2006).

Attempts have been made using molecular sequences to facilitate species authentication, taking advantage of the polymorphism in genomic and chloroplast DNA between species. Chan (2004) applied 5S rRNA spacer sequences for the differentiation of five *Stemona* and an *Asparagus* species. These include *S. japonica*, *S. parviflora*, *S. sessilifolia*, *S. shandongensis*, *S. tuberosa*, and *Asparagus filicinus* Ham. ex D. Don collected from China. The size of PCR product from *A. filicinus* was 600 bp, while that of *Stemona* was only 500 bp. From the sequence alignment, it showed that each *Stemona* species had unique insertion and deletion sections or unique sequences, and also confirmed that *S. shandongensis* is conspecific with *S. sessilifolia* (Li and Fu, 2007).

Chan (2004) and Jiang *et al.* (2006a) tried the *trn*L intron but found this region unhelpful in differentiating the three pharmacopoeial species. Nevertheless, the *trn*L intron sequences shed light on the phylogenetic relationship among *Stemona*, *Croomia*, *Pentastemona* and *Stichoneuron*, and favoured grouping them together in Stemonaceae (Fig. 2), although there are suggestions to segregate them into separate families (Stemonaceae, Croomiaceae, and Pentastemonaceae) (Duyfjes, 1991, 1992; Inthachub *et al.*, 2009; Thorne and Reveal, 2007). It also suggested closer relationship with Pandanaceae

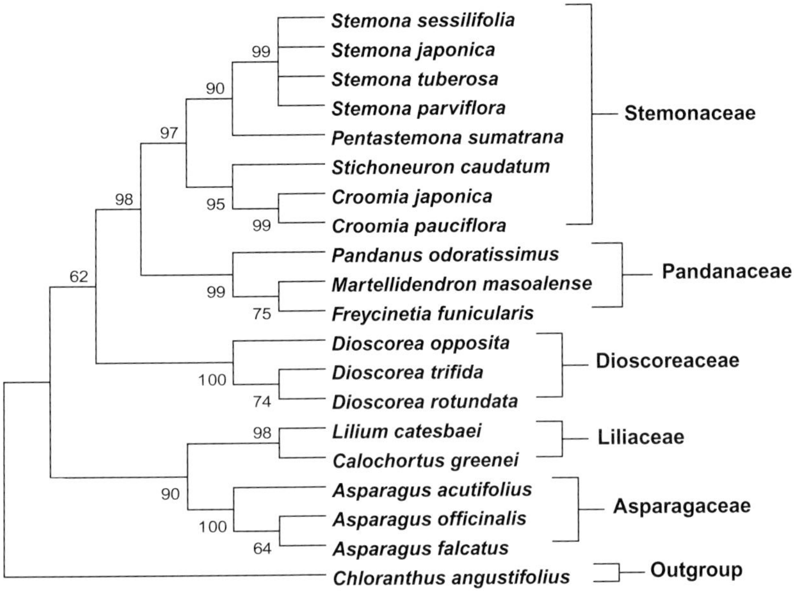


Fig. 2. Consensus parsimony tree based on *trn*L intron of plants from Stemonaceae, Pandanaceae, Dioscoreaceae, Liliaceae, and Asparagaceae (20 sequences). The tree is rooted through an outgroup *Chloranthus angustifolius*. Bootstrap values based on 1000 replications are shown on branches.

than with Dioscoreaceae (Caddick *et al.*, 2002; Greger *et al.*, 2009). The *trn*L-F region, however, was found useful in differentiating the species in a related genus *Croomia* (Li *et al.*, 2008). Vongsak *et al.* (2008) looked into chloroplast *trn*H-*psb*A region and found partial success in differentiating *S. tuberosa* from four other Thai *Stemona* species.

Fan *et al.* (2009a), after preliminary trials on *trn*L intron sequences, concluded that this region was unable to differentiate *Stemona* species. They expanded their investigations into four chloroplast regions (*trn*L-*trn*F, *trn*H-*psb*A, *pet*B-*pet*D, and *trn*K-*rps*16) in four *Stemona* species (*S. japonica*, *S. sessilifolia*, *S. parviflora,* and *S. tuberosa*). Relatively higher substitution rates were found in the sequences of *pet*B-*pet*D and *trn*K-*rps*16 regions than those of *trn*L-*trn*F and *trn*H-*sb*A regions. These sequences provided useful information to discriminate the four species from one another. Fan *et al.* (2009b) went further to analyse the restriction maps of the *trn*L-*trn*F and *pet*B-*pet*D sequences. Their results showed that restriction enzymes *Mwo*I, *Aci*I, and *Xmn*I were able to recognize specific sequence sites in the *trn*L-*trn*F region, while *Bcl*I, *Hin*cII, and *Bsl*I could recognize specific sequence sites in the *pet*B-*pet*D region. This combination of polymerase chain

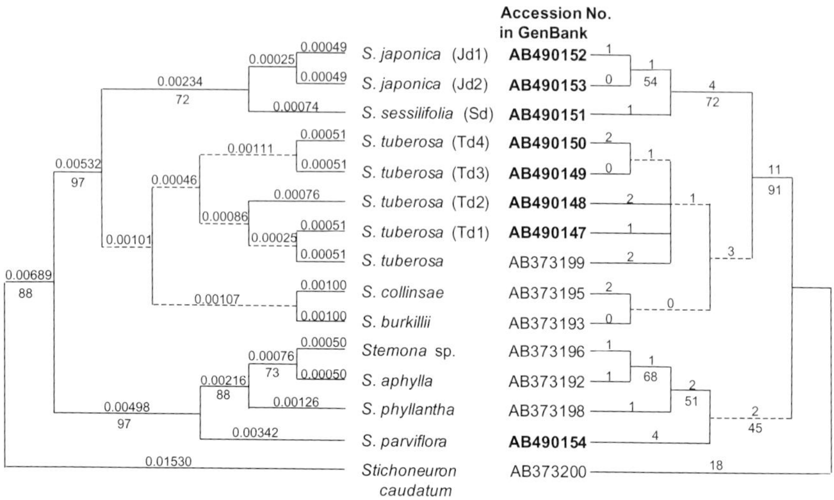


Fig. 3. Phylogenetic tree based on sequences of cpDNA *trn*H-*psb*A region (courtesy of Prof Katsuko Komatsu and Prof Shao-Qing Cai). Left: The phylogenetic tree reconstructed by UPGMA method with distance calculated by Kimura's two parameters. Right: The 50% major-rule consensus tree obtained by maximum parsimonious analysis. Tree length_62, CI_0.7419, RI_0.8095, RC_0.6006. Number above line is branch length and number below line is the bootstrap value with 1000 replicates. Bootstrap values were obtained from 1000 replications. The broken line indicates the clade with bootstrap value less than 50%. The accession number in bold face is investigated and registered in GenBank by the present study and the remaining is obtained from GenBank.

reaction–restriction fragment length polymorphism (PCR-RFLP) methods showed excellent results in discriminating the four *Stemona* species from one another (Figs. 3–5).

## C. SUBSTITUTES FROM *ASPARAGUS* SPECIES

Pharmacognostic authentication of Radix Stemonae does not stop with the three *Stemona* species. In southwest China, the fusiform tuberous roots of *A. filicinus* and related species (Asparagaceae) are harvested and sold as Xiao-Baibu or Tu-Baibu (Jiangsu New Medical College, 1977). It is used as a substitute of Radix Stemonae for treating cough (Ding and Yang, 1990; Li *et al.*, 2005). *Asparagus* plants are readily distinguishable from *Stemona* species by the fascicles of needle-shaped 'leaves', which are actually modified stems (cladodes), and the trimerous flowers composed of six perianth parts, six stamens and a superior ovary with three chambers. However, the dried and sliced tuberous roots of *Stemona* and *Asparagus* look very similar.

An incident of misidentifying the tuberous roots of *Stemona collinsae* as from *Asparagus racemosus* led to publishing new compounds from the former species but naming them after the latter (Greger, 2006; Kostecki *et al.*, 2004; Sekine *et al.*, 1995, 1997). In the pursuit for standardization of Radix Stemonae, it is, therefore, necessary to cover not only the *Stemona* species but also the *Asparagus* species.

*1. Pharmacognostic differences*

Herb samples from *Asparagus* species can be distinguished from those of *Stemona* species by anatomical and chemical characters.

Anatomical studies showed that *Asparagus* roots differ from *Stemona* roots in having bundles of raphide needles in parenchyma cells in the root cortex (Cong and Xu, 1997; Kauff *et al.*, 2000).

Chemically, differentiation between *Asparagus* and *Stemona* species is generally simple, as no alkaloid is found in *Asparagus* species (Cong and Xu, 1997; Cong *et al.*, 1992). There are, however, confusing reports in the literature about alkaloids in *Asparagus*, which interpretation might possibly have involved misidentification of the experimental material or false positive chemical reactions (Potduang *et al.*, 2008; Sekine *et al.*, 1995, 1997).

Various DNA regions were reported useful in distinguishing *Asparagus* from *Stemona*. The variation in the 5S rRNA sequences was high enough to differentiate *Stemona* species from one another and from *A. filicinus* (Chan 2004). In studying the applications of DNA sequences from four chloroplast DNA regions and subsequent RFLP analyses, Fan *et al.* (2009a,b) included an *Asparagus* species (*Asparagus cochinchinensis* (Lour.) Merrill). Their results also showed that this species stayed in a very distinct clade from those of the four *Stemona* species and could be easily separated from them by PCR-RFLP analyses. Obviously, the DNA polymorphism between *Asparagus* and *Stemona* provide excellent markers for distinguishing them apart from each other (Figs. 4 and 5). Ling (2008), on the other hand, revisited the *trn*L-*trn*F region and covered a wider collection of both *Stemona* and *Asparagus* species in China. With proper alignment, this region was found to be useful for segregating the many species included in his study and may possibly offer phylogenetic interpretations (Fig. 6).

## III. *STEMONA* ALKALOIDS

*Stemona* alkaloids are very unique; they possess an exposed or a hidden pyrrolo- or pyrido[1,2-a]azepine nucleus as the smallest common structural core (Kaltenegger *et al.*, 2003; Pilli *et al.*, 2010). So far, such alkaloids are not

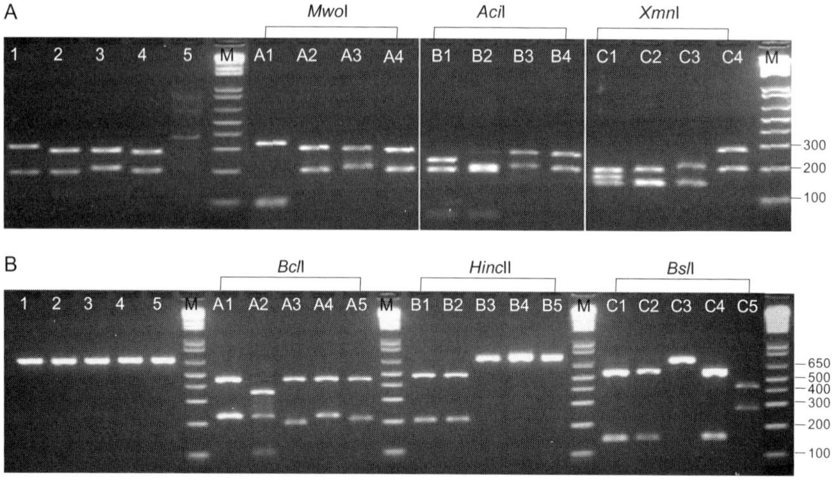


Fig. 4. Electrophoretograms of PCR-RFLP analysis (courtesy of Prof Katsuko Komatsu and Prof Shao-Qing Cai). (A) *trn*L-*trn*F region. Lanes 1–5: intact PCR products of *S. japonica*, *S. sessilifolia*, *S. tuberosa*, *S. parviflora* and *Asparagus cochinchinensis*, respectively; Lanes A1, B1, and C1 are PCR-RFLP analysis results of *S. japonica* by using restriction enzymes *Mwo*I, *Aci*I, and *Xmn*I, respectively; Lanes A2, B2, and C2 are those of *S. sessilifolia*; Lanes A3, B3, and C3 are those of *S. tuberosa*; Lanes A4, B4, and C4 are those of *S. parviflora*. (B) *pet*B-*pet*D region. Lanes 1–5: intact PCR products of *S. japonica*, *S. sessilifolia*, *S. tuberosa*, *S. parviflora*, and *A. cochinchinensis*, respectively; Lanes A1, B1, and C1 are PCR-RFLP analysis results of *S. japonica* by using restriction enzymes *Bcl*I, *Hinc*II, and *Bsl*I, respectively; Lanes A2, B2, and C2 are those of *S. sessilifolia*; Lanes A3, B3, and C3 are those of *S. tuberosa*; Lanes A4, B4, and C4 are those of *S. parviflora*; Lanes A5, B5, and C5 are those of *A. cochinchinensis*. M: 1 kb plus DNA ladder.

found elsewhere in the plant kingdom but the four genera in Stemonaceae. The same structural nucleus has been found in the skin of the Colombian poison frog *Dendrobates lehmanni* (Garraffo *et al.*, 2001); but the auxiliary ring systems and substitution patterns are significantly different. Greger (2006), based on biosynthetic considerations with reference to the differences in the carbon chains attached to C-9 of the pyrrolo[1,2-a]azepine nucleus, suggested classifying *Stemona* alkaloids into three skeletal types: the sticho-neurine type (tuberostemonine type), protostemonine type, and croomine type. Pilli *et al.* (2010), on the other hand, organized them into eight groups: stenine, stemoamide, tuberostemospironine, stemonamine, parvistemoline, stemofoline (all of which contain the pyrrolo[1,2-a]azepine core characteristic of the majority of the *Stemona* alkaloids), stemocurtisine displaying the pyrido[1,2-a]azepine nucleus, and a miscellaneous group without the structural motifs of the other seven groups.

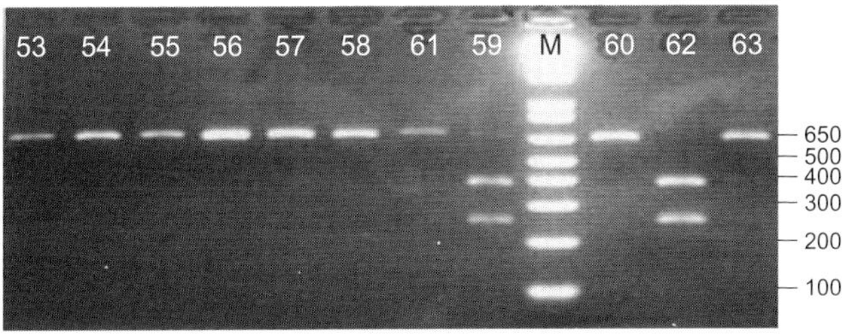


Fig. 5. PCR-RFLP analysis of crude drugs by using restriction enzyme *Bsl*I to digest partial fragment of chloroplast *pet*B-*pet*D region (courtesy of Prof Katsuko Komatsu and Prof Shao-Qing Cai). Lane number represents batch number of crude drug. M: 1 kb plus DNA ladder.

The complexity and biological properties of this type of alkaloids have attracted much interest from researchers. The inventory of *Stemona* alkaloids has kept growing from only 7 by 1975 (Götz and Strunz, 1975) to 42 by 1998 (Pilli and de Oliveira, 2000), 68 by 2004 (Pilli *et al.*, 2005), about 100 by 2007 (Schinnerl *et al.*, 2007), and 139 by 2009 (Pilli *et al.*, 2010). Synthetic ones have further enriched the library of alkaloids with the pyrrolo[1,2-a]azepine nucleus (Frankowski *et al.*, 2008, 2011).

## A. CHEMICAL DIVERSITY IN *S. TUBEROSA*

Significant chemical variation was observed in *S. tuberosa* regarding the alkaloidal composition and corresponding contents (Jiang *et al.*, 2006b). The HPLC chromatograms of the total alkaloids of this herb from 13 places in China were compared (Fig. 7A–D). Since most of the *Stemona* alkaloids are not sensitive under direct UV detection due to the absence of conjugated functional groups, evaporative light scattering detector (ELSD) depending on the size, shape, and number of eluate particles was hyphenated with HPLC. The HPLC profiles generated from these samples revealed the presence of four chemical types A–D. Type A had croomine (**1**) of the tuberostemospironine group as the major component (Fig. 7A, $Rt = 9.0$ min). Type B presented stemoninine (**3**) of the stemoamide group as the major component (Fig. 7B, $Rt = 15.7$ min). Type C carried neotuberostemonine (**4**) of the stenine group as the major component (Fig. 7C, $Rt = 19.0$ min). Type D appeared more common, with tuberostemonine (**5**) also of the stenine group as the major component (Fig. 7D, $Rt = 19.7$ min). The structures of these alkaloids are shown in Scheme 1.

The chemical diversity among different samples of *S. tuberosa* is an interesting phenomenon. Phytochemists have been amazed by this observation

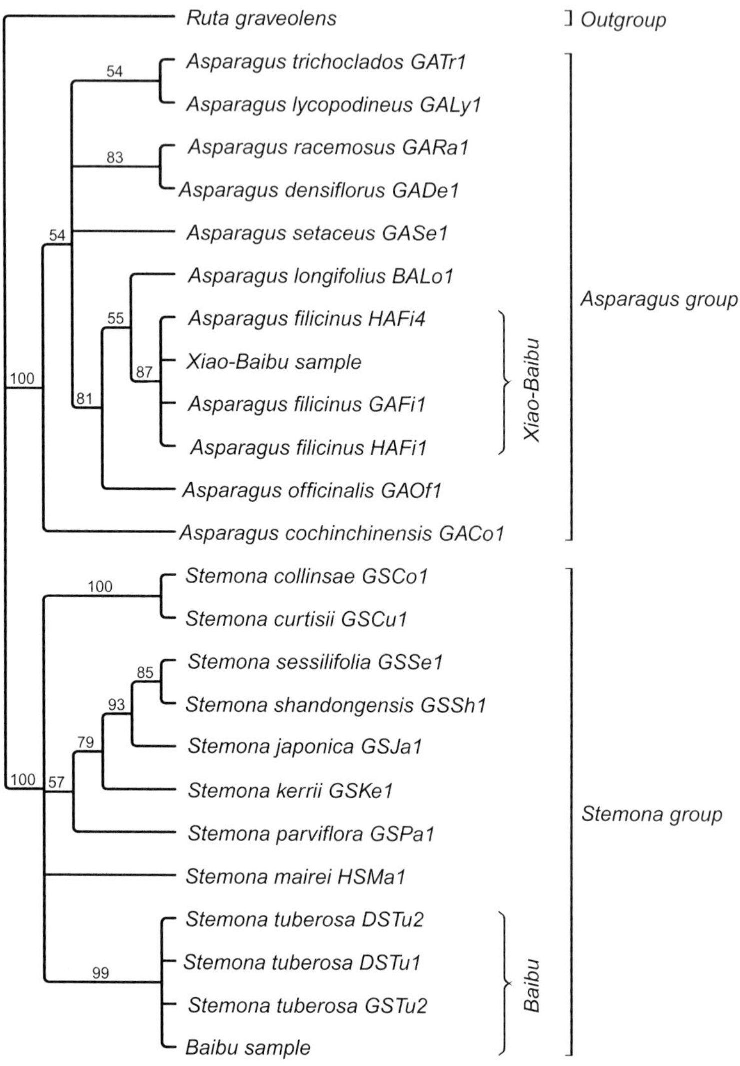


Fig. 6. The *trnL-trnF* region (both spacer and intron) neighbour-joining tree with bootstrap values more than 50% indicated at branches.

(H. Greger, personal communication). Through phytochemical studies and comparative HPLC and LC–MS tandem analyses, the chemical profiles of *S. tuberosa* and the range of their variations were further characterized (Jiang *et al.*, 2002, 2006a,b, 2010; Zhou *et al.*, 2006). Apparently, one possibility of this variation is the results of genetic parameters. Not a few plant species, such as *Pogostemon cablin* (Luo *et al.*, 2003) and *Callitris columellaris*

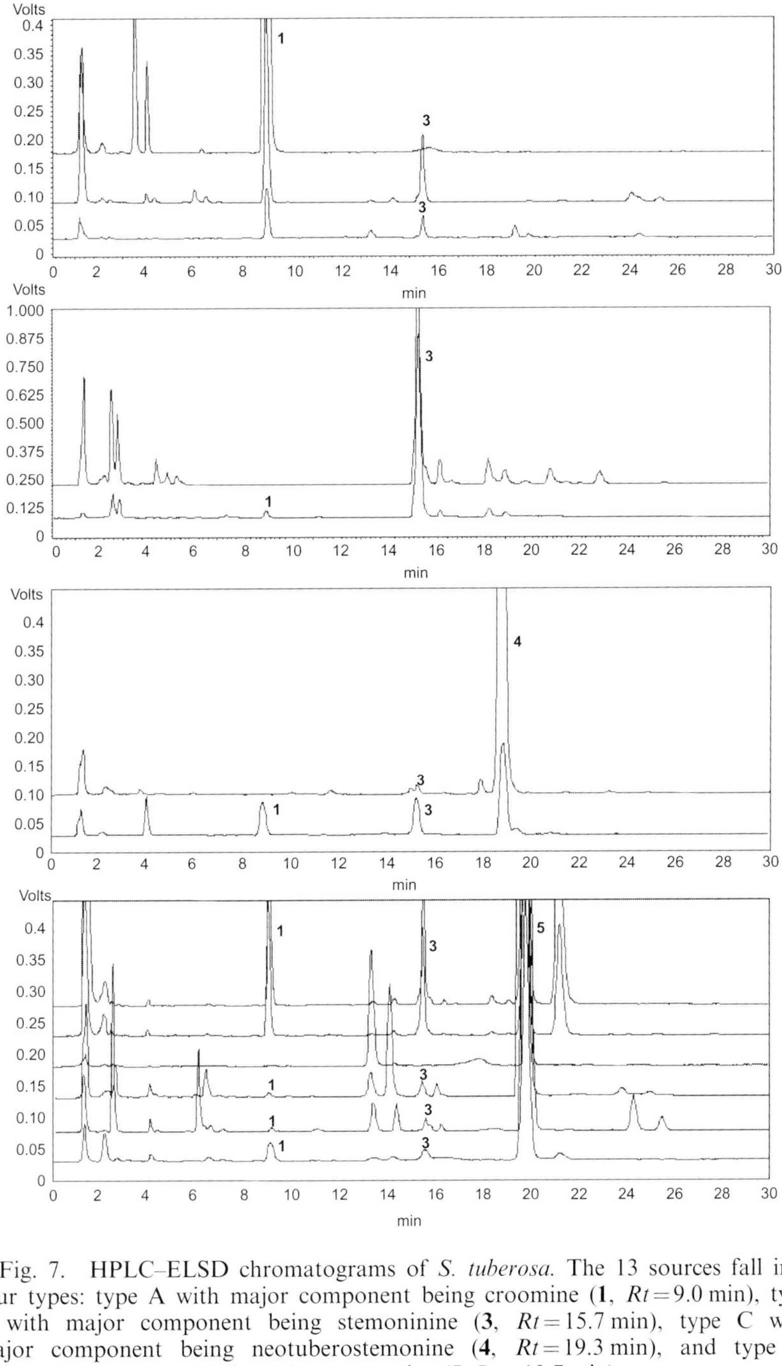


Fig. 7. HPLC–ELSD chromatograms of *S. tuberosa*. The 13 sources fall into four types: type A with major component being croomine (**1**, $Rt = 9.0$ min), type B with major component being stemoninine (**3**, $Rt = 15.7$ min), type C with major component being neotuberostemonine (**4**, $Rt = 19.3$ min), and type D with major component being tuberostemonine (**5**, $Rt = 19.7$ min).

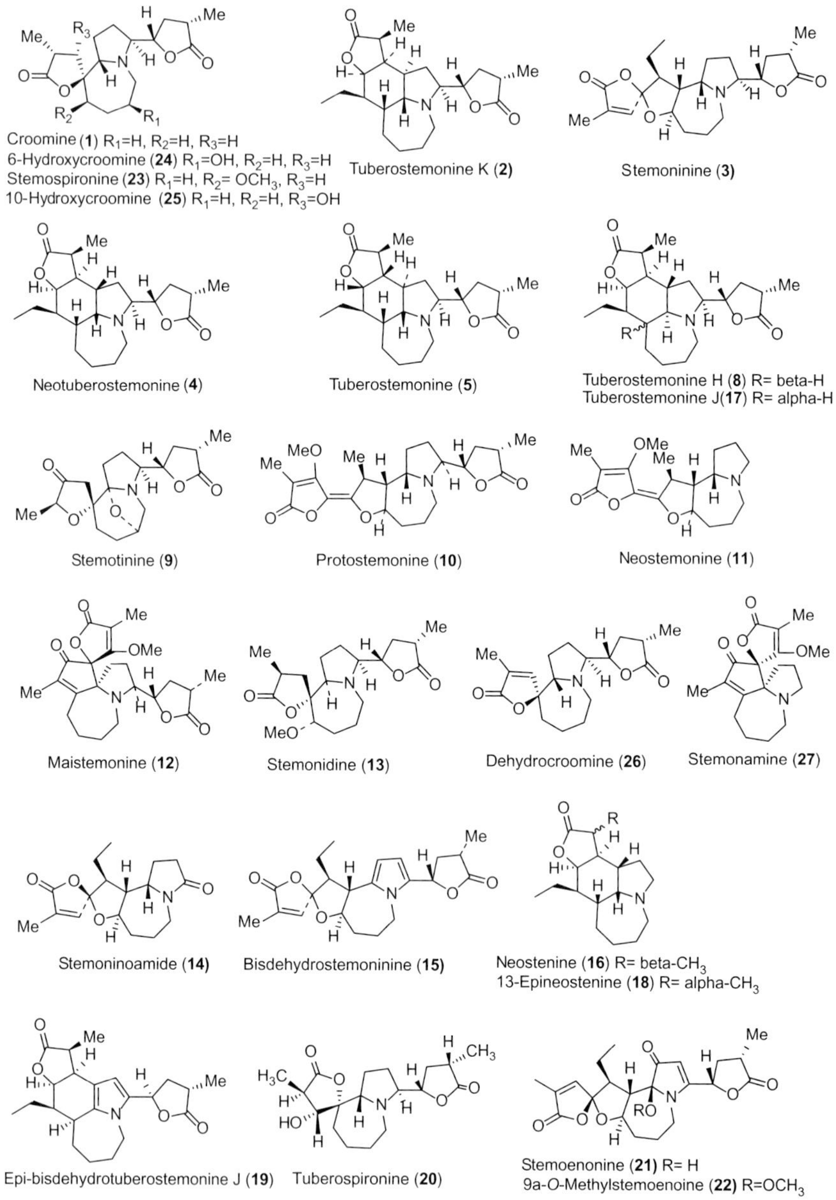


Scheme 1. Structural formulae of *Stemona* alkaloids identified in the extracts of *S. tuberosa*, *S. japonica* and *S. sessilifolia*.

(Ogunwande *et al.*, 2005), are known to have chemical races. Another possibility is environmental factors, such as differences in age, seasons of collection and geographic distributions. Naturally, the two possibilities are not mutually exclusive, and the interplay between genetic and environmental factors may have resulted in such chemical variations. This kind of chemical diversity may also reflect the active role of plants against pests, as illustrated by the increase in the contents of stilbenoids in *S. collinsae* in response to fungal attack (Pacher *et al.*, 2002).

### B. CHEMICAL DIVERSITY AMONG *STEMONA* SPECIES

Most *Stemona* alkaloids are not sensitive to direct UV detection due to the absence of any conjugated functional groups, but they are easily detectable by ESI–MS. Since the tuberous roots of three species of *Stemona* are recognized as genuine sources of the herb Radix Stemonae by China Pharmacopoeia, it is necessary to identify herbs to specific species. Online HPLC–ESI/MS and detailed tandem mass spectrometric analyses were used to study the *Stemona* alkaloids from the three pharmacopoeial species. Six alkaloids [croomine (**1**), tuberostemonine K (**2**), stemoninine (**3**), neotuberostemonine (**4**), tuberostemonine (**5**), and tuberostemonine H (**8**)] from *S. tuberosa* were unequivocally identified by comparing mass spectrometry and retention times with those of authentic samples. The established chromatographic profiles of the six selected samples indicated that different collections of *S. tuberosa* exhibited significant chemical variations in these alkaloids (Fig. 8).

Using the same LC–MS method, croomine (**1**), stemotinine (**9**), protostemonine (**10**), neostemonine (**11**), maistemonine (**12**), and stemonidine (**13**) were identified from the three samples of *S. japonica*, but the latter four compounds were not found in *S. tuberosa*. Similarly, croomine (**1**), protostemonine (**10**), and maistemonine (**12**) were also identified from *S. sessilifolia*; however, stemonidine (**13**) was found only in *S. sessilifolia* and thus could be selected as a marker to differentiate it from *S. japonica* and *S. tuberosa* (Fig. 9).

Based on phytochemical analyses of properly identified experimental materials, it is possible to generate chemotaxonomic interpretations on the distribution of representative types of *Stemona* alkaloids in the three pharmacopoeial species. The formation of tuberostemonine and its closely related stichoneurine-type (stenine type) derivatives with a saturated tricyclic pyrrolobenzazepine nucleus is restricted to the *S. tuberosa* group, and so is stemoninine representing another stichoneurine-type alkaloid with a spiro system (stemoamide type). On the other hand, protostemonine is only observed in *S. japonica* and *S. sessilifolia* (Kongkiatpaiboon *et al.*, 2011).

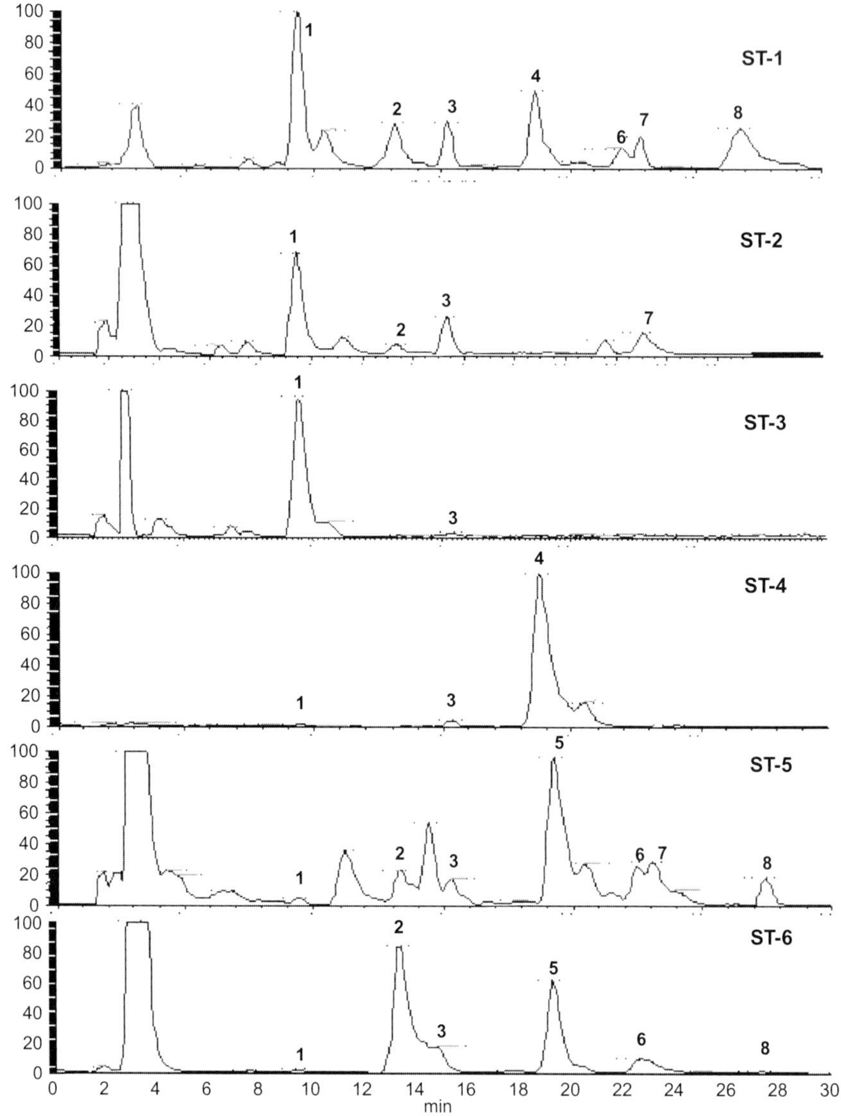


Fig. 8. Base peak traces of LC–MS analyses of the total alkaloids from six samples of *S. tuberosa* (ST). The TIC chromatograms refer to positive ion in ESI mode, and compound identities were assigned on the basis of $MS^n$ ($n=2$–4) analyses and their structures are shown in Scheme 1. Peaks 6 and 7 were unknown.

It is noteworthy that no bisdehydro derivatives with a stable aromatic pyrrole system could be found with HPLC/ELSD or with TLC/Dragendorff detection. Very possibly, bisdehydro alkaloids are artefacts only produced during storage, extraction, and/or fractionation (Greger, 2006; Schinnerl *et al.*, 2007).

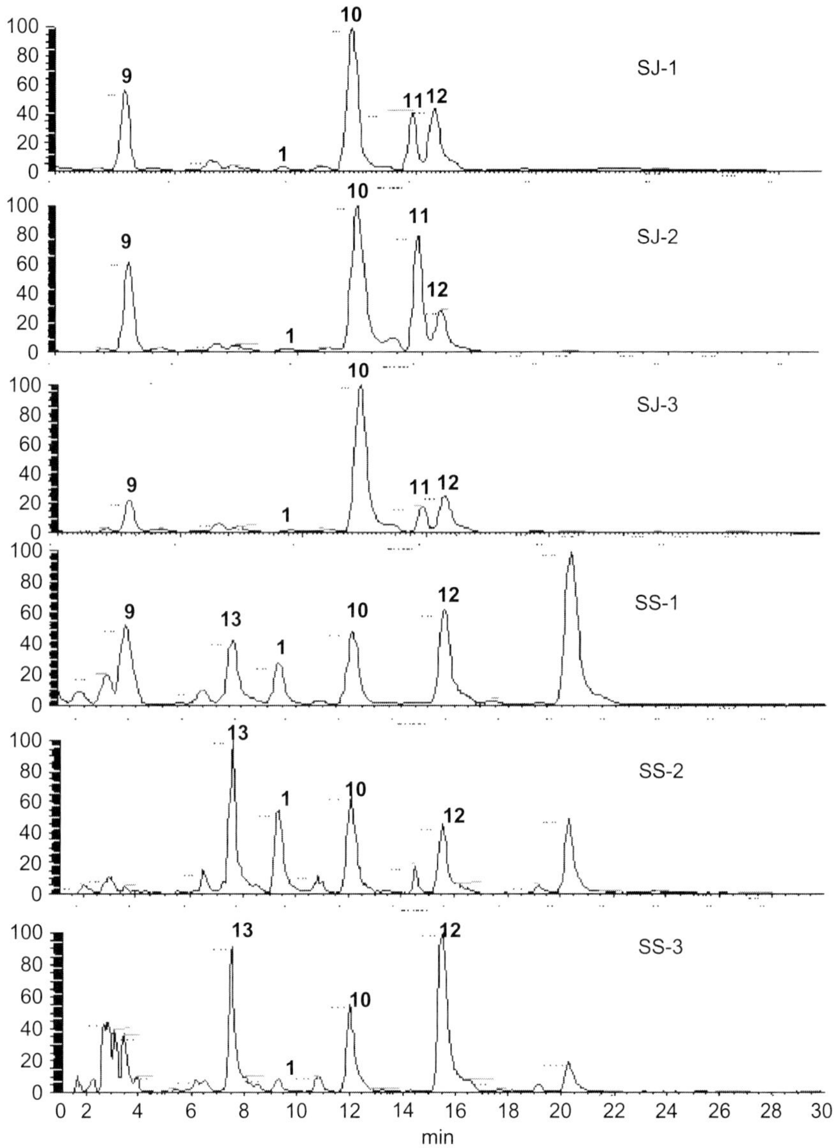


Fig. 9. Base peak traces of LC/MS analyses of the total alkaloids from three samples of *S. japonica* (SJ1–SJ3) and three samples of *S. sessilifolia* (SS1–SS3); The TIC chromatograms refer to positive ion in ESI mode and compound identities were assigned on the basis of $MS^n$ ($n=2\sim4$) analyses. The structures are shown in Scheme 1. (For colour version of this figure, the reader is referred to the online version of this chapter.)

## C. ANTITUSSIVE COMPONENTS

Although Radix Stemonae has been used as an antitussive in traditional Chinese medicine for over 15 centuries (Jiangsu New Medical College, 1977), no scientific documentation was available until rather recently. As summarized by Wei (1986), a 1954 paper reported that intragastric administration of an aqueous extract of Radix Stemonae at 0.6–0.9 g/kg did not demonstrate antitussive action on cough induced by intrapleural injection of 1% iodine solution in cats, while a 1973 report found that intraperitoneal (i.p.) injection of the 100% alkaloid extract at the dose of 10 ml/kg failed to suppress cough induced by sulphur dioxide in mice. Cong and Xu (1997) tested the 95% ethanol extracts of Chinese species of *Stemona* and also the polysaccharide, total saponin, total alkaloid and lipid-soluble fractions of *S. sessilifolia*. All the extracts and fractions were found to have significant, though of different potency, antitussive effects against cough induced by ammonia aerosol in mice. However, their results could only be taken with reservation because mice do not have rapidly adapting receptors, which play an important role in the cough reflex. Moreover, mice lack intraepithelial nerve endings and thus are thought to be devoid of any cough reflex (Belvisi and Hele, 2003). A similar study on the three pharmacopoeial species suffered the same problem (Hu *et al.*, 2009); the authors tested the aqueous extracts, alkaloid fractions and non-alkaloid fractions of herb samples of the three pharmacopoeial *Stemona* species. Their results suggested that all of them, especially at high dosages, were effective in reducing cough in mice induced with ammonia aerosol. Not to say that this model does not reflect the cough mechanism in man, it remains difficult to interpret their results, as no details were given on the selection of mice and rationales for the selected aerosol concentrations and duration of exposure. Moreover, the numbers of cough in mice in 2 min, after aerosol exposure for 17 s, were rather high, up to $164.1 \pm 12$. It is not clear if sneezes were distinguished from coughs, if any.

In the absence of convincing information on the antitussive properties of Radix Stemonae, colleagues in The Chinese University of Hong Kong (CUHK) looked into this puzzle by adopting a guinea pig model with citric acid aerosol for cough induction. This animal model is the most generally accepted one, because cough reflexes in guinea pigs and humans are similar (Belvisi and Hele, 2003).

Initial tests at CUHK of the total alkaloid fraction of Radix Stemonae purchased from herb shop revealed significant antitussive effects in guinea pigs via i.p. administration; but the second batch of herb purchased from the same herb shop showed no effect (Jiang *et al.*, 2006a). Thin layer chromatography revealed substantial chemical differences between the two batches and was used to guide the procurement of subsequent batches. Their efforts led to the discovery that neotuberostemonine and other stenine-type alkaloids exhibited antitussive potency comparable to codeine but not involving the opioid receptor (Chung *et al.*, 2003). This antitussive property of neotuberostemonine was further confirmed by intragastric administration 400 μmol/kg in guinea pigs. Structure–activity relationship analyses revealed that the saturated tricyclic pyrrolo[3,2,1-*jk*][1]benzazepine nucleus is the primary key structure contributing to the antitussive function and all *cis*-configurations at the three ring junctions are the optimal structure for the antitussive activity of stenine-type *Stemona* alkaloids (Chung *et al.*, 2003).

After the report by Chung *et al.* (2003), further pursuit continued at CUHK on the identities and quality of Radix Stemonae, and oral absorption, action pathways and safety of *Stemona* alkaloids. Extensive collections of the many taxa in China were made, and some were grown to flowering and fruiting in greenhouse for confirmation of their identities (Chan, 2004; Jiang *et al.*, 2006a,b; Xu *et al.*, 2006). Anatomical and molecular markers were identified to facilitate phytochemical analyses (Ling, 2008; Xu *et al.*, 2006).

Over 30 herb samples belonging to *S. tuberosa* from different parts of China were tested by HPLC and LC–MS. The results showed that *S. tuberosa* is chemically a very variable species. The antitussive neotuberostemonine (**4**) was not universally present in Radix Stemonae, but only present in two of the 30 samples. A natural step, in view of such chemical variations, was to check whether or not the samples without neotuberostemonine had any antitussive effects. The total alkaloid fractions of samples representing the four chemical types A–D of Radix Stemonae were subjected to a guinea pig assay with cough induced by citric acid aerosol. Results indicated that, although quite variable in chemical contents among the four samples, all their alkaloid fractions showed antitussive effects (Fig. 10). The samples with croomine (**1**, tuberostemospironine type) or stemoninine (**3**, stemoamide type) as the main ingredient showed stronger antitussive functions, indicating that non-stenine-type alkaloids also contributed to the antitussive functions (Xu *et al.*, 2006). The antitussive effects were further confirmed by intraperitoneal and intragastric administrations of croomine, stemoninine, and stemoninoamide (**14**) in guinea pigs (Lin *et al.*, 2008a,b). Intraperitoneal and intragastric administration of tuberostemonine and neotuberostemonine showed antitussive effects in guinea pigs, but tuberostemonine H (**8**) did so only in intragastric applications (Zhou *et al.*, 2009). Bisdehydrostemoninine (**15**) was also reported to have significant antitussive effects in guinea pigs (Lin *et al.*, 2006); but this compound is probably an artefact (Greger, 2006; Schinnerl *et al.*, 2007). In summary, so far, antitussive effects of 21 *Stemona* alkaloids have been tested (Table I). Thirteen of them, covering stenine, stemoamide, tuberostemospirnine, and stemonamine types, showed significant antitussive effects.

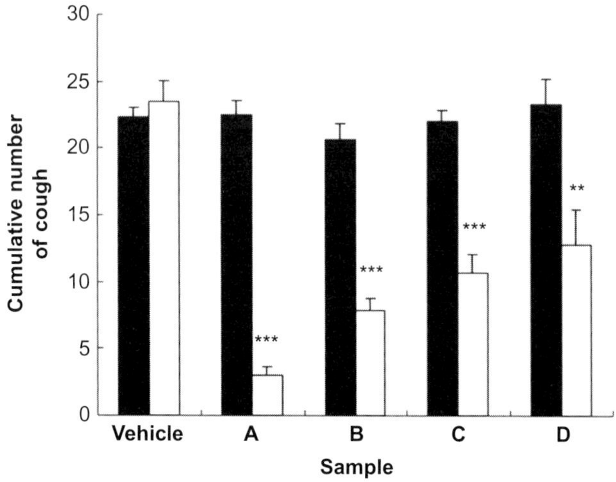


Fig. 10. Antitussive effects of total alkaloids from different *Stemona* samples at 100 mg/kg i.p. Major ingredient in samples A–D being croomine, stemoninine, neotuberostemonine, and tuberostemonine. White columns represent first challenge with citric acid aerosol, black columns represent second challenge with citric acid aerosol after administration with test samples ($n=6$). ***$P<0.001$ in A–C, **$P<0.01$ in D.

The intestinal absorption of five stenine-type alkaloids, neostenine **(16)**, neotuberostemonine, tuberostemonine, and tuberostemonines H and J **(17)** was investigated using a 21-day cultured Caco-2 monolayer model. Cell integrity was confirmed with transepithelial electric resistance (TEER) values higher than 600 $\Omega$ cm$^2$, and the samples were analysed by HPLC–evaporative light scattering detector (HPLC–ELSD). All of them exhibited different degrees of absorptive permeability, indicating that they are likely to be well absorbed and orally active. Further, neostenine and neotuberostemonine were identified as the substrates of P-glycoprotein and have a transport preference from the basolateral to apical direction with efflux ratios between 2 and 3. Cyclosporin A dose-dependently inhibited the secretory permeability of these alkaloids and abolished their active efflux transport (Leung *et al.*, 2006; Zhou *et al.*, 2009).

## D. ANTITUSSIVE PATHWAYS AND MECHANISMS OF ACTION

Electrical stimulation of the superior laryngeal nerve in guinea pigs was adapted to differentiate peripherally from centrally acting antitussives. The results showed that reproducible cough responses could be induced by the constant stimulation voltages at $103.3 \pm 2.1$ mV after operation during a period of 60 min in the control group (Fig. 11A). After separate administration of neotuberostemonine, tuberostemonine, or stemoninine 100 mg/kg i.p.,

TABLE I

*Antitussive Effects of* Stemona *Alkaloids (Routes of Administration in Different Papers Varied from Intragastric to Intraperitoneal or Intracerebroventricular)*

| Type of alkaloid | Name | Antitussive effect | References |
|---|---|---|---|
| Stenine | Epi-bisdehydro-tuberostemonine J (**19**) | n.s. | Chung *et al.* (2003) |
| | Neostenine (**16**) | s | Chung *et al.* (2003) |
| | Neotuberostemonine (**4**) | s | Chung *et al.* (2003), Xu *et al.* (2010), Zhou *et al.* (2009) |
| | Tuberostemonine (**5**) | s | Xu *et al.* (2010), Zhou *et al.* (2009) |
| | Tuberostemonine H (**8**) | s | Chung *et al.* (2003), Zhou *et al.* (2009) |
| | Tuberostemonine J (**17**) | s | Chung *et al.* (2003) |
| | Tuberospironine (**20**) | n.s. | Lin *et al.* (2008a) |
| Stemoamide | Bisdehydrostemoninine (**15**) | s | Lin *et al.* (2006) |
| | 9a-*O*-Methylstemoenoine (**22**) | n.s | Lin *et al.* (2008b) |
| | Protostemonine (**10**) | s | Yang *et al.* (2009) |
| | Stemoenonine (**21**) | n.s. | Lin *et al.* (2008b) |
| | Stemoninine (**3**) | s | Lin *et al.* (2008b), Xu *et al.* (2010) |
| | Stemoninoamide (**14**) | s | Lin *et al.* (2008b) |
| Tuberostemospironine | Croomine (**1**) | s | Lin *et al.* (2008a), Xu *et al.* (2010) |
| | Dehyrocroomine (**26**) | n.s | Lin *et al.* (2008a) |
| | 6-Hydroxycroomine (**24**) | n.s. | Lin *et al.* (2008a) |
| | 10-Hydroxycroomine (**25**) | n.s. | Lin *et al.* (2008a) |
| | Stemospironine (**23**) | s | Yang *et al.* (2009) |
| Stemonamine | Maistemonine (**12**) | s | Yang *et al.* (2009) |
| | Neostemonine (**11**) | n.s. | |
| | Stemonamine (**27**) | s | |

*Note*: s, significant; n.s., not significant.

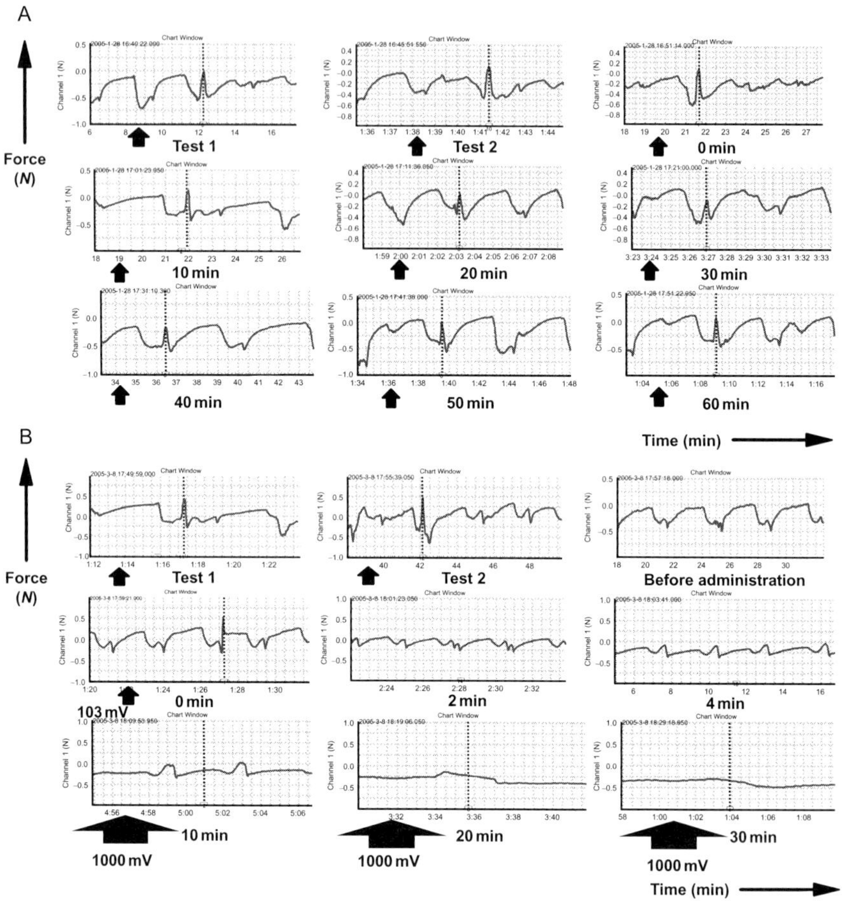


Fig. 11. Cough responses on the animal model of electrical stimulation on superior laryngeal nerve in guinea pigs ($n=6$). Note: The labelled sharp peaks represent cough responses or where cough responses are expected. Test samples 100 mg/kg (i.p.) were administered at 0 min. Arrow heads (⬆) represents application of electric stimuli. (A) Reproducible cough responses in the control or after administration of neotuberostemonine (purity $\geq$ 98%). The same pattern was observed after separate administration of tuberostemonine or stemoninine. (B) Response to administration of croomine. In the chart, coughing was induced by afferent electrical stimulation of the superior laryngeal nerve by a stimulator. Tests 1 and 2 were used to determine the thresholds of reproducible coughs in the animals. Electrical stimuli with the same voltage threshold were also given at 0, 10, 20, 30, 40, 50, and 60 min after dosing in (A). In (B), the responses to croomine, even by increasing the electrical stimuli, were different from those in (A). (For colour version of this figure, the reader is referred to the online version of this chapter.)

the results indicated that the stimulation threshold of reproducible cough responses was equal to the control group at 103.3 ± 2.1 mV during a period of 60 min and showed the same patterns of recording as in Fig. 11A. These results suggested that all three compounds acted on the peripheral system. Croomine (**1**), however, exhibited obviously different activities from the other three alkaloids. Cough responses could be induced by the stimulation voltage at 103.3 ± 2.1 mV but not after administration of croomine 100 mg/kg i.p. at 10 min even when the electrical stimulation voltage was increased to 1000 mV (Fig. 11B). The results suggested that the threshold of cough reflex increased to a value higher than 1000 mV, which was significantly different from the value 103.3 ± 2.1 mV in the control group. This marked difference suggests that croomine is a centrally acting antitussive. Obviously, this result has important implications, particularly the fact that the antitussive effects of some *Stemona* alkaloids could be mediated through the central route while some others through the peripheral pathway. It also may offer explanation to adverse reactions and also hints for future approaches to development of therapeutic agents from *Stemona* for relieving cough.

So far, several studies have looked into the mechanisms of action of *Stemona* alkaloids. Chung *et al.* (2003) studied the antitussive mechanisms of neotuberostemonine (**4**) by *in vitro* receptor binding assays and demonstrated that this alkaloid did not show affinity to opioid, 5-$HT_{1A}$, $GABA_B$, and dopamine $D_2$ receptors. Meanwhile, the insecticidal action mechanisms of tuberostemonine (**5**) were studied by electrophysiological methods, and the results indicated that it inhibited excitatory transmission of neuromuscular junction as a partial open-channel blocker in crayfish (Shinozaki and Ishida, 1985). In screening against 40 G protein-coupled receptor and other molecular targets, synthetic (±)-neostenine (**16**) was found binding only to the muscarinic M5 receptor, whereas (±)-13-epineostenine (**18**) to sigma receptors (Frankowski *et al.*, 2011). Further, in-depth investigation is needed to clarify action targets and mechanisms of action. An expanded study on 104 synthetic analogues based on the pyrrolo[1,2-a]azepine nucleus demonstrated their affinity to multiple classes of G protein-coupled receptor targets, including adrenergic, muscarinic, serotonin and dopaminergic classes, and sigma receptors (Frankowski *et al.*, 2011). In-depth investigation will be rewarding in further analyses of the targets and mechanisms of action of *Stemona* alkaloids.

## E. ADVERSE REACTIONS

When first appeared in Mingyi Bielu, an herbal compiled in around 500 A.D., Radix Stemonae was noted to have mild toxicity. Consumption of preparations containing Radix Stemonae is noted to have caused heartburn, dryness of the mouth, nose and pharynx, dizziness, chest discomfort, shortness of breath, anorexia, abdominal pain, diarrhoea, and epistaxis, whereas over dosage may cause paralysis of the respiratory centre (Wei, 1986).

In the study on the pathways of antitussive actions using an electrical stimulation model, intraperitoneal administration of croomine (**1**) 100 mg/kg, resulted in obvious decrease in both the respiratory rate and the amplitude of respiratory movement (Fig. 11B). This observation led to the question whether or not the respiratory depressant effect was caused by side effects of high dose croomine. To test this hypothesis, the effects of croomine at high and low doses were compared. Using the same recording system without electrical stimulation, croomine at 100 mg/kg i.p. was administered to guinea pigs. The amplitude of respiratory movement decreased obviously in 2 min and stayed low for the next 60 min. The respiratory rate showed two phases. At first, it demonstrated a slight increase in 2 min, but then gradually dropped to a slower rate for 60 min (Fig. 12A). On the other hand, after administration of croomine 25 mg/kg i.p., the amplitude of respiratory movement dropped from 4 to 20 min but then began to recover and reached a steady level at 30 min. The respiratory rate also dropped but did not show a fast phase (Fig. 12B). These results indicated that both doses decreased the amplitude of respiratory movement and respiratory rate. This is a good piece of evidence that croomine can account for part of the adverse reactions reported in literature.

## IV. QUALITY ASSESSMENT

Phytochemical studies have identified a rich collection of over 130 *Stemona* alkaloids from the four genera in Stemonaceae (Pilli *et al.*, 2010). Significant intra- and inter-specific chemical diversities are found among the three pharmacopoeial species (Schinnerl *et al.*, 2007). Such results have led to complicated but interesting issues in the use of botanicals. One issue is the heterogeneous and variable nature in the chemical constitution of plants, which would cause difficulties in the quality control of herbs. This is more obvious, when there are more than one bioactive components, which are sometimes even mutually exclusive as in the case of the isomers tuberostemonine and neotuberostemonine. The problem is more complicated in traditional Chinese medicine as more than one species are regarded as genuine source materials of Radix Stemonae (Pharmacopoeia Commission of People's Republic of China, 2010). When several species are regarded as alternative sources of an herb, the variability among them increases substantially. These situations would make it extremely difficult to maintain quality

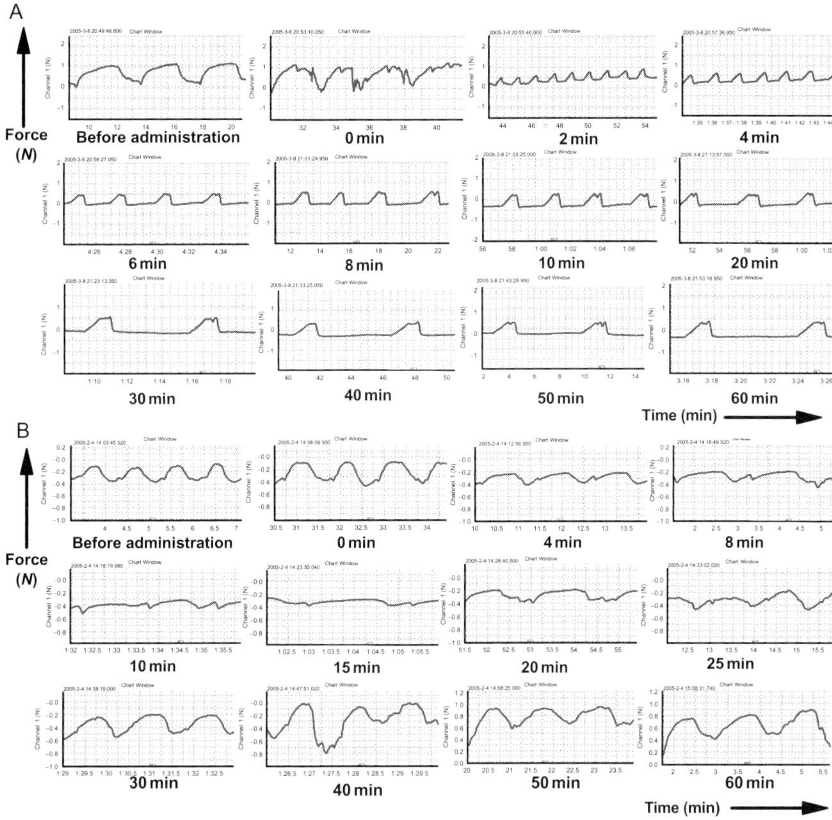


Fig. 12. Respiratory movements after croomine administration at 100 mg/kg i.p. (A) ($n = 3$) and 25 mg/kg i.p. (B) ($n = 3$), respectively. (For colour version of this figure, the reader is referred to the online version of this chapter.)

control for botanical products. The problem of chemical variations among herb samples would also cause difficulties for herbal practitioners and suppliers of herbal materials. For herbal practitioners, it is a nightmare in clinical practices. Although they are sophisticated and experienced in diagnosing an illness and prescribing herbal prescriptions, there is no guarantee for the consistence and efficacy of the herbs dispensed in different cases. Meanwhile, in view of the need for modernization and harmonization of herbal products, suppliers of crude botanicals and manufacturers in sourcing raw materials are urged to commit to more vigorous selection of the proper materials in order to meet the objective of consistence and standardization of their products. Another issue is that more than one chemical component or even several types of components in an herb may independently or collectively contribute to a desired biological effect. This variability is a challenge in the selection of markers for quality control and in the understanding of the pharmacology and pharmacokinetics of botanicals. The pursuit for standardization and harmonization of such variable botanical products would remain an uncertainty, if not an impossible dream.

Bioassays conducted on guinea pigs with induction of cough using citric acid aerosol have shown that antitussive actions come from various *Stemona* alkaloids. Both neotuberostemonine (**4**) and tuberostemonine (**5**) were reported to possess potent biological activities. However, these two isomers are not universally present in all samples of *S. tuberosa*. The chemical diversity and variability of this herb suggest that careful chemical authentication is needed to provide guarantee for the consistence and efficacy in its ethnomedical applications.

Various approaches have been developed with reference to the centrally and peripherally acting *Stemona* alkaloids. The previously described on-line HPLC–ESI/MS and detailed tandem mass spectrometric analyses are able to monitor the *Stemona* alkaloids from the three pharmacopoeial species (Zhou *et al.*, 2006).

A HPLC method coupled with diode array detection and evaporative light scattering detection (HPLC–DAD–ELSD) was developed to simultaneously quantify six major bioactive alkaloids belonging to different structure types in Radix Stemonae (Li *et al.*, 2007). Diode array detector (DAD) with the wavelengths at 307 and 260 nm was used to monitor the conjugated system of protostemonine (**10**) and maistemonine (**12**), respectively, whereas evaporative light scattering detector (ELSD) was employed to detect croomine (**1**), stemoninine (**3**), neotuberostemonine (**4**), and tuberostemonine (**5**), the other four analytes with no or poor chromophores. The assay was validated to be sensitive, precise and accurate, with a detection limit of 3.64–0.04 μg/mL depending on the individual analytes. The overall intra- and inter-day variations were less than 9.3%, and the overall recoveries higher than 91.2%, respectively. The correlation coefficients of the calibration curves were better than 0.996 for all analytes. This simple, rapid, low-cost, and reliable method is suitable for the routine quality control of herbal medicines containing bioactive components with different structure types such as Radix Stemonae (Li *et al.*, 2007)

A HPLC method was developed and validated for the determination of stemoninine in plasma after oral and intravenous administration of the extract of the roots of *S. tuberosa* to rats. Plasma samples were analysed on a Waters reversed-phase C18 column using a gradient mobile-phase of eluent A (water containing 0.1% formic acid and 0.2% triethylamine, pH 3.68) and eluent B (acetonitrile–water, 50:50, v/v).The flow rate was 1.0 mL/min and the detector wavelength was 210 nm. The Waters Oasis solid-phase extraction cartridge was applied for the preparation of plasma samples with high recovery. A good linear relationship was obtained in the concentration range of 1.55–124 mg/mL ($r = 0.9995$). The limits of quantification and detection were 1.55 and 0.42 mg/mL, respectively. The average recoveries ranged from 91% to 96% in plasma at stemoninine concentrations of 3.10, 62.0, and 99.2 mg/mL. Intra- and inter-batch coefficients of variations were 3.3–5.4% and 2.5–3.9%, respectively. This method was successfully applied to pharmacokinetic studies after oral and intravenous administration of *S. tuberosa* extract in rats (Jiang *et al.*, 2011).

A HPLC method with UV detection at 305 nm was reported for the quantitative analyses of protostemonine in rat plasma, which has a chromophore and is present in samples derived from *S. japonica* and *S. sessilifolia* (Zhang *et al.*, 2010).

Besides the HPLC-based quantification methods, a capillary electrophoresis method for the separation, identification, and quantification of the pyrrolo- and pyrido[1,2-a]azepine derivatives in *Stemona curtisii, S. collinsae* and *S. tuberosa* was developed with a NACE-ESI-IT-MS method, utilizing an electrolyte of 50 mM ammonium acetate, 1 M acetic acid and 10% methanol in ACN and a separation voltage of 30 kV. Samples were injected voltage-assisted with 20 kV for 1 s. Isopropanol:water (1:1) was used as ESI sheath liquid at a flow rate of 3 mL/min. For unambiguous peak assignment of more than 40 unidentified alkaloids MS/MS experiments were performed and fragmentation patterns studied. Subsequently, the method was validated for the quantitative determination of four selected derivatives (RSD inter- and intra-day; 6%, LODs; 7.5 mg/mL, LOQs; 25.0 mg/mL, for all analytes; recovery rates 98.9%) in several *Stemona* extracts (Sturm *et al.*, 2008).

## V. CONCLUSIONS

Cough is a normal protective reflex for keeping the airways free of obstruction and harmful substances. It is also the most common symptom for which medical advice is sought (Reynolds *et al.*, 2003, 2004). Antitussives are among the most widely used drugs in the world, although reviews of randomized controlled clinical trials either concluded that over-the-counter (OTC) antitussive drugs possess little clinically relevant efficacy (Schroeder and Fahey, 2002) or admitted that there is no good evidence for or against the effectiveness of OTC medicines in acute cough (Smith *et al.*, 2008). There are two sites for antitussive actions, central, and peripheral. Centrally acting antitussives act within the CNS to suppress the central cough pathway and comprise the majority of currently used antitussives. However, many of these drugs, including the opioids, which have been the antitussives of choice for decades, are limited by their many side effects. Peripherally acting antitussives generally inhibit the responsiveness of airway nerve subtypes that evoke cough, although these drugs are not used widely. However, as a better understanding of the cough pathway emerges, many new peripherally acting antitussives are being developed to avoid unwanted CNS effects (Reynolds *et al.*, 2004).

To relieve coughing, it is natural that many botanicals have been tried down the ages in different cultures (Gairola *et al.*, 2010; Rodriguez-Fragoso *et al.*, 2008). Most of them await concrete scientific evaluation and are generally viewed with suspicion or reservation (Widdicombe and Ernst, 2009; Ziment, 2002). The need and wait for better antitussives continues.

Radix Stemonae can be a good candidate. Bioassays in guinea pigs and double-blind clinical studies (Wong *et al.*, 2006) have documented its antitussive properties and, in turn, confirmed the claims of its traditional uses in Chinese medicine. Phytochemical analyses have revealed the presence of antitussive *Stemona* alkaloids in all three pharmacopoeial *Stemona* species and thus justified their inclusion in China Pharmacopoeia (Pharmacopoeia Commission, 2010). However, total alkaloid contents, chosen as a quality marker in China Pharmacopoeia, has limited value as an indicator, especially when high concentrations of some components such as croomine (**1**) can lead to respiratory depression (Xu *et al.*, 2010). On the other hand, through recent pharmacognostic, phytochemical, and pharmacological studies, a clearer picture on Radix Stemonae becomes available. A battery of morphological, anatomical, chemical, and molecular markers allows proper identification of the source plants for this herb and provides measures for prevention of substitutes. In fact, as it is now known that croomine (**1**) accounts for much of the adverse reactions and varies in contents in different populations of Radix Stemonae, it would be strategically possible to select samples high in *Stemona* alkaloids with obvious antitussive value but low or devoid of croomine, especially for GAP (good agricultural practice) farms and GMP (good manufacturing practice) manufacturers. The HPLC–DAD–ELSD, LC–MS, and HPCE methods, found in literature, can help monitor the contents of *Stemona* alkaloids for quality assurance purposes.

## ACKNOWLEDGEMENT

Assistance from Yiu-Man Chan, Song-lin Li, Ka-Ho Ling, and Yan Zhou is deeply appreciated.

# REFERENCES

Belvisi, M. G. and Hele, D. J. (2003). Animal models of cough. *In* Cough: Causes, Mechanisms and Therapy, (F. Chung, J. Widdicombe and H. Boushey, eds.), pp. 217–222. Blackwell Publishing, Oxford.

But, P. P. H. (1995). Attitudes and approaches of traditional Chinese medicine to herbal toxicity. *Journal of Natural Toxins* **4**, 207–217.

Caddick, L. R., Rudall, P. J., Wilkin, P., Hedderson, T. A. J. and Chase, M. W. (2002). Phylogenetics of Dioscoreales based on combined analyses of morphological and molecular data. *Botanical Journal of the Linnean Society* **138**, 123–144.

Chan, Y. M. (2004). Molecular authentication and taxonomy of Radix Stemonae. Master of Philosophy Thesis, The Chinese University of Hong Kong, Hong Kong.

Chung, H. S., Hon, P. M., Lin, G., But, P. P. H. and Dong, H. (2003). Antitussive activity of *Stemona* alkaloids from *Stemona tuberosa*. *Planta Medica* **69**, 914–920.

Cong, X. D. and Xu, G. J. (1997). Studies on Radix Stemonae. *In* Species Systematization and Quality Evaluation of Commonly Used Chinese Traditional Drugs, (G. J. Xu and G. S. Xu, eds.), Vol. 2, pp. 55–108. Fujian Science and Technology Publisher, Fuzhou.

Cong, X. D., Xu, G. J., Jin, R. L. and Zhi, H. J. (1992). Pharmacognostic studies on Baibu Radix Stemonae and allied drugs. IX. Determination and evaluation of total alkaloid content in the roots of Chinese Stemona spp. *Acta Pharmaceutica Sinica* **27**, 556–560.

Ding, Y. and Yang, C. R. (1990). Steroidal saponins from *Asparagus filicinus*. *Acta Academiae Medicinae Shanghai* **25**, 509–514.

Duyfjes, B. E. E. (1991). Stemonaceae and Pentastemonaceae; with miscellaneous notes on members of both families. *Blumea* **36**, 239–252.

Duyfjes, B. E. E. (1992). Formal description of the family Pentastemonaceae with some additional notes on Pentastemonaceae and Stemonaceae. *Blumea* **36**, 551–552.

Fan, L. L., Zhu, S., Che, H. B., Yang, D. H., Cai, S. Q. and Komatsu, K. (2009a). Molecular analysis of *Stemona* plants in China based on sequences of four chloroplast DNA regions. *Biological and Pharmaceutical Bulletin* **32**, 1439–1446.

Fan, L. L., Zhu, S., Che, H. B., Yang, D. H., Cai, S. Q. and Komatsu, K. (2009b). Identification of the botanical source of stemonae radix based on polymerase chain reaction with specific primers and polymerase chain reaction-restriction fragment length polymorphism. *Biological & Pharmaceutical Bulletin* **32**, 1624–1627.

Frankowski, K. J., Neuenswander, B. and Aubé, J. (2008). Explorations of *Stemona* alkaloid-inspired analogues: Skeletal modification and functional group diversification. *Journal of Combinatorial Chemistry* **10**, 721–725.

Frankowski, K. J., Setola, V., Evans, J. M., Neuenswander, B., Roth, B. L. and Aubé, J. (2011). Synthesis and receptor profiling of *Stemona* alkaloid analogues reveal a potent class of sigma ligands. *Proceedings of the National Academy of Sciences* **108**, 6727–6732.

Gairola, S., Gupta, V., Bansal, P., Singh, R. and Maithani, M. (2010). Herbal antitussives and expectorants—A review. *International Journal of Pharmaceutical Sciences Review and Research* **5**(2), 5–9.

Garraffo, H. M., Jain, P., Spande, T. F., Daly, J. W., Jones, T. H., Smith, L. J. and Zottig, V. E. (2001). Structure of alkaloid 275A, a novel 1-azabicyclo[5.3.0] decane from a dendrobatid frog, *Dendrobates lehmanni*: Synthesis of the tetrahydrodiastereomers. *Journal of Natural Products* **64**, 421–427.

Götz, M. and Strunz, G. M. (1975). Tuberostemonine and related compounds: The chemistry of the *Stemona* alkaloids. *In* MTP, International Review of Sciences, Organic Chemistry, (K. Wiesner, ed.), Vol. 9, pp. 143–160. Butterworths, LondonSeries One.

Greger, H. (2006). Structural relationships, distribution and biological activities of *Stemona* alkaloids. *Planta Medica* **72**, 99–113.

Greger, H., Schinnerl, J., Vajrodaya, S., Brecker, L. and Hofer, O. (2009). *Pandanus* alkaloids in Stemonaceae: Finding of a plausible biogenetic origin of *Stemona* alkaloids. *Journal of Natural Products* **72**, 1708–1711.

Hu, J. P., Zhang, N., Mao, Y. Q., Yand, D. H., Tan, H. R., Siran, N. I., Liu, X. Q. and Cai, S. Q. (2009). Antitussive activity comparison of three kinds of Stemonae Radix. *China Journal of Chinese Materia Medica* **34**, 3096–3104.

Inthachub, P., Vajrodaya, S. and Duyfjes, B. E. E. (2009). Review of the genus *Stichoneuron*. *Edinburgh Journal of Botany* **66**, 213–228.

Ji, Z. H. and Duyfjes, B. E. E. (2000). Stemonaceae. *In* Flora of China, (Z. Y. Wu and P. H. Raven, eds.), Vol. 24. Science Press/Missouri Botanical Garden Press, Beijing/St. Louis.

Jiang, R. W., Hon, P. M., But, P. P. H., Chung, H. S., Lin, G., Ye, W. C. and Mak, T. C. W. (2002). Isolation and stereochemistry of two new alkaloids from *Stemona tuberosa* Lour. *Tetrahedron* **58**, 6705–6712.

Jiang, R. W., Hon, P. M., Xu, Y. T., Chan, Y. W., Xu, H. X., Shaw, P. C. and But, P. P. H. (2006a). Isolation and chemotaxonomic significance of tuberostemospironine-type alkaloids from *Stemona tuberosa*. *Phytochemistry* **67**, 52–57.

Jiang, R. W., Hon, P. M., Zhou, Y., Chan, Y. W., Xu, Y. T., Xu, H. X., Greger, H., Shaw, P. C. and But, P. P. H. (2006b). Alkaloids and chemical diversity of *Stemona tuberosa*. *Journal of Natural Products* **69**, 749–754.

Jiang, R. W., Ye, W. C., Shaw, P. S., But, P. P. H. and Mak, C. W. (2010). Absolute configuration of neostenine. *Journal of Molecular Structure* **966**, 18–22.

Jiang, D., Yang, X., Liu, H. and Guan, H. (2011). Development and validation of a high-performance liquid chromatographic method for the determination of stemoninine in rat plasma after administration of *Stemona tuberosa* extracts. *Biomedical Chromatography* **25**, 498–502.

Jiangsu New Medical College (1977). Encyclopedia of Chinese Materia Medica. Shanghai People's Press, Shanghai, 1, pp. 858–861.

Kaltenegger, E., Brem, B., Mereiter, K., Kalchhauser, H., Kählig, H., Hofer, O., Vajrodaya, S. and Greger, H. (2003). Insecticidal pyrido[1,2-*a*]azepine alkaloids and related derivatives from *Stemona* species. *Phytochemistry* **63**, 803–816.

Kauff, F., Rudall, P. J. and Conran, J. G. (2000). Systematic root anatomy of Asparagales and other monocotyledons. *Plant Systematics and Evolution* **223**, 139–154.

Kongkiatpaiboon, S., Keeratinijakal, V. and Gritsanapan, W. (2010). Microscopic characterization as a tool for separation of *Stemona* groups. *Pharmacognosy Journal* **17**, 1–4.

Kongkiatpaiboon, S., Schinnerl, J., Felsinger, S., Keeratinijakal, V., Vajrodaya, S., Gritsanapan, W., Brecker, L. and Greger, H. (2011). Structural relationships of *Stemona* alkaloids: Assessment of species-specific accumulation trends for exploiting their biological activities. *Journal of Natural Products* **74**, 1931–1938.

Kostecki, K., Engelmeier, D., Pacher, T., Hofer, O., Vajrodaya, S. and Greger, H. (2004). Dihydrophenanthrenes and their antifungal stilbenoids from *Stemona cf. pierrei*. *Phytochemistry* **65**, 99–106.

Leung, P. H. H., Zhang, L., Zuo, Z. and Lin, G. (2006). Intestinal absorption of *Stemona* alkaloids in a Caco-2 cell model. *Planta Medica* **71**, 211–216.

Li, E. X. and Fu, C. X. (2007). A new synonym of *Stemona sessilifolia* (Stemonaceae). *Acta Phytotaxonomica Sinica* **45**, 399–402.

Li, Y. F., Hu, L. H., Lou, F. C., Hong, J. R., Li, J. and Shen, Q. (2005). Furostanoside from *Asparagus filicinus*. *Journal of Asian Natural Products Research* **7**, 43–47.

Li, S. L., Jiang, R. W., Hon, P. M., Cheng, L., Xu, H. X., Greger, H., But, P. P. H. and Shaw, P. C. (2007). Quality evaluation of Radix Stemonae through simultaneous quantification of bioactive alkaloids by high-performance liquid chromatography coupled with diode array and evaporative light scattering detectors. *Biomedical Chromatography* **21**, 1088–1094.

Li, E. X., Yi, S., Qiu, Y. X., Guo, J. T., Comes, H. P. and Fu, C. X. (2008). Phylogeography of two East Asian species in Croomia (Stemonaceae) inferred from chloroplast DNA and ISSR fingerprinting variation. *Molecular Phylogenetics and Evolution* **49**, 702–714.

Li, J. S., Zhang, T., Pu, Y. Q. and Wu, T. (2011). Identification of the three source species of Radix Stemonae in the Pharmacopoeia of the People's Republic of China. *Journal of Chinese Medicinal Materials* **34**, 209–210.

Lin, L. G., Zhong, Q. X., Cheng, T. Y., Tang, C. P., Ke, C. Q., Lin, G. and Ye, Y. (2006). Stemoninines from the roots of *Stemona tuberosa*. *Journal of Natural Products* **69**, 1051–1054.

Lin, L. G., Leung, H. P. H., Zhu, J. Y., Tang, C. P., Ke, C. Q., Rudd, J. A., Lin, G. and Ye, Y. (2008a). Croomine- and tuberostemonine-type alkaloids from roots of *Stemona tuberosa* and their antitussive activity. *Tetrahedron* **64**, 10155–10161.

Lin, L. G., Li, K. M., Tang, C. P., Ke, C. Q., Rudd, J. A., Lin, G. and Ye, Y. (2008b). Antitussive stemoninine alkaloids from the roots of *Stemona tuberosa*. *Journal of Natural Products* **71**, 1107–1110.

Ling, K. H. (2008). 'Authentication of *Stemona* Root, Oilfish, Crocodile Meat and Frog Oviduct'. Master of Philosophy Thesis, The Chinese University of Hong Kong, Hong Kong.

Luo, J. P., Liu, Y. P., Feng, Y. F., Guo, X. L. and Cao, H. (2003). Two chemotypes of *Pogostemon cablin* and influence of region of cultivation and harvesting time on volatile oil composition. *Acta Pharmaceutica Sinica* **38**, 307–310.

Ogunwande, I. A., Olawore, N. O., Adeleke, K. A. and Konig, W. A. (2005). Analyses of the volatile compounds of *Callitris columellaris* F. Muell. needles from two different regions of Nigeria. *Journal of Essential Oil Research* **17**, 44–46.

Pacher, T., Seger, C., Engelmeier, D., Vajrodaya, S., Hofer, O. and Greger, H. (2002). Antifungal stilbenoids from *Stemona collinsae*. *Journal of Natural Products* **65**, 820–827.

Pharmacopoeia Commission (2010). Pharmacopoeia of the People's Republic of China. Chemistry Industry Publisher, Beijing, p. 100.

Pilli, R. A. and de Oliveira, M. C. F. (2000). Recent progress in the chemistry of the *Stemona* alkaloids. *Natural Product Reports* **17**, 117–127.

Pilli, R. A., Rossoa, G. B. and de Oliveira, M. C. F. (2005). The *Stemona* alkaloids. *In* The Alkaloids: Chemistry and Biology, (G. A. Cordell, ed.), Vol. 62, pp. 77–173. Elsevier, New York.

Pilli, R. A., Rossoa, G. B. and de Oliveira, M. C. F. (2010). The chemistry of Stemona alkaloids: An update. *Natural Product Reports* **27**, 1108–1937.

Potduang, B., Meeploy, M., Giwanon, R., Benmart, Y., Kaewduang, M. and Supatanakul, W. (2008). Biological activities of Asparagus racemosus. *African Journal of Traditional, Complementary, and Alternative Medicines* **5**, 230–237.

Reynolds, S. M., Spina, D. and Page, C. P. (2003). Pharmacology of peripherally acting antitussives. *In* Cough: Causes, Mechanisms and Therapy, (K. F. Chung, J. G. Widdicombe and H. A. Boushey, eds.), pp. 237–245. Blackwell Publishing, Oxford.

Reynolds, S. M., Mackenzie, A. J., Spina, D. and Page, C. P. (2004). The pharmacology of cough. *Trends in Pharmacological Sciences* **25**, 569–576.

Rodriguez-Fragoso, L., Reyes-Esparza, J., Burchiel, S. W., Herrera-Ruiz, R. and Torres, E. (2008). Risks and benefits of commonly used herbal medicines in Mexico. *Toxicology and Applied Pharmacology* **227**, 125–135.

Rudall, P. J., Cunniff, J., Wilkin, P. and Caddick, L. R. (2005). Evolution of dimery, pentamery and the monocarpellary condition in the monocot family Stemonaceae (Pandanales). *Taxon* **54**, 701–711.

Schinnerl, J., Brem, B., But, P. P. H., Vajrodaya, S., Hofer, O. and Greger, H. (2007). Pyrrolo- and pyridoazepine alkaloids as chemical markers in *Stemona* species. *Phytochemistry* **68**, 1417–1427.

Schroeder, K. and Fahey, T. (2002). Systematic review of randomized controlled trials of over the counter cough medicines for acute cough in adults. *British Medical Journal* **324**, 329–331.

Sekine, T., Ikegami, F., Fukasawa, N., Kashiwagi, Y., Aizawa, T., Fujii, Y., Ruangrungsi, N. and Murakoshi, I. (1995). Structure and relative stereochemistry of a new polycyclic alkaloid, asparagamine A, showing anti-oxytocin activity, isolated from *Asparagus racemosus*. *Journal of the Chemical Society. Perkin Transactions 1* **1995**, 391–393.

Sekine, T., Fukasawa, N., Murakoshi, I. and Ruangrungsi, N. (1997). A 9,10-dihydrophenanthrene from *Asparagus racemosus*. *Phytochemistry* **44**, 763–764.

Shinozaki, H. and Ishida, M. (1985). Inhibitory actions of tuberostemonine on the excitatory tansmission at the craryfish neuromuscular junction. *Brain research* **334**, 33–40.

Smith, S. M., Schroeder, K. and Fahey, T. (2008). Over-the-counter medications for acute cough in children and adults in ambulatory settings. *Cochrane Database Systematic Reviews* (1), CD001831.

Sturm, S., Schinnerl, J., Greger, H. and Stuppner, H. (2008). Nonaqueous capillary electrophoresis-electrospray ionization-ion trap-mass spectrometry analysis of pyrrolo- and pyrido[1,2-*a*]azepine alkaloids in *Stemona*. *Electrophoresis* **29**, 2079–2087.

Thorne, R. and Reveal, J. R. (2007). An updated classification of the class Magnoliopsida ("Angiospermae"). *The Botanical Review* **73**, 67–181.

Tie, Q. S. (2010). Clinical observations on the therapeutic values of Baibu Cough Syrup in treating cough. *Journal of Clinical Medicine in Practice* **14**(9), 82–83.

Vongsak, B., Kengtong, S., Vajrodaya, S. and Sukrong, S. (2008). Sequencing analysis of the medicinal plant *Stemona tuberosa* and five related species existing in Thailand based on *trn*H-*psb*A chloroplast DNA. *Planta Medica* **74**, 1764–1766.

Wang, X. X. (2011). Quality Evaluation Study of Stemona tuberosa Lour. Unpublished PhD thesis, Chengdu University of Traditional Chinese Medicine, Chengdu, pp. 154.

Wei, Y. (1986). Baibu. (H. M. Chang and P. P. H. But, eds.), *In* Vol. 1, pp. 451–455. World Scientific, Singapore.

Wen, R. L. (2002). *Zhang Yao Cai Cai Se Xian Wei Tu Jian*. Beijing, Chinese Medicine Science and Technology Publishing House, pp. 111–112.

Widdicombe, J. G. and Ernst, E. (2009). Clinical Cough V: Complementary and alternative medicine: Therapy of cough. *In* Pharmacology and Therapeutics of Cough, (K. F. Chung and J. G. Widdicombe, eds.), **187**, pp. 321–342. Springer-Verlag, Berlin Heidelberg.

Wong, W. C., Lee, A., Lam, A. T., Li, K. T., Leung, C. Y., Leung, P. C., Wong, E. L. and Tang, J. L. (2006). Effectiveness of a Chinese herbal medicine preparation in the treatment of cough in uncomplicated upper respiratory tract infection: A randomized double-blinded placebo-control trial. *Cough* **2**, 5.

Xu, Y. T., Hon, P. M., Jiang, R. W., Cheng, L., Chan, Y. P., Li, S. H., Xu, H. X., Shaw, P. C. and But, P. P. H. (2006). Antitussive effects of *Stemona tuberosa* with different chemical profiles. *Journal of Ethnopharmacology* **108**, 46–53.

Xu, Y. T., Shaw, P. C., Jiang, R. W., Hon, P. M., Chan, Y. M. and But, P. P. H. (2010). Antitussive and central respiratory depressant effects of *Stemona tuberosa*. *Journal of Ethnopharmacology* **128**, 679–684.

Yang, X. Z., Zhu, J. Y., Tang, C. P., Ke, C. Q., Lin, G., Chang, T. Y., Rudd, J. A. and Ye, Y. (2009). Alkaloids from roots of *Stemona sessilifolia* and their antitussive activity. *Planta Medica* **75**, 174–177.

Zhang, Y. Z., Tong, Z. and Tao, J. S. (2010). Study on pharmacokinetics of protostemonine in rats. *Li Shizhen Medicine and Materia Medica Research* **21**, 1364–1365.

Zhao, Z. Z. (2005) *An Illustrated Microscopic Identification of Chinese Materia Medica*. Hong Kong, Chung Hwa Book Co., pp. 281–283.

Zhou, Y., Jiang, R. W., Hon, P. M., Xu, Y. T., Chan, Y. M., Chan, T. W. D., Xu, H. X., Ding, L. S., But, P. P. H. and Shaw, P. C. (2006). Analyses of *Stemona* alkaloids in *Stemona tuberosa* by liquid chromatography/tandem mass spectrometry. *Rapid Communication in Mass Spectrometry* **20**, 1030–1038.

Zhou, X., Leung, P. H., Li, N., Ye, Y., Zhang, L., Zuo, Z. and Lin, G. (2009). Oral absorption and antitussive activity of tuberostemonine alkaloids from the roots of *Stemona tuberosa*. *Planta Medica* **75**, 575–580.

Ziment, I. (2002). Herbal antitussives. *Pulmonary Pharmacology & Therapeutics* **15**, 327–333.

# *In Vitro* Propagation and Approaches for Metabolites Production in Medicinal Plants

SUSHIM KUMAR GUPTA,* CHAO-LIN KUO,† HUNG-CHI CHANG,* HSIAO-SUNG CHAN,* EMILY CHIN-FUN CHEN,* FU-SHIN CHUEH‡ AND HSIN-SHENG TSAY*,§,[1]

**Institute of Biochemical Sciences and Technology, Chaoyang University of Technology, Taichung, Taiwan*

*†School of Chinese Pharmaceutical Sciences and Chinese Medicine Resources, China Medical University, Taichung, Taiwan*

*‡Department of Health and Nutrition Biotechnology, Asia University, Taichung, Taiwan*

*§Department of Agronomy, National Chung-Hsing University, Taichung, Taiwan*



[1]Corresponding author. E-mail: hstsay@cyut.edu.tw

Advances in Botanical Research, Vol. 62 0065-2296/12 $35.00

DOI: 10.1016/B978-0-12-394591-4.00002-7

## ABSTRACT

There has been a renewed interest in the use of herbal medicines throughout the world due to toxicities and health hazards associated with synthetic drugs and antibiotics. The World Health Organization estimates that up to 80% of people still rely mainly on traditional remedies such as herbs for their medicine, resulting in the increasing demand for medicinal plants. About 85% of traditional medicines involve the use of plant extracts. However, a large number of medicinal plants are needed to be investigated for their possible pharmaceutical value. Many pharmaceutical industries harness wildly growing plant populations for the supply of raw materials for extraction of medicinally important compounds. Many of the medicinal plants are severely threatened owing to illicit and indiscriminate collection and destruction of natural habitat. Advanced biotechnological methods of culturing plant cells and tissue provide an alternative means for rapid propagation and conservation of rare and endangered and/or commercially important medicinal plants. This chapter reviews the work carried out by our group in Taiwan on *in vitro* propagation of *Pinellia ternata*, *Momordica charantia*, *Gentiana scabra* var Bunge and *Taraxacum formosanum* and production of secondary metabolites from the callus of *Stephania tetrandra*, hairy root of *Salvia miltiorrhiza* and cell suspension culture of *Gentiana davidii* and *Dioscorea doryophora*.

## I. INTRODUCTION

Millions of people in the third world countries prefer herbal medicines, because they believe in them and regard it as their medicines in contrast to western allopathic prescriptions. Over the past 2500 years, there have been very strong traditional systems of medicines, such as Chinese, Ayurvedic and Unani, born and practised, more in the eastern continent. These traditions have been flourishing since then; approximately 80% of populations in the developing countries still rely on these systems of medicines for their primary health care needs. Now, even in western countries, there is an increased awareness about herbal medicines, largely due to powerful synthetic agents used in the western medicines which exert more unwanted side effects. About 85% of traditional medicines involve the use of plant extracts (Vieira and Skorupa, 1993). Many plant species processing medicinally important compounds are disappearing at an alarming rate because of destructions of their natural habitats owing to rapid agricultural development, urbanization, indiscriminate deforestation and uncontrolled collection of plant materials.

Taiwan is situated in tropical and subtropical zones, but it also possesses a temperate zone based on the ecological system of its mountains. The central mountain range in Taiwan is the hot spot for many highly valued medicinal herbs. Indeed, surveys have shown that almost 7000 species of vascular plants are grown in Taiwan, including at least 4477 native species and 2500 introduced species. Moreover, one-fourth of native species are indigenous and many of these plants have medicinal uses. Like many other parts in the world, in Taiwan too, medicinal herbs are being indiscriminately collected in large quantities from the wild to meet the ever-increasing demand for traditional crude drugs. Plant biotechnology can play a significant role in increasing plant population of locally available medicinally important plant species and in the conservation and maintenance of biodiversity. The National Science Council (NSC) of Taiwan has been promoting research on traditional Chinese medicinal plants since 1988. The main objectives of these research programmes are (i) to collect information about important and rare traditional medicinal herbs, (ii) to develop a simple method for identification of medicinal herbs, (iii) to develop methods for mass propagation of medicinal herbs through *in vitro* methodologies, (iv) to study active principles and pharmacology for their safer use and (v) to promote export of traditional medicinal herbs. On the basis of initial exciting findings for the production of medicinal compounds using *in vitro* cultures, an array of research investigations has been undertaken in the area of plant, cell, tissue and organ cultures to produce valuable metabolites from traditional medicinal plants collected from Taiwan. We have successfully propagated many medicinally important species through shoot morphogenesis and/or somatic embryogenesis (Liu *et al.*, 2011; Nalawade and Tsay, 2004; Nalawade *et al.*, 2003). Also, we have been successful in establishing callus and cell suspension culture of several plant species and extracting medicinally important compounds (Gupta *et al.*, 2011; Kuo *et al.*, 2011; Mulabagal and Tsay, 2004). This chapter reviews the work carried out by our group in Taiwan on *in vitro* propagation of *Pinellia ternate*, *Gentiana scabra* var Bunge, *Momordica charantia* and *Taraxacum formosanum* and production of secondary metabolites from callus of *Stephania tetrandra*, hairy root of *Salvia miltiorrhiza* and cell suspension culture of *Gentiana davidii* and *Dioscorea doryophora*.

# II. *IN VITRO* PROPAGATION OF MEDICINAL PLANTS

## A. *PINELLIA TERNATE*

*P. ternate* belonging to family Araceae grows wildly in Japan and China. It is a perennial herb. Tubers of *P. ternate* contain homogentisic acid, its glucoside, 3,4-dihydroxybenzaldehyde, its diglucoside and ephedrine (Shoyama *et al.*, 1983). The species has been used in Chinese medicines to prevent vomiting and for analgesic and sedative effects. The traditional Japanese herbal (Kampo) medicine 'Sho-seiryu-to' that is used for the treatment of cold syndromes (Nagai and Yamada, 1994) contains tubers of *P. ternate* as one of the components. Pinellic acid from the tubers had shown oral adjuvant activity for nasal influenza HA vaccine (Nagai *et al.*, 2002). Presently, the species is not under cultivation owing to the limited planting materials in the wild. Hence, tissue culture studies were carried out with a view to establish mass propagation method for this important medicinal plant.

An efficient method for plant regeneration via adventitious bud or protocorm-like body formation directly from the bulbils, leaf blade and petiole explants without intervening callus has been developed by Tsay *et al.* (1989). The explants developed adventitious buds when cultured on MS basal medium supplemented with BA (1.0–5.0 mg/l) and NAA (0.0–0.2 mg/l). The maximum number of adventitious buds (73 per explant) was induced from protocorm-like body cultured on the MS basal medium supplemented with BA alone, without any NAA (Table I). It was observed that incorporation of NAA in the medium at all levels of BA had adverse effect on the proliferation rate. However, a different trend was observed with adventitious buds as an explant. At higher levels of BA (5.0 mg/l), proliferation rate was drastically improved on inclusion of NAA in the proliferation medium (Table I).

The regeneration efficiency varied with the explant type. Maximum response was observed in bulbils, followed by leaf blades and petiole explants. The protocorm-like bodies induced on the medium containing BA and NAA multiplied at a prolific rate and were then cut into parts and transferred to a liquid MS basal medium supplemented with BA (1.0–5.0 mg/l) and NAA

TABLE I

*Effect of BA and NAA on Shoot Bud Proliferation in Adventitious Bud and Protocorm like Bodies of* Pinellia ternate[a]

| BA/NAA (mg/l) | | Number of buds proliferated per protocorms like body | Number of buds proliferated per adventitious buds |
|---|---|---|---|
| 1 | 0 | 55.5 ± 6.3 | 15.0 ± 3.1 |
| 1 | 0.2 | 46.0 ± 5.7 | 12.1 ± 2.0 |
| 2 | 0 | 73.0 ± 6.8 | 21.8 ± 3.6 |
| 2 | 0.2 | 60.2 ± 6.1 | 23.0 ± 3.4 |
| 5 | 0 | 70.0 ± 7.2 | 11.7 ± 2.6 |
| 5 | 0.2 | 56.1 ± 5.9 | 21.5 ± 2.9 |

Incubation period: 30 days; No. of explants per treatment: 50.

[a]Culture medium: MS basal medium with agar (1%), sucrose (3%).

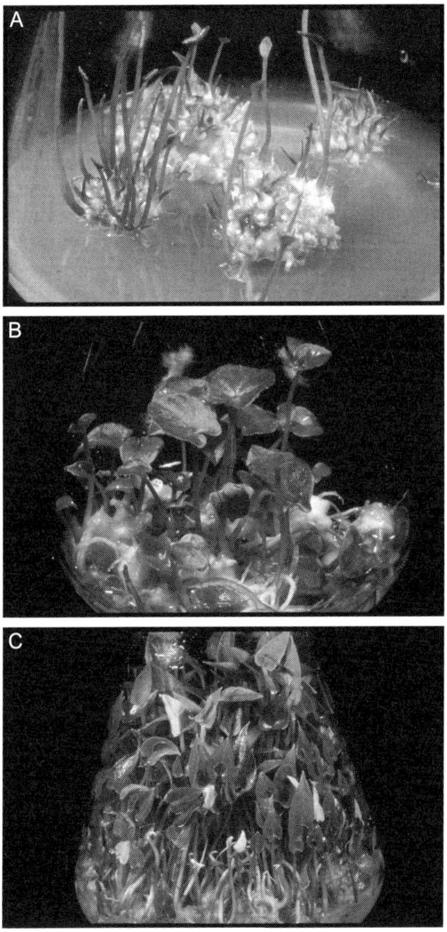


Fig. 1. (A) Multiple shoots induction in protocorm-like bodies in *P. ternata* (2 weeks after culture in MS basal agar medium). (B) Shoot regeneration in protocorm-like bodies in *Pinellia ternata* (4 weeks after culture in MS basal liquid medium). (C) Shoot elongation in *P. ternata* (4 weeks after culture in MS basal, agar medium). (See color plate section at the back of the book.)

(0.2 mg/l) or 2,4-D (0.2 mg/l). The regenerative potential of the culture could be maintained by culturing them alternately in solid and liquid medium (Fig. 1A–C). The plant that was regenerated from protocorm-like bodies was observed after continuous culture in liquid medium (Fig. 1A). Rooting was achieved in half strength MS basal medium supplemented with NAA (0.1 mg/l). A high survival rate of plants (96%) was achieved on their transfer to a mixture of vermiculite:loam soil:peat moss potting substrates.

No morphological abnormalities in tissue-culture-raised plants were observed when compared to plants grown in field. By using this *in vitro* procedure, it is estimated that about $1.7 \times 10^{27}$ plantlets from single bulbils can be produced in 1 year (Tsay and Agrawal, 2005).

### B. *MOMORDICA CHARANTIA*

*M. charantia* (commonly known as bitter gourd) is a member of the family Cucurbitaceae, cultivated in tropical and subtropical regions. *M. charantia* is grown for its fruits, young shoots and edible flowers. The immature fruits are a good source of phosphorus, iron, and vitamins A and C (Grover and Yadav, 2004). Its notable bitterness is attributed to the non-toxic glycoside momordicine, which has been used for centuries as traditional medicine. Bitter melon has been the subject of countless studies and has demonstrated significant pharmacological activity in a variety of experimental models. Bitter melon increases the renewal of beta-cells in the pancreas, or is involved in the recovery of partially destroyed beta-cells (Ahmed *et al.,* 1998), and improves uptake of glucose (Welihinda *et al.,* 1986). Several constituents of bitter melon (e.g. alpha- and beta-momorcharin, lectin and MAP 30) have demonstrated *in vitro* antiviral activity (Beloin *et al.,* 2005; Foa-Tomasi *et al.,* 1982; Grover and Yadav, 2004). Broad-spectrum antibacterial activity has been demonstrated for the leaf extracts (Grover and Yadav, 2004).

The newly developed variety of *M. charantia* L. Hualien No.1 is fast growing, attains early maturity, has short lifespan (110–130 days), and is high yielding and disease resistant. Since most of the bitter gourd species are prone to interspecific hybridization and seeds are produced in the natural habitat, there are chances of inbreeding depression. Thus, micropropagation may be a significant approach to conserve the genotype. Attempts have been already made in cross-pollinated wood apple (*Aegle marmelos* L.) to conserve the genotype through multiple shoot regeneration (Das *et al*., 2008). An efficient method of regeneration through shoot apex and nodal segment has been developed by Liu *et al.* (2011) (Fig. 2A and B). The explants developed adventitious shoots when cultured on MS medium supplemented with 3% sucrose and 1.0 mg/l BA. The shoot apex and nodal segment became unresponsive to adventitious bud formation when 1.0 mg/l TDZ was used with 0.1 mg/l NAA. Regenerated shoots (~1 cm) were excised and transferred into a 500-ml conical flask containing 100 ml ½ × MS medium supplemented with different concentrations of IBA (indole butyric acid) and NAA. Rooting was achieved in half strength MS basal medium supplemented with 1.0 mg/l IBA (Fig. 2C). A high survival rate of plants (100%) was achieved on their transfer to a mixture containing turf soil:pearl stone:vermiculite (3:1:1),

Fig. 2. *In vitro* regeneration of *Momordica charantia* Hualien No 1. Shoot formation from (A) shoot apex, (B) nodal segment, (C) root formation, (D) *ex vivo* acclimation and (E) acclimatized plants in clay pot. (For colour version of this figure, the reader is referred to the online version of this chapter.)

and the plants were kept in the growth chamber under high light intensity of 100 μmol/m$^2$ s for 16 h photoperiod at 22/16 °C light/dark with 80% relative humidity for 2 weeks for hardening (Fig. 2D). Acclimatized hardened plants were later transferred to greenhouse for further growth (Fig. 2E). The established *in vitro* regeneration system may be used in the future to conserve the genotype of this newly developed variety of bitter gourd.

### C. *GENTIANA SCABRA* VAR BUNGE

The genus *Gentiana* belongs to the family Gentianaceae and comprises over 400 species that are widely distributed in alpine habitats of temperate regions of Asia, Europe and the Americas. *Gentiana* is one of the important medicinal herbs in traditional Chinese medicine, commonly known as 'longdan', and it has been used for over 2000 years as an analgesic, anti-inflammatory, antipyretic, antirheumatic, diuretic, febrifuge, hypoglycaemic and for treating hypotensive rheumatic pains, fevers and allergic inflammations (Tang and Eisenbrand, 1992a,b). The dried roots and rhizomes from *G. manshurica* Kitag, *G. scabra* Bunge, *G. triflora* Pall and G. *rigescens* Franch are officially listed in the Chinese Pharmacopoeia under the name of *Radix gentianae*, commonly known as Longdan.

*G. scabra*, colloquially called Japanese Gentian or Long Dan Cao, is a semi-evergreen flowering plant native to Japan. The delicate *G. scabra* produces star-shaped flowers in a range of vivid blues and purples. An efficient method of plant regeneration using shoot apex as explants has been developed recently in our laboratory. The maximum regeneration frequency was obtained when cultured on ½ × MS basal medium supplemented with 3% sucrose, 0.3% gelrite, 0.1 mg/l BA and 0.1 mg/l NAA (Fig. 3A). It was observed that higher concentration of BA and NAA had reduced total regeneration frequency. The higher regeneration frequency was obtained with 0.3% gelrite compared to 0.9% agar; thus, 0.3% gelrite was the preferred gelling agent for regeneration (Fig. 3B). The higher *in vitro* plant growth response in all aspects of the measurements was obtained, when containers were enclosed with two layers of non-permeable aluminium foil and sealed with parafilm for first 4 weeks followed by their enclosure in four layers of dispense paper sealed from side for another 4 weeks. The survival frequency increased further, when plantlets were directly transferred to green house without prior hardening in the growth chamber (Fig. 3C). The developed methodology can be implemented for rapid generation of healthy plantlets with the highest survival frequency.

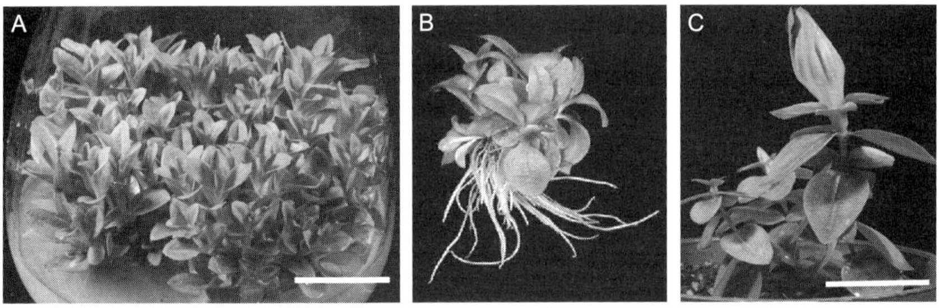


Fig. 3. *In vitro* regeneration of *G. scabra*. (A) Regenerated plantlet in 1/2 × MS major salts and vitamins, 3% sucrose, 1 mg/l BA, 0.9% Bacto-agar for 8 weeks. (B) *In vitro* regenerated plantlets of *G. scabra*. (C) Acclimatized plant in greenhouse. (For colour version of this figure, the reader is referred to the online version of this chapter.)

## D. *TARAXACUM FORMOSANUM*

*T. formosanum* is a member of the family Asteraceae, subfamily Cichorioideae and tribe Lactuceae, which is distributed mainly in the littoral areas of northern Taiwan. It is an endemic species in Taiwan but is also distributed in Japan, South Korea and Eastern China. This perennial herbaceous species has been known since ancient times for its curative properties, and it is being utilized either alone or in combination with other herbs (collectively called dandelion) for the treatment of various ailments such as mastitis, tonsillitis, cholestasis, hepatitis, bronchitis or pneumonia and also used to enhance immune response to upper respiratory tract infections, enhance urinary secretion and clean the waste of the liver (Bluementhal *et al.*, 1998; Leu *et al.*, 2005; Sweeney *et al.*, 2005). The inflorescences, leaves and roots of dandelion are used as an active ingredient in different processed food products. Tender leaves of cultivated or wild species are consumed fresh as salad, whereas roots are roasted and utilized as herbal drink (Leung and Foster 1996).

Owing to limited geographical distribution, indiscriminate collection and less frequency of self-pollination, *T. formosanum* is now rarely found in its natural habitat and listed in second grade endangered species in Taiwan (IUCN, 1994). An efficient method of plant regeneration from leaf and root was developed recently in our laboratory. Shoot induction was initiated within a week directly from the cut surfaces of explants on ½ × MS basal medium supplemented with 3% sucrose. However, shooting rate was significantly affected by the type of primary explant used and higher shoot induction was recorded from the leaf (1.9 ± 2.6 shoots per explant). Thus, different segments of the leaf (Fig. 4) were used for multiple shooting in subsequent

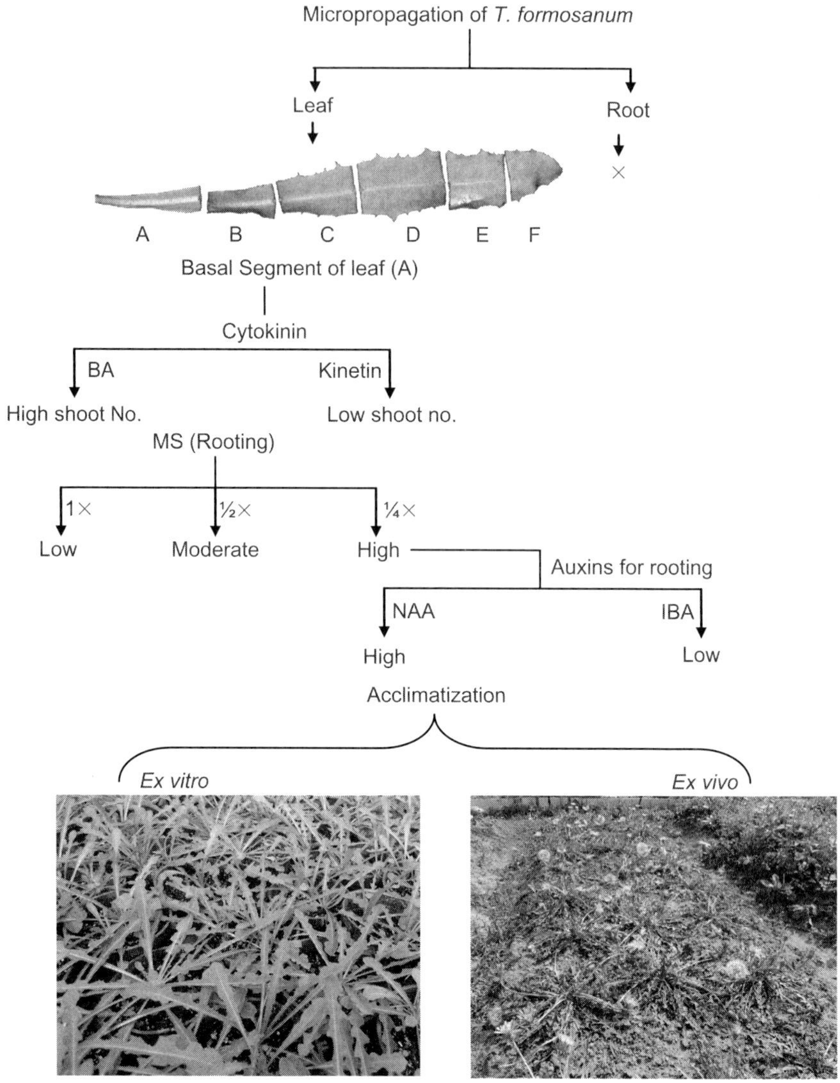


Fig. 4. Schematic presentation of *in vitro* methodology applied for micropropagation of *T. formosanum*. (For colour version of this figure, the reader is referred to the online version of this chapter.)

experiments. The basal part of the leaf (petiole) designated as A responded better to shoot multiplication (Fig. 4). The number of shoots generated per explant improved with the addition of BA on ½ × MS, whereas a different concentration of kinetin was ineffective in enhancing shooting. An average of 3.4 shoots was obtained with BA at 0.3 mg/l and approximately 92.3%

explants responded to shooting with 13-fold increase in fresh weight, whereas the multiplication frequency reduced further with the increased BA concentration (0.5 mg/l) and minimum shoots were observed at higher BA concentration (2.0 mg/l). The influence of the strengths of MS on rooting was investigated to determine the optimal nutrient medium for rooting. The best result was obtained on ¼ × MS medium, with a mean of 3.3 roots. The rooting frequency was further enhanced by the addition of different concentrations of NAA and IBA (0.1–0.5 mg/l). NAA at low concentrations favoured rooting compared to high concentrations, and an average of 5.2 roots was obtained with 0.1 mg/l NAA. IBA induced less roots compared to NAA. The benefit of any micropropagation system can be fully realized by the successful acclimatization of plantlets from tissue-culture vessels to the ambient conditions found *ex vitro*. *In vitro* raised plantlets with well-developed shoots and roots were transferred to pots containing sterile soilrite in the growth chamber with 100% acclimatization. The schematic description of the methodology followed for plant regeneration and acclimation is presented in Fig. 4. Acclimatized plants were transferred to the greenhouse and open field, where they grew well and their phenotypic variation was recorded after 3 months.

## III. PRODUCTION OF SECONDARY METABOLITES

With an ever-increasing global inclination towards herbal medicine, there is an obligatory demand not only for a huge raw material of medicinal plants but also on right stage when active principles are available in optimum quantities at the requisite time for standardization of herbal preparation. Based on initial exciting findings for the production of medicinal compounds using *in vitro* cultures, an array of research investigations has been undertaken in the area of plant, cell, tissue and organ cultures to produce valuable metabolites from traditional medicinal plants collected from Taiwan.

### A. *IN VITRO* PRODUCTION OF FANGOCHINOLINE AND TETRANDRINE FROM *S. TETRANDRA*

*S. tetrandra* S. Moore (Fan fang ji) belongs to the family Menispermaceae. In traditional Chinese medicines, *S. tetrandra* is prescribed for its diuretic, antiphlogistic and antirheumatic properties (Huang, 1999). In the Chinese Pharmacopoeia, the herb is recommended for treating general symptoms such as inflammation, fever, pain, oedema and stagnant bronchial mucous and for detoxification. The use of Fan fang ji roots has been limited to China. Many of its pharmacological properties are relatively well described including its hypotensive activity (Kawashima *et al.*, 1990), its use as a vasodilator (Kim *et al.*, 1997) and an anti-fibrotic (Nan *et al.*, 2000), and its use in coronary disease (Yu *et al.*, 2001) and retinopathy (Liang *et al.*, 2002).

Plant secondary metabolites have enormous potential for research and new drug development. Many secondary metabolites have complex and unique structure and their production can be enhanced by introducing different types of additives into the basal media. An efficient *in vitro* callus induction system in *S. tetrandra* was established on MS medium supplemented with 3% sucrose and different concentrations (0.5 to 2.0 mg/l) of plant growth regulators, that is, auxins (2,4-D, IAA and NAA) and cytokinins (BA, kinetin, TDZ and zeatin) in the dark. Different plant parts (leaf, stem, petiole and root) were used as explants for the callus induction, and 100% callus induction was obtained from the leaves. Approximately, 200 mg of fresh callus was used as the initial material and transferred to MS media supplemented with 3% sucrose, 0.9% agar, 1 mg/l BA and 0.5 mg/l TDZ. Half of the replicates were grown under continuous cool white fluorescent lamps (100 $\mu E/m^2$ s), and the other half of the calluses were grown in complete darkness at $25 \pm 1$ °C. Although callus mass increased sevenfold in the dark as well as under light after 6 weeks, dark condition promoted fangochinoline (Fan) and tetrandrine (Tet) accumulation (Table II). Different auxins and cytokinins were examined to study their effects on Fan and Tet accumulation. No Fan and Tet were detected in the auxin-derived callus, whereas cytokinin-derived callus favoured Fan and Tet accumulation. Higher amounts of Fan and Tet were obtained with 1 mg/l TDZ and BA, respectively (Table II).

TABLE II

*Effect of Light and Different Growth Regulators on Fangchinoline and Tetrandrine Accumulation in Callus Culture of* Stephania tetrandra

| | Different conditions* | Fan** (mg/g) | Tet (mg/g) |
|---|---|---|---|
| Physical condition | Light | $0.10^b$ | 0.0 |
| | Dark | $0.203^a$ | $0.09^a$ |
| Auxin (1 mg/l) | 2-4D | 0.0 | 0.0 |
| | IAA | 0.0 | 0.0 |
| | NAA | 0.0 | 0.0 |
| Cytokinins (1 mg/l) | BA | $0.192^b$ | $0.087^c$ |
| | Kinetin | $0.088^{e, d}$ | 0.0 |
| | TDZ | $0.191^a$ | $0.083^c$ |
| | Zeatin | $0.181^c$ | $0.087^b$ |

*Data collected from the 200 mg of the callus were cultured for 45 days with varying concentrations of different constituents.

**Different letter within a column indicate significant differences at 5% level by LSD (least significant difference) test.

### B. ENHANCED TANSHINONE PRODUCTION IN HAIRY ROOTS OF '*S. MILTIORRHIZA* BUNGE'

*S. miltiorrhiza* Bunge (Lamiaceae) is a well-known Chinese herb and the rhizome is known as 'tanshen' or 'danshen', an ancient drug in traditional Chinese medicine (Duke and Avensu, 1985). It has been used widely to treat coronary heart disease, menstrual disorders, miscarriage, hypertension and viral hepatitis (Chang and But, 1986). Organized cultures, especially root cultures, can make a significant contribution to secondary metabolite production. *A. rhizogenes* causes hairy-root disease in plants. The neoplastic (cancerous) roots produced by *A. rhizogenes* infection are characterized by high growth rate, genetic stability and growth in hormone-free media (Srivastava and Srivastava, 2007). These genetically transformed root cultures can produce higher levels of secondary metabolites comparable to that of intact plants. It is a valuable tool for plant biotechnology and a promising tissue culture method for mass production of useful plant secondary metabolites (Giri and Narasu, 2000). The greatest advantage of hairy roots is that they often exhibit similar or greater biosynthetic capacity for secondary metabolite production as compared to their mother plants (Kim *et al.*, 2002a,b). *S. miltiorrhiza* roots contain two major classes of chemicals, a class of lipid-soluble diterpene quinone pigments, generally known as tanshinone, and a class of water-soluble phenolic acids (Tang and Eisenbrand, 1992a,b). Among these tanshinones, cryptotanshinone is a typical compound having an *ortho*-quinone skeleton. Cryptotanshinone acts as an effective topoisomerase I inhibitor (Gantchev and Hunting, 1997) and exhibits significant cytotoxicity against a number of cultured human tumour cell lines (Ryu *et al.*, 1997).

Leaves were cut into 0.25 $cm^2$ pieces and a small cut was made perpendicular to the midrib into the centre of the squared leaf. Explants were precultured on MS basal medium for 1 day prior to infection. *A. rhizogenes* strain BCRC15010 (Food Industry Development Institute, Taiwan) was grown on BEP (beef extract and peptone) medium at 28 °C in the dark overnight. The cultures were inoculated with fresh BEP media and grown for 8–10 h. Cells were collected by centrifugation and resuspended in liquid MS medium at an $OD_{580}$ of 0.4. Precultured explants were submerged into the bacterial suspension for 30 min by shaking. After blotting off the excess bacterial suspension, discs were transferred to MS basal medium and co-cultivated for 2 days. They were then cultured on MS basal medium supplemented with 3% sucrose and cefotaxime (200 mg/l) and gelled with 0.9% agar (w/v). Hairy root initiations were observed in 75–80% infected leaves after 3 weeks. The hairy roots, which developed from the infected areas, were individually excised after 3 weeks and transferred to B5 medium (Gamborg *et al.*, 1968) containing 200 mg/l of cefotaxime. After 4–5 subcultures, the roots were able to grow without *A. rhizogenes* in the absence of cefotaxime and were negatively geotropic. Hairy roots obtained from a single clone were transferred to 70 ml B5 liquid medium in 250-ml flasks and kept at 100 rpm. Stable transgenic lines with high multiplication rates were selected for analysis. We investigated the effects of exogenous PGR on the growth and tanshinone accumulation in hairy roots of *S. miltiorrhiza.* Hairy roots were cultured in a medium containing various concentrations of PGR, and the results are summarized in Table III. There was no significant difference in hairy root multiplication with ABA, IAA, NAA and kinetin (101.67, 103.00, 116.67 and 118.33 mg) compared to the control (98.67 mg). However, a significant difference in root multiplication was observed with IBA, BA and TDZ (137.00, 139.33 and 185.00 mg, respectively). The multiplication of hairy roots was in direct proportion to the increased rate of root branching and elongation. Although ABA had minimal influence on hairy root multiplication, it caused 2-, 3- and 3-fold increase in tanshinone I, tanshinone IIA and cryptotanshinone contents, respectively. On other hand, TDZ both influenced hairy root multiplication and caused 2-, 1.5- and 3.5-fold increase in tanshinone I, tanshinone IIA and cryptotanshinone content, respectively (Table III).

TABLE III

*Influence of Different PGR on Tanshinones Accumulation from the Hairy Root of* Salvia miltiorrhiza *in Liquid Medium**

| PGR constituents | Average dry weight after 12 weeks** (mg) | Weight gain (folds) | TI[a] (mg/g of dw) | TII[b] (mg/g of dw) | CT[c] (mg/g of dw***) |
|---|---|---|---|---|---|
| 0 | $98.67 \pm 7.22^{b}$ | 18 | $0.216 \pm 0.06$ | $0.120 \pm 0.02$ | $0.374 \pm 0.06$ |
| NAA | $116.67 \pm 8.95^{a,b}$ | 21 | ND | $0.069 \pm 0.01$ | $0.158 \pm 0.04$ |
| IAA | $103.00 \pm 10.07^{b}$ | 19 | $0.134 \pm 0.05$ | $0.097 \pm 0.02$ | $0.249 \pm 0.05$ |
| IBA | $137.00 \pm 10.15^{a}$ | 25 | ND | $0.069 \pm 0.02$ | $0.195 \pm 0.02$ |
| BA | $139.33 \pm 8.76^{a}$ | 25 | $0.430 \pm 0.06$ | $0.300 \pm 0.04$ | $0.772 \pm 0.09$ |
| Kinetin | $118.33 \pm 11.02^{a,b}$ | 22 | $0.316 \pm 0.06$ | $0.210 \pm 0.02$ | $0.534 \pm 0.07$ |
| TDZ | $185.00 \pm 8.14^{a}$ | 34 | $0.393 \pm 0.03$ | $0.195 \pm 0.03$ | $1.310 \pm 0.17$ |
| ABA | $101.67 \pm 6.69^{b}$ | 19 | $0.451 \pm 0.05$ | $0.341 \pm 0.02$ | $1.038 \pm 0.11$ |

ND, Not detected.

*B5 basal salts supplemented with 3% sucrose. Observations were recorded after 12 weeks of culture.

**Value are means ± standard error, $n = 3$. Different letter within a column indicate significant differences at 5% level by LSD (least significant difference) test.

***Dry weight.

[a]Tanshinone I.

[b]Tanshinone IIA.

[c]Cryptotanshinone.

### C. GENTIOPICROSIDE AND SWERTIAMARIN PRODUCTION FROM *G. DAVIDII* VAR. *FORMOSANA* (HAYATA)

*G. davidii* var. *formosana* (Hayata) commonly known as longdan in Chinese belongs to the family Gentianaceae. The genus *Gentiana* comprises over 400 species that are widely distributed in alpine habitats of temperate regions of Asia, Europe and the Americas. Some species also occur in northwest Africa, eastern Australia and New Zealand (Georgieva *et al.*, 2005). Members of the family are glabrous with opposite, simple leaves and often lack stipulus. Eleven species and two additional varieties of *Gentiana* have been recorded from Taiwan (Chen and Wang, 1999; Wang and Chen, 1998). The dried roots and rhizomes, which contain bitter-tasting secoiridoid glucosides, have been used in the Chinese system of traditional medicine for over 2000 years as an analgesic, anti-inflammatory, antipyretic, antirheumatic, diuretic, febrifuge and hypoglycaemic and for treating hypotensive rheumatic pains, fevers and allergic inflammations (Tang and Eisenbrand, 1992a,b).

Continuous collection of plant material from the natural habitat has led to the depletion of *G. davidii* population. *G. davidii* comes under the list of protected plants. This section of the book provides details of the cell suspension cultures of *G. davidii*. The initiation of callus is the first step for the suspension culture. The method for callus induction from different tissues is essentially the same as conventional procedure, viz, sterilization of tissue (5–10 min in 0.5% (w/v) solution of sodium hypochlorite after dipping into 70% (v/v) ethanol for 30 s), rinsing with sterile water and inoculation of plants with the media. The stem of *G. davidii* was used as an explant for callus induction and callus was induced on MS basal medium supplemented with naphthalene acetic acid (NAA) and kinetin. The cultures were incubated at $25 \pm 1$ °C in the dark for 5–7 weeks. The fragile callus was cut into small pieces and cultured in approximately 20 ml liquid MS medium supplemented with NAA and kinetin in a 125-ml Erlenmeyer flask for the establishment of primary cell suspension cultures.

Gentiopicroside and swertiamarin are the two important secoiridoid glucosides found in *G. davidii*. Gentiopicroside is found to be capable of suppressing chemically and immunologically induced hepatic injuries (Kondo *et al.*, 1994). The aglucons of gentiopicroside and swertiamarin showed inhibitory actions against *Staphylococcus aureus* (Isiguro *et al.*, 1982). The gentiopicroside and swertiamarin have been successfully isolated from the cell suspension culture of the *G. davidii* (Chueh *et al.*, 2000). The samples were freeze-dried in a lyophilizer. The dried samples were crushed into fine powder and ultra-sonicated for 10 min in methanol (Branson Ultrasonic Cleaner, Branson Cleaning Equipment Co., Shelton, CT, USA). The supernatant was collected after centrifugation and the process repeated three times for each sample. The combined methanol extracts were evaporated to dryness in a rotary evaporator. The residue was dissolved in 10 ml methanol and filtered through 0.45–0.20 μm (Nalgene®, New York, USA) membrane filter and the filtrate was diluted appropriately for HPLC analysis. The gentiopicroside and swertiamarin contents were analysed in the suspension culture of *G. davidii*, which was harvested every fourth day up to 36 days. The swertiamarin contents started accumulating after 4 days and reached a maximum at 12 days and then started decreasing again (Fig. 5). The gentiopicroside contents started accumulating after 16 days and reached a maximum at 24 days and then started decreasing again (Fig. 5). Thus, to obtain maximum swertiamarin and gentiopicroside from the suspension culture, the culture can be harvested after 12 days and 24 days, respectively. Thus, to obtain maximum gentiopicroside and from the suspension culture, the culture may be harvested 24 days after culture. These contents in aerial as well as underground part of *G.davidii* were analysed. The gentiopicroside and swertiamarin contents varied in different plant parts of *G. davidii* depending on the age of plants (Table IV). Higher gentiopicroside content was found in the aerial part of *G. davidi* compared to that in the underground part, and it increased with age. However, the higher swertiamarin content was found in the underground part of the plant, compared to that in the aerial parts. The swertiamarin contents decreased with advanced age in underground parts, while it increased in the aerial parts of *G. davidii* (Table IV).

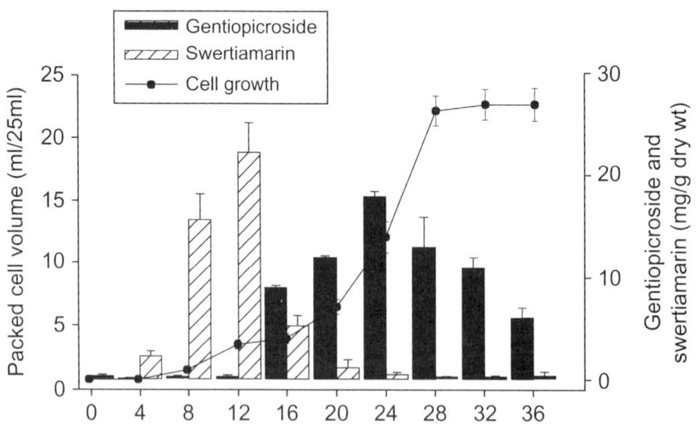


Fig. 5. Time course on the production of gentiopicroside and swertiamarin and the suspension cell growth of *G. davidii* var. *formosana*. The basal medium was MS salts supplemented with 0.2 mg/l kinetin, 3% sucrose, pH 5.2, initial PCV 1 ml and incubated under low light intensity (2.33 μE $m^{-2}$ $s^{-1}$) at 80 rev/min shaker speed.

TABLE IV

*Analysis of Gentiopicroside and Swertiamarin in Aerial and Underground Part of Wild and* In Vitro *Propagated Plants of* Gentiana davidii

| Plant materials | | Gentiopicroside (mg/g dry weight) | | Swertiamarin (mg/g dry weight) | |
|---|---|---|---|---|---|
| | | Underground part | Aerial parts | Underground part | Aerial parts |
| *G. davidii* | A | 7.60 ± 2.74 | 44.75 ± 3.42 | 0.77 ± 0.21 | 0.31 ± 0.02 |
| | B | 28.68 ± 7.93 | 52.40 ± 5.71 | 0.48 ± 0.06 | 0.29 ± 0.02 |
| | C | 52.53 ± 5.68 | 78.35 ± 10.37 | 0.66 ± 0.18 | 0.79 ± 0.18 |

A, *in vitro* rooted plant; B, Six months old acclimatized plant; C, Wild plant; NA, Not analyzed; ±, Standard deviation.

## D. DIOSGENIN PRODUCTION FROM THE CELL SUSPENSION CULTURE OF *D. DORYOPHORA* HANCE

*Dioscorea* belongs to the family Dioscoreaceae, which are abundant in nature and distributed widely in the tropics. The rhizomes of some of the species are commonly used as traditional Chinese medicines (TCMs; Pharmacopoeia Committee of P. R. China, 2010). *D. doryophora* Hance, a Taiwanese species whose tuber possesses high quality and higher quantity of the active component, is being wildly propagated by farmers in Taiwan (Huang *et al.*, 1993). The tuber is in great demand as it is used both as a crude drug and as food. The most important active principle discovered in tubers is diosgenin, which can be used as a precursor of many important medicinal steroids such as prednisolone, dexamethasone, norethisterone and metenolone (Tsukamoto *et al.*, 1936). More than 50 compounds of steroidal saponins from plants of *Dioscorea* L. have been isolated (Tang *et al.*, 2007). Extensive phytochemical and pharmacological studies have shown that steroidal saponins have biological activities contributing to the efficacy of the herb (Liu *et al.*, 2010).

The cell suspension culture was established by Yeh *et al.*, 1994 in *D. doryophora* for the production of diosgenin. A plant cell suspension culture is a sterile (closed) system normally initiated by aseptically placing friable callus fragments into a suitable sterile liquid medium (Dixon and Paiva, 1995). Callus tissue in *D. doryophora* was induced from the surface of both microtuber and stem node. The explants are placed on different solid growth media; in successful media, callus tissue will appear on the explants in the course of 2–6 weeks. A callus is subsequently cut from the explant and subcultured further. Callus material can be inoculated in a liquid medium (in flasks), and through continuous shaking cell suspension cultures are obtained. Finally, dispersed cell suspension cultures were obtained from both microtuber- and stem node-derived callus in liquid culture medium supplemented with 2,4-D (0.1 mg/l) and sucrose (3%) by incubation on a rotator shaker at 120 rpm. Although 6% sucrose was found to be optimum for the growth of cell suspension culture, it was observed that cells cultured in 3% sucrose containing medium produced higher level of diosgenin. Analysis by HPLC revealed that both microtuber- and stem node-derived suspension culture contained diosgenin. The diosgenin contents were analysed from the different plant parts (leaf, node, root, seed and stem) and higher level of diosgenin was obtained from the root (Table V). An approximately similar level of diosgenin was obtained in field grown and *in vitro* grown tubers (Chen, 1985; Table V). Diosgenin content as high as 3.3% per gram dry weight was obtained in microtuber-derived cell suspension culture compared to the stem node-derived callus cell suspension, which contained only 0.3% diosgenin (Table VI). The higher diosgenin contents and maximum biomass was obtained from the developed tuber-derived cell suspension culture of *D. doryophora*.

TABLE V

*Diosgenin Content in Callus Obtained from Different Explants of* Dioscorea *(Tsay and Agrawal, 2005)*

| Source of calllus | Diosgenin content* (%) |
|---|---|
| Seed | $0.4^{c,d}$ |
| Leaf | $0.6^{c}$ |
| Root | $1.57^{b}$ |
| Stem | 0.0 |
| Node | $0.3^{d}$ |
| Microtubers (*In vitro*) | $3.3^{a}$ |
| Microtubers (Field grown) | $3.2^{a}$ |

*(a, b, c & d) Different letters within a column indicate significant differences at P = 0.05 by Duncan's multiple range test.

TABLE VI

*Diosgenin Concentration in Cell Suspension from Two Different Explants of* Dioscorea doryophora

| Source of cell suspension | Diosgenin content* (%) |
|---|---|
| Node | $0.3^{b}$ |
| Microtubers (*In vitro*) | $3.3^{a}$ |

*(a, b) Different letters within a column indicate significant differences at P = 0.05 by Duncan's multiple range test.

## REFERENCES


Ahmed, I., Adeghate, E., Sharma, A. K., Pallot, D. J. and Singh, J. (1998). Effects of *Momordica charantia* fruit juice on islet morphology in the pancreas of the streptozotocin-diabetic rat. *Diabetes Research and Clinical Practice* **40**, 145–151.

Beloin, N., Gbeassorb, M., Akpaganab, K., Hudsonc, J., Soussab, K. D., Koumaglob, K. and Arnasona, J. T. (2005). Ethnomedicinal uses of *Momordica charantia* (Cucurbitaceae) in Togo and relation to its phytochemistry and biological activity. *Journal of Ethnopharmacology* **96**, 49–55.

Blumenthal, M., Busse, W. R., Goldberg, A., Gruenwald, J., Hall, T., Riggins, C. W. and Rister, R. S. (1998). "Dandelion herb" and "Dandelion root with herb". *In* The complete german commission E monographs (eds.). Therapeutic Guide to Herbal Medicines, pp. 118–120. American Botanical Council, Austin, Texas.

Chang, H. M. and But, P. (1986). *In* Pharmacology and Applications of Chinese Materia Medica. Vol. 1, pp. 255–268. World Science, Singapore.

Chen, A. H. (1985). Study on application of diosgenin-I. Analysis of Diosgenin constituents of plants from Taiwan. *Science Monthly* **43**, 79–85.

Chen, C. H. and Wang, J. C. (1999). Revision of the genus *Gentiana* L. (Gentianaceae) in Taiwan. *Botanical Bulletin of Academia Sinica* **40**, 9–38.

Chueh, F. S., Chen, C. C. and Tsay, H. S. (2000). Studies on factors effecting the establishment of *Gentiana davidii* var. *formosana* (Hayata) T. N. Ho cell suspension cultures. *Journal of Food and Drug Analysis* **8**, 297–303.

Das, R., Hasan, M. F., Rahman, M. S., Rashid, M. H. and Rahman, M. (2008). Study on *in vitro* propagation through multiple shoot proliferation in wood apple (*Aegle marmelos* L.). *International Journal of Sustainable Crop Production* **3**, 16–20.

Dixon, R. A. and Paiva, N. L. (1995). Stressed-induced phenylpropanoid metabolism. *The Plant Cell* **7**, 1085–1097.

Duke, J. A. and Avensu, E. S. (1985). *In* Medicinal Plants of China. Vol. 2, p. 381. Reference, Algonac, MI.

Foa-Tomasi, L., Campadelli-Fiume, G., Barbieri, L. and Stirpe, F. (1982). Effect of ribosome-inactivating proteins on virus-infected cells. Inhibition of virus multiplication and protein synthesis. *Archives of Virology* **71**, 323–332.

Gamborg, O. L., Miller, R. A. and Ojima, K. (1968). Nutrient requirements of suspension cultures of soyabean root cells. *Experimental Cell Research* **50**, 151–158.

Gantchev, T. G. and Hunting, D. J. (1997). Biochemical inhibition of the topoisomerase II-DNA cleavable complex by the *ortho*-quinone derivative of the antitumor drug etoposide (VP-16). *Biochemical and Biophysical Research Communications* **237**, 24–27.

Georgieva, E., Handjieva, N., Popov, S. and Evstatieva, L. (2005). Comparative analysis of the volatiles from flowers and leaves of three *Gentiana* species. *Biochemical Systematics and Ecology* **33**, 938–947.

Giri, A. and Narasu, L. (2000). Transformed hairy roots: Recent trends and applications. *Biotechnology Advances* **18**, 1–22.

Grover, J. K. and Yadav, S. P. (2004). Pharmacological actions and potential uses of *Momordica charantia*: A review. *Journal of Ethnopharmacology* **93**, 123–132.

Gupta, S. K., Liu, R. B., Liaw, S. Y., Chan, S. S. and Tsay, H. S. (2011). Enhanced production of tanshinone by hairy roots of '*Salvia miltiorrhiza* Bunge' under the influence of plant growth regulators in liquid culture. *Botanical Studies* **52**, 435–443.

Huang, K. C. (1999). The Pharmacology of Chinese Herbs. CRC, Boca Raton5.

Huang, W. W., Cheng, C. C., Yeh, F. T. and Tsay, H. S. (1993). Tissue culture of *Dioscorea doryophora* Hance I. Callus induction from different sources of organs and the measurement of Diosgenin content. *China Medical College Journal* **2**, 151–160.

Isiguro, K., Yamaki, M. and Takagi, S. (1982). Studies on the iridoid related compounds. On the antimicrobial activity of aucubigenin and certain iridoid aglycons. *Yakugaku Zassi* **102**, 755–759.

IUCN. International Union for Conservation of Nature and Ziranziyuan (1994). http://econgis.forest.gov.tw/rareplant/species.asp?id=t00371.

Kawashima, K., Hayakawa, T. and Miwa, Y. (1990). Structure and hypotensive activity relationships of tetrandrine derivatives in stroke-prone spontaneously hypertensive rats. *General Pharmacology* **21**, 343–348.

Kim, H. S., Zhang, Y. H., Oh, K. W. and Ahn, H. Y. (1997). Vasodilating and hypotensive effects of fangchinoline and tetrandrine on the rat aorta and the stroke-prone spontaneously hypertensive rat. *Journal of Ethnopharmacology* **58**(2), 117–123.

Kim, Y. J., Weathers, P. J. and Wyslouzil, B. E. (2002a). The growth of *Artemisia annua* hairy roots in liquid and gas phase reactors. *Biotechnology and Bioengineering* **80**, 454–464.

Kim, Y. J., Wyslouzil, B. E. and Weathers, P. J. (2002b). Invited review: Secondary metabolism of hairy root cultures in bioreactors. *In Vitro Cellular & Developmental Biology—Plant* **38**, 1–10.

Kondo, Y., Takano, F. and Hojo, H. (1994). Suppression of chemically and immunologically induced hepatic injuries by gentionpicroside in mice. *Planta Medica* **60**, 414–416.

Kuo, C. L., Chang, J. Y., Chang, H. C., Gupta, S. K., Chan, H. S., Chen, E. C. F. and Tsay, H. S. (2011). *In vitro* production of benzylisoquinoline from *Stephania tetrandra* through callus culture under the influence of different additives. *Botanical Studies* **52**, 285–294.

Leu, Y. L., Wang, Y. L., Huang, S. H. and Shi, L. S. (2005). Chemical constituents from roots of *Taraxacum formosanum*. *Chemical and Pharmaceutical Bulletin* **53**, 853–855.

Leung, A. Y. and Foster, S. (1996). Dandelion root. Encyclopedia of Common Natural Ingredients Used in Food Drugs and Cosmetics. John Wiley & Sons, New York, pp. 205–207.

Liang, X., Hagino, N., Guo, S., Tsutsumi, T. and Kobayashi, S. (2002). Therapeutic efficacy of *Stephania tetrandra* S. Moore for treatment of neovascularization of retinal capillary (retinopathy) in diabetes *In vitro* study. *Phytomedicine* **9**, 377–384.

Liu, J. Q., Wang, C. F., Qiu, M. H. and Hu, W. X. (2010). Sterodial saponins from flowers of *Hosta plantaginea* and their antitumor activities. *Chinese Traditional and Herbal Drugs* **41**, 520–526.

Liu, R. B., Gupta, S. K., Liaw, S. Y., Chien, C. H. and Tsay, H. S. (2011). *In vitro* studies on newly developed variety of *Momordica charantia* L. Hualien No.1 and its metabolites analysis. *International Journal of Integrative Biology* **11**, 155–160.

Mulabagal, V. and Tsay, H. S. (2004). Plant cell cultures—An alternative and efficient source for the production of biologically important secondary metabolites. *International Journal of Applied Science and Engineering* **1**, 29–48.

Nagai, T. and Yamada, H. (1994). *In vivo* anti-influenza virus activity of Kampo (Japnese herbal) medicine "Sho-seiryu-to" and its mode of action. *International Journal of Immunopharmacology* **16**, 605–613.

Nagai, T., Kiyohara, H., Munakata, K., Shirahata, T., Sunazuka, T., Harigaya, Y. and Yamada, H. (2002). Pinellic acid from the tuber of *Pinellia ternata*

Breitenbach as an effective oral adjuvant for nasal influenza vaccine. *International Journal of Immunopharmacology* **2**, 1183–1193.

Nalawade, S. M. and Tsay, H. S. (2004). *In vitro* propagation of some important Chinese medicinal plants and their sustainable usage. *In Vitro Cellular & Developmental Biology* **40**, 143–154.

Nalawade, S. M., Sagar, A. P., Lee, C. Y., Kao, C. L. and Tsay, H. S. (2003). Studies on tissue culture of Chinese medicinal plant resources in Taiwan and their sustainable utilization. *Botanical Bulletin of Academia Sinica* **44**, 79–98.

Nan, J. X., Park, E. J., Lee, S. H., Park, P. H., Young, K. J., Ko, G. and Sohn, D. H. (2000). Antifibrotic effect of *Stephania tetrandra* on experimental liver fibrosis induced by bile duct ligation and scission in rats. *Archives of Pharmacal Research* **23**(5), 501–506.

Ryu, S. Y., Lee, C. O. and Choi, A. U. (1997). *In vitro* Cytotoxicity of tanshinones from Salvia miltiorrhiza. *Planta Medica* **63**, 339–342.

Shoyama, Y., Hatano, K. and Nishioka, I. (1983). Clonal multiplication of *Pinellia ternata* by tissue culture. *Planta Medica* **49**, 14–16.

Srivastava, S. and Srivastava, A. K. (2007). Hairy root culture for mass-production of high-value secondary metabolites. *Critical Reviews in Biotechnology* **27**, 29–43.

Sweeney, B., Vora, M., Ulbricht, C. and Basch, E. (2005). Evidence-based systematic review of dandelion (*Taraxacum officinale*) by natural standard research collaboration. *Journal of Herbal Pharmacotherapy* **5**, 79–93.

Tang, W. and Eisenbrand, G. (1992a). *Gentiana* spp. Chinese drugs of Plant Origin, Chemistry, Pharmacology, and Use in Traditional and Modern Medicine. Springer-Verlag, Berlin, pp. 549–553.

Tang, W. and Eisenbrand, G. (1992b). Chinese Drugs and Plant Origin, Chemistry, Pharmacology, and Use in Traditional and Modern Medicine. Springer-Verlag, Berlin, p. 481.

Tang, S. R., Yang, R. T. and Pan, F. S. (2007). Steroidal saponin and steroidal sapogenin in Chinese *Dioscorea* L. *Journal of Plant Resources and Environment* **16**, 64–72.

Tsay, H. S. and Agrawal, D. C. (2005). Tissue culture technology of Chinese medicinal plant resources in Taiwan and their sustainable utilization. *International Journal of Applied Science and Engineering* **3**, 215–223.

Tsay, H. S., Gau, T. G. and Chen, C. C. (1989). Rapid clonal propagation of *Pinellia ternata* by tissue culture. *Plant Cell Reports* **8**, 450–454.

Tsukamoto, T., Yeno, Y. and Ohta, Z. (1936). Glycosides of *Dioscorea tokoro* I. Dioscin, dioscoerasapotoxin and Diosgenin. *Journal Pharmacology Society of Japan* **56**, 135.

Vieira, R. F. and Skorupa, L. A. (1993). Brazilian medicinal plants gene bank. *Acta Horticulturae* **330**, 51–58.

Wang, J. C. and Chen, C. H. (1998). *Gentiana*ceae. 2nd edn. (T. C. Huang *et al.*, eds.), Vol. 4, pp. 154–191. Editorial Committee of the Flora of Taiwan, Taipei.

Welihinda, J., Karunanayake, E. H., Sheriff, M. H. and Jayasinghe, K. S. (1986). Effect of *Momordica charantia* on the glucose tolerance in maturity onset diabetes. *Journal of Ethnopharmacology* **17**, 277–282.

Yeh, F. T., Huang, W. W., Cheng, C. C., Na, C. and Tsay, H. S. (1994). Tissue culture of *Dioscorea doryophora* Hance. II. Establishment of suspension culture and the measurement of diosgenin content. *Chinese Agronomy Journal* **4**, 257–268.

Yu, X. C., Wu, S., Wang, G. Y., Shan, J., Wong, T. M., Chen, C. F. and Pang, K. T. (2001). Cardiac effects of the extract and active components of radix *Stephaniae tetrandrae*, II. Myocardial infarct, arrhythmia, coronary arterial flow and heart rate in the isolated perfused rat heart. *Life Sciences* **68**, 2863–2872.

# Regulation of Inflammation-Mediated Chronic Diseases by Botanicals

BOKYUNG SUNG, SAHDEO PRASAD, SUBASH C. GUPTA, SRIDEVI PATCHVA AND BHARAT B. AGGARWAL[1]

*Cytokine Research Laboratory, Department of Experimental Therapeutics, The University of Texas MD Anderson Cancer Center, Houston, Texas, USA*



[1]Corresponding author: E-mail: aggarwal@mdanderson.org

Advances in Botanical Research, Vol. 62 
0065-2296/12 $35.00
DOI: 10.1016/B978-0-12-394591-4.00003-9

## ABSTRACT

Extensive research over the past several years has indicated a close association between chronic inflammation and chronic diseases. Chronic inflammation has now been shown to be involved in the onset and development of numerous chronic diseases, including cancer, neurological diseases, cardiovascular diseases, hypertension, blood pressure, atherosclerosis, diabetes, obesity, respiratory disorders, musculoskeletal disorders, gastro-intestinal disorders, and autoimmune disorders. An interesting fact that has emerged over the years is that most chronic diseases are caused by lifestyle factors such as stress, toxicants, tobacco, alcohol, infectious agents, and radiation. How chronic inflammation contributes to chronic diseases has also been elucidated over the years. The discovery of transcription factors such as NF-κB, STAT3, AP-1, NRF2, PPAR-γ, β-catenin/Wnt, HIF-1α, and Hedgehog, as well as the signalling molecules regulating these transcription factors has provided a molecular link between chronic inflammation and chronic diseases. Thus agents that can modulate the expression of these transcription factors might be useful against chronic diseases. Because of limited efficacy and high toxicity, mono-targeted anti-inflammatory agents have little effect against chronic diseases. Agents derived from natural sources called botanicals have gained particular attention for their anti-inflammatory activity, not only because they are multi-targeted but also because they are safe, cost effective, and readily available. How transcription factors contribute to the development of chronic diseases is the focus of this review. Additionally, we also describe various botanicals and the inflammatory transcription factors that they modulate.

## I. INTRODUCTION

Chronic diseases are major health concerns worldwide. According to one report, every second American has at least one chronic disease. The most common chronic diseases that are known to affect the human population are cancer, neurological diseases, cardiovascular diseases, hypertension, atherosclerosis, diabetes, obesity, respiratory disorders, musculoskeletal disorders, gastro-intestinal disorders, and immune system disorders (Table I). The past half-century has seen major advances in our understanding of the pathogenesis of chronic diseases. The discovery of transcription factors and signalling molecules regulating these factors has provided a molecular basis for chronic diseases. We now know that most of these diseases are caused by dysregulation of inflammatory pathways, leading in turn to chronic inflammation. Chronic inflammation is now known to be involved in the onset and development of numerous chronic diseases, including cardiovascular diseases, cancer, diabetes, obesity, arthritis, neurologic diseases, pulmonary diseases, psychological diseases, and autoimmune diseases (Aggarwal *et al.*, 2006; Dantzer *et al.*, 2008;

TABLE 1
*List of Chronic Diseases*

*Central Nervous System*
- Alzheimer's disease
- Attention deficit hyperactivity disorder
- Bipolar mood disorder
- Chronic fatigue syndrome
- Dementia
- Depression
- Epilepsy
- Generalized anxiety disorder
- Migraine
- Multiple sclerosis
- Parkinson's disease
- Schizophrenia
- Stroke

*Cardiovascular system*
- Angina
- Cardiac arrythmias
- Coronary artery disease
- Endocarditis
- Heart failure and cardiomyopathy
- Hyperlipidemia
- Hypertension
- Peripheral vascular disease

*Blood*
- Anemia
- Cryoglobulinemia
- Deep vein thrombosis
- Hemophilia
- Thalassemia
- Thromboangitis obliterans
- Thrombocytopenia
- Wegener's granulomatosis

*Respiratory system*
- Asthma
- Bronchiectasis
- Chronic obstructive airways disease
- Cystic fibrosis
- Pulmonary tuberculosis

*Endocrine system*
- Addison's disease
- Congenital adrenal hyperplasia
- Cushing's syndrome
- Diabetes mellitus/ Diabetes insipidus
- Hypoparathyroidism
- Pheochromocytoma
- Pituitary adenomas
- Polycystic ovarian syndrome
- Thyroid disorder

*Musculoskeletal system*
- Ankylosing spondylitis
- Dystonia
- Gout / Hyperuricemia
- Motor neuron disease
- Myasthenia gravis
- Rheumatoid arthritis / Osteoarthritis
- Osteoporosis
- Paget's disease
- Paraplegia
- Sjogren's disease
- Systemic lupus erythematosus

*Ear, Nose and Throat*
- Allergic rhinitis
- Tonsillitis
- Adenoiditis
- Chronic serous otitis media

*Gastrointestinal system*
- Chronic cholecystitis / Chronic pancreatitis
- Cirrhosis
- Gastro-esophageal reflux disorder
- Hemorrhoids
- Hepatitis
- Inflammatory bowel disease (Crohn's disease, ulcerative colitis)

*Skin*
- Dermatomyositis
- Psoriasis
- Pemiphigus
- Scleroderma

*Eye*
- Dry eye syndrome
- Glaucoma

*Genitourinary system*
- Benign prostate hypertrophy
- Chronic renal failure
- Chronic pyelonephritis

*Immune system*
- Allergies
- DiGeorge syndrome

*Miscellaneous*
- AIDS
- Cancer
- Obesity

Hansson *et al.*, 2006; Hold and El-Omar, 2008; Khanna *et al.*, 2007; Odrowaz-Sypniewska, 2007; Packard and Libby, 2008).

Reports indicate that most chronic diseases are interlinked and are caused by dysregulation of multiple genes (Argiles *et al.*, 2005). Numerous inflammatory transcription factors have been found to play a role in the pathogenesis of chronic disease. The most common transcription factors linked with chronic diseases are nuclear factor-κB (NF-κB), activator protein-1 (AP-1), signal transducer and activator of transcription-3 (STAT3), nuclear factor erythroid 2-related factor (Nrf2), peroxisome proliferator-activated receptor-γ (PPARγ), β-catenin/Wnt, hypoxia inducible factor-1 (HIF-1), and Hedgehog (Hh). An interesting fact that has emerged is that most chronic diseases are caused by lifestyle factors such as stress, toxicants, tobacco, alcohol, infectious agents and radiation (Fig. 1). How lifestyle factors contribute to chronic diseases has also been investigated over the years. These factors have been shown to induce chronic inflammation through modulation of pro-inflammatory molecules including chemokines, cytokines, pro-inflammatory transcription factors, enzymes, and other factors (Aggarwal *et al.*, 2006). According to one report, 15–20% of smokers with bronchitis have a tendency to develop lung cancer (Wingo *et al.*, 1999). Similarly, people

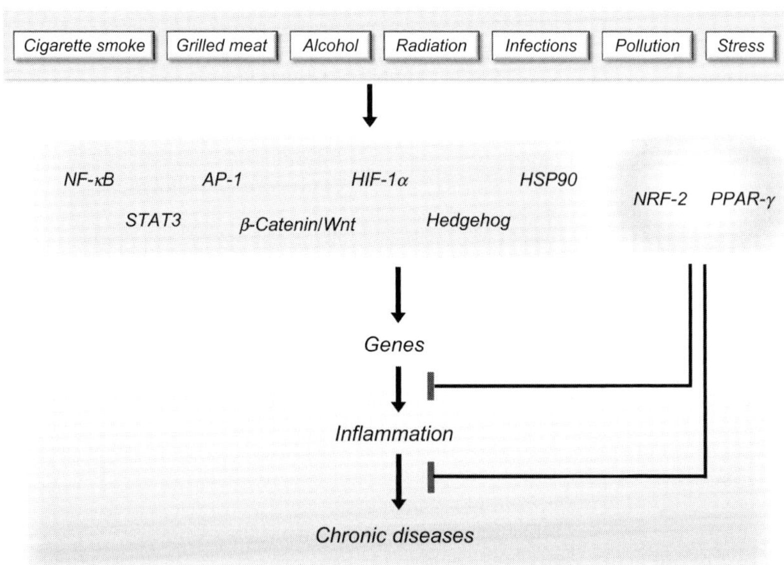


Fig. 1. List of transcription factors known to modulate chronic diseases. (See color plate section at the back of the book.).

who have colitis are at high risk of developing colon cancer (Itzkowitz and Yio, 2004). Infection with *Helicobacter pylori* has been shown to induce gastritis, which in its chronic form can lead to gastric cancer (Peter and Beglinger, 2007). Similarly, cigarette smoke is likely to account for 80–90% of the cases of pulmonary disease in the United States (Sethi and Rochester, 2000). Lifestyle factors are also known to generate reactive oxygen species (ROS) that in turn can induce inflammation (Dickinson and Chang, 2011; Mantovani, 2005). For example, cyclooxygenase-2 (COX-2), inflammatory cytokines, chemokines, and pro-inflammatory transcription factors are all known to be regulated by ROS (Hussain and Harris, 2007). These statements imply that agents with potential to down-modulate pro-inflammatory transcription factors and the genes regulated by them may have potential efficacy against such chronic diseases.

Although chronic diseases are caused by dysregulation of multiple genes, many modern medicines are based on the modulation of a single target and therefore are less likely to be effective. In addition, these medicines often produce numerous side effects and cannot be consumed over long periods of time. Therefore, the current paradigm for the prevention and treatment of chronic diseases is either to combine multiple single-targeted agents or to design a molecule that can target multiple pathways (Aggarwal *et al.*, 2009; Keith *et al.*, 2005; Mencher and Wang, 2005). Agents called nutraceuticals derived from the Mother Nature have gained considerable attention not only because they have multi-targeting properties but also because they are cost effective and immediately available. In addition, these nutraceuticals are safe and can be taken over extended periods of time. According to one estimate, more than 63% of the anti-cancer drugs introduced over the past 25 years have been natural products or can be traced back to natural sources (Newman and Cragg, 2007). In addition, some dietary agents have shown potential to inactivate inflammatory molecules by direct binding. For example, curcumin, one of the most widely studied dietary agents, has been found to bind to a number of inflammatory molecules (Gupta *et al.*, 2011).

How transcription factors contribute to chronic inflammation and chronic diseases is the focus of this review. We provide evidence supporting transcription factors as potential targets for the prevention and treatment of chronic diseases. Finally, we discuss the efficacy of common plant-derived nutraceuticals against chronic diseases. Given the large of nutraceuticals identified to date, we here focus on some of the more promising nutraceuticals, including triterpenoids, chalcones, flavonoids, chavicols, quinones, and xanthones.

# II. TRANSCRIPTION FACTORS AS MEDIATORS OF CHRONIC DISEASES

Extensive research over the past several years has indicated that pro-inflammatory genes, oncogenes, and tumour-suppressor genes are regulated by transcription factors. Transcription factors act as drivers to control gene expression and to regulate signalling pathways. Dysregulation of these transcription factors has been shown to induce chronic inflammation and chronic diseases. In this section, we will discuss how transcription factors such as NF-κB, AP-1, STAT3, Nrf2, PPARγ, β-catenin/Wnt, HIF-1, and Hh are connected to chronic diseases (Fig. 1). In addition, we will also describe the molecular chaperone, heat shock protein-90 (Hsp90), which is known to regulate numerous signalling pathways and thus has emerged as an attractive target.

## A. NUCLEAR FACTOR-κB

NF-κB is a pleiotropic transcription factor found in nearly all animal cell types. In the cell's normal state, NF-κB stays in an inactive state in the cytoplasm as a heterotrimer consisting of the subunits p50, p65, and the inhibitory subunit IκBα. In response to inflammatory stimuli, IκBα is phosphorylated at $Ser^{32}$ and $Ser^{36}$ by IκB kinase (IKK), polyubiquitinated, and subsequently proteolytically degraded (Hayden and Ghosh, 2004). The active p65–p50 dimer is then released and translocated to the nucleus, where it binds to a specific DNA sequence and activates the transcription of NF-κB-dependent genes. Since its discovery in 1986, NF-κB has been shown to regulate the expression of over 500 different genes (Fig. 2; Baker *et al.*, 2011; Gupta *et al.*, 2010; Sen and Baltimore, 1986).

Aberrant regulation of NF-κB is associated with a number of chronic diseases including cancer, AIDS, atherosclerosis, asthma, arthritis, diabetes, inflammatory bowel disease, stroke, muscle wasting, and viral infections (Aggarwal, 2004; Balkwill *et al.*, 2005; Clevers, 2006; Haefner, 2002; Karin and Greten, 2005; Karin *et al.*, 2002; Li *et al.*, 2005; Richmond, 2002). One of the common mediators for most of these diseases is inflammation, and NF-κ B has been shown to act as a link between inflammation and chronic diseases. For instance, the role of NF-κB in development of insulin resistance and obesity was demonstrated more than a decade ago (Yuan *et al.*, 2001). In that study, the authors found that mice homozygous for IKKβ ($IKK\beta^{+/-}$) are protected against insulin resistance from both diet-induced and genetic obesity (Yuan *et al.*, 2001). Insulin resistance was also effectively reversed in murine models of obesity by high-dose salicylates, which are NF-κB inhibitors (Yuan *et al.*, 2001). Salicylates have also been shown to inhibit

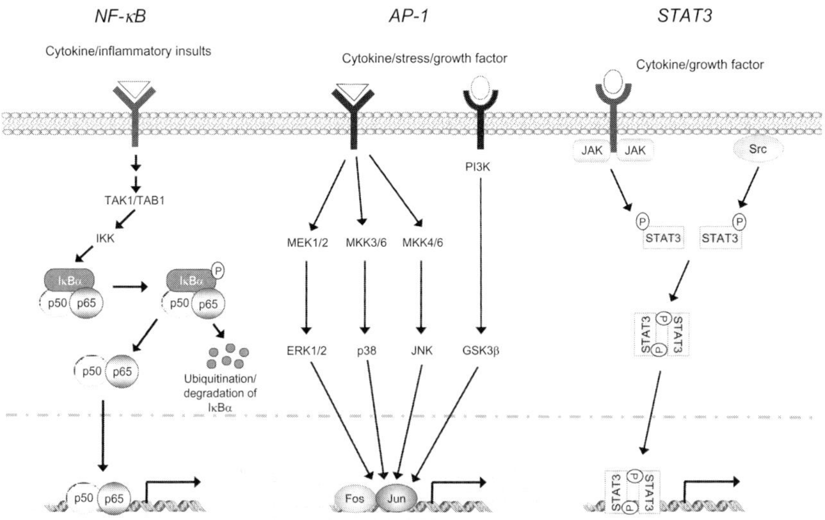


Fig. 2. Regulation of NF-κB, AP-1, and STAT3. (For colour version of this figure, the reader is referred to the online version of this chapter.).

fat-induced insulin resistance in the skeletal muscle of mice (Kim *et al.*, 2001; Yuan *et al.*, 2001). The role of NF-κB in the development of atherosclerosis has been demonstrated in a number of studies (Brand *et al.*, 1996; Ferreira *et al.*, 2007; Gareus *et al.*, 2008; Kanters *et al.*, 2004). NF-κB has been shown to control the expression of genes directing the initiation and progression of atherosclerosis, including cytokines (tumour necrosis factor [TNF]-α, interleukin [IL]-1β, IL-6), chemokine (MCP-1), and the adhesion molecule (ICAM-1).

NF-κB activation has also been shown to play a major role in numerous neurodegenerative diseases including Huntington's disease, Parkinson's disease, stroke, multiple sclerosis, and Alzheimer's disease. Activated NF-κB has been found predominantly in neurons and glial cells in beta-amyloid (Aβ) plaque surrounding areas in the brains of patients with Alzheimer's disease (Boissiere *et al.*, 1997; Kaltschmidt *et al.*, 1997; Lukiw and Bazan, 1998; Terai *et al.*, 1996). The role of NF-κB in the pathogenesis of Alzheimer's disease is further supported by the potential of NF-κB inhibitors (flurbiprofen and indomethacin) to reduce the amyloid load *in vitro* and also in transgenic mice (Eriksen *et al.*, 2003; Sung *et al.*, 2004). NF-κB-regulated genes are often found upregulated in brain tissues from patients with multiple sclerosis

(Achiron and Gurevich, 2006; Chabas *et al.*, 2001; Lindberg *et al.*, 2004; Lock *et al.*, 2002; Tajouri *et al.*, 2003). Similarly, mice lacking functional NF-κB have been found to be protected from the development of allergic airway disease (Donovan *et al.*, 1999; Yang *et al.*, 1998)

The role of NF-κB in mediating cancer progression is evident from numerous studies (Aggarwal, 2004). This transcription factor has now been shown to regulate the expression of genes associated with tumour cell development, including those associated with transformation (Balmain and Pragnell, 1983), survival (Kreuz *et al.*, 2001; Tamatani *et al.*, 1999; Wang *et al.*, 1998), proliferation (Ahn and Aggarwal, 2005; Habib *et al.*, 2001; Mukhopadhyay *et al.*, 2002; Romashkova and Makarov, 1999), invasion, angiogenesis, and metastasis (Wang *et al.*, 1999).

### B. ACTIVATOR PROTEIN-1

AP-1 was first identified as a transcription factor that binds to an essential *cis*-element of the human metallothionein lla promoter (Lee *et al.*, 1987). AP-1 is mainly composed of the Jun (c-Jun, JunB, JunD) and Fos (c-Fos, FosB, Fra-1, Fra-2) subfamilies. AP-1 is activated by numerous stimuli including pro-inflammatory cytokines, stress, and growth factors (Fig. 2). This pleiotropic transcription factor plays a role in a wide range of cellular processes, including inflammation, cell proliferation, survival, and differentiation and is considered a key player in inducing chronic diseases in humans. AP-1 has been shown to play a role more specifically in the development of inflammatory bowel disease, chronic obstructive pulmonary disease, rheumatoid arthritis, and psoriasis (Rannou *et al.*, 2006; Shimizu *et al.*, 2006).

Elevated AP-1 activity has been detected in a variety of cancers and tumour cell lines, suggesting a role for AP-1 in tumour progression (Young *et al.*, 2003). Bernstein and Colburn were the first to show the role of AP-1 in tumour promotion (Bernstein and Colburn, 1989), reporting that transformation-resistant JB6 cells failed to activate AP-1 in response to such tumour promoters as 12-*O*-tetradecanoylphorbol 13-acetate and epithermal growth factor (EGF), whereas the AP-1 response was intact in transformation-sensitive JB6 cells. AP-1 regulates the expression of genes known to mediate proliferation and angiogenesis, such as c-Myc and fos, COX-2, urokinase-type plasminogen activator (uPA), matrix metallopeptidase (MMP)-9, cyclin D1, and vascular endothelial growth factor (VEGF; Shaulian and Karin, 2001, 2002). This transcription factor also represses tumour-suppressor genes such as p53, p21cip1/waf-1, and p16 (Eferl and Wagner, 2003).

C. SIGNAL TRANSDUCER AND ACTIVATOR OF TRANSCRIPTION-3

STAT3 is one of six members of a family of transcription factors. It was first identified as a DNA-binding factor that selectively binds to the IL-6-responsive element in the promoter of acute-phase genes from IL-6-stimulated hepatocytes (Akira *et al.*, 1994). STAT3 is normally present in the cytoplasm of most cells and is activated by inflammatory stimuli and growth factors. STAT3 is activated by phosphorylation at $Tyr^{705}$ and $Ser^{727}$. The phosphorylation of STAT3 in the cytoplasm leads to its dimerization, translocation into the nucleus, DNA binding, and gene transcription (Fig. 2). Several protein kinases are known to cause specific phosphorylation of STAT3, including Janus-activated kinases (Lutticken *et al.*, 1994), EGF receptor (EGFR) kinase (Garcia *et al.*, 1997), Src (Yu *et al.*, 1995), and extracellular signal-regulated kinase (ERK; Megeney *et al.*, 1996).

Aberrant regulation of STAT3 has been shown to be associated with the development of numerous chronic diseases, especially cancer. This transcription factor is now known to promote every step of tumour development. For example, tumour cell transformation has been shown to be regulated by STAT3 (Yoshida *et al.*, 2002). Similarly, STAT3 plays a role in the regulation of genes associated with tumour cell survival (Catlett-Falcone *et al.*, 1999; Zushi *et al.*, 1998). STAT3 activation has also been shown to promote proliferation of certain tumour cells. The ability of STAT3 to promote proliferation depends upon its ability to induce expression of cyclin D1 (Masuda *et al.*, 2002). Other growth-promoting genes known to be regulated by STAT3 include c-Myc (Kiuchi *et al.*, 1999) and pim-1 (Shirogane *et al.*, 1999). STAT3 has also been reported to play a major role in tumour cell invasion, angiogenesis, and metastasis through numerous mechanisms (Xiong *et al.*, 2008; Zhao *et al.*, 2008). The role of STAT3 in tumour metastasis is supported by observations that blockage of activated STAT3 in highly metastatic cells significantly suppresses the invasiveness of the tumour cells and prevents metastasis in nude mice. Furthermore, overexpression of activated STAT3 correlates with the invasion and metastasis of cutaneous squamous cell carcinoma (Suiqing *et al.*, 2005). STAT3 also controls the expression of the *MUC1* gene, which can mediate tumour invasion (Gaemers *et al.*, 2001). STAT3 has also been involved in the regulation of VEGF and tumour angiogenesis (Niu *et al.*, 2002). STAT3 has also been shown to regulate TWIST, another mediator of tumour metastasis (Cheng *et al.*, 2008).

STAT3-regulated pro-inflammatory cytokines have also been shown to play a pivotal role in the pathogenesis of asthma (Litonjua *et al.*, 2005). STAT3 has been shown to play a role in the development of idiopathic pulmonary fibrosis (Bonniaud *et al.*, 2003, 2004; Lasky *et al.*, 1998; Pan *et al.*, 2001; Wu *et al.*, 2006). STAT3 activation has also been implicated in the pathogenesis of renal diseases (Arakawa *et al.*, 2008). Although STAT3 is active in most chronic diseases, in Alzheimer's disease, STAT3 was reported to be inactive (Chiba *et al.*, 2009).

### D. β-CATENIN/WNT

The Wnt and Frizzled (Fz) family genes were first discovered in *Drosophila*. In mammalian systems, around 19 Wnt ligands and more than 12 Fz receptors have been identified (Cadigan and Nusse, 1997; Logan and Nusse, 2004). The Wnt signalling is mediated through binding of Wnt ligands to Fz receptors. The Wnt-Fz complex inactivates cytoplasmic proteins such as adenomatous polyposis coli (APC) and axin, which otherwise degrade β-catenin. This results in the cytoplasmic accumulation and nuclear localization of β-catenin. In the nucleus, β-catenin interacts with T-cell factor/lymphoid enhancer factors (TCF/LEF) to regulate the expression of target genes (Fig. 3).

Dysregulation or mutations in Wnt have been implicated in a variety of chronic human diseases, including cancer (Clevers, 2006). Wnt and Fz

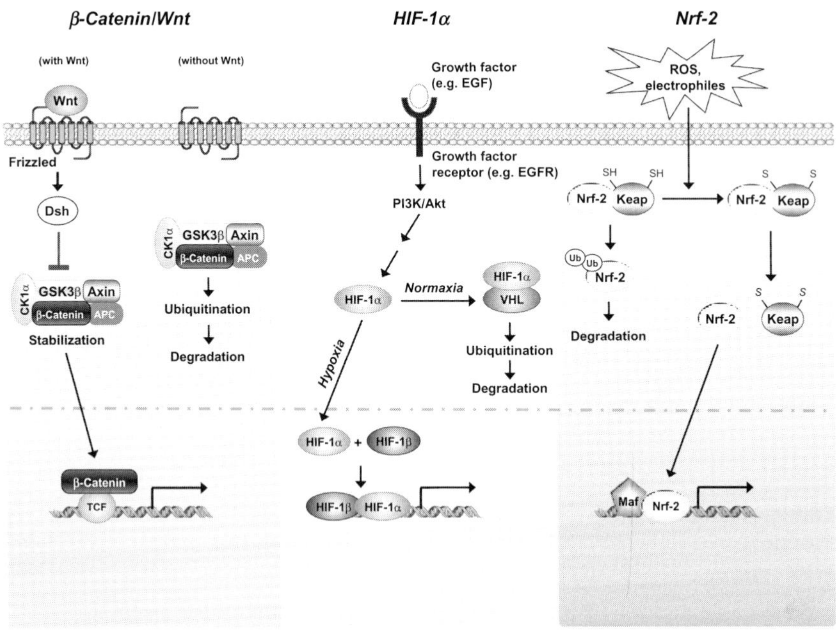


Fig. 3. Regulation of Wnt/β-catenin, HIF-1, and Nrf-2. (For colour version of this figure, the reader is referred to the online version of this chapter.).

homologues have been found to be highly expressed in the fibroblast-like synoviocytes isolated from the rheumatoid arthritis synovium (Sen *et al.*, 2000). An anti-Wnt1 antibody or an excess of the Wnt1 antagonist has been shown to reduce both cartilage degradation and invasion of fibroblast-like synoviocytes into cartilage (Sen *et al.*, 2002). The β-catenin/Wnt pathway has been implicated in gastric cancer (Oshima *et al.*, 2006), and the overexpression of COX-2, which increases prostaglandin $E_2$ ($PGE_2$) production, is critical in gastric tumourigenesis. Infection from *H. pylori* has also been shown to activate the Wnt pathway (Chang *et al.*, 2004; Franco *et al.*, 2005). Oshima *et al.* found an association between Wnt and COX-2 overexpression and development of gastric tumours in mice (Oshima *et al.*, 2006). The inflammatory cytokine TNF-α has also been shown to activate β-catenin/Wnt through the inhibition of GSK3β, which also contributes to tumour progression (Oguma *et al.*, 2008). Hyperactive β-catenin turns on a genetic programme sufficient to initiate the development of a multitude of different tumour types, primarily those of gastro-intestinal origin. One of the best-known examples is familial adenomatous polyposis (FAP), an inherited disease in which patients have numerous polyps in the colon and rectum. For example, truncations in APC have been shown to promote aberrant activation of the Wnt pathway, leading to adenomatous lesions in up to 70% of colorectal cancer cases (Nishisho *et al.*, 1991).

### E. HYPOXIA INDUCIBLE FACTOR-1

HIF-1 is a major transcription factor that is induced in response to hypoxia and has been shown to regulate the expression of more than 70 genes linked with cellular adaptation and survival under hypoxic stress (Semenza, 1998). HIF-1 is a member of the PAS [per/aryl-hydrocarbon-receptor nuclear translocator (ARNT)/Sim] family of basic helix–loop–helix transcription factors. These transcription factors consist of an oxygen-sensitive α-subunit and a constitutively expressed β-unit, also known as the ARNT or simply HIF-1β (Fig. 3; Wenger, 2002).

HIF-1 is considered to be a crucial transcription factor involved in the progression of a broad range of human malignancies (Giaccia *et al.*, 2003; Maxwell *et al.*, 2001; Semenza, 2003). Solid tumours often have high levels of HIF-1α (Bertout *et al.*, 2008). HIF can directly upregulate a number of genes linked with tumour cell proliferation (c-myc), angiogenesis (*VEGF*, *PDGF*), apoptosis/autophagy (*NDRG2*, *BNIP3*), extracellular matrix remodelling (*LOX*, *MMP1*), and cell migration and invasion (*CXCR4*, *SDF1*). HIF-1 has also been shown to promote tumour metastasis through regulation of E-cadherin, TWIST, and SNAIL (Erler *et al.*, 2009; Esteban *et al.*, 2006;

Mak *et al.*, 2010; Yang *et al.*, 2008). In obesity, the imbalance of oxygen tension in enlarged adipocytes causes hypoxia and an increase in inflammatory adipokines in fat. The resultant infiltration by macrophages and chronic low-grade systemic inflammation promote insulin resistance (Ye, 2009). The elevated HIF-1 in experimental animal models of diabetic nephropathy (Edlund *et al.*, 2009) has been correlated with the severity of the disease (Higgins *et al.*, 2007). In cases of inflammatory bowel disease, not only does the entire mucosa become more hypoxic, but surgical specimens of the inflamed intestine contain elevated levels of HIF-1$\alpha$ (Giatromanolaki *et al.*, 2003; Karhausen *et al.*, 2004). Because hypoxia and inflammatory conditions are closely intertwined, targeting hypoxia-dependent signalling pathways could help to attenuate chronic inflammatory disorders.

### F. Nuclear Factor Erythroid 2-Related Factor

Nrf2 is a potent transcription factor that plays a major role in inducing cytoprotective genes in response to oxidative and electrophilic stresses (Itoh *et al.*, 1997). Nrf2 has been shown to regulate the expression of approximately 100 cytoprotective genes, including glutathione S-transferases (GSTs), heme oxygenase-1 (HO-1), NAD(P)H:quinone oxidoreductase 1 (NQO1), and glutamate-cysteine ligase, by binding to the anti-oxidant response element (ARE), a DNA regulatory element. This transcription factor is normally retained in the cytoplasm by binding with its negative regulator Keap1 (Dinkova-Kostova *et al.*, 2002). Once dissociated from Keap1, Nrf2 translocates to the nucleus, heterodimerizes with small Maf, and binds to ARE, resulting in the expression of the Nrf2 responsive gene (Itoh *et al.*, 1999; Fig. 3).

Nrf2 plays a crucial role in combating various oxidative stress-mediated diseases, including chronic diseases (Mariani *et al.*, 2005). Ramsey *et al.* found a reduction in nuclear Nrf2 in patients with Alzheimer's disease, which suggests that Nrf2 signalling may be actively involved in early-stage pathogenesis of this disease (Ramsey *et al.*, 2007). Nrf2 has also been shown to protect endothelial cells from development of atherosclerotic plaques and inflammation in response to low shear stress (Zakkar *et al.*, 2009). Compared with wild-type mice, *nrf*-deficient mice ($nrf2^{-/-}$) show increased susceptibility to cigarette smoke-induced pulmonary emphysema, indicating that Nrf2 protects against the development of emphysema (Iizuka *et al.*, 2005; Rangasamy *et al.*, 2004).

Several lines of evidence have indicated the role of Nrf2 in increasing susceptibility to carcinogens. This transcription factor has been shown to protect normal cells from malignancy, but it also promotes the survival of malignant cells by enhancing drug resistance, leading to an increased risk of cancer. The gene deletion of *nrf2* has been shown to increase the incidence, multiplicity, and size of colorectal tumours in mice exposed to azoxymethane–dextran sodium sulphate (AOM/DSS; Khor *et al.*, 2008). In addition, Nrf2-deficient mice are also more susceptible to cancers of skin (Xu *et al.*, 2006), lung (Aoki *et al.*, 2001), stomach (Ramos-Gomez *et al.*, 2001), and urinary bladder (Iida *et al.*, 2004). The mutations in *KEAP1* and Nrf2 have also been observed in human cancers (Hayes and McMahon, 2009). For example, *KEAP1* mutation (C23Y), found in tumours from breast cancer patients, has been associated with impaired ubiquitination of Nrf2 (Nioi and Hayes, 2004), and recurrent *KEAP1* gene alterations were observed in gallbladder cancer with a frequency of 30% (Shibata *et al.*, 2008).

### G. PEROXISOME PROLIFERATOR-ACTIVATED RECEPTOR

PPARs are members of the nuclear hormone receptor superfamily, which includes receptors for steroids, thyroid hormone, vitamin D, and retinoic acid. PPARs are best understood as regulators of lipid metabolism. Three isotypes of PPARs (α, β/δ, and γ) have been described (Ricote and Glass, 2007). Among these three isotypes, PPARγ plays an important role in the regulation of glucose and lipid metabolism and has been implicated in several pathological conditions, including diabetes, cardiovascular diseases, cancer, and inflammation (Belfiore *et al.*, 2009; Fig. 4). A growing body of evidence indicates that PPARγ plays a crucial role in the control of the inflammatory response by inhibiting pro-inflammatory gene expression (Rizzo and Fiorucci, 2006). Treatment with PPARγ has been shown to reduce a wide variety of inflammatory markers in several animal models of osteoarthritis, rheumatoid arthritis, sepsis, pancreatitis, atherosclerosis, ulcerative colitis, chronic asthma, and neurodegenerative diseases (Belvisi *et al.*, 2006; Moraes *et al.*, 2006).

The somatic mutations in PPARγ have been found in sporadic colorectal carcinomas, thus emphasizing the role of PPARγ as a tumour suppressor (Kinzler and Vogelstein, 1996). However, results with several murine models have suggested that, under certain circumstances, PPARγ ligands may stimulate cancer formation (Koeffler, 2003). The correlation between dysregulation of PPARγ and carcinogenesis is well documented in head and neck, colorectal, and bladder cancers and in thyroid follicular carcinomas (Hamakawa *et al.*, 2008; Kroll *et al.*, 2000; Sarraf *et al.*, 1999; Yoshimura *et al.*, 2003).

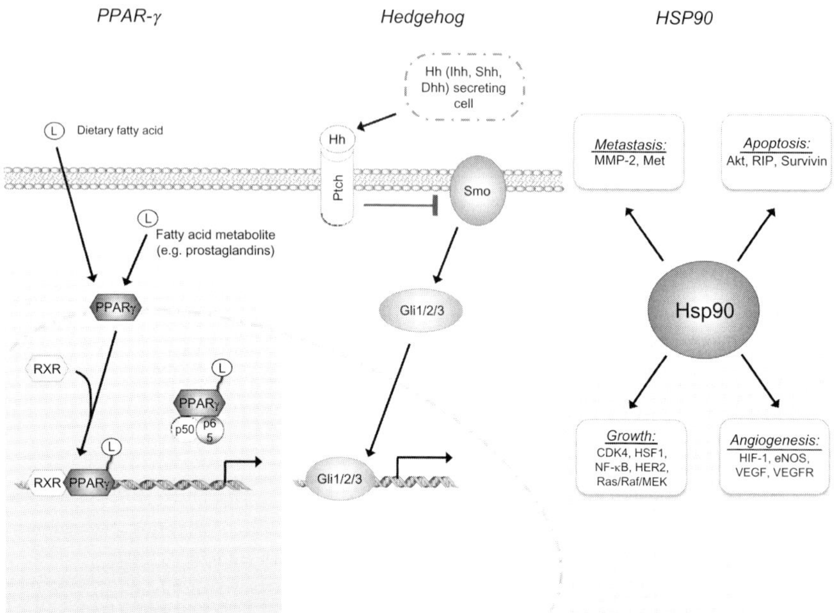


Fig. 4. Regulation of PPARγ, Hedgehog, and HSP90. (For colour version of this figure, the reader is referred to the online version of this chapter.).

## H. HEDGEHOG

Hh was discovered in 1980 in *Drosophila* embryos (Nusslein-Volhard and Wieschaus, 1980). The name of Hedgehog came from the "spiny" phenotype of the embryos, which resembled a hedgehog. Three vertebrate Hh homologues have been identified: Sonic hedgehog (Shh), Indian hedgehog (Ihh), and Desert hedgehog (Dhh). Shh is one of the most widely studied Hh pathways in vertebrates. Key regulatory components of the Hh signalling include Smoothened (SMO), a seven-transmembrane domain cell surface protein essential to pathway activation, and PTCH1, a cell surface receptor protein that serves as a primary repressor of SMO. Interaction of any of the Hh ligands with PTCH1 relieves PTCH1 repression of SMO, leading to downstream pathway activation (Fig. 4; Yang *et al.*, 2010).

Shh signalling has been associated with numerous inflammatory diseases (Lowrey *et al.*, 2002; Stewart *et al.*, 2002, 2003; Wang *et al.*, 2003). Stewart *et al.* found an association between Shh signalling and chronic lung inflammation in human T cells (Stewart *et al.*, 2002). These authors found that Shh expression was confined to the epithelium, with damage at sites of tissue remodelling and fibrosis but not in normal lung tissue in models of fibrotic disease.

The dysregulation of Hh signalling has been associated with the induction of numerous cancer types, including medulloblastomas, leukaemia, glioma, and cancers of the pancreas, lung, ovary, breast, prostate, skin, and colon (Guo *et al.*, 2011; Jagani *et al.*, 2010; Li *et al.*, 2011; Mantamadiotis and Taraviras, 2011; Morris *et al.*, 2010; Yang *et al.*, 2010). Activation of the Hh pathway is also found in poorly differentiated and more aggressive tumours (Fukaya *et al.*, 2006; Wang *et al.*, 2006b). Blockage of Hh signalling has been shown to inhibit epithelial-to-mesenchymal transition (EMT) and metastases in pancreatic cancer cell lines (Feldmann *et al.*, 2007). Thus, inhibitors of the Hh pathway might be useful against inflammatory diseases and for prevention of inflammation-related tumour progression.

### I. HEAT SHOCK PROTEIN 90

Hsp90 is the most abundant cytosolic heat shock protein family in eukaryotic cells and ranges in molecular weight from 82 to 90 kDa. Under physiological conditions, Hsp90 has been found in association with several intracellular proteins including calmodulin, actin, tubulin, kinases, and some receptor proteins (Csermely *et al.*, 1998). Hsp90 is highly conserved and is expressed in a variety of different organisms from bacteria to mammals. In mammalian cells, two or more genes encode Hsp90 homologues (Chen *et al.*, 2006). Basically, Hsp90 has a role in assisting in the folding, intracellular transport, maintenance, and degradation of proteins as well as in facilitating cell signalling. Hsp90 has also been shown to suppress the aggregation of a wide range of "client" or "substrate" proteins and hence acts as a general protective chaperone (Fig. 4; Jakob *et al.*, 1995; Miyata and Yahara, 1992; Wiech *et al.*, 1992). In addition, Hsp90 stabilizes various growth factor receptors (Sawai *et al.*, 2008) and some signalling molecules including PI3K and AKT proteins; hence inhibition of Hsp90 may induce apoptosis in cancer cells (Mohsin *et al.*, 2005; Stebbins *et al.*, 1997). Another important role of Hsp90 in cancer is the stabilization of mutant proteins such as v-Src, the fusion oncogene Bcr/Abl, and mutant forms of p53 that appear during cell transformation.

## III. ROLE OF BOTANICALS AGAINST CHRONIC DISEASES

### A. ROLE OF CHALCONES AGAINST CHRONIC DISEASES

Chalcones belong to the flavonoid family and are often responsible for the yellow pigmentation in plants. Chalcones have a variety of biological properties, including analgesic, anti-oxidant, anti-fungal (Opletalova and Sedivy, 1999), anti-bacterial, anti-protozoal (Herencia *et al.*, 1999; Hsieh *et al.*, 1998), gastric protectant, anti-mutagenic, anti-tumourogenic (Makita *et al.*, 1996), and anti-inflammatory properties (Ban *et al.*, 2004; Herencia *et al.*, 1999). Figure 5 shows the chemical structures of chalcones that exhibit anti-inflammatory activity. Among the numerous chalcones that have shown potential against various chronic diseases are butein, cardamonin, and xanthohumol.

*1. Butein*

Butein (3,4,2′,4′-tetrahydroxychalcone) is obtained from stem bark of cashews (*Semecarpus anacardium*) and the heartwood of *Dalbergia odorifera*. Butein has been shown to exhibit anti-inflammatory, anti-cancer, and anti-fibrogenic activities and to inhibit NO production and iNOS expression. It has also been shown to inhibit activation of pro-inflammatory transcription factor NF-κB (Lee *et al.*, 2004b; Pandey *et al.*, 2007a). Butein has also been shown to inhibit the proliferation, metastasis, and invasion of multiple myeloma and leukaemia cells through inhibition of the STAT3 and NF-κB pathways (Pandey *et al.*, 2007a, 2009). Butein has shown efficacy against colon (Yit and Das, 1994), melanoma (Iwashita *et al.*, 2000), osteosarcoma (Jang *et al.*, 2005), lymphoma (Lee *et al.*, 2004a), breast (Samoszuk *et al.*, 2005), and bladder cancer cells (Zhang *et al.*, 2008).

Butein has also shown promise for the prevention and treatment of diabetic complications. Butein has been found to inhibit cytokine-induced β-cell damage, which can be helpful in the prevention of type 1 diabetes mellitus (Jeong *et al.*, 2011). In rats, butein was shown to inhibit angiotensin converting enzyme in a dose-dependent manner (Kang *et al.*, 2003).

*2. Cardamonin*

Cardamonin is isolated from the fruits of *Alpinia rafflesiana* and has demonstrated anti-inflammatory activity in cellular models of inflammation. Anti-inflammatory activity of cardamonin is likely to be mediated through inhibition of NO and $PGE_2$ synthesis and the key molecules in the NF-κB activation pathway. In one study, cardamonin suppressed the production of NO and $PGE_2$ in IFN-γ- and lipopolysaccharide (LPS)-induced RAW 264.7 cells (Israf *et al.*, 2007). In colon cancer cells, cardamonin was shown to enhance the apoptotic effects of TRAIL through CHOP-mediated upregulation of death receptors (DRs; Yadav *et al.*, 2012a).

Protective effects of cardamonin against cardiovascular disease have been observed in experimental models (Wang *et al.*, 2001). Cardamonin has also been shown to suppress systemic hypertension in rat artery myocytes (Fusi *et al.*, 2010). Cardamonin has also been shown to play a role in ameliorating insulin resistance and smooth muscle hyperplasia of major vessels in rats with

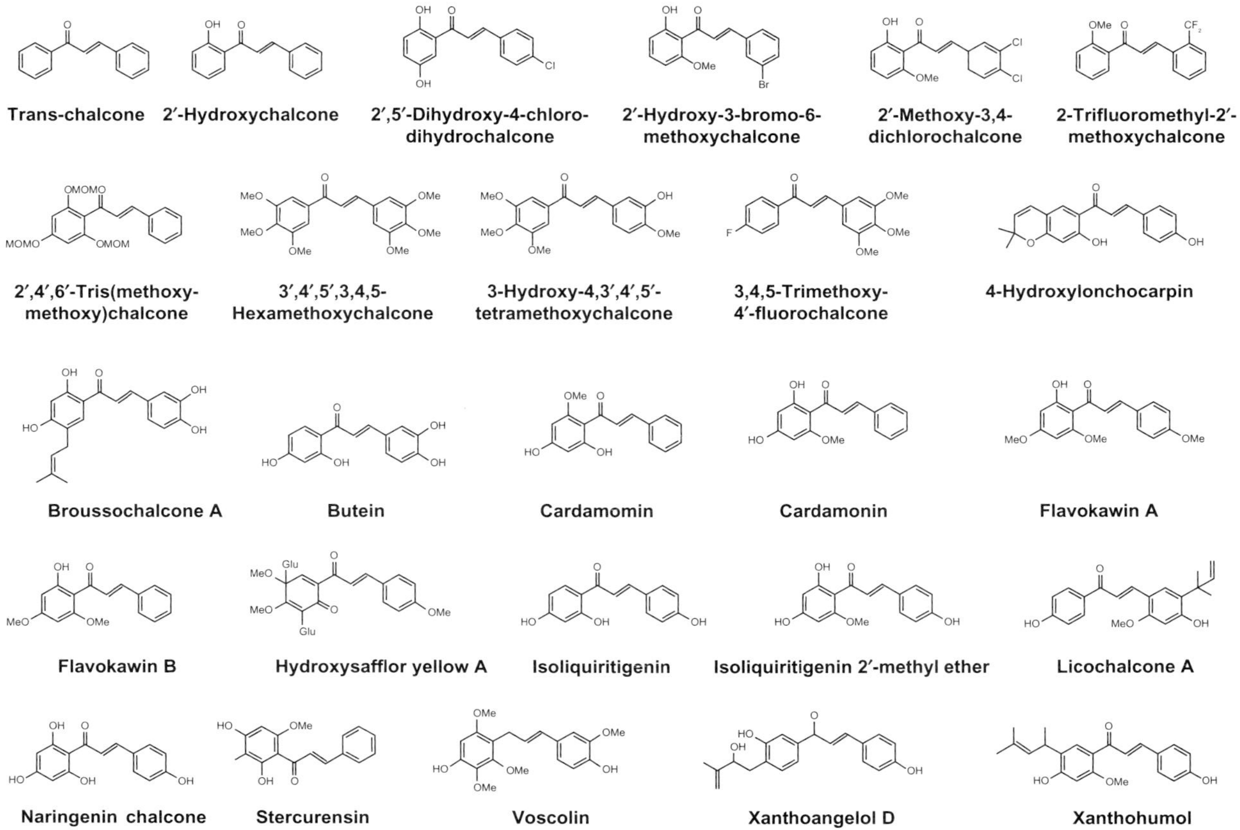


Fig. 5. Chemical structures of chalcones that exhibit anti-inflammatory activity.

fructose-induced diabetes, possibly via a mechanism that involves the modulation of insulin/mTOR signalling (Liao *et al.*, 2010).

*3. Xanthohumol*

Xanthohumol is a prenylated chalcone and is isolated from the hop plant. It has demonstrated potential as an anti-inflammatory, anti-proliferative, and anti-angiogenic agent. Xanthohumol has been shown to inhibit IL-12 production in macrophages stimulated by LPS/interferon-γ and to attenuate skin inflammation (Cho *et al.*, 2010) and to inhibit both constitutive and inducible NF-κB activation (Harikumar *et al.*, 2009). Xanthohumol inhibited the growth and proliferation of breast (Miranda *et al.*, 1999), ovarian (Drenzek *et al.*, 2011), colon (Goncalves *et al.*, 2011), and prostate (Deeb *et al.*, 2010) cancer cells.

Xanthohumol has shown potential against other chronic diseases as well. In one study, xanthohumol-fed diabetic mice had lowered levels of plasma glucose and hepatic triglyceride. Xanthohumol-fed diabetic mice also had less white adipose tissue and higher levels of plasma adiponectin (Nozawa, 2005). Xanthohumol has also been shown to inhibit the proliferation and to induce apoptosis in adipocytes (Rayalam *et al.*, 2009). Xanthohumol reduced differentiation of adipocytes, as shown by decreased lipid content and PPARγ expression in 3T3-L1 cells (Mendes *et al.*, 2008).

## B. ROLE OF FLAVONOIDS AGAINST CHRONIC DISEASES

Flavonoids are plant-based anti-inflammatory agents with potential against numerous chronic diseases. Figure 6 shows the chemical structures of flavonoids. In this section, we discuss some of the promising flavonoids such as fisetin, deguelin, gossypin, and morin.

*1. Fisetin*

Fisetin (3,7,3′,4′-tetrahydroxyflavone) is most commonly found in the smoke tree (*Cotinus coggygria*) and is also widely distributed in fruits and vegetables (Arai *et al.*, 2008). This flavonoid exhibits a wide variety of activities, including neurotrophic (Maher, 2006), anti-oxidant (Hanneken *et al.*, 2006), anti-inflammatory (Higa *et al.*, 2003), and anti-angiogenic (Fotsis *et al.*, 1998) activities. In a mouse model of LPS-induced acute pulmonary inflammation, fisetin significantly reduced the levels of myeloperoxidase, IL-6, TNF-α, IL-1β, MIP-1α, and MIP-2 (Geraets *et al.*, 2009).

Fisetin has shown efficacy against numerous cancer types including prostate cancer (Haddad *et al.*, 2006), liver cancer (Chen *et al.*, 2002), colon cancer (Lu *et al.*, 2005), and leukaemia (Lee *et al.*, 2002). Fisetin has also

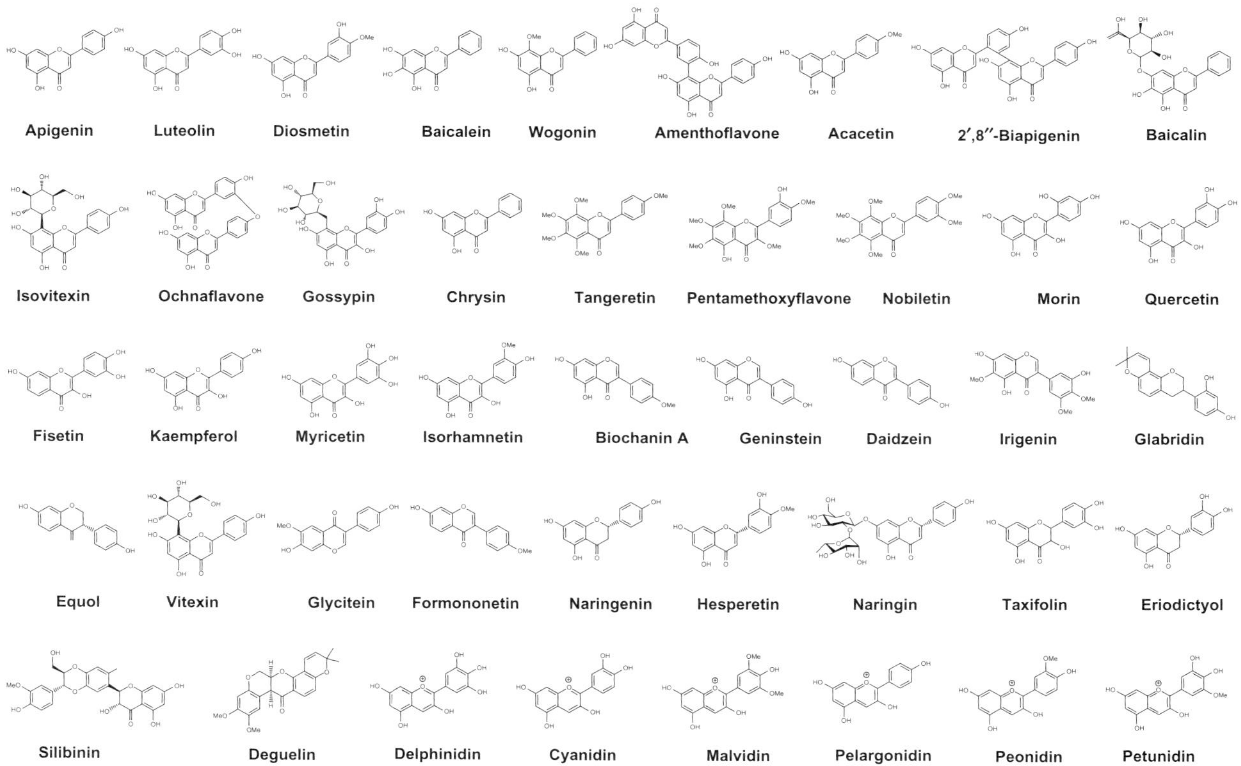


Fig. 6. Chemical structures of flavonoids.

been shown to possess anti-inflammatory activity by inhibiting NF-κB activity (Sung *et al.*, 2007).

Fisetin has also shown potential against other chronic diseases. For example, in ovalbumin-induced asthma, fisetin attenuated lung inflammation and airway hyper-responsiveness and decreased eosinophils and lymphocytes in bronchoalveolar fluid. Fisetin reduced the expression of the key initiators of allergic airway inflammation (eotaxin-1 and TSLP), Th2-associated cytokines (IL-4, IL-5, and IL-13) in lungs, and Th2-predominant transcription factor GATA-3 and cytokines in thoracic lymph node cells and splenocytes (Wu *et al.*, 2011). This flavonoid has shown cardioprotective properties. In isolated rat aorta, fisetin showed concentration-dependent relaxant effects against $K^+$- and phenylephrine-induced contractions (Ajay *et al.*, 2003).

*2. Deguelin*

Deguelin is isolated from the African plant *Mundulea sericea* (Leguminosae) and has been identified as a potent chemopreventive and chemotherapeutic agent against inflammatory diseases. It has been shown to inhibit iNOS expression and activity in LPS-treated endothelial NO synthase (eNOS) knockout mice (Connelly *et al.*, 2005). Deguelin has shown activity against preneoplastic lesions in a mammary organ culture and also inhibited papillomas in a two-stage mouse skin carcinogenesis model (Gerhauser *et al.*, 1995; Gills *et al.*, 2005; Udeani *et al.*, 1997). Deguelin has also been found to suppress the formation of carcinogen-induced aberrant crypt foci in mouse colons (Murillo *et al.*, 2003). This flavonoid was shown to suppress cigarette smoke-induced lung carcinogenesis (Lee *et al.*, 2005; Yan *et al.*, 2005). Additionally, deguelin enhanced the sensitivity of leukaemia cells to chemotherapeutic agents (Bortul *et al.*, 2005). Deguelin also exhibited a potent anti-proliferative effect via the downregulation of survivin expression and STAT3 phosphorylation in HTLV-1-transformed cells (Ito *et al.*, 2010). Deguelin has also been shown to suppress NF-κB activation (Nair *et al.*, 2006).

*3. Gossypin*

Gossypin is isolated from *Hibiscus vitifolius* (tropical rose mallow) and *H. furcatus*. Gossypin has exhibited anti-inflammatory activity by preventing carrageenin-induced paw edema in mice and has also inhibited arachidonic acid metabolism (Ferrandiz and Alcaraz, 1991). Gossypin has been shown to inhibit growth of the L929, HT29, and K562 tumour cell lines in culture. In mice, gossypin inhibited the growth and angiogenesis of Ehrlich's ascites carcinoma. Gossypin also exhibited anti-carcinogenic activity against

DMBA-induced papilloma in rodents. Gossypin's anti-tumoural effects have been partly ascribed to its ability to inhibit topoisomerase I and II (Babu *et al.*, 2003).

Gossypin has shown potent anti-diabetic activity in streptozotocin-induced experimental diabetes in rats. The extracts of these plants are traditionally used for the treatment of diabetes, jaundice, and inflammation (Vijayan *et al.*, 2004). Oral administration of gossypin to these diabetic rats resulted in improved glucose tolerance. The higher levels of blood glucose and glycosylated haemoglobin and lower levels of plasma insulin and haemoglobin observed in diabetic rats significantly returned to near normal after oral administration of gossypin. Gossypin also improved the levels of glycogen content in liver and muscles in diabetic rats (Venkatesan and Sorimuthu Pillai, 2012). Gossypin significantly reduced the total cholesterol concentration in HepG2 cells, thus exhibiting its efficacy against cardiovascular diseases (Lu *et al.*, 2008b).

*4. Morin*

Morin is isolated from members of the Moraceae family, such as mulberry figs and other Chinese herbs. Morin had anti-inflammatory activity in a rat colitis model (Ocete *et al.*, 1998), inhibited AOM-induced aberrant crypt foci in rats (Tanaka *et al.*, 1999), and exhibited chemopreventive effects against chemical-induced rat tongue carcinogenesis (Kawabata *et al.*, 1999).

Morin suppressed the proliferation of a wide variety of tumour cells, including oral squamous cell carcinoma (Brown *et al.*, 2003), leukaemia (Krol *et al.*, 2002), and colon cancer cell lines (Ranelletti *et al.*, 1992). Morin inhibited the growth of COLO205 cells in nude mice (Chen *et al.*, 2004) and induced differentiation of keratinocytes (Thuillier *et al.*, 2002). Morin has been shown to inhibit NF-κB activation induced by inflammatory agents, carcinogens, and tumour promoters. This flavone also suppressed the expression of NF-κB-regulated gene products involved in cell survival, proliferation, and invasion (Manna *et al.*, 2007).

LDL glycation has been shown to contribute to the increased atherosclerotic risk of patients with diabetes, and morin was found to ameliorate this risk (Ghaffari and Mojab, 2007). Morin has also demonstrated cardioprotective activity against oxyradical damage. Morin has been shown to protect three types of human cells (ventricular myocytes, saphenous vein endothelial cells, and erythrocytes) against damage by oxyradicals generated *in situ*, indicating its efficacy in the cardiovascular system (Wu *et al.*, 1994).

## C. Role of Alkaloids Against Chronic Diseases

Alkaloids are natural products that contain a nitrogen atom in their heterocyclic ring. Alkaloid-containing plants have been known for their medicinal properties since ancient times (Hughes and Shanks, 2002). Vinblastine was the first natural alkaloid found to have anti-proliferative activity and was discovered by Noble and colleagues from the leaves of *Vinca rosea* in 1958. So far, more than 12,000 alkaloids have been identified (Fig. 7A). In this section, we provide experimental evidence supporting the role of most common alkaloids such as berberine, evodiamine, noscapine, sanguinarine, and indirubin against chronic diseases.

### *1. Berberine*

Berberine is the major bioactive constituent of *Rhizoma coptidis*, a popular traditional Chinese medication used to treat diabetes and infections. There are a substantial number of clinical reports about the hypoglycemic action of berberine in the Chinese literature (Yin *et al.*, 2008). Berberine has been reported to reduce gain in body weight, to enhance insulin sensitivity, and to decrease blood glucose in animal models of type 2 diabetes (Zhou *et al.*, 2009). The administration of berberine has been shown to attenuate cardiac dysfunction in hyperglycemic and hypercholesterolemic rats (Dong *et al.*, 2011). Furthermore, berberine significantly increased cardiac fatty acid transport protein-1, fatty acid β-oxidase, glucose transporter-4, and PPARγ in hyperglycemic and hypercholesterolemic rats.

Berberine has been shown to possess anti-cancer activity as well. In a rat model, berberine inhibited AOM-induced aberrant crypt foci (ACF) formation (Fukutake *et al.*, 1998). In a mouse model, berberine inhibited teleocidin-induced tumourigenesis in the skin (Nishino *et al.*, 1986). Berberine suppressed the growth of a wide variety of tumour cells, including leukaemia (Lin *et al.*, 2006a), melanoma (Letasiova *et al.*, 2006), epidermoid carcinoma (Kettmann *et al.*, 2004), hepatoma (Hwang *et al.*, 2006), oral carcinoma (Kuo *et al.*, 2005), glioblastoma (Sanders *et al.*, 1998), prostate carcinoma (Mantena *et al.*, 2006), and gastric carcinoma (Lin *et al.*, 2006b). Berberine has also been shown to exert an inhibitory effect on AP-1 activity in human hepatoma cells (Fukuda *et al.*, 1999). Results from our group showed berberine's potential in suppressing NF-κB activity in Jurkat cells (Pandey *et al.*, 2008).

### *2. Evodiamine*

Evodiamine, a quinolone alkaloid isolated from the fruit of *Evodia rutaecarpa*, is used to treat numerous inflammation-related disorders such as eczema, ulcerative stomatitis, and others (Eisenbrand *et al.*, 1982).

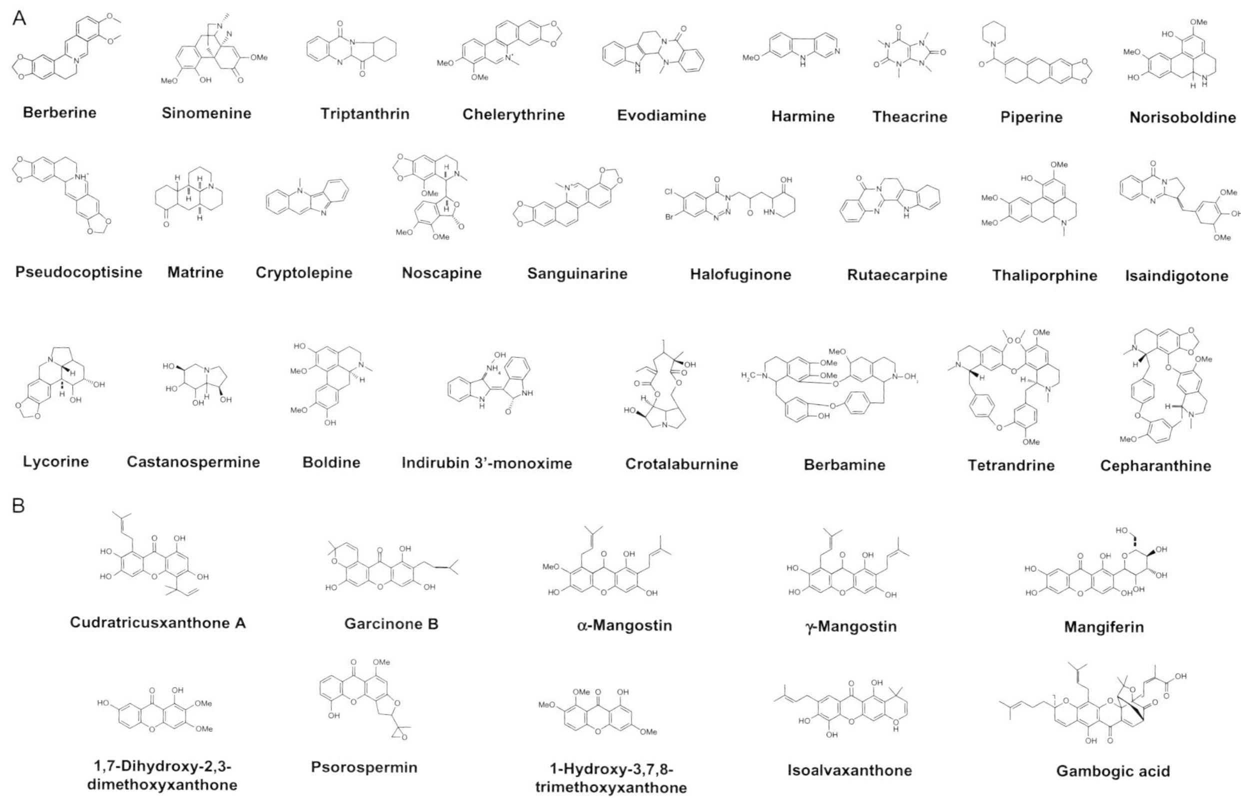


Fig. 7. Chemical structures of alkaloids (A) and xanthones (B).

Evodiamine has been shown to inhibit $PGE_2$ production and COX-2 induction in LPS-stimulated RAW 264.7 cells (Choi *et al.*, 2006). The anti-inflammatory activities of evodiamine are supported by *in vivo* studies as well. For example, in a transgenic mouse model, evodiamine significantly alleviated impairments in learning ability and memory (Yuan *et al.*, 2011). In addition, evodiamine also reversed the inhibition of glucose uptake due to the development of Alzheimer's disease traits in mice. Furthermore, expressions of IL-1β, IL-6, TNF-α, and COX-2 were also significantly decreased by evodiamine treatment.

Evodiamine has been shown to suppress the proliferation of a wide variety of tumour cells, including prostate cancer (Kan *et al.*, 2004), leukaemia (Fei *et al.*, 2003; Huang *et al.*, 2004), melanoma (Fei *et al.*, 2003; Zhang *et al.*, 2003), cervical cancer (Fei *et al.*, 2003), and fibrosarcoma cells (Fei *et al.*, 2003). Further, evodiamine has no apparent toxicity against normal peripheral blood mononuclear cells (Fei *et al.*, 2003). In human leukaemia cells, evodiamine induced apoptosis in both a caspase-dependent and a caspase-independent manner (Lee *et al.*, 2006). In addition to its anti-proliferative and apoptotic effects, evodiamine has been shown to suppress the invasion and migration of human colon carcinoma cells and melanoma cells to the lung (Ogasawara and Suzuki, 2004; Ogasawara *et al.*, 2001). Our group found that evodiamine has the ability to abrogate both inducible and constitutive NF-κB activation in human cancer cells and to decrease the expression of NF-κB-regulated gene products linked with tumourigenesis. In human liver cancer cells, evodiamine exhibited anti-cancer activity through inhibition of transcription factors NF-κB and AP-1, and it also suppressed hepatocellular transformation (Chao *et al.*, 2011).

### 3. *Noscapine*

Noscapine is a phthalideisoquinoline alkaloid and constitutes 1–10% of the total alkaloid content of opium (poppy, *Papaver somniferum*). Noscapine has been widely used as a cough suppressant in humans as well as in experimental models. The anti-tumour activity of noscapine was first reported in 1998 (Ye *et al.*, 1998). Its anti-inflammatory activity is evident from a recent study by Zughaier *et al.* (2010), who found that the brominated noscapine analogues have the potential to inhibit production of cytokines and chemokines from macrophages without affecting cell viability.

Noscapine has been shown to inhibit the growth of numerous cancer types including T-cell lymphoma, melanoma, leukaemia, myeloma, glioblastoma, and breast, ovarian, and colorectal cancer cells (Aneja *et al.*, 2006; Ke *et al.*, 2000; Landen *et al.*, 2002; Verma *et al.*, 2006; Zhou *et al.*, 2002). Noscapine can significantly reduce the tumour volumes in mice transplanted with human glioblastoma, non-small cell lung cancer, and breast cancer cells (Aneja *et al.*, 2006; Jackson *et al.*, 2008b; Verma *et al.*, 2006).

In some cases, noscapine has been shown to enhance the efficacy of conventional chemotherapeutic agents. For instance, in a xenograft model of human lung cancer, noscapine was shown to enhance the efficacy of cisplatin (Chougule *et al.*, 2011a). This alkaloid has also been shown to potentiate the effects of doxorubicin in a synergistic manner in an animal model bearing triple negative breast cancer through inactivation of the NF-κB pathway (Chougule *et al.*, 2011b). In our laboratory, noscapine inhibited proliferation of leukaemia cells and sensitized them to chemotherapeutic agents by suppressing the NF-κB activation pathway (Sung *et al.*, 2010).

#### 4. *Sanguinarine*

Sanguinarine, a benzophenanthridine alkaloid derived from *Sanguinaria canadensis* and poppy *Fumaria* species (Shamma and Guinaudeau, 1986), has been shown to exhibit anti-microbial, anti-oxidant, and anti-inflammatory activities (Walterova *et al.*, 1995). This alkaloid has also shown anti-tumour activity (Agarwal *et al.*, 1991; Lenfeld *et al.*, 1981; Walterova *et al.*, 1995) and shown potential against cardiovascular diseases as well (Jeng *et al.*, 2007). Sanguinarine is also known to inhibit the activation of transcription factor NF-κB (Chaturvedi *et al.*, 1997).

Sanguinarine has been shown to induce death in cancer cells by numerous mechanisms. First, this alkaloid induced apoptosis via modulation of Bcl-2 family proteins in a variety of cancer cells including leukaemia, keratinocyte, oral squamous cell carcinoma, pancreas, and breast cancer cells (Adhami *et al.*, 2003; Ahsan *et al.*, 2007; Han *et al.*, 2008; Kim *et al.*, 2008; Tsukamoto *et al.*, 2011; Weerasinghe *et al.*, 2001). Second, sanguinarine has been shown to induce ROS generation and glutathione depletion that led to cell death (Choi *et al.*, 2008; Debiton *et al.*, 2003; Kim *et al.*, 2008). Third, sanguinarine induced apoptosis in human cancer cells in both a p53-dependent and a p53-independent manner (Ahsan *et al.*, 2007). Sanguinarine has been shown to inactivate STAT3 and to downregulate the expression of cell survival proteins in human prostate cancer cells (Sun *et al.*, 2011).

*5. Indirubin*

Indirubin was originally identified as the active principle of a traditional Chinese medicinal preparation *Dang Gui Long Hui Wan*, which is used to treat chronic diseases. This alkaloid is also used in Asia as a systemic treatment for psoriasis. Indirubin has anti-inflammatory activity, as indicated by its ability to inhibit the 2,4,6-trinitro-1-chlorobenzene-induced inflammatory reaction in mice (Kunikata *et al.*, 2000) and to suppress the expression of the influenza virus-induced chemokine RANTES in human bronchial epithelial cells (Mak *et al.*, 2004).

Indirubin inhibited the activities of cyclin-dependent kinases (CDKs) by direct binding to the ATP binding site of enzymes (Hoessel *et al.*, 1999). Indirubin is also a highly effective inhibitor of GSK3β, an essential element of the β-catenin/Wnt signalling pathway (Meijer *et al.*, 2003; Skardelly *et al.*, 2011; Williams *et al.*, 2011). Indirubin has also been reported to suppress the NF-κB signalling pathway and to inhibit expression of NF-κB-dependent gene products linked with tumourigenesis (Sethi *et al.*, 2006). Indirubin has been shown to inhibit STAT3 activation in cancer cells as well (Schwaiberger *et al.*, 2010; Zhang *et al.*, 2011).

## D. ROLE OF XANTHONES AGAINST CHRONIC DISEASES

Xanthones belong to the class of oxygenated heterocycles and is found in some higher plants, fungi, and lichens. Depending on the chemical nature of the substituents in the dibenzo-γ-pirone scaffold, xanthones can be classified into simple oxygenated xanthones, glycosylated xanthones, and prenylated xanthones (Pinto *et al.*, 2005). The chemical structure of some of the xanthones is presented in Fig. 7B.

*1. γ-Mangostin*

γ-Mangostin (γ-MG) is a xanthone derived from *Garcinia mangostana*, used in folk medicine for treatment of abdominal pain, diarrhoea, dysentery, wound infections, and chronic ulcers (Suksamrarn *et al.*, 2006). This xanthone has been shown to inhibit activities of COX-1 and COX-2 and to reduce $PGE_2$ release in C6 rat glioma cells (Nakatani *et al.*, 2002, 2004). In addition, this xanthone has also been shown to directly inhibit the activity of IKK, the enzyme responsible for NF-κB activation (Nakatani *et al.*, 2002, 2004). γ-MG has also been shown to block LPS-induced insulin resistance by attenuating the activation of MAPK, NF-κB, and AP-1 activities in primary cultures of human adipocytes (Bumrungpert *et al.*, 2009). Garcinone B is a structural analogue of γ-MG and has been reported to possess anti-inflammatory activity by suppressing $PGE_2$ release and inactivation of NF-κB activity in C6 rat glioma cells (Yamakuni *et al.*, 2006).

*2. Mangiferin*

Mangiferin is isolated from mango fruit (*Mangifera indica*), *Salacia oblonga*, and *Salacia reticulata*. One study indicated the anti-inflammatory activity of mangiferin in UVB-induced skin damage in SKH-1 mice (Petrova *et al.*, 2011). Mangiferin has been suggested to be a potent chemopreventive agent, owing to its ability to inhibit AOM-induced colon carcinogenesis (Yoshimi *et al.*, 2001) and benzo(a)pyrene-induced lung carcinogenesis (Rajendran *et al.*, 2008a,b). In addition, mangiferin has also been shown to inhibit the NF-κB activation pathway by inhibiting IKK activity in human breast cancer cells (Garcia-Rivera *et al.*, 2011). It has also shown anti-proliferative and pro-apoptotic effects (Peng *et al.*, 2004; Shoji *et al.*, 2011; Yao *et al.*, 2010). This xanthone was shown to possess anti-growth effects in ascitic fibrosarcoma by both *in vitro* and *in vivo* studies (Guha *et al.*, 1996).

Mangiferin has also been shown to have anti-diabetic properties. For example, in a streptozotocin-induced rat diabetic model, mangiferin prevented diabetic nephropathy, hyperglycemia, and atherogenicity and improved renal function (Li *et al.*, 2010c; Muruganandan *et al.*, 2005). In another rat model, mangiferin was shown to attenuate diabetic renal fibrosis by suppressing angiotensin II/AT1 signalling (He *et al.*, 2009).

*3. Gambogic acid*

Gambogic acid is derived from *Garcinia morella* and *Garcinia hurbury* and has been extensively used in traditional medicine (Han *et al.*, 2005). Gambogic acid has inhibited the growth of a broad range of tumour cells, including human hepatoma (Guo *et al.*, 2006), breast cancer (Zhang *et al.*, 2004), gastric carcinoma (Li *et al.*, 2010c; Zhao *et al.*, 2004), and lung carcinoma (Wu *et al.*, 2004). Several mechanisms have been proposed for the anti-cancer activity of gambogic acid. For example, gambogic acid downregulated telomerase and telomerase reverse transcriptase (hTERT) expression (Wu *et al.*, 2004). Gambogic acid has also been shown to inactivate CDC2/p34 by inhibition of cyclin-dependent kinase 7 (CDK7) activity (Yu *et al.*, 2007). We showed that gambogic acid has the potential to inhibit the expression of NF-κB-regulated gene products involved in cell survival, proliferation, invasion, and angiogenesis (Pandey *et al.*, 2007b). Gambogic acid has also inhibited the NF-κB activation induced by LPS in macrophages (Palempalli *et al.*, 2009). Gambogic acid has been shown to physically bind with Hsp90, to inhibit Hsp90 ATPase activity, and to degrade Hsp90 client proteins (i.e. Akt, IKK) in HeLa cells (Zhang *et al.*, 2010a). In addition, gambogic acid has been shown to induce apoptosis in multiple myeloid cells through inactivation of STAT3 and downregulation of STAT3-regulated gene products linked with tumour cell survival (Prasad *et al.*, 2011).

## E. ROLE OF TRITERPENOIDS AGAINST CHRONIC DISEASES

Triterpenoids are obtained from plants such as onion, ginseng, brahmi, shallaki, salai guggal, lei gong teng, licorice, mango, olive, bearberry, Chinese bellflower, sickle-leaf, tulsi, and ashwagandha. Triterpenoids are chemically diverse (Fig. 8) and have demonstrated potential against numerous chronic diseases.

More than 20,000 triterpenoids have been identified to date. In this section, we discuss some of the most promising triterpenoids, such as betulinic acid, celastrol, ursolic acid (UA), pristimerin, AKBA, diosgenin, and escin.

### *1. Betulinic acid*

Betulinic acid is one of the major triterpenes and is obtained from numerous plants including *Triphyophyllum peltatum*, *Ancistrocladus heyneanus*, *Zizyphus joazeiro*, *Diospyros leucomelas*, *Tetracera boliviana*, *Syzygium formosanum*, and *Betula pubescens*. Betulinic acid has demonstrated potential against numerous chronic diseases. For example, in carrageenan- and serotonin-induced rat paw edema, betulinic acid exhibited potent anti-inflammatory activity (Mukherjee *et al.*, 1997). Betulinic acid has been shown to inhibit NF-κB activation and NF-κB-regulated gene products linked with tumour development (Takada and Aggarwal, 2003). Betulinic acid has also been shown to suppress STAT3 activity (Pandey *et al.*, 2010b).

Betulinic acid has been shown to possess anti-cancer activity in neuroblastomas (Schmidt *et al.*, 1997), glioblastomas (Fulda *et al.*, 1999), gliomas (Rzeski *et al.*, 2006), prostate cancers (Chintharlapalli *et al.*, 2007), and leukaemia and multiple myeloma cells (Pandey *et al.*, 2010b; Rzeski *et al.*, 2006). In one study, betulinic acid inhibited the growth of melanoma in an athymic nude mice model (Pisha *et al.*, 1995). The anti-cancer activity of betulinic acid is evident from clinical studies as well (Salti *et al.*, 2001).

Cardiovascular risk factors such as diabetes mellitus, hypertension, hypercholesterolemia, and cigarette smoking reduce the bioactive NO in the body, which is generated by endothelial NO synthase (eNOS). Betulinic acid has been shown to enhance eNOS expression (Forstermann and Li, 2011). Betulinic acid has been shown to possess anti-diabetic activity as well (Singab *et al.*, 2005). The oral administration of Egyptian *Morus alba* root bark extract containing betulinic acid to streptozotocin-induced diabetic rats

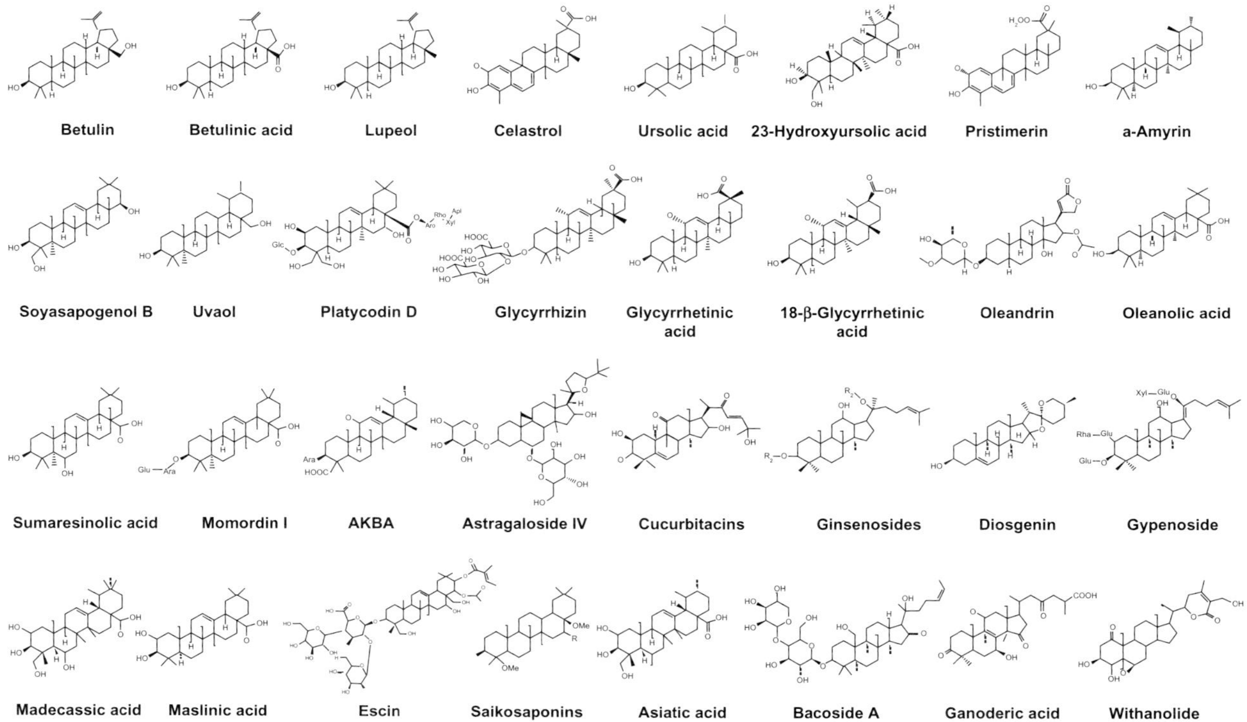


Fig. 8. Chemical structures of triterpenoids.

was associated with a significant decrease in lipid peroxides. Administration of the extract for 10 days (600 mg/kg/day) also significantly reduced the glucose level and significantly increased the insulin level (Singab *et al.*, 2005).

*2. Celastrol*

Celastrol is a natural triterpenoid found in the thunder god vine. It has demonstrated potential as an anti-inflammatory and anti-oxidant and has shown efficacy against numerous chronic diseases. In nanomolar concentrations, celastrol has been shown to suppress production of pro-inflammatory cytokines by human monocytes and macrophages (Allison *et al.*, 2001) and to suppress NF-κB activation in melanoma cells (Chen *et al.*, 2009). Celastrol has been shown to inhibit proliferation of numerous other cancer types including multiple myeloma, hepatocellular carcinoma, gastric cancer, prostate cancer, renal cell carcinoma, head and neck carcinoma, non-small cell lung carcinoma, melanoma, glioma, and breast cancer (Kannaiyan *et al.*, 2011).

Deposition of Aβ proteins is a neuropathological hallmark of Alzheimer's disease, and celastrol is known to inhibit Aβ deposition (Paris *et al.*, 2010). Celastrol was found to inhibit Aβ1–40 and Aβ1–42 production by reducing the beta-cleavage of amyloid precursor protein (APP) that leads to a decrease in APP-CTFβ and APPsβ (Paris *et al.*, 2010). Celastrol has also shown potential against Crohn's disease (Pinna *et al.*, 2004).

*3. Ursolic acid*

UA is found in numerous medicinal plants, including rosemary (*Rosmarinus officinalis*), apple (*Malus domestica*), cranberry (*Vaccinium macrocarpon*), beefsteak (*Perilla frutescens*), pear (*Pyrus pyrifolia*), plum (*Prunus domestica*), bearberry (*Arctostaphylos alpina*), loquat (*Eriobotrya japonica*), scotch heather (*Calluna vulgaris*), basil (*Ocimum tenuiflorum*), and jamun (*Eugenia jambolana*) (Liu, 1995). Ursolic acid has exhibited anti-cancer activity both *in vitro* and *in vivo*. Ursolic acid has been shown to suppress proliferation and induce apoptosis in numerous cancer types, including breast cancer (Es-Saady *et al.*, 1996), colon cancer (Andersson *et al.*, 2003), lung cancer (Hsu *et al.*, 2004), cervical cancer (Yim *et al.*, 2006), multiple myeloma (Pathak *et al.*, 2007), pancreatic cancer (Chadalapaka *et al.*, 2008), melanoma (Harmand *et al.*, 2005), and prostate cancer (Zhang *et al.*, 2010b). Ursolic acid has been shown to inhibit tumour invasion and metastasis *in vivo* (Yamai *et al.*, 2009). Ursolic acid exerts its anti-cancer activity by numerous mechanisms, including caspase activation (Choi *et al.*, 2000b; Harmand *et al.*, 2005), inhibition of DNA replication (Kim *et al.*, 2000), downregulation of the expression of cell survival proteins (Kassi *et al.*, 2009; Shishodia *et al.*, 2003), and inhibition of the activities of STAT3 (Pathak *et al.*, 2007) and NF-κB (Shishodia *et al.*, 2003). Ursolic acid has shown potential against arthritis (Kang *et al.*, 2008), chronic bronchitis (Huang *et al.*, 1994), and Alzheimer's disease (Chung *et al.*, 2001).

*4. Pristimerin*

Pristimerin is found in many species of the Celastraceae family, such as *Pristimera indica*, *Catha edulis*, *Schaefferia cuneifolia*, and *Maytenus heterophylla*. It has demonstrated potential as an anti-inflammatory agent against numerous chronic diseases (Dirsch *et al.*, 1997). Pristimerin has also been shown to inhibit NF-κB activation by inhibiting IKK activation and degradation of ubiquitinated IκB. Pristimerin has also shown potential to selectively kill primary myeloma cells and to inhibit xenografted plasmacytoma tumours in mice (Tiedemann *et al.*, 2009).

*5. Acetyl-11-keto-β-boswellic acid*

Acetyl-11-keto-β-boswellic acid (AKBA) is a derivative of boswellic acid and is a potent anti-inflammatory agent, known to bind and inhibit 5-LOX activity (Sailer *et al.*, 1996). AKBA has shown anti-growth activity in colon, prostate, and pancreatic tumour cells both *in vitro* and *in vivo* (Lu *et al.*, 2008a; Park *et al.*, 2011a; Yadav *et al.*, 2012b). Expressions of cyclin D1, cyclin E, CDK2, CDK4, and phosphorylated retinoblastoma protein have been found to be downregulated by AKBA treatment (Liu *et al.*, 2006).

The cytostatic and apoptosis-inducing activities of boswellic acid against malignant cell lines have been demonstrated by *in vitro* studies (Hostanska *et al.*, 2002). In one study, boswellic acid was found to trigger apoptosis in colon cancer cells in a caspase-8-dependent manner (Liu *et al.*, 2002). AKBA has also been shown to inhibit NF-κB activation (Takada *et al.*, 2006) and STAT3 activation (Kunnumakkara *et al.*, 2009).

AKBA has also been found effective against arthritis. For instance, 5-Loxin, which contains 30% AKBA, significantly lessened pain and improved physical function in patients with osteoarthritis (Sengupta *et al.*, 2008). In addition, a significant reduction in the MMP-3 content of synovial fluid was observed in patients supplemented with 5-Loxin compared with placebo control (Sengupta *et al.*, 2008). Sphingomyelinase (SMase) is found to be upregulated in several inflammation-related diseases, such as inflammatory bowel disease, atherosclerosis, and diabetes. In one study, AKBA decreased SMase activity in intestinal cell lines (Zhang and Duan, 2009).

*6. Diosgenin*

Diosgenin, a triterpenoid having two pentacyclic rings, is present in *Trigonella foenum-graecum*. Diosgenin has been shown to suppress inflammation and proliferation and to induce apoptosis in a variety of tumour cells. Diosgenin has also demonstrated inhibitory effects on Akt, IKK, NF-κB, and NF-κB-regulated gene expression. Diosgenin has been shown to suppress NF-κB through direct DNA binding (Shishodia and Aggarwal, 2006). In addition, diosgenin has the potential to suppress STAT3 (Li *et al.*, 2010a) activity and to induce DR5 (Lepage *et al.*, 2011).

Diosgenin is also effective in reducing plasma glucose levels in diabetic rats. Diosgenin decreased the activities of diabetes-associated enzymes (such as ATP-citrate lyase, pyruvate kinase, and glucose-6-phosphate dehydrogenase) in the liver of diabetic rats (McAnuff *et al.*, 2005). Diosgenin obtained from fenugreek has also been found to ameliorate diabetes in mice fed a high-fat diet (Uemura *et al.*, 2010).

Expression of adhesion molecules on vascular smooth muscle cells (VSMCs) has been shown to contribute to the pathogenesis of atherosclerosis. In one study, diosgenin inhibited the expression of adhesion molecules induced by TNF-α in the cultured mouse VSMC cell line, MOVAS-1 (Choi *et al.*, 2000a). Diosgenin has shown potential against rheumatoid arthritis as well (Liagre *et al.*, 2004).

*7. Escin*

Escin is a pentacyclic triterpene, existing in α and β forms, and is isolated from the seeds of the horse chestnut (*Aesculus hippocastanum*). The β form of escin has demonstrated potential as an anti-inflammatory agent (Matsuda *et al.*, 1998; Rothkopf and Vogel, 1976). β-Escin also exhibits anti-edema, hypoglycemic (Kimura *et al.*, 2006), and anti-obesity (Hu *et al.*, 2008) activities. In one study, escin inhibited acute inflammation induced by acetic acid in mice and by histamine in rats (Matsuda *et al.*, 1998). This triterpene has also been shown to inhibit chronic aberrant foci formation in rats and to induce apoptosis in human colon cancer HT29 cells (Patlolla *et al.*, 2006). Escin also inhibited NF-κB activation through inhibition of IKK activation (Harikumar *et al.*, 2010). Escin was also shown to inhibit STAT3 activation (Tan *et al.*, 2010). Escin has clinical efficacy in patients with HIV-1 (Grases *et al.*, 2004), for the treatment of blunt impact injuries (Wetzel *et al.*, 2002), and for cutaneous pruritus (Li *et al.*, 2004). Escin also manifests hypoglycemic activity. In one study, escin and its derivatives at a dose of 100 mg/kg effectively attenuated blood glucose levels in mice, as measured by oral glucose tolerance test (Kimura *et al.*, 2006). The anti-diabetic activity of escin was demonstrated in another study in which this triterpene decreased leptin levels in mice fed a high-fat diet (Avci *et al.*, 2010).

## F. ROLE OF CHAVICOLS AGAINST CHRONIC DISEAESES

Chavicols (also known as *p*-allylphenol) are a type of organic compounds consisting of a benzene ring substituted with a hydroxy group and a propenyl group (Fig. 8A). These natural compounds are found in betel leaf oils and in oil of bay. In this section, we discuss the potential of some common chavicols such as eugenol, hydroxychavicol, and 1′-acetoxychavicol acetate against chronic diseases.

### *1. Eugenol*

Eugenol, obtained from *Syzygium aromaticum* (commonly known as clove), is commonly used as a flavouring agent in food products, in cosmetics, and particularly in dentistry. Both *S. aromaticum* and its major component eugenol have been shown to possess anti-ulcer activities in an indomethacin-induced and ethanol/HCl-induced ulcer model (Santin *et al.*, 2011). Eugenol has also been reported to suppress LPS-induced iNOS and COX-2 expression by downregulating NF-κB and AP-1 activity (Kim *et al.*, 2003; Yeh *et al.*, 2011). The inhibition of NF-κB by eugenol was correlated with a reduction in LPS-induced inflammatory cytokine production, such as TNF-α and IL-1β (Yeh *et al.*, 2011). In a mouse model of LPS-induced lung inflammation, eugenol was found to possess anti-inflammatory activity (Magalhaes *et al.*, 2010).

Eugenol has also demonstrated potential as an anti-cancer agent. In studies with human cancer cells, this chavicol induced apoptosis in human colorectal and melanoma cells (Ghosh *et al.*, 2005; Jaganathan *et al.*, 2011). In another studies, eugenol suppressed N-methyl-N′-nitro-N-nitrosoguanidine-induced gastric carcinogenesis in rats and DMBA-induced skin carcinogenesis in mice through inhibition of the NF-κB signalling pathway (Kaur *et al.*, 2010; Manikandan *et al.*, 2011).

### *2. Hydroxychavicol*

Hydroxychavicol is a major phenolic compound present in the aqueous extract of the betel leaf (*Piper betle*), which is used in a number of traditional medicines. Hydroxychavicol has been reported to have anti-oxidant, anti-cancer, and anti-inflammatory activities (Jeng *et al.*, 2004). In a rat model of Alzheimer's disease, hydroxychavicol improved cognitive impairment and attenuated elevated levels of pro-inflammatory cytokines (Pandey and Bani, 2010). The potential of hydroxychavicol against chronic diseases such as atherosclerosis, cardiovascular diseases (Chang *et al.*, 2007), and arthritis (Pandey *et al.*, 2010a) has also been reported. Hydroxychavicol has also been shown to possess anti-cancer activity, as demonstrated by its potential to inhibit DMBA-induced skin tumours in rats (Azuine *et al.*, 1991), tobacco-specific carcinogen-induced tumourigenesis in mice (Amonkar *et al.*, 1989; Bhide *et al.*, 1991a), and benzo[a]pyrene-induced forestomach tumours in mice (Bhide *et al.*, 1991b).

#### *3. 1′-Acetoxychavicol acetate*

1′-Acetoxychavicol acetate (ACA) is derived from the rhizomes of an Asian ginger, *Alpinia galanga* (also known as *Languas galanga* Stuntz). Numerous researchers have demonstrated the anti-inflammatory activity of ACA. Ohata *et al.* were the first to show the inhibition of LPS-induced iNOS expression by ACA through suppression of NF-κB, AP-1, and STAT1 in murine macrophage cells (Ohata *et al.*, 1998).

ACA has also been investigated for its chemopreventive effect in various animal models. Ohnishi *et al.* showed the anti-cancer activity of ACA against 4-nitroquinoline 1-oxide-induced oral carcinogenesis in a rat model (Ohnishi *et al.*, 1996). The inhibition of tumour promotion by topical application of ACA in TPA-induced skin carcinogenesis in a mouse model has also been demonstrated (Murakami *et al.*, 1996). Tanaka *et al.* studied the effect of ACA on the development of colonic ACF induced by AOM in a rat model (Tanaka *et al.*, 1997). This group showed that feeding ACA reduced the incidence of colonic carcinoma through elevation in the activities of phase II enzymes. In another study, the chemopreventive effect of ACA was associated with upregulation of phase II enzymes and Nrf2 activation in IEC6, a rat intestine epithelial cell line (Yaku *et al.*, 2011). Our group showed that inhibition of NF-κB activity by ACA was associated with downregulation of NF-κB-regulated proliferative, anti-apoptotic, and metastatic gene products in human leukaemia cells (Ichikawa *et al.*, 2005). The inhibition of NF-κB by ACA has also been shown to inhibit osteoclastogenesis induced by RANKL and human cancer cells, thus suggesting the potential of ACA as a therapeutic agent for osteoporosis and cancer-associated bone loss (Ichikawa *et al.*, 2006).

### G. ROLE OF QUINONES AGAINST CHRONIC DISEASES

Quinones are a group of organic compounds that have been known for medicinal properties since ancient times. In this section, we discuss the efficacy of the most promising quinones: thymoquinone, embelin, capsaicin, plumbagin, and emodin (Fig. 8B).

*1. Thymoquinone*

Thymoquinone, the predominant bioactive component of black seed oil (*Nigella sativa)*, has shown potential against lung disease, arthritis, and hypercholesterolemia. Thymoquinone has been reported to exert an anti-inflammatory effect, as shown by the significant inhibition of acetic acid-induced colitis in rats (Mahgoub, 2003), TNF-α production in murine septic peritonitis (Haq *et al.*, 1999), adjuvant-induced rheumatoid arthritis in rats (Tekeoglu *et al.*, 2007), and carrageenan-induced paw edema in rats (Hajhashemi *et al.*, 2004). In an experimental asthma model, thymoquinone significantly reduced ovalbumin-induced lung inflammation (El Gazzar *et al.*, 2006; El Mezayen *et al.*, 2006). Thymoquinone administration has been shown to improve experimental allergic encephalomyelitis, presumably due to inhibition of oxidative stress (Mohamed *et al.*, 2003).

Several investigators have shown the chemopreventive effect of thymoquinone *in vivo*. In a DMBA-induced buccal pouch carcinogenesis model, thymoquinone reduced tumour formation (Rajkamal *et al.*, 2010). Thymoquinone has also been reported as a powerful chemopreventive agent against 20-methylcholanthrene-induced fibrosarcoma (Badary and Gamal El-Din, 2001) and benzo(a)pyrene-induced forestomach carcinogenesis (Hajhashemi *et al.*, 2004). Kaseb *et al.* examined the effect of thymoquinone on androgen-dependent LNCaP prostate cancer cells (Kaseb *et al.*, 2007) and found that thymoquinone caused cell cycle arrest at G1/S phase, with concomitant reduction in androgen receptor (AR), E2F-1, and the E2F-1-regulated proteins essential for cell cycle progression.

Thymoquinone has been shown to downregulate the expression of numerous genes linked to tumour development, such as Bcl-xL, COX-2, iNOS, 5-LOX, TNF, and cyclin D1. Thymoquinone has also been shown to inhibit NF-κB activation (Sethi *et al.*, 2008) and STAT3 activation (Li *et al.*, 2010b).

*2. Embelin*

Embelin (2,5-dihydroxy-3-undecyl-1,4-benzoquinone), a derivative from *Embelia ribes* Burm (Myrsinaceae), has been reported to be an effective analgesic and anti-fertility agent (Gupta *et al.*, 1976; Radhakrishnan and Alam, 1975). There is now ample evidence to suggest anti-inflammatory activity for embelin. The anti-inflammatory activity of embelin is mediated through its inhibitory effects on IL-1β and TNF-α production, as was shown recently (Kalyan Kumar *et al.*, 2011). Embelin has also demonstrated inhibitory action in the alloxan-induced rat diabetes model (Mahendran *et al.*, 2010).

The chemopreventive activities of embelin have been demonstrated by numerous studies. For example, embelin has been shown to inhibit *N*-nitrosodiethylamine-initiated and phenobarbital-promoted hepatocarcinogenesis (Poojari *et al.*, 2010; Tabakoff *et al.*, 1991). In another study, embelin inhibited 1,2-dimethylhydrazine dihydrochloride-induced colon carcinogenesis (Thippeswamy *et al.*, 2011). In one study, the anti-tumour activity of embelin was attributed to its ability to bind to and inhibit XIAP (Nikolovska-Coleska *et al.*, 2004). Embelin has also been reported to inhibit NF-κB activation (Ahn *et al.*, 2007a) and STAT3 activation (Heo *et al.*, 2011) in cancer cells.

*3. Capsaicin*

Capsaicin (*trans*-8-methyl-*N*-vanillyl-6-nonenamide) is a principal pungent ingredient of hot and red chili peppers belonging to the genus *Capsicum*. It has demonstrated potential against neuropathic pain, itching, and numerous cancer types. One epidemiological study showed that chili consumers were at greater risk of developing stomach cancer than non-consumers (Lopez-Carrillo *et al.*, 1994). In contrast to this result, however, some evidence suggests that capsaicin has chemopreventive and chemotherapeutic activities against numerous cancers. For example, in a rat model of diethylnitrosamine-induced hepatocarcinogenesis, capsaicin treatment significantly inhibited the formation of preneoplastic foci (Jang *et al.*, 1991). In the AOM-induced rat colon carcinogenesis model, capsaicin significantly reduced the incidence of colonic adenocarcinoma (Yoshitani *et al.*, 2001). Capsaicin has also been shown to exhibit chemopreventive potential in an animal model of chronic pancreatitis (Bai *et al.*, 2011).

The anti-cancer and anti-proliferative effects of capsaicin may be linked to its ability to suppress activation of transcription factors. Our group examined the effect of capsaicin on the activation of NF-κB (Singh *et al.*, 1996). In human myeloid cells, capsaicin abrogated TNF-induced NF-κB activation. In another study, capsaicin was found to suppress TPA-stimulated NF-κB activation in human promyelocytic leukaemia cells (Han *et al.*, 2002). Capsaicin has also been shown to inhibit the STAT3 activation pathway (Bhutani *et al.*, 2007). This group found that capsaicin inhibited constitutive activation of STAT3 in multiple myeloma cells, with a minimal effect on STAT5. Joung *et al.* showed activation of the Nrf2 signalling pathway by capsaicin (Joung *et al.*, 2007). Recently, capsaicin was found to repress transcriptional activity of β-catenin in human colorectal cancer cells (Lee *et al.*, 2011). In addition, capsaicin also suppressed TCF-4 expression and disrupted the interaction of TCF-4 with β-catenin.

*4. Plumbagin*

The naphthoquinone plumbagin is a naturally occurring yellow pigment found in the roots, leaves, bark, and wood of plants of the Plumbaginaceae, Droseraceae, Ancistrocladaceae, and Dioncophyllaceae families. The root of *Plumbago zeylanica* (also called Chitrak), a major source of plumbagin, has been used in traditional Indian medicine since 750 BC as an anti-atherogenic, cardiotonic, hepatoprotective, and neuroprotective agent. *P. zeylanica* has also been used to treat rheumatoid arthritis, dysmenorrhea, injury by bumping, and cancer.

The anti-inflammatory and analgesic activities of plumbagin were recently examined (Sheeja *et al.*, 2010). Plumbagin inhibited carrageenan-induced rat hind paw edema, and it shortened the duration of pain response in formalin-induced nociception in mice. Luo *et al.* also investigated the anti-inflammatory and anti-analgesic effects of plumbagin in carrageenan-induced rat paw edema (Luo *et al.*, 2010). Plumbagin significantly reduced the size of paw edema and various pro-inflammatory mediators, including histamine, serotonin, bradykinin, and $PGE_2$.

The neuroprotective activity of plumbagin against cerebral ischemia was investigated in another study (Son *et al.*, 2010). Plumbagin upregulated the expression of transcription factor Nrf2 in neuroblastoma cells. *In vivo*, administration of plumbagin significantly reduced brain damage and ameliorated the associated neurological deficits in a mouse model of focal ischemic stroke. Plumbagin has also been shown to exert anti-arthritic activity (Jackson *et al.*, 2008a).

Plumbagin has been shown to possess anti-cancer, anti-proliferative, chemopreventive, radiosensitizing, anti-angiogenic, and anti-metastatic activities (Devi *et al.*, 1998; Lai *et al.*, 2011; Manu *et al.*, 2011; Prasad *et al.*, 1996; Sugie *et al.*, 1998; Wang *et al.*, 2008). Our group showed that plumbagin has the potential to suppress NF-κB activity (Sandur *et al.*, 2006) and STAT3 activity (Sandur *et al.*, 2010).

*5. Emodin*

Emodin (1,3,8-trihydroxy-6-methylanthraquinone) is a naturally occurring anthraquinone present in numerous plants, moulds, and lichens. Emodin is an active ingredient of such Chinese medicinal plants as *Rheum officinale*, *Cassia occidentalis* L., and *Polygonum cuspidatum*. *P. cuspidatum* has been traditionally used in China for skin burns, hepatitis, gallstone, inflammation, and osteomyelitis. Emodin has shown anti-inflammatory effects in various experimental models. Various chronic diseases in which emodin has shown potential include atherosclerosis, ulcer, colitis, glomerulonephritis, pancreatitis, and hepatitis, as evident from experimental studies (Ding *et al.*, 2008;

Goel *et al.*, 1991; Hei *et al.*, 2006; Park *et al.*, 2011b; Wang *et al.*, 2007; Xia *et al.*, 2010; Yamada *et al.*, 2005).

Emodin has shown potential for the treatment of metabolic diseases. In one study, emodin ameliorated renal dysfunction in rats with diabetic nephropathy, probably owing to its ability to inhibit p38 MAPK activation and to downregulate the expression of fibronectin (Wang *et al.*, 2006a). The ability of emodin to act as a PPAR$\gamma$ ligand has been suggested as one of the mechanisms for its anti-diabetic effects in mice (Xue *et al.*, 2010).

Emodin has been described as a potent anti-cancer agent. For example, this anthraquinone was shown to inhibit proliferation of breast, lung, cervical, colorectal, and prostate cancer cells (Cha *et al.*, 2005; Chan *et al.*, 1993; Huang *et al.*, 2007; Zhang *et al.*, 1995). In one study, the effects of emodin on cell cycle progression were examined in hepatoma cells. Treatment of cells with emodin resulted in G2/M cell cycle arrest by upregulating the expression of p53 and p21 (Shieh *et al.*, 2004).

In addition, many studies have shown that emodin targets transcription factors that are involved in the progression of chronic diseases and cancer. For example, the potential of emodin to target NF-$\kappa$B, AP-1 and STAT3 has been described in numerous studies (Kumar *et al.*, 1998; Liu *et al.*, 2011).

HSP90 has been identified as a target of emodin. Zhang *et al.* found that emodin inhibited the growth of HER-2/neu-overexpressing cancer cells through its action as a tyrosine kinase inhibitor (Zhang *et al.*, 1995). The suppression of HER-2/neu by emodin resulted in the blockage of metastasis, growth inhibition, and sensitization of cancer cells to chemotherapy (Zhang and Hung, 1996; Zhang *et al.*, 1995, 1998). In another study, Yan *et al.* demonstrated that the downregulation of HER-2/neu by emodin was mediated through dissociation of HSP90 from HER-2/neu and through activation of the proteosomal degradation pathway (Yan *et al.*, 2011). Emodin has also been shown to target AR directly and suppress prostate cancer cell growth (Cha *et al.*, 2005). The downregulation of AR was mediated through dissociation of AR and HSP90.

## H. ROLE OF POLYPHENOLS AGAINST CHRONIC DISEASES

Extensive research over the past several years has shown the anti-inflammatory and chemopreventive potential of polyphenols. The mechanism of action of polyphenols has also been delineated over the years. Two of the most extensively studied polyphenols are curcumin and $\gamma$-tocotrienol (Fig. 8C). In this section, we provide evidence from *in vitro*, *in vivo*, and clinical studies for the role of curcumin and $\gamma$-tocotrienol against chronic diseases.

*1. Curcumin*

Curcumin (diferuloylmethane) is a yellow colouring agent derived from turmeric (*Curcuma longa*), a golden spice. Turmeric is used in Indian cooking to add colour and flavour and as a preservative. Turmeric has also been used in the Ayurvedic system of medicine against various ailments, including stomach upset, dysentery, ulcers, jaundice, arthritis, sprains, wounds, acne, and skin and eye infections (Singh, 2007; Fig. 9).

Curcumin has demonstrated potential against a wide range of chronic diseases, including cancer, lung diseases, neurological diseases, liver diseases, metabolic diseases, autoimmune diseases, cardiovascular diseases, and various other inflammatory diseases. At the molecular level, curcumin has been shown to modulate a wide range of signalling molecules. The molecular targets of curcumin fall into two categories: those to which curcumin binds directly and those whose activity curcumin modulates indirectly (Aggarwal and Sung, 2009; Gupta *et al.*, 2011). Curcumin has been shown to activate various transcription factors, including PPARγ, p53, Nrf2, C/EBP homologous protein, and activating transcription factor 3. Curcumin has also been shown to downregulate various transcription factors (e.g. NF-κB, HIF-1α, AP-1, STAT3, and β-catenin), chemokines, chemokine receptors, anti-apoptotic proteins, cell cycle regulatory proteins, invasion and angiogenesis biomarkers, and inflammatory molecules. A comprehensive review of the disease targets of curcumin and the molecular mechanisms involved can be found in numerous articles by our laboratory (Aggarwal, 2010; Anand *et al.*, 2008; Kunnumakkara *et al.*, 2008). The chemopreventive potential of curcumin might be due to its ability to induce apoptosis by numerous mechanisms. How curcumin controls different genes or gene products involved in cell death pathways has also been reviewed by our group (Ravindran *et al.*, 2009).

*2. Tocotrienol*

Vitamin E is composed of tocopherols and tocotrienols and was first discovered in 1938. Tocopherols and tocotrienols further consist of α, β, δ, and γ analogues. Although tocopherols have been extensively studied, tocotrienols have gained considerable attention only during the past decade. Tocotrienols have shown great potential against such human diseases as diabetes, cardiovascular diseases, Parkinson's disease, and cancer. Tocotrienols have been shown to target several transcription factors linked with inflammatory conditions and to suppress the expression of various inflammatory cytokines and inflammatory mediators such as iNOS and COX-2 (Aggarwal *et al.*, 2010; Miyazawa *et al.*, 2009; Sen *et al.*, 2004, 2007; Theriault *et al.*, 1999). In the

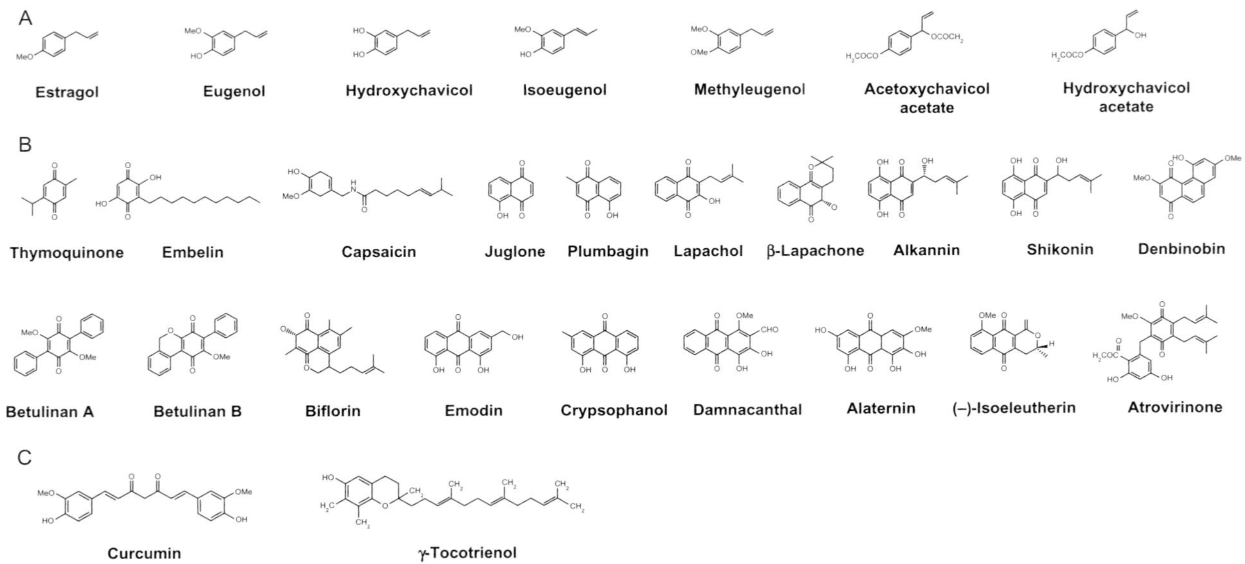


Fig. 9. Chemical structures of chavicols (A), quinones (B), and other phytochemicals (C).

paragraphs to follow, we provide evidence for the role of γ-tocotrienol against chronic diseases.

γ-Tocotrienol has been shown to inhibit NF-κB activity (Shah and Sylvester, 2005). Our group elucidated the molecular mechanism by which γ-tocotrienol inhibits NF-κB activation (Ahn *et al.*, 2007b). We found that γ-tocotrienol inhibited NF-κB activation by inhibiting the IKK activity that leads to suppression of proliferation, anti-apoptosis, and metastasis in tumour cells. Similar results were also observed in breast cancer cells (Yap *et al.*, 2010). Our group recently found that γ-tocotrienol has the potential to augment the effect of gemcitabine in pancreatic cancer cells both *in vitro* and *in vivo* (Kunnumakkara *et al.*, 2010). γ-Tocotrienol also inhibited NF-κB activation and NF-κB-dependent gene products linked with tumourigenesis.

γ-Tocotrienol has been also found to modulate the activation of other transcription factors such as STAT3, HIF-1α, PPARγ, and Nrf2. Our group reported that γ-tocotrienol, but not γ-tocopherol, inhibited constitutive activation of STAT3 in multiple myeloma cells (Kannappan *et al.*, 2010). γ-Tocotrienol inhibited STAT3 activation through suppression of upstream kinases such as Src, JAK1, and JAK2. γ-Tocotrienol has also been shown to induce Nrf2 expression (Hsieh *et al.*, 2010). γ-Tocotrienol has been found to downregulate the expression of HIF-1α and the paracrine secretion of VEGF under both normoxic and hypoxic conditions in human gastric adenocarcinoma cells (Bi *et al.*, 2010). Recently, Campbell *et al.* found that γ-tocotrienol has the potential to inhibit the growth of prostate cancer cells in a PPARγ-dependent manner (Campbell *et al.*, 2011).

# IV. CONCLUSIONS

It is clear from the above discussion that chronic inflammation plays a major role in the pathogenesis of chronic diseases and that the transcription factors are the major mediators. The role of plant-based nutraceuticals against chronic diseases seems promising. However, most of the known activities of these botanicals are based only on *in vitro* and *in vivo* studies, and only limited clinical data are available. None of the botanicals has been approved for human use. We also need to compare these nutraceuticals with steroids, nonsteroidal anti-inflammatory agents, anti-diabetic drugs such as metformin, cholesterol-lowering drugs such as statins and other synthetic compounds known to exhibit antinflammatory activities and have been approved for human use. Therefore, more extensive and well-controlled human studies are required to demonstrate the safety and efficacy of these botanicals. Future research should be focused on bringing these fascinating botanicals to the forefront of therapeutic agents for the treatment of human diseases.

## ACKNOWLEDGEMENTS

Dr. Aggarwal is the Ransom Horne, Jr., Professor of Cancer Research. This work was supported by a program project grant from National Institutes of Health (NIH CA-124787-01A2) and a grant from the Center for Targeted Therapy of The University of Texas MD Anderson Cancer Center. This work was also supported by Malaysian Palm Oil Board, Kuala Lumpur, Malaysia.

## REFERENCES

Achiron, A. and Gurevich, M. (2006). Peripheral blood gene expression signature mirrors central nervous system disease: The model of multiple sclerosis. *Autoimmunity Reviews* **5**, 517–522.

Adhami, V. M., Aziz, M. H., Mukhtar, H. and Ahmad, N. (2003). Activation of prodeath Bcl-2 family proteins and mitochondrial apoptosis pathway by sanguinarine in immortalized human HaCaT keratinocytes. *Clinical Cancer Research* **9**, 3176–3182.

Agarwal, S., Reynolds, M. A., Pou, S., Peterson, D. E., Charon, J. A. and Suzuki, J. B. (1991). The effect of sanguinarine on human peripheral blood neutrophil viability and functions. *Oral Microbiology and Immunology* **6**, 51–61.

Aggarwal, B. B. (2004). Nuclear factor-kappaB: The enemy within. *Cancer Cell* **6**, 203–208.

Aggarwal, B. B. (2010). Targeting inflammation-induced obesity and metabolic diseases by curcumin and other nutraceuticals. *Annual Review of Nutrition* **30**, 173–199.

Aggarwal, B. B. and Sung, B. (2009). Pharmacological basis for the role of curcumin in chronic diseases: An age-old spice with modern targets. *Trends in Pharmacological Sciences* **30**, 85–94.

Aggarwal, B. B., Shishodia, S., Sandur, S. K., Pandey, M. K. and Sethi, G. (2006). Inflammation and cancer: How hot is the link? *Biochemical Pharmacology* **72**, 1605–1621.

Aggarwal, B. B., Danda, D., Gupta, S. and Gehlot, P. (2009). Models for prevention and treatment of cancer: Problems vs promises. *Biochemical Pharmacology* **78**, 1083–1094.

Aggarwal, B. B., Sundaram, C., Prasad, S. and Kannappan, R. (2010). Tocotrienols, the vitamin E of the 21st century: Its potential against cancer and other chronic diseases. *Biochemical Pharmacology* **80**, 1613–1631.

Ahn, K. S. and Aggarwal, B. B. (2005). Transcription factor NF-kappaB: A sensor for smoke and stress signals. *Annals of the New York Academy of Sciences* **1056**, 218–233.

Ahn, K. S., Sethi, G. and Aggarwal, B. B. (2007a). Embelin, an inhibitor of X chromosome-linked inhibitor-of-apoptosis protein, blocks nuclear factor-kappaB (NF-kappaB) signaling pathway leading to suppression of NF-kappaB-regulated antiapoptotic and metastatic gene products. *Molecular Pharmacology* **71**, 209–219.

Ahn, K. S., Sethi, G., Krishnan, K. and Aggarwal, B. B. (2007b). Gamma-tocotrienol inhibits nuclear factor-kappaB signaling pathway through inhibition of receptor-interacting protein and TAK1 leading to suppression of antiapoptotic gene products and potentiation of apoptosis. *The Journal of Biological Chemistry* **282**, 809–820.

Ahsan, H., Reagan-Shaw, S., Breur, J. and Ahmad, N. (2007). Sanguinarine induces apoptosis of human pancreatic carcinoma AsPC-1 and BxPC-3 cells via modulations in Bcl-2 family proteins. *Cancer Letters* **249**, 198–208.

Ajay, M., Gilani, A. U. and Mustafa, M. R. (2003). Effects of flavonoids on vascular smooth muscle of the isolated rat thoracic aorta. *Life Sciences* **74**, 603–612.

Akira, S., Nishio, Y., Inoue, M., Wang, X. J., Wei, S., Matsusaka, T., Yoshida, K., Sudo, T., Naruto, M. and Kishimoto, T. (1994). Molecular cloning of APRF, a novel IFN-stimulated gene factor 3 p91-related transcription factor involved in the gp130-mediated signaling pathway. *Cell* **77**, 63–71.

Allison, A. C., Cacabelos, R., Lombardi, V. R., Alvarez, X. A. and Vigo, C. (2001). Celastrol, a potent antioxidant and anti-inflammatory drug, as a possible treatment for Alzheimer's disease. *Progress in Neuro-Psychopharmacology & Biological Psychiatry* **25**, 1341–1357.

Amonkar, A. J., Padma, P. R. and Bhide, S. V. (1989). Protective effect of hydroxychavicol, a phenolic component of betel leaf, against the tobacco-specific carcinogens. *Mutation Research* **210**, 249–253.

Anand, P., Sundaram, C., Jhurani, S., Kunnumakkara, A. B. and Aggarwal, B. B. (2008). Curcumin and cancer: An "old-age" disease with an "age-old" solution. *Cancer Letters* **267**, 133–164.

Andersson, D., Liu, J. J., Nilsson, A. and Duan, R. D. (2003). Ursolic acid inhibits proliferation and stimulates apoptosis in HT29 cells following activation of alkaline sphingomyelinase. *Anticancer Research* **23**, 3317–3322.

Aneja, R., Zhou, J., Zhou, B., Chandra, R. and Joshi, H. C. (2006). Treatment of hormone-refractory breast cancer: Apoptosis and regression of human tumors implanted in mice. *Molecular Cancer Therapeutics* **5**, 2366–2377.

Aoki, Y., Sato, H., Nishimura, N., Takahashi, S., Itoh, K. and Yamamoto, M. (2001). Accelerated DNA adduct formation in the lung of the Nrf2 knockout mouse exposed to diesel exhaust. *Toxicology and Applied Pharmacology* **173**, 154–160.

Arai, M. A., Tateno, C., Hosoya, T., Koyano, T., Kowithayakorn, T. and Ishibashi, M. (2008). Hedgehog/GLI-mediated transcriptional inhibitors from Zizyphus cambodiana. *Bioorganic & Medicinal Chemistry* **16**, 9420–9424.

Arakawa, T., Masaki, T., Hirai, T., Doi, S., Kuratsune, M., Arihiro, K., Kohno, N. and Yorioka, N. (2008). Activation of signal transducer and activator of transcription 3 correlates with cell proliferation and renal injury in human glomerulonephritis. *Nephrology, Dialysis, Transplantation* **23**, 3418–3426.

Argiles, J. M., Busquets, S., Felipe, A. and Lopez-Soriano, F. J. (2005). Molecular mechanisms involved in muscle wasting in cancer and ageing: Cachexia versus sarcopenia. *The International Journal of Biochemistry & Cell Biology* **37**, 1084–1104.

Avci, G., Kucukkurt, I., Kupeli Akkol, E. and Yesilada, E. (2010). Effects of escin mixture from the seeds of Aesculus hippocastanum on obesity in mice fed a high fat diet. *Pharmaceutical Biology* **48**, 247–252.

Azuine, M. A., Amonkar, A. J. and Bhide, S. V. (1991). Chemopreventive efficacy of betel leaf extract and its constituents on 7,12-dimethylbenz(a)anthracene induced carcinogenesis and their effect on drug detoxification system in mouse skin. *Indian Journal of Experimental Biology* **29**, 346–351.

Babu, B. H., Jayram, H. N., Nair, M. G., Ajaikumar, K. B. and Padikkala, J. (2003). Free radical scavenging, antitumor and anticarcinogenic activity of gossypin. *Journal of Experimental & Clinical Cancer Research* **22**, 581–589.

Badary, O. A. and Gamal El-Din, A. M. (2001). Inhibitory effects of thymoquinone against 20-methylcholanthrene-induced fibrosarcoma tumorigenesis. *Cancer Detection and Prevention* **25**, 362–368.

Bai, H., Li, H., Zhang, W., Matkowskyj, K. A., Liao, J., Srivastava, S. K. and Yang, G. Y. (2011). Inhibition of chronic pancreatitis and pancreatic intraepithelial neoplasia (PanIN) by capsaicin in LSL-KrasG12D/Pdx1-Cre mice. *Carcinogenesis* **32**, 1689–1696.

Baker, R. G., Hayden, M. S. and Ghosh, S. (2011). NF-kappaB, inflammation, and metabolic disease. *Cell Metabolism* **13**, 11–22.

Balkwill, F., Charles, K. A. and Mantovani, A. (2005). Smoldering and polarized inflammation in the initiation and promotion of malignant disease. *Cancer Cell* **7**, 211–217.

Balmain, A. and Pragnell, I. B. (1983). Mouse skin carcinomas induced in vivo by chemical carcinogens have a transforming Harvey-ras oncogene. *Nature* **303**, 72–74.

Ban, H. S., Suzuki, K., Lim, S. S., Jung, S. H., Lee, S., Ji, J., Lee, H. S., Lee, Y. S., Shin, K. H. and Ohuchi, K. (2004). Inhibition of lipopolysaccharide-induced expression of inducible nitric oxide synthase and tumor necrosis factor-alpha by 2′-hydroxychalcone derivatives in RAW 264.7 cells. *Biochemical Pharmacology* **67**, 1549–1557.

Belfiore, A., Genua, M. and Malaguarnera, R. (2009). PPAR-gamma agonists and their effects on IGF-I receptor signaling: Implications for cancer. *PPAR Research* **2009**, 830501.

Belvisi, M. G., Hele, D. J. and Birrell, M. A. (2006). Peroxisome proliferator-activated receptor gamma agonists as therapy for chronic airway inflammation. *European Journal of Pharmacology* **533**, 101–109.

Bernstein, L. R. and Colburn, N. H. (1989). AP1/jun function is differentially induced in promotion-sensitive and resistant JB6 cells. *Science* **244**, 566–569.

Bertout, J. A., Patel, S. A. and Simon, M. C. (2008). The impact of O2 availability on human cancer. *Nature Reviews. Cancer* **8**, 967–975.

Bhide, S. V., Padma, P. R. and Amonkar, A. J. (1991a). Antimutagenic and anticarcinogenic effects of betel leaf extract against the tobacco-specific nitrosamine 4-(N-nitrosomethylamino)-1-(3-pyridyl)-1-butanone (NNK). *IARC Scientific Publications* 520–524.

Bhide, S. V., Zariwala, M. B., Amonkar, A. J. and Azuine, M. A. (1991b). Chemopreventive efficacy of a betel leaf extract against benzo[a]pyrene-induced forestomach tumors in mice. *Journal of Ethnopharmacology* **34**, 207–213.

Bhutani, M., Pathak, A. K., Nair, A. S., Kunnumakkara, A. B., Guha, S., Sethi, G. and Aggarwal, B. B. (2007). Capsaicin is a novel blocker of constitutive and interleukin-6-inducible STAT3 activation. *Clinical Cancer Research* **13**, 3024–3032.

Bi, S., Liu, J. R., Li, Y., Wang, Q., Liu, H. K., Yan, Y. G., Chen, B. Q. and Sun, W. G. (2010). gamma-Tocotrienol modulates the paracrine secretion of VEGF induced by cobalt(II) chloride via ERK signaling pathway in gastric adenocarcinoma SGC-7901 cell line. *Toxicology* **274**, 27–33.

Boissiere, F., Hunot, S., Faucheux, B., Duyckaerts, C., Hauw, J. J., Agid, Y. and Hirsch, E. C. (1997). Nuclear translocation of NF-kappaB in cholinergic neurons of patients with Alzheimer's disease. *Neuroreport* **8**, 2849–2852.

Bonniaud, P., Margetts, P. J., Kolb, M., Haberberger, T., Kelly, M., Robertson, J. and Gauldie, J. (2003). Adenoviral gene transfer of connective tissue growth factor in the lung induces transient fibrosis. *American Journal of Respiratory and Critical Care Medicine* **168**, 770–778.

Bonniaud, P., Martin, G., Margetts, P. J., Ask, K., Robertson, J., Gauldie, J. and Kolb, M. (2004). Connective tissue growth factor is crucial to inducing a profibrotic environment in "fibrosis-resistant" BALB/c mouse lungs. *American Journal of Respiratory Cell and Molecular Biology* **31**, 510–516.

Bortul, R., Tazzari, P. L., Billi, A. M., Tabellini, G., Mantovani, I., Cappellini, A., Grafone, T., Martinelli, G., Conte, R. and Martelli, A. M. (2005). Deguelin, A PI3K/AKT inhibitor, enhances chemosensitivity of leukaemia cells with an active PI3K/AKT pathway. *British Journal of Haematology* **129**, 677–686.

Brand, K., Page, S., Rogler, G., Bartsch, A., Brandl, R., Knuechel, R., Page, M., Kaltschmidt, C., Baeuerle, P. A. and Neumeier, D. (1996). Activated transcription factor nuclear factor-kappa B is present in the atherosclerotic lesion. *The Journal of Clinical Investigation* **97**, 1715–1722.

Brown, J., O'Prey, J. and Harrison, P. R. (2003). Enhanced sensitivity of human oral tumours to the flavonol, morin, during cancer progression: Involvement of the Akt and stress kinase pathways. *Carcinogenesis* **24**, 171–177.

Bumrungpert, A., Kalpravidh, R. W., Chitchumroonchokchai, C., Chuang, C. C., West, T., Kennedy, A. and McIntosh, M. (2009). Xanthones from mangosteen prevent lipopolysaccharide-mediated inflammation and insulin resistance in primary cultures of human adipocytes. *The Journal of Nutrition* **139**, 1185–1191.

Cadigan, K. M. and Nusse, R. (1997). Wnt signaling: A common theme in animal development. *Genes & Development* **11**, 3286–3305.

Campbell, S. E., Rudder, B., Phillips, R. B., Whaley, S. G., Stimmel, J. B., Leesnitzer, L. M., Lightner, J., Dessus-Babus, S., Duffourc, M., Stone, W. L., Menter, D. G., Newman, R. A. *et al.* (2011). gamma-Tocotrienol induces growth arrest through a novel pathway with TGFbeta2 in prostate cancer. *Free Radical Biology & Medicine* **50**, 1344–1354.

Catlett-Falcone, R., Landowski, T. H., Oshiro, M. M., Turkson, J., Levitzki, A., Savino, R., Ciliberto, G., Moscinski, L., Fernandez-Luna, J. L., Nunez, G., Dalton, W. S. and Jove, R. (1999). Constitutive activation of Stat3 signaling confers resistance to apoptosis in human U266 myeloma cells. *Immunity* **10**, 105–115.

Cha, T. L., Qiu, L., Chen, C. T., Wen, Y. and Hung, M. C. (2005). Emodin down-regulates androgen receptor and inhibits prostate cancer cell growth. *Cancer Research* **65**, 2287–2295.

Chabas, D., Baranzini, S. E., Mitchell, D., Bernard, C. C., Rittling, S. R., Denhardt, D. T., Sobel, R. A., Lock, C., Karpuj, M., Pedotti, R., Heller, R., Oksenberg, J. R. *et al.* (2001). The influence of the proinflammatory cytokine, osteopontin, on autoimmune demyelinating disease. *Science* **294**, 1731–1735.

Chadalapaka, G., Jutooru, I., McAlees, A., Stefanac, T. and Safe, S. (2008). Structure-dependent inhibition of bladder and pancreatic cancer cell growth by 2-substituted glycyrrhetinic and ursolic acid derivatives. *Bioorganic & Medicinal Chemistry Letters* **18**, 2633–2639.

Chan, T. C., Chang, C. J., Koonchanok, N. M. and Geahlen, R. L. (1993). Selective inhibition of the growth of ras-transformed human bronchial epithelial cells by emodin, a protein-tyrosine kinase inhibitor. *Biochemical and Biophysical Research Communications* **193**, 1152–1158.

Chang, Y. J., Wu, M. S., Lin, J. T., Sheu, B. S., Muta, T., Inoue, H. and Chen, C. C. (2004). Induction of cyclooxygenase-2 overexpression in human gastric epithelial cells by Helicobacter pylori involves TLR2/TLR9 and c-Src-dependent nuclear factor-kappaB activation. *Molecular Pharmacology* **66**, 1465–1477.

Chang, M. C., Uang, B. J., Tsai, C. Y., Wu, H. L., Lin, B. R., Lee, C. S., Chen, Y. J., Chang, C. H., Tsai, Y. L., Kao, C. J. and Jeng, J. H. (2007). Hydroxychavicol, a novel betel leaf component, inhibits platelet aggregation by suppression of cyclooxygenase, thromboxane production and calcium mobilization. *British Journal of Pharmacology* **152**, 73–82.

Chao, D. C., Lin, L. J., Kao, S. T., Huang, H. C., Chang, C. S., Liang, J. A., Wu, S. L., Hsiang, C. Y. and Ho, T. Y. (2011). Inhibitory effects of Zuo-Jin-Wan and its alkaloidal ingredients on activator protein 1, nuclear factor-kappaB, and cellular transformation in HepG2 cells. *Fitoterapia* **82**, 696–703.

Chaturvedi, M. M., Kumar, A., Darnay, B. G., Chainy, G. B., Agarwal, S. and Aggarwal, B. B. (1997). Sanguinarine (pseudochelerythrine) is a potent inhibitor of NF-kappaB activation, IkappaBalpha phosphorylation, and degradation. *The Journal of Biological Chemistry* **272**, 30129–30134.

Chen, Y. C., Shen, S. C., Lee, W. R., Lin, H. Y., Ko, C. H., Shih, C. M. and Yang, L. L. (2002). Wogonin and fisetin induction of apoptosis through activation of caspase 3 cascade and alternative expression of p21 protein in hepatocellular carcinoma cells SK-HEP-1. *Archives of Toxicology* **76**, 351–359.

Chen, Y. C., Shen, S. C., Chow, J. M., Ko, C. H. and Tseng, S. W. (2004). Flavone inhibition of tumor growth via apoptosis in vitro and in vivo. *International Journal of Oncology* **25**, 661–670.

Chen, B., Zhong, D. and Monteiro, A. (2006). Comparative genomics and evolution of the HSP90 family of genes across all kingdoms of organisms. *BMC Genomics* **7**, 156.

Chen, M., Rose, A. E., Doudican, N., Osman, I. and Orlow, S. J. (2009). Celastrol synergistically enhances temozolomide cytotoxicity in melanoma cells. *Molecular Cancer Research* **7**, 1946–1953.

Cheng, G. Z., Zhang, W. Z., Sun, M., Wang, Q., Coppola, D., Mansour, M., Xu, L. M., Costanzo, C., Cheng, J. Q. and Wang, L. H. (2008). Twist is transcriptionally induced by activation of STAT3 and mediates STAT3 oncogenic function. *The Journal of Biological Chemistry* **283**, 14665–14673.

Chiba, T., Yamada, M., Sasabe, J., Terashita, K., Shimoda, M., Matsuoka, M. and Aiso, S. (2009). Amyloid-beta causes memory impairment by disturbing the JAK2/STAT3 axis in hippocampal neurons. *Molecular Psychiatry* **14**, 206–222.

Chintharlapalli, S., Papineni, S., Ramaiah, S. K. and Safe, S. (2007). Betulinic acid inhibits prostate cancer growth through inhibition of specificity protein transcription factors. *Cancer Research* **67**, 2816–2823.

Cho, Y. C., You, S. K., Kim, H. J., Cho, C. W., Lee, I. S. and Kang, B. Y. (2010). Xanthohumol inhibits IL-12 production and reduces chronic allergic contact dermatitis. *International Immunopharmacology* **10**, 556–561.

Choi, B. M., Park, R., Pae, H. O., Yoo, J. C., Kim, Y. C., Jun, C. D., Jung, B. H., Oh, G. S., So, H. S., Kim, Y. M. and Chung, H. T. (2000a). Cyclic adenosine monophosphate inhibits ursolic acid-induced apoptosis via activation of protein kinase A in human leukaemic HL-60 cells. *Pharmacology & Toxicology* **86**, 53–58.

Choi, Y. H., Baek, J. H., Yoo, M. A., Chung, H. Y., Kim, N. D. and Kim, K. W. (2000b). Induction of apoptosis by ursolic acid through activation of caspases and down-regulation of c-IAPs in human prostate epithelial cells. *International Journal of Oncology* **17**, 565–571.

Choi, Y. H., Shin, E. M., Kim, Y. S., Cai, X. F., Lee, J. J. and Kim, H. P. (2006). Anti-inflammatory principles from the fruits of Evodia rutaecarpa and their cellular action mechanisms. *Archives of Pharmacal Research* **29**, 293–297.

Choi, W. Y., Kim, G. Y., Lee, W. H. and Choi, Y. H. (2008). Sanguinarine, a benzophenanthridine alkaloid, induces apoptosis in MDA-MB-231 human breast carcinoma cells through a reactive oxygen species-mediated mitochondrial pathway. *Chemotherapy* **54**, 279–287.

Chougule, M., Patel, A. R., Sachdeva, P., Jackson, T. and Singh, M. (2011a). Anticancer activity of Noscapine, an opioid alkaloid in combination with Cisplatin in human non-small cell lung cancer. *Lung Cancer* **71**, 271–282.

Chougule, M. B., Patel, A. R., Jackson, T. and Singh, M. (2011b). Antitumor activity of Noscapine in combination with Doxorubicin in triple negative breast cancer. *PLoS One* **6**, e17733.

Chung, Y. K., Heo, H. J., Kim, E. K., Kim, H. K., Huh, T. L., Lim, Y., Kim, S. K. and Shin, D. H. (2001). Inhibitory effect of ursolic acid purified from Origanum majorana L on the acetylcholinesterase. *Molecules and Cells* **11**, 137–143.

Clevers, H. (2006). Wnt/beta-catenin signaling in development and disease. *Cell* **127**, 469–480.

Connelly, L., Madhani, M. and Hobbs, A. J. (2005). Resistance to endotoxic shock in endothelial nitric-oxide synthase (eNOS) knock-out mice: A pro-inflammatory role for eNOS-derived no in vivo. *The Journal of Biological Chemistry* **280**, 10040–10046.

Csermely, P., Schnaider, T., Soti, C., Prohaszka, Z. and Nardai, G. (1998). The 90-kDa molecular chaperone family: Structure, function, and clinical applications. A comprehensive review. *Pharmacology & Therapeutics* **79**, 129–168.

Dantzer, R., O'Connor, J. C., Freund, G. G., Johnson, R. W. and Kelley, K. W. (2008). From inflammation to sickness and depression: When the immune system subjugates the brain. *Nature Reviews Neuroscience* **9**, 46–56.

Debiton, E., Madelmont, J. C., Legault, J. and Barthomeuf, C. (2003). Sanguinarine-induced apoptosis is associated with an early and severe cellular glutathione depletion. *Cancer Chemotherapy and Pharmacology* **51**, 474–482.

Deeb, D., Gao, X., Jiang, H., Arbab, A. S., Dulchavsky, S. A. and Gautam, S. C. (2010). Growth inhibitory and apoptosis-inducing effects of xanthohumol, a prenylated chalone present in hops, in human prostate cancer cells. *Anticancer Research* **30**, 3333–3339.

Devi, P. U., Rao, B. S. and Solomon, F. E. (1998). Effect of plumbagin on the radiation induced cytogenetic and cell cycle changes in mouse Ehrlich ascites carcinoma in vivo. *Indian Journal of Experimental Biology* **36**, 891–895.

Dickinson, B. C. and Chang, C. J. (2011). Chemistry and biology of reactive oxygen species in signaling or stress responses. *Nature Chemical Biology* **7**, 504–511.

Ding, Y., Zhao, L., Mei, H., Zhang, S. L., Huang, Z. H., Duan, Y. Y. and Ye, P. (2008). Exploration of Emodin to treat alpha-naphthylisothiocyanate-induced cholestatic hepatitis via anti-inflammatory pathway. *European Journal of Pharmacology* **590**, 377–386.

Dinkova-Kostova, A. T., Holtzclaw, W. D., Cole, R. N., Itoh, K., Wakabayashi, N., Katoh, Y., Yamamoto, M. and Talalay, P. (2002). Direct evidence that sulfhydryl groups of Keap1 are the sensors regulating induction of phase 2 enzymes that protect against carcinogens and oxidants. *Proceedings of the National Academy of Sciences of the United States of America* **99**, 11908–11913.

Dirsch, V. M., Kiemer, A. K., Wagner, H. and Vollmar, A. M. (1997). The triterpenoid quinonemethide pristimerin inhibits induction of inducible nitric oxide synthase in murine macrophages. *European Journal of Pharmacology* **336**, 211–217.

Dong, S. F., Hong, Y., Liu, M., Hao, Y. Z., Yu, H. S., Liu, Y. and Sun, J. N. (2011). Berberine attenuates cardiac dysfunction in hyperglycemic and hypercholesterolemic rats. *European Journal of Pharmacology* **660**, 368–374.

Donovan, C. E., Mark, D. A., He, H. Z., Liou, H. C., Kobzik, L., Wang, Y., De Sanctis, G. T., Perkins, D. L. and Finn, P. W. (1999). NF-kappa B/Rel transcription factors: c-Rel promotes airway hyperresponsiveness and allergic pulmonary inflammation. *Journal of Immunology* **163**, 6827–6833.

Drenzek, J. G., Seiler, N. L., Jaskula-Sztul, R., Rausch, M. M. and Rose, S. L. (2011). Xanthohumol decreases Notch1 expression and cell growth by cell cycle arrest and induction of apoptosis in epithelial ovarian cancer cell lines. *Gynecologic Oncology* **122**, 396–401.

Edlund, J., Hansell, P., Fasching, A., Liss, P., Weis, J., Glickson, J. D. and Palm, F. (2009). Reduced oxygenation in diabetic rat kidneys measured by T2* weighted magnetic resonance micro-imaging. *Advances in Experimental Medicine and Biology* **645**, 199–204.

Eferl, R. and Wagner, E. F. (2003). AP-1: A double-edged sword in tumorigenesis. *Nature Reviews. Cancer* **3**, 859–868.

Eisenbrand, G., Heydt, G. and Preussmann, R. (1982). Metabolism of isomeric N-nitroso-N-methylaminopyridines by rat liver microsomes. *IARC Scientific Publications* 509–516.

El Gazzar, M., El Mezayen, R., Nicolls, M. R., Marecki, J. C. and Dreskin, S. C. (2006). Downregulation of leukotriene biosynthesis by thymoquinone attenuates airway inflammation in a mouse model of allergic asthma. *Biochimica et Biophysica Acta* **1760**, 1088–1095.

El Mezayen, R., El Gazzar, M., Nicolls, M. R., Marecki, J. C., Dreskin, S. C. and Nomiyama, H. (2006). Effect of thymoquinone on cyclooxygenase expression and prostaglandin production in a mouse model of allergic airway inflammation. *Immunology Letters* **106**, 72–81.

Eriksen, J. L., Sagi, S. A., Smith, T. E., Weggen, S., Das, P., McLendon, D. C., Ozols, V. V., Jessing, K. W., Zavitz, K. H., Koo, E. H. and Golde, T. E. (2003). NSAIDs and enantiomers of flurbiprofen target gamma-secretase and lower Abeta 42 in vivo. *The Journal of Clinical Investigation* **112**, 440–449.

Erler, J. T., Bennewith, K. L., Cox, T. R., Lang, G., Bird, D., Koong, A., Le, Q. T. and Giaccia, A. J. (2009). Hypoxia-induced lysyl oxidase is a critical mediator of bone marrow cell recruitment to form the premetastatic niche. *Cancer Cell* **15**, 35–44.

Es-Saady, D., Simon, A., Jayat-Vignoles, C., Chulia, A. J. and Delage, C. (1996). MCF-7 cell cycle arrested at G1 through ursolic acid, and increased reduction of tetrazolium salts. *Anticancer Research* **16**, 481–486.

Esteban, M. A., Tran, M. G., Harten, S. K., Hill, P., Castellanos, M. C., Chandra, A., Raval, R., O'Brien, T. S. and Maxwell, P. H. (2006). Regulation of E-cadherin expression by VHL and hypoxia-inducible factor. *Cancer Research* **66**, 3567–3575.

Fei, X. F., Wang, B. X., Li, T. J., Tashiro, S., Minami, M., Xing, D. J. and Ikejima, T. (2003). Evodiamine, a constituent of Evodiae Fructus, induces anti-proliferating effects in tumor cells. *Cancer Science* **94**, 92–98.

Feldmann, G., Dhara, S., Fendrich, V., Bedja, D., Beaty, R., Mullendore, M., Karikari, C., Alvarez, H., Iacobuzio-Donahue, C., Jimeno, A., Gabrielson, K. L., Matsui, W. *et al.* (2007). Blockade of hedgehog signaling inhibits pancreatic cancer invasion and metastases: A new paradigm for combination therapy in solid cancers. *Cancer Research* **67**, 2187–2196.

Ferrandiz, M. L. and Alcaraz, M. J. (1991). Anti-inflammatory activity and inhibition of arachidonic acid metabolism by flavonoids. *Agents and Actions* **32**, 283–288.

Ferreira, V., van Dijk, K. W., Groen, A. K., Vos, R. M., van der Kaa, J., Gijbels, M. J., Havekes, L. M. and Pannekoek, H. (2007). Macrophage-specific inhibition of NF-kappaB activation reduces foam-cell formation. *Atherosclerosis* **192**, 283–290.

Forstermann, U. and Li, H. (2011). Therapeutic effect of enhancing endothelial nitric oxide synthase (eNOS) expression and preventing eNOS uncoupling. *British Journal of Pharmacology* **164**, 213–223.

Fotsis, T., Pepper, M. S., Montesano, R., Aktas, E., Breit, S., Schweigerer, L., Rasku, S., Wahala, K. and Adlercreutz, H. (1998). Phytoestrogens and inhibition of angiogenesis. *Baillière's Clinical Endocrinology and Metabolism* **12**, 649–666.

Franco, A. T., Israel, D. A., Washington, M. K., Krishna, U., Fox, J. G., Rogers, A. B., Neish, A. S., Collier-Hyams, L., Perez-Perez, G. I., Hatakeyama, M., Whitehead, R., Gaus, K. *et al.* (2005). Activation of beta-catenin by carcinogenic Helicobacter pylori. *Proceedings of the National Academy of Sciences of the United States of America* **102**, 10646–10651.

Fukaya, M., Isohata, N., Ohta, H., Aoyagi, K., Ochiya, T., Saeki, N., Yanagihara, K., Nakanishi, Y., Taniguchi, H., Sakamoto, H., Shimoda, T., Nimura, Y. *et al.* (2006). Hedgehog signal activation in gastric pit cell and in diffuse-type gastric cancer. *Gastroenterology* **131**, 14–29.

Fukuda, K., Hibiya, Y., Mutoh, M., Koshiji, M., Akao, S. and Fujiwara, H. (1999). Inhibition of activator protein 1 activity by berberine in human hepatoma cells. *Planta Medica* **65**, 381–383.

Fukutake, M., Yokota, S., Kawamura, H., Iizuka, A., Amagaya, S., Fukuda, K. and Komatsu, Y. (1998). Inhibitory effect of Coptidis Rhizoma and Scutellariae Radix on azoxymethane-induced aberrant crypt foci formation in rat colon. *Biological & Pharmaceutical Bulletin* **21**, 814–817.

Fulda, S., Jeremias, I., Steiner, H. H., Pietsch, T. and Debatin, K. M. (1999). Betulinic acid: A new cytotoxic agent against malignant brain-tumor cells. *International Journal of Cancer* **82**, 435–441.

Fusi, F., Cavalli, M., Mulholland, D., Crouch, N., Coombes, P., Dawson, G., Bova, S., Sgaragli, G. and Saponara, S. (2010). Cardamonin is a bifunctional vasodilator that inhibits Ca(v)1.2 current and stimulates K(Ca)1.1 current in rat tail artery myocytes. *The Journal of Pharmacology and Experimental Therapeutics* **332**, 531–540.

Gaemers, I. C., Vos, H. L., Volders, H. H., van der Valk, S. W. and Hilkens, J. (2001). A stat-responsive element in the promoter of the episialin/MUC1 gene is involved in its overexpression in carcinoma cells. *The Journal of Biological Chemistry* **276**, 6191–6199.

Garcia, R., Yu, C. L., Hudnall, A., Catlett, R., Nelson, K. L., Smithgall, T., Fujita, D. J., Ethier, S. P. and Jove, R. (1997). Constitutive activation of

Stat3 in fibroblasts transformed by diverse oncoproteins and in breast carcinoma cells. *Cell Growth & Differentiation* **8**, 1267–1276.

Garcia-Rivera, D., Delgado, R., Bougarne, N., Haegeman, G. and Berghe, W. V. (2011). Gallic acid indanone and mangiferin xanthone are strong determinants of immunosuppressive anti-tumour effects of Mangifera indica L. bark in MDA-MB231 breast cancer cells. *Cancer Letters* **305**, 21–31.

Gareus, R., Kotsaki, E., Xanthoulea, S., van der Made, I., Gijbels, M. J., Kardakaris, R., Polykratis, A., Kollias, G., de Winther, M. P. and Pasparakis, M. (2008). Endothelial cell-specific NF-kappaB inhibition protects mice from atherosclerosis. *Cell Metabolism* **8**, 372–383.

Geraets, L., Haegens, A., Brauers, K., Haydock, J. A., Vernooy, J. H., Wouters, E. F., Bast, A. and Hageman, G. J. (2009). Inhibition of LPS-induced pulmonary inflammation by specific flavonoids. *Biochemical and Biophysical Research Communications* **382**, 598–603.

Gerhauser, C., Mar, W., Lee, S. K., Suh, N., Luo, Y., Kosmeder, J., Luyengi, L., Fong, H. H., Kinghorn, A. D., Moriarty, R. M. *et al.* (1995). Rotenoids mediate potent cancer chemopreventive activity through transcriptional regulation of ornithine decarboxylase. *Nature Medicine* **1**, 260–266.

Ghaffari, M. A. and Mojab, S. (2007). Influence of flavonols as in vitro on low density lipoprotein glycation. *Iranian Biomedical Journal* **11**, 185–191.

Ghosh, R., Nadiminty, N., Fitzpatrick, J. E., Alworth, W. L., Slaga, T. J. and Kumar, A. P. (2005). Eugenol causes melanoma growth suppression through inhibition of E2F1 transcriptional activity. *The Journal of Biological Chemistry* **280**, 5812–5819.

Giaccia, A., Siim, B. G. and Johnson, R. S. (2003). HIF-1 as a target for drug development. *Nature Reviews. Drug Discovery* **2**, 803–811.

Giatromanolaki, A., Sivridis, E., Maltezos, E., Papazoglou, D., Simopoulos, C., Gatter, K. C., Harris, A. L. and Koukourakis, M. I. (2003). Hypoxia inducible factor 1alpha and 2alpha overexpression in inflammatory bowel disease. *Journal of Clinical Pathology* **56**, 209–213.

Gills, J. J., Kosmeder, J., 2nd, Moon, R. C., Lantvit, D. D. and Pezzuto, J. M. (2005). Effect of deguelin on UVB-induced skin carcinogenesis. *Journal of Chemotherapy* **17**, 297–301.

Goel, R. K., Das Gupta, G., Ram, S. N. and Pandey, V. B. (1991). Antiulcerogenic and anti-inflammatory effects of emodin, isolated from Rhamnus triquerta wall. *Indian Journal of Experimental Biology* **29**, 230–232.

Goncalves, P., Araujo, J. R., Pinho, M. J. and Martel, F. (2011). In vitro studies on the inhibition of colon cancer by butyrate and polyphenolic compounds. *Nutrition and Cancer* **63**, 282–294.

Grases, F., Garcia-Gonzalez, R., Redondo, E., Costa-Bauza, A., Simonet, B. M., Sartini, R. P., Bassa, A., Gallegos, C., Payeras, A., Saro, D., Martinez, A. I., Cifuentes, C. *et al.* (2004). Effects of escin on indinavir crystallization time in the urine of patients with HIV-I infection: A multicenter, randomized, open-label, controlled, four-period crossover trial. *Clinical Therapeutics* **26**, 2045–2055.

Guha, S., Ghosal, S. and Chattopadhyay, U. (1996). Antitumor, immunomodulatory and anti-HIV effect of mangiferin, a naturally occurring glucosylxanthone. *Chemotherapy* **42**, 443–451.

Guo, Q. L., Lin, S. S., You, Q. D., Gu, H. Y., Yu, J., Zhao, L., Qi, Q., Liang, F., Tan, Z. and Wang, X. (2006). Inhibition of human telomerase reverse transcriptase gene expression by gambogic acid in human hepatoma SMMC-7721 cells. *Life Sciences* **78**, 1238–1245.

Guo, S., Liu, M. and Gonzalez-Perez, R. R. (2011). Role of Notch and its oncogenic signaling crosstalk in breast cancer. *Biochimica et Biophysica Acta* **1815**, 197–213.

Gupta, O. P., Anand, K. K., Ghatak, B. J., Atal, C. K. and Ali, M. (1976). In vitro anthelmintic activity of disalts of embelin. *Indian Journal of Experimental Biology* **14**, 356–357.

Gupta, S. C., Kim, J. H., Prasad, S. and Aggarwal, B. B. (2010). Regulation of survival, proliferation, invasion, angiogenesis, and metastasis of tumor cells through modulation of inflammatory pathways by nutraceuticals. *Cancer Metastasis Reviews* **29**, 405–434.

Gupta, S. C., Prasad, S., Kim, J. H., Patchva, S., Webb, L. J., Priyadarsini, I. K. and Aggarwal, B. B. (2011). Multitargeting by curcumin as revealed by molecular interaction studies. *Natural Product Reports* **28**, 1937–1955.

Habib, A. A., Chatterjee, S., Park, S. K., Ratan, R. R., Lefebvre, S. and Vartanian, T. (2001). The epidermal growth factor receptor engages receptor interacting protein and nuclear factor-kappa B (NF-kappa B)-inducing kinase to activate NF-kappa B. Identification of a novel receptor-tyrosine kinase signalosome. *The Journal of Biological Chemistry* **276**, 8865–8874.

Haddad, A. Q., Venkateswaran, V., Viswanathan, L., Teahan, S. J., Fleshner, N. E. and Klotz, L. H. (2006). Novel antiproliferative flavonoids induce cell cycle arrest in human prostate cancer cell lines. *Prostate Cancer and Prostatic Diseases* **9**, 68–76.

Haefner, B. (2002). NF-kappa B: Arresting a major culprit in cancer. *Drug Discovery Today* **7**, 653–663.

Hajhashemi, V., Ghannadi, A. and Jafarabadi, H. (2004). Black cumin seed essential oil, as a potent analgesic and antiinflammatory drug. *Phytotherapy Research* **18**, 195–199.

Hamakawa, H., Nakashiro, K., Sumida, T., Shintani, S., Myers, J. N., Takes, R. P., Rinaldo, A. and Ferlito, A. (2008). Basic evidence of molecular targeted therapy for oral cancer and salivary gland cancer. *Head & Neck* **30**, 800–809.

Han, S. S., Keum, Y. S., Chun, K. S. and Surh, Y. J. (2002). Suppression of phorbol ester-induced NF-kappaB activation by capsaicin in cultured human promyelocytic leukemia cells. *Archives of Pharmacal Research* **25**, 475–479.

Han, Q. B., Cheung, S., Tai, J., Qiao, C. F., Song, J. Z. and Xu, H. X. (2005). Stability and cytotoxicity of gambogic acid and its derivative, gambogoic acid. *Biological & Pharmaceutical Bulletin* **28**, 2335–2337.

Han, M. H., Yoo, Y. H. and Choi, Y. H. (2008). Sanguinarine-induced apoptosis in human leukemia U937 cells via Bcl-2 downregulation and caspase-3 activation. *Chemotherapy* **54**, 157–165.

Hanneken, A., Lin, F. F., Johnson, J. and Maher, P. (2006). Flavonoids protect human retinal pigment epithelial cells from oxidative-stress-induced death. *Investigative Ophthalmology & Visual Science* **47**, 3164–3177.

Hansson, G. K., Robertson, A. K. and Soderberg-Naucler, C. (2006). Inflammation and atherosclerosis. *Annual Review of Pathology* **1**, 297–329.

Haq, A., Lobo, P. I., Al-Tufail, M., Rama, N. R. and Al-Sedairy, S. T. (1999). Immunomodulatory effect of Nigella sativa proteins fractionated by ion exchange chromatography. *International Journal of Immunopharmacology* **21**, 283–295.

Harikumar, K. B., Kunnumakkara, A. B., Ahn, K. S., Anand, P., Krishnan, S., Guha, S. and Aggarwal, B. B. (2009). Modification of the cysteine residues in IkappaBalpha kinase and NF-kappaB (p65) by xanthohumol leads to suppression of NF-kappaB-regulated gene products and potentiation of apoptosis in leukemia cells. *Blood* **113**, 2003–2013.

Harikumar, K. B., Sung, B., Pandey, M. K., Guha, S., Krishnan, S. and Aggarwal, B. B. (2010). Escin, a pentacyclic triterpene, chemosensitizes human tumor cells through inhibition of nuclear factor-kappaB signaling pathway. *Molecular Pharmacology* **77**, 818–827.

Harmand, P. O., Duval, R., Delage, C. and Simon, A. (2005). Ursolic acid induces apoptosis through mitochondrial intrinsic pathway and caspase-3 activation in M4Beu melanoma cells. *International Journal of Cancer* **114**, 1–11.

Hayden, M. S. and Ghosh, S. (2004). Signaling to NF-kappaB. *Genes & Development* **18**, 2195–2224.

Hayes, J. D. and McMahon, M. (2009). NRF2 and KEAP1 mutations: Permanent activation of an adaptive response in cancer. *Trends in Biochemical Sciences* **34**, 176–188.

He, L., Qi, Y., Rong, X., Jiang, J., Yang, Q., Yamahara, J., Murray, M. and Li, Y. (2009). The Ayurvedic medicine Salacia oblonga attenuates diabetic renal fibrosis in rats: Suppression of angiotensin II/AT1 signaling. *Evidence-Based Complementary and Alternative Medicine*.

Hei, Z. Q., Huang, H. Q., Tan, H. M., Liu, P. Q., Zhao, L. Z., Chen, S. R., Huang, W. G., Chen, F. Y. and Guo, F. F. (2006). Emodin inhibits dietary induced atherosclerosis by antioxidation and regulation of the sphingomyelin pathway in rabbits. *Chinese Medical Journal* **119**, 868–870.

Heo, J. Y., Kim, H. J., Kim, S. M., Park, K. R., Park, S. Y., Kim, S. W., Nam, D., Jang, H. J., Lee, S. G., Ahn, K. S., Kim, S. H., Shim, B. S. *et al.* (2011). Embelin suppresses STAT3 signaling, proliferation, and survival of multiple myeloma via the protein tyrosine phosphatase PTEN. *Cancer Letters* **308**, 71–80.

Herencia, F., Ferrandiz, M. L., Ubeda, A., Guillen, I., Dominguez, J. N., Charris, J. E., Lobo, G. M. and Alcaraz, M. J. (1999). Novel anti-inflammatory chalcone derivatives inhibit the induction of nitric oxide synthase and cyclooxygenase-2 in mouse peritoneal macrophages. *FEBS Letters* **453**, 129–134.

Higa, S., Hirano, T., Kotani, M., Matsumoto, M., Fujita, A., Suemura, M., Kawase, I. and Tanaka, T. (2003). Fisetin, a flavonol, inhibits TH2-type cytokine production by activated human basophils. *The Journal of Allergy and Clinical Immunology* **111**, 1299–1306.

Higgins, D. F., Kimura, K., Bernhardt, W. M., Shrimanker, N., Akai, Y., Hohenstein, B., Saito, Y., Johnson, R. S., Kretzler, M., Cohen, C. D., Eckardt, K. U., Iwano, M. *et al.* (2007). Hypoxia promotes fibrogenesis in vivo via HIF-1 stimulation of epithelial-to-mesenchymal transition. *The Journal of Clinical Investigation* **117**, 3810–3820.

Hoessel, R., Leclerc, S., Endicott, J. A., Nobel, M. E., Lawrie, A., Tunnah, P., Leost, M., Damiens, E., Marie, D., Marko, D., Niederberger, E., Tang, W. *et al.* (1999). Indirubin, the active constituent of a Chinese antileukaemia medicine, inhibits cyclin-dependent kinases. *Nature Cell Biology* **1**, 60–67.

Hold, G. L. and El-Omar, E. M. (2008). Genetic aspects of inflammation and cancer. *The Biochemical Journal* **410**, 225–235.

Hostanska, K., Daum, G. and Saller, R. (2002). Cytostatic and apoptosis-inducing activity of boswellic acids toward malignant cell lines in vitro. *Anticancer Research* **22**, 2853–2862.

Hsieh, H. K., Lee, T. H., Wang, J. P., Wang, J. J. and Lin, C. N. (1998). Synthesis and anti-inflammatory effect of chalcones and related compounds. *Pharmaceutical Research* **15**, 39–46.

Hsieh, T. C., Elangovan, S. and Wu, J. M. (2010). Differential suppression of proliferation in MCF-7 and MDA-MB-231 breast cancer cells exposed to alpha-, gamma- and delta-tocotrienols is accompanied by altered expression of oxidative stress modulatory enzymes. *Anticancer Research* **30**, 4169–4176.

Hsu, Y. L., Kuo, P. L. and Lin, C. C. (2004). Proliferative inhibition, cell-cycle dysregulation, and induction of apoptosis by ursolic acid in human non-small cell lung cancer A549 cells. *Life Sciences* **75**, 2303–2316.

Hu, J. N., Zhu, X. M., Han, L. K., Saito, M., Sun, Y. S., Yoshikawa, M., Kimura, Y. and Zheng, Y. N. (2008). Anti-obesity effects of escins extracted from the seeds of Aesculus turbinata BLUME (Hippocastanaceae). *Chemical & Pharmaceutical Bulletin (Tokyo)* **56**, 12–16.

Huang, M. T., Ho, C. T., Wang, Z. Y., Ferraro, T., Lou, Y. R., Stauber, K., Ma, W., Georgiadis, C., Laskin, J. D. and Conney, A. H. (1994). Inhibition of skin tumorigenesis by rosemary and its constituents carnosol and ursolic acid. *Cancer Research* **54**, 701–708.

Huang, Y. C., Guh, J. H. and Teng, C. M. (2004). Induction of mitotic arrest and apoptosis by evodiamine in human leukemic T-lymphocytes. *Life Sciences* **75**, 35–49.

Huang, Q., Lu, G., Shen, H. M., Chung, M. C. and Ong, C. N. (2007). Anti-cancer properties of anthraquinones from rhubarb. *Medicinal Research Reviews* **27**, 609–630.

Hughes, E. H. and Shanks, J. V. (2002). Metabolic engineering of plants for alkaloid production. *Metabolic Engineering* **4**, 41–48.

Hussain, S. P. and Harris, C. C. (2007). Inflammation and cancer: An ancient link with novel potentials. *International Journal of Cancer* **121**, 2373–2380.

Hwang, J. M., Kuo, H. C., Tseng, T. H., Liu, J. Y. and Chu, C. Y. (2006). Berberine induces apoptosis through a mitochondria/caspases pathway in human hepatoma cells. *Archives of Toxicology* **80**, 62–73.

Ichikawa, H., Takada, Y., Murakami, A. and Aggarwal, B. B. (2005). Identification of a novel blocker of I kappa B alpha kinase that enhances cellular apoptosis and inhibits cellular invasion through suppression of NF-kappa B-regulated gene products. *Journal of Immunology* **174**, 7383–7392.

Ichikawa, H., Murakami, A. and Aggarwal, B. B. (2006). 1′-Acetoxychavicol acetate inhibits RANKL-induced osteoclastic differentiation of RAW 264.7 monocytic cells by suppressing nuclear factor-kappaB activation. *Molecular Cancer Research* **4**, 275–281.

Iida, K., Itoh, K., Kumagai, Y., Oyasu, R., Hattori, K., Kawai, K., Shimazui, T., Akaza, H. and Yamamoto, M. (2004). Nrf2 is essential for the chemopreventive efficacy of oltipraz against urinary bladder carcinogenesis. *Cancer Research* **64**, 6424–6431.

Iizuka, T., Ishii, Y., Itoh, K., Kiwamoto, T., Kimura, T., Matsuno, Y., Morishima, Y., Hegab, A. E., Homma, S., Nomura, A., Sakamoto, T., Shimura, M. *et al.* (2005). Nrf2-deficient mice are highly susceptible to cigarette smoke-induced emphysema. *Genes to Cells* **10**, 1113–1125.

Israf, D. A., Khaizurin, T. A., Syahida, A., Lajis, N. H. and Khozirah, S. (2007). Cardamonin inhibits COX and iNOS expression via inhibition of p65NF-kappaB nuclear translocation and Ikappa-B phosphorylation in RAW 264.7 macrophage cells. *Molecular Immunology* **44**, 673–679.

Ito, S., Oyake, T., Murai, K. and Ishida, Y. (2010). Deguelin suppresses cell proliferation via the inhibition of survivin expression and STAT3 phosphorylation in HTLV-1-transformed T cells. *Leukemia Research* **34**, 352–357.

Itoh, K., Chiba, T., Takahashi, S., Ishii, T., Igarashi, K., Katoh, Y., Oyake, T., Hayashi, N., Satoh, K., Hatayama, I., Yamamoto, M. and Nabeshima, Y.

(1997). An Nrf2/small Maf heterodimer mediates the induction of phase II detoxifying enzyme genes through antioxidant response elements. *Biochemical and Biophysical Research Communications* **236**, 313–322.

Itoh, K., Wakabayashi, N., Katoh, Y., Ishii, T., Igarashi, K., Engel, J. D. and Yamamoto, M. (1999). Keap1 represses nuclear activation of antioxidant responsive elements by Nrf2 through binding to the amino-terminal Neh2 domain. *Genes & Development* **13**, 76–86.

Itzkowitz, S. H. and Yio, X. (2004). Inflammation and cancer IV. Colorectal cancer in inflammatory bowel disease: The role of inflammation. *American Journal of Physiology. Gastrointestinal and Liver Physiology* **287**, G7–G17.

Iwashita, K., Kobori, M., Yamaki, K. and Tsushida, T. (2000). Flavonoids inhibit cell growth and induce apoptosis in B16 melanoma 4A5 cells. *Bioscience, Biotechnology, and Biochemistry* **64**, 1813–1820.

Jackson, J. K., Higo, T., Hunter, W. L. and Burt, H. M. (2008a). Topoisomerase inhibitors as anti-arthritic agents. *Inflammation Research* **57**, 126–134.

Jackson, T., Chougule, M. B., Ichite, N., Patlolla, R. R. and Singh, M. (2008b). Antitumor activity of noscapine in human non-small cell lung cancer xenograft model. *Cancer Chemotherapy and Pharmacology* **63**, 117–126.

Jaganathan, S. K., Mazumdar, A., Mondhe, D. and Mandal, M. (2011). Apoptotic effect of eugenol in human colon cancer cell lines. *Cell Biology International* **35**, 607–615.

Jagani, Z., Dorsch, M. and Warmuth, M. (2010). Hedgehog pathway activation in chronic myeloid leukemia. *Cell Cycle* **9**, 3449–3456.

Jakob, U., Lilie, H., Meyer, I. and Buchner, J. (1995). Transient interaction of Hsp90 with early unfolding intermediates of citrate synthase. Implications for heat shock in vivo. *The Journal of Biological Chemistry* **270**, 7288–7294.

Jang, J. J., Cho, K. J., Lee, Y. S. and Bae, J. H. (1991). Different modifying responses of capsaicin in a wide-spectrum initiation model of F344 rat. *Journal of Korean Medical Science* **6**, 31–36.

Jang, H. S., Kook, S. H., Son, Y. O., Kim, J. G., Jeon, Y. M., Jang, Y. S., Choi, K. C., Kim, J., Han, S. K., Lee, K. Y., Park, B. K., Cho, N. P. *et al.* (2005). Flavonoids purified from Rhus verniciflua Stokes actively inhibit cell growth and induce apoptosis in human osteosarcoma cells. *Biochimica et Biophysica Acta* **1726**, 309–316.

Jeng, J. H., Wang, Y. J., Chang, W. H., Wu, H. L., Li, C. H., Uang, B. J., Kang, J. J., Lee, J. J., Hahn, L. J., Lin, B. R. and Chang, M. C. (2004). Reactive oxygen species are crucial for hydroxychavicol toxicity toward KB epithelial cells. *Cellular and Molecular Life Sciences* **61**, 83–96.

Jeng, J. H., Wu, H. L., Lin, B. R., Lan, W. H., Chang, H. H., Ho, Y. S., Lee, P. H., Wang, Y. J., Wang, J. S., Chen, Y. J. and Chang, M. C. (2007). Antiplatelet effect of sanguinarine is correlated to calcium mobilization, thromboxane and cAMP production. *Atherosclerosis* **191**, 250–258.

Jeong, G. S., Lee, D. S., Song, M. Y., Park, B. H., Kang, D. G., Lee, H. S., Kwon, K. B. and Kim, Y. C. (2011). Butein from Rhus verniciflua protects pancreatic beta cells against cytokine-induced toxicity mediated by inhibition of nitric oxide formation. *Biological & Pharmaceutical Bulletin* **34**, 97–102.

Joung, E. J., Li, M. H., Lee, H. G., Somparn, N., Jung, Y. S., Na, H. K., Kim, S. H., Cha, Y. N. and Surh, Y. J. (2007). Capsaicin induces heme oxygenase-1 expression in HepG2 cells via activation of PI3K-Nrf2 signaling: NAD(P)H: quinone oxidoreductase as a potential target. *Antioxidants & Redox Signaling* **9**, 2087–2098.

Kaltschmidt, B., Uherek, M., Volk, B., Baeuerle, P. A. and Kaltschmidt, C. (1997). Transcription factor NF-kappaB is activated in primary neurons by amyloid beta peptides and in neurons surrounding early plaques from patients with Alzheimer disease. *Proceedings of the National Academy of Sciences of the United States of America* **94**, 2642–2647.

Kalyan Kumar, G., Dhamotharan, R., Kulkarni, N. M., Mahat, M. Y., Gunasekaran, J. and Ashfaque, M. (2011). Embelin reduces cutaneous TNF-alpha level and ameliorates skin edema in acute and chronic model of skin inflammation in mice. *European Journal of Pharmacology* **662**, 63–69.

Kan, S. F., Huang, W. J., Lin, L. C. and Wang, P. S. (2004). Inhibitory effects of evodiamine on the growth of human prostate cancer cell line LNCaP. *International Journal of Cancer* **110**, 641–651.

Kang, D. G., Kim, Y. C., Sohn, E. J., Lee, Y. M., Lee, A. S., Yin, M. H. and Lee, H. S. (2003). Hypotensive effect of butein via the inhibition of angiotensin converting enzyme. *Biological & Pharmaceutical Bulletin* **26**, 1345–1347.

Kang, S. Y., Yoon, S. Y., Roh, D. H., Jeon, M. J., Seo, H. S., Uh, D. K., Kwon, Y. B., Kim, H. W., Han, H. J., Lee, H. J. and Lee, J. H. (2008). The anti-arthritic effect of ursolic acid on zymosan-induced acute inflammation and adjuvant-induced chronic arthritis models. *The Journal of Pharmacy and Pharmacology* **60**, 1347–1354.

Kannaiyan, R., Manu, K. A., Chen, L., Li, F., Rajendran, P., Subramaniam, A., Lam, P., Kumar, A. P. and Sethi, G. (2011). Celastrol inhibits tumor cell proliferation and promotes apoptosis through the activation of c-Jun N-terminal kinase and suppression of PI3 K/Akt signaling pathways. *Apoptosis* **16**, 1028–1041.

Kannappan, R., Yadav, V. R. and Aggarwal, B. B. (2010). gamma-Tocotrienol but not gamma-tocopherol blocks STAT3 cell signaling pathway through induction of protein-tyrosine phosphatase SHP-1 and sensitizes tumor cells to chemotherapeutic agents. *The Journal of Biological Chemistry* **285**, 33520–33528.

Kanters, E., Gijbels, M. J., van der Made, I., Vergouwe, M. N., Heeringa, P., Kraal, G., Hofker, M. H. and de Winther, M. P. (2004). Hematopoietic NF-kappaB1 deficiency results in small atherosclerotic lesions with an inflammatory phenotype. *Blood* **103**, 934–940.

Karhausen, J., Furuta, G. T., Tomaszewski, J. E., Johnson, R. S., Colgan, S. P. and Haase, V. H. (2004). Epithelial hypoxia-inducible factor-1 is protective in murine experimental colitis. *The Journal of Clinical Investigation* **114**, 1098–1106.

Karin, M. and Greten, F. R. (2005). NF-kappaB: Linking inflammation and immunity to cancer development and progression. *Nature Reviews Immunology* **5**, 749–759.

Karin, M., Cao, Y., Greten, F. R. and Li, Z. W. (2002). NF-kappaB in cancer: From innocent bystander to major culprit. *Nature Reviews. Cancer* **2**, 301–310.

Kaseb, A. O., Chinnakannu, K., Chen, D., Sivanandam, A., Tejwani, S., Menon, M., Dou, Q. P. and Reddy, G. P. (2007). Androgen receptor and E2F-1 targeted thymoquinone therapy for hormone-refractory prostate cancer. *Cancer Research* **67**, 7782–7788.

Kassi, E., Sourlingas, T. G., Spiliotaki, M., Papoutsi, Z., Pratsinis, H., Aligiannis, N. and Moutsatsou, P. (2009). Ursolic acid triggers apoptosis and Bcl-2 downregulation in MCF-7 breast cancer cells. *Cancer Investigation* **27**, 723–733.

Kaur, G., Athar, M. and Alam, M. S. (2010). Eugenol precludes cutaneous chemical carcinogenesis in mouse by preventing oxidative stress and inflammation and by inducing apoptosis. *Molecular Carcinogenesis* **49**, 290–301.

Kawabata, K., Tanaka, T., Honjo, S., Kakumoto, M., Hara, A., Makita, H., Tatematsu, N., Ushida, J., Tsuda, H. and Mori, H. (1999). Chemopreventive effect of dietary flavonoid morin on chemically induced rat tongue carcinogenesis. *International Journal of Cancer* **83**, 381–386.

Ke, Y., Ye, K., Grossniklaus, H. E., Archer, D. R., Joshi, H. C. and Kapp, J. A. (2000). Noscapine inhibits tumor growth with little toxicity to normal tissues or inhibition of immune responses. *Cancer Immunology, Immunotherapy* **49**, 217–225.

Keith, C. T., Borisy, A. A. and Stockwell, B. R. (2005). Multicomponent therapeutics for networked systems. *Nature Reviews. Drug Discovery* **4**, 71–78.

Kettmann, V., Kosfalova, D., Jantova, S., Cernakova, M. and Drimal, J. (2004). In vitro cytotoxicity of berberine against HeLa and L1210 cancer cell lines. *Die Pharmazie* **59**, 548–551.

Khanna, D., Sethi, G., Ahn, K. S., Pandey, M. K., Kunnumakkara, A. B., Sung, B., Aggarwal, A. and Aggarwal, B. B. (2007). Natural products as a gold mine for arthritis treatment. *Current Opinion in Pharmacology* **7**, 344–351.

Khor, T. O., Huang, M. T., Prawan, A., Liu, Y., Hao, X., Yu, S., Cheung, W. K., Chan, J. Y., Reddy, B. S., Yang, C. S. and Kong, A. N. (2008). Increased susceptibility of Nrf2 knockout mice to colitis-associated colorectal cancer. *Cancer Prevention Research (Philadelphia, Pa.)* **1**, 187–191.

Kim, D. K., Baek, J. H., Kang, C. M., Yoo, M. A., Sung, J. W., Chung, H. Y., Kim, N. D., Choi, Y. H., Lee, S. H. and Kim, K. W. (2000). Apoptotic activity of ursolic acid may correlate with the inhibition of initiation of DNA replication. *International Journal of Cancer* **87**, 629–636.

Kim, J. K., Kim, Y. J., Fillmore, J. J., Chen, Y., Moore, I., Lee, J., Yuan, M., Li, Z. W., Karin, M., Perret, P., Shoelson, S. E. and Shulman, G. I. (2001). Prevention of fat-induced insulin resistance by salicylate. *The Journal of Clinical Investigation* **108**, 437–446.

Kim, S. S., Oh, O. J., Min, H. Y., Park, E. J., Kim, Y., Park, H. J., Nam Han, Y. and Lee, S. K. (2003). Eugenol suppresses cyclooxygenase-2 expression in lipopolysaccharide-stimulated mouse macrophage RAW264.7 cells. *Life Sciences* **73**, 337–348.

Kim, S., Lee, T. J., Leem, J., Choi, K. S., Park, J. W. and Kwon, T. K. (2008). Sanguinarine-induced apoptosis: Generation of ROS, down-regulation of Bcl-2, c-FLIP, and synergy with TRAIL. *Journal of Cellular Biochemistry* **104**, 895–907.

Kimura, H., Ogawa, S., Jisaka, M., Kimura, Y., Katsube, T. and Yokota, K. (2006). Identification of novel saponins from edible seeds of Japanese horse chestnut (Aesculus turbinata Blume) after treatment with wooden ashes and their nutraceutical activity. *Journal of Pharmaceutical and Biomedical Analysis* **41**, 1657–1665.

Kinzler, K. W. and Vogelstein, B. (1996). Lessons from hereditary colorectal cancer. *Cell* **87**, 159–170.

Kiuchi, N., Nakajima, K., Ichiba, M., Fukada, T., Narimatsu, M., Mizuno, K., Hibi, M. and Hirano, T. (1999). STAT3 is required for the gp130-mediated full activation of the c-myc gene. *The Journal of Experimental Medicine* **189**, 63–73.

Koeffler, H. P. (2003). Peroxisome proliferator-activated receptor gamma and cancers. *Clinical Cancer Research* **9**, 1–9.

Kreuz, S., Siegmund, D., Scheurich, P. and Wajant, H. (2001). NF-kappaB inducers upregulate cFLIP, a cycloheximide-sensitive inhibitor of death receptor signaling. *Molecular and Cellular Biology* **21**, 3964–3973.

Krol, W., Dworniczak, S., Pietsz, G., Czuba, Z. P., Kunicka, M., Kopacz, M. and Nowak, D. (2002). Synthesis and tumoricidal activity evaluation of new morin and quercetin sulfonic derivatives. *Acta Poloniae Pharmaceutica* **59**, 77–79.

Kroll, T. G., Sarraf, P., Pecciarini, L., Chen, C. J., Mueller, E., Spiegelman, B. M. and Fletcher, J. A. (2000). PAX8-PPARgamma1 fusion oncogene in human thyroid carcinoma [corrected]. *Science* **289**, 1357–1360.

Kumar, A., Dhawan, S. and Aggarwal, B. B. (1998). Emodin (3-methyl-1,6,8-trihydroxyanthraquinone) inhibits TNF-induced NF-kappaB activation, IkappaB degradation, and expression of cell surface adhesion proteins in human vascular endothelial cells. *Oncogene* **17**, 913–918.

Kunikata, T., Tatefuji, T., Aga, H., Iwaki, K., Ikeda, M. and Kurimoto, M. (2000). Indirubin inhibits inflammatory reactions in delayed-type hypersensitivity. *European Journal of Pharmacology* **410**, 93–100.

Kunnumakkara, A. B., Anand, P. and Aggarwal, B. B. (2008). Curcumin inhibits proliferation, invasion, angiogenesis and metastasis of different cancers through interaction with multiple cell signaling proteins. *Cancer Letters* **269**, 199–225.

Kunnumakkara, A. B., Nair, A. S., Sung, B., Pandey, M. K. and Aggarwal, B. B. (2009). Boswellic acid blocks signal transducers and activators of transcription 3 signaling, proliferation, and survival of multiple myeloma via the protein tyrosine phosphatase SHP-1. *Molecular Cancer Research* **7**, 118–128.

Kunnumakkara, A. B., Sung, B., Ravindran, J., Diagaradjane, P., Deorukhkar, A., Dey, S., Koca, C., Yadav, V. R., Tong, Z., Gelovani, J. G., Guha, S., Krishnan, S. *et al.* (2010). {Gamma}-tocotrienol inhibits pancreatic tumors and sensitizes them to gemcitabine treatment by modulating the inflammatory microenvironment. *Cancer Research* **70**, 8695–8705.

Kuo, C. L., Chi, C. W. and Liu, T. Y. (2005). Modulation of apoptosis by berberine through inhibition of cyclooxygenase-2 and Mcl-1 expression in oral cancer cells. *In Vivo* **19**, 247–252.

Lai, L., Liu, J., Zhai, D., Lin, Q., He, L., Dong, Y., Zhang, J., Lu, B., Chen, Y., Yi, Z. and Liu, M. (2011). Plumbagin inhibits tumor angiogenesis and tumor growth through VEGFR2-mediated Ras signaling pathway. *British Journal of Pharmacology* **165**, 1084–1096.

Landen, J. W., Lang, R., McMahon, S. J., Rusan, N. M., Yvon, A. M., Adams, A. W., Sorcinelli, M. D., Campbell, R., Bonaccorsi, P., Ansel, J. C., Archer, D. R., Wadsworth, P. *et al.* (2002). Noscapine alters microtubule dynamics in living cells and inhibits the progression of melanoma. *Cancer Research* **62**, 4109–4114.

Lasky, J. A., Ortiz, L. A., Tonthat, B., Hoyle, G. W., Corti, M., Athas, G., Lungarella, G., Brody, A. and Friedman, M. (1998). Connective tissue growth factor mRNA expression is upregulated in bleomycin-induced lung fibrosis. *The American Journal of Physiology* **275**, L365–L371.

Lee, W., Haslinger, A., Karin, M. and Tjian, R. (1987). Activation of transcription by two factors that bind promoter and enhancer sequences of the human metallothionein gene and SV40. *Nature* **325**, 368–372.

Lee, W. R., Shen, S. C., Lin, H. Y., Hou, W. C., Yang, L. L. and Chen, Y. C. (2002). Wogonin and fisetin induce apoptosis in human promyeloleukemic cells, accompanied by a decrease of reactive oxygen species, and activation of caspase 3 and Ca(2+)-dependent endonuclease. *Biochemical Pharmacology* **63**, 225–236.

Lee, J. C., Lee, K. Y., Kim, J., Na, C. S., Jung, N. C., Chung, G. H. and Jang, Y. S. (2004a). Extract from Rhus verniciflua Stokes is capable of inhibiting the growth of human lymphoma cells. *Food and Chemical Toxicology* **42**, 1383–1388.

Lee, S. H., Seo, G. S. and Sohn, D. H. (2004b). Inhibition of lipopolysaccharide-induced expression of inducible nitric oxide synthase by butein in RAW 264.7 cells. *Biochemical and Biophysical Research Communications* **323**, 125–132.

Lee, H. Y., Oh, S. H., Woo, J. K., Kim, W. Y., Van Pelt, C. S., Price, R. E., Cody, D., Tran, H., Pezzuto, J. M., Moriarty, R. M. and Hong, W. K. (2005). Chemopreventive effects of deguelin, a novel Akt inhibitor, on tobacco-induced lung tumorigenesis. *Journal of the National Cancer Institute* **97**, 1695–1699.

Lee, T. J., Kim, E. J., Kim, S., Jung, E. M., Park, J. W., Jeong, S. H., Park, S. E., Yoo, Y. H. and Kwon, T. K. (2006). Caspase-dependent and caspase-independent apoptosis induced by evodiamine in human leukemic U937 cells. *Molecular Cancer Therapeutics* **5**, 2398–2407.

Lee, S. H., Richardson, R. L., Dashwood, R. H. and Baek, S. J. (2011). Capsaicin represses transcriptional activity of beta-catenin in human colorectal cancer cells. *The Journal of Nutritional Biochemistry* (in press).

Lenfeld, J., Kroutil, M., Marsalek, E., Slavik, J., Preininger, V. and Simanek, V. (1981). Antiinflammatory activity of quaternary benzophenanthridine alkaloids from Chelidonium majus. *Planta Medica* **43**, 161–165.

Lepage, C., Leger, D. Y., Bertrand, J., Martin, F., Beneytout, J. L. and Liagre, B. (2011). Diosgenin induces death receptor-5 through activation of p38 pathway and promotes TRAIL-induced apoptosis in colon cancer cells. *Cancer Letters* **301**, 193–202.

Letasiova, S., Jantova, S., Cipak, L. and Muckova, M. (2006). Berberine-antiproliferative activity in vitro and induction of apoptosis/necrosis of the U937 and B16 cells. *Cancer Letters* **239**, 254–262.

Li, F. L., Xu, R., Zhou, R., Fan, B., Zhou, M., Xu, W. B., Gu, D. Q., Zhang, M. and Li, B. (2004). Clinical effect of escin on patients with cutaneous pruritus caused by blood stasis and wind-dryness. *Zhong Xi Yi Jie He Xue Bao* **2**, 426–428.

Li, H. L., Chen, H. L., Li, H., Zhang, K. L., Chen, X. Y., Wang, X. W., Kong, Q. Y. and Liu, J. (2005). Regulatory effects of emodin on NF-kappaB activation and inflammatory cytokine expression in RAW 264.7 macrophages. *International Journal of Molecular Medicine* **16**, 41–47.

Li, F., Fernandez, P. P., Rajendran, P., Hui, K. M. and Sethi, G. (2010a). Diosgenin, a steroidal saponin, inhibits STAT3 signaling pathway leading to suppression of proliferation and chemosensitization of human hepatocellular carcinoma cells. *Cancer Letters* **292**, 197–207.

Li, F., Rajendran, P. and Sethi, G. (2010b). Thymoquinone inhibits proliferation, induces apoptosis and chemosensitizes human multiple myeloma cells through suppression of signal transducer and activator of transcription 3 activation pathway. *British Journal of Pharmacology* **161**, 541–554.

Li, X., Cui, X., Sun, X., Zhu, Q. and Li, W. (2010c). Mangiferin prevents diabetic nephropathy progression in streptozotocin-induced diabetic rats. *Phytotherapy Research* **24**, 893–899.

Li, C., Chi, S. and Xie, J. (2011). Hedgehog signaling in skin cancers. *Cellular Signalling* **23**, 1235–1243.

Liagre, B., Vergne-Salle, P., Corbiere, C., Charissoux, J. L. and Beneytout, J. L. (2004). Diosgenin, a plant steroid, induces apoptosis in human rheumatoid

arthritis synoviocytes with cyclooxygenase-2 overexpression. *Arthritis Research & Therapy* **6**, R373–R383.
Liao, Q., Shi, D. H., Zheng, W., Xu, X. J. and Yu, Y. H. (2010). Antiproliferation of cardamonin is involved in mTOR on aortic smooth muscle cells in high fructose-induced insulin resistance rats. *European Journal of Pharmacology* **641**, 179–186.
Lin, C. C., Lin, S. Y., Chung, J. G., Lin, J. P., Chen, G. W. and Kao, S. T. (2006a). Down-regulation of cyclin B1 and up-regulation of Wee1 by berberine promotes entry of leukemia cells into the G2/M-phase of the cell cycle. *Anticancer Research* **26**, 1097–1104.
Lin, J. P., Yang, J. S., Lee, J. H., Hsieh, W. T. and Chung, J. G. (2006b). Berberine induces cell cycle arrest and apoptosis in human gastric carcinoma SNU-5 cell line. *World Journal of Gastroenterology* **12**, 21–28.
Lindberg, R. L., De Groot, C. J., Certa, U., Ravid, R., Hoffmann, F., Kappos, L. and Leppert, D. (2004). Multiple sclerosis as a generalized CNS disease—Comparative microarray analysis of normal appearing white matter and lesions in secondary progressive MS. *Journal of Neuroimmunology* **152**, 154–167.
Litonjua, A. A., Tantisira, K. G., Lake, S., Lazarus, R., Richter, B. G., Gabriel, S., Silverman, E. S. and Weiss, S. T. (2005). Polymorphisms in signal transducer and activator of transcription 3 and lung function in asthma. *Respiratory Research* **6**, 52.
Liu, J. (1995). Pharmacology of oleanolic acid and ursolic acid. *Journal of Ethnopharmacology* **49**, 57–68.
Liu, J. J., Nilsson, A., Oredsson, S., Badmaev, V., Zhao, W. Z. and Duan, R. D. (2002). Boswellic acids trigger apoptosis via a pathway dependent on caspase-8 activation but independent on Fas/Fas ligand interaction in colon cancer HT-29 cells. *Carcinogenesis* **23**, 2087–2093.
Liu, J. J., Huang, B. and Hooi, S. C. (2006). Acetyl-keto-beta-boswellic acid inhibits cellular proliferation through a p21-dependent pathway in colon cancer cells. *British Journal of Pharmacology* **148**, 1099–1107.
Liu, A., Chen, H., Tong, H., Ye, S., Qiu, M., Wang, Z., Tan, W., Liu, J. and Lin, S. (2011). Emodin potentiates the antitumor effects of gemcitabine in pancreatic cancer cells via inhibition of nuclear factor-kappaB. *Molecular Medicine Reports* **4**, 221–227.
Lock, C., Hermans, G., Pedotti, R., Brendolan, A., Schadt, E., Garren, H., Langer-Gould, A., Strober, S., Cannella, B., Allard, J., Klonowski, P., Austin, A. *et al.* (2002). Gene-microarray analysis of multiple sclerosis lesions yields new targets validated in autoimmune encephalomyelitis. *Nature Medicine* **8**, 500–508.
Logan, C. Y. and Nusse, R. (2004). The Wnt signaling pathway in development and disease. *Annual Review of Cell and Developmental Biology* **20**, 781–810.
Lopez-Carrillo, L., Hernandez Avila, M. and Dubrow, R. (1994). Chili pepper consumption and gastric cancer in Mexico: A case-control study. *American Journal of Epidemiology* **139**, 263–271.
Lowrey, J. A., Stewart, G. A., Lindey, S., Hoyne, G. F., Dallman, M. J., Howie, S. E. and Lamb, J. R. (2002). Sonic hedgehog promotes cell cycle progression in activated peripheral CD4(+) T lymphocytes. *Journal of Immunology* **169**, 1869–1875.
Lu, X., Jung, J., Cho, H. J., Lim, D. Y., Lee, H. S., Chun, H. S., Kwon, D. Y. and Park, J. H. (2005). Fisetin inhibits the activities of cyclin-dependent kinases leading to cell cycle arrest in HT-29 human colon cancer cells. *The Journal of Nutrition* **135**, 2884–2890.

Lu, M., Xia, L., Hua, H. and Jing, Y. (2008a). Acetyl-keto-beta-boswellic acid induces apoptosis through a death receptor 5-mediated pathway in prostate cancer cells. *Cancer Research* **68**, 1180–1186.

Lu, N., Li, Y., Qin, H., Zhang, Y. L. and Sun, C. H. (2008b). Gossypin up-regulates LDL receptor through activation of ERK pathway: A signaling mechanism for the hypocholesterolemic effect. *Journal of Agricultural and Food Chemistry* **56**, 11526–11532.

Lukiw, W. J. and Bazan, N. G. (1998). Strong nuclear factor-kappaB-DNA binding parallels cyclooxygenase-2 gene transcription in aging and in sporadic Alzheimer's disease superior temporal lobe neocortex. *Journal of Neuroscience Research* **53**, 583–592.

Luo, P., Wong, Y. F., Ge, L., Zhang, Z. F., Liu, Y., Liu, L. and Zhou, H. (2010). Anti-inflammatory and analgesic effect of plumbagin through inhibition of nuclear factor-kappaB activation. *The Journal of Pharmacology and Experimental Therapeutics* **335**, 735–742.

Lutticken, C., Wegenka, U. M., Yuan, J., Buschmann, J., Schindler, C., Ziemiecki, A., Harpur, A. G., Wilks, A. F., Yasukawa, K., Taga, T. *et al.* (1994). Association of transcription factor APRF and protein kinase Jak1 with the interleukin-6 signal transducer gp130. *Science* **263**, 89–92.

Magalhaes, C. B., Riva, D. R., DePaula, L. J., Brando-Lima, A., Koatz, V. L., Leal-Cardoso, J. H., Zin, W. A. and Faffe, D. S. (2010). In vivo anti-inflammatory action of eugenol on lipopolysaccharide-induced lung injury. *Journal of Applied Physiology* **108**, 845–851.

Mahendran, S., Badami, S. and Maithili, V. (2010). Evaluation of antidiabetic effect of embelin from Embelia ribes in alloxan induced diabetes in rats. *Biomedicine & Pharmacotherapy* (in press).

Maher, P. (2006). A comparison of the neurotrophic activities of the flavonoid fisetin and some of its derivatives. *Free Radical Research* **40**, 1105–1111.

Mahgoub, A. A. (2003). Thymoquinone protects against experimental colitis in rats. *Toxicology Letters* **143**, 133–143.

Mak, N. K., Leung, C. Y., Wei, X. Y., Shen, X. L., Wong, R. N., Leung, K. N. and Fung, M. C. (2004). Inhibition of RANTES expression by indirubin in influenza virus-infected human bronchial epithelial cells. *Biochemical Pharmacology* **67**, 167–174.

Mak, P., Leav, I., Pursell, B., Bae, D., Yang, X., Taglienti, C. A., Gouvin, L. M., Sharma, V. M. and Mercurio, A. M. (2010). ERbeta impedes prostate cancer EMT by destabilizing HIF-1alpha and inhibiting VEGF-mediated snail nuclear localization: Implications for Gleason grading. *Cancer Cell* **17**, 319–332.

Makita, H., Tanaka, T., Fujitsuka, H., Tatematsu, N., Satoh, K., Hara, A. and Mori, H. (1996). Chemoprevention of 4-nitroquinoline 1-oxide-induced rat oral carcinogenesis by the dietary flavonoids chalcone, 2-hydroxychalcone, and quercetin. *Cancer Research* **56**, 4904–4909.

Manikandan, P., Vinothini, G., Vidya Priyadarsini, R., Prathiba, D. and Nagini, S. (2011). Eugenol inhibits cell proliferation via NF-kappaB suppression in a rat model of gastric carcinogenesis induced by MNNG. *Investigational New Drugs* **29**, 110–117.

Manna, S. K., Aggarwal, R. S., Sethi, G., Aggarwal, B. B. and Ramesh, G. T. (2007). Morin (3,5,7,2′,4′-Pentahydroxyflavone) abolishes nuclear factor-kappaB activation induced by various carcinogens and inflammatory stimuli, leading to suppression of nuclear factor-kappaB-regulated gene expression and up-regulation of apoptosis. *Clinical Cancer Research* **13**, 2290–2297.

Mantamadiotis, T. and Taraviras, S. (2011). Self-renewal mechanisms in neural cancer stem cells. *Frontiers in Bioscience* **16**, 598–607.

Mantena, S. K., Sharma, S. D. and Katiyar, S. K. (2006). Berberine, a natural product, induces G1-phase cell cycle arrest and caspase-3-dependent apoptosis in human prostate carcinoma cells. *Molecular Cancer Therapeutics* **5**, 296–308.

Mantovani, A. (2005). Cancer: Inflammation by remote control. *Nature* **435**, 752–753.

Manu, K. A., Shanmugam, M. K., Rajendran, P., Li, F., Ramachandran, L., Hay, H. S., Kannaiyan, R., Nanjunda Swamy, S., Vali, S., Kapoor, S., Ramesh, B., Bist, P. *et al.* (2011). Plumbagin inhibits invasion and migration of breast and gastric cancer cells by downregulating the expression of chemokine receptor CXCR4. *Molecular Cancer* **10**, 107.

Mariani, E., Polidori, M. C., Cherubini, A. and Mecocci, P. (2005). Oxidative stress in brain aging, neurodegenerative and vascular diseases: An overview. *Journal of Chromatography. B, Analytical Technologies in the Biomedical and Life Sciences* **827**, 65–75.

Masuda, M., Suzui, M., Yasumatu, R., Nakashima, T., Kuratomi, Y., Azuma, K., Tomita, K., Komiyama, S. and Weinstein, I. B. (2002). Constitutive activation of signal transducers and activators of transcription 3 correlates with cyclin D1 overexpression and may provide a novel prognostic marker in head and neck squamous cell carcinoma. *Cancer Research* **62**, 3351–3355.

Matsuda, H., Murakami, T., Li, Y., Yamahara, J. and Yoshikawa, M. (1998). Mode of action of escins Ia and IIa and E, Z-senegin II on glucose absorption in gastrointestinal tract. *Bioorganic & Medicinal Chemistry* **6**, 1019–1023.

Maxwell, P. H., Pugh, C. W. and Ratcliffe, P. J. (2001). Activation of the HIF pathway in cancer. *Current Opinion in Genetics & Development* **11**, 293–299.

McAnuff, M. A., Omoruyi, F. O., Morrison, E. Y. and Asemota, H. N. (2005). Changes in some liver enzymes in streptozotocin-induced diabetic rats fed sapogenin extract from bitter yam (Dioscorea polygonoides) or commercial diosgenin. *The West Indian Medical Journal* **54**, 97–101.

Megeney, L. A., Perry, R. L., LeCouter, J. E. and Rudnicki, M. A. (1996). bFGF and LIF signaling activates STAT3 in proliferating myoblasts. *Developmental Genetics* **19**, 139–145.

Meijer, L., Skaltsounis, A. L., Magiatis, P., Polychronopoulos, P., Knockaert, M., Leost, M., Ryan, X. P., Vonica, C. A., Brivanlou, A., Dajani, R., Crovace, C., Tarricone, C. *et al.* (2003). GSK-3-selective inhibitors derived from Tyrian purple indirubins. *Chemical Biology* **10**, 1255–1266.

Mencher, S. K. and Wang, L. G. (2005). Promiscuous drugs compared to selective drugs (promiscuity can be a virtue). *BMC Clinical Pharmacology* **5**, 3.

Mendes, V., Monteiro, R., Pestana, D., Teixeira, D., Calhau, C. and Azevedo, I. (2008). Xanthohumol influences preadipocyte differentiation: Implication of antiproliferative and apoptotic effects. *Journal of Agricultural and Food Chemistry* **56**, 11631–11637.

Miranda, C. L., Stevens, J. F., Helmrich, A., Henderson, M. C., Rodriguez, R. J., Yang, Y. H., Deinzer, M. L., Barnes, D. W. and Buhler, D. R. (1999). Antiproliferative and cytotoxic effects of prenylated flavonoids from hops (Humulus lupulus) in human cancer cell lines. *Food and Chemical Toxicology* **37**, 271–285.

Miyata, Y. and Yahara, I. (1992). The 90-kDa heat shock protein, HSP90, binds and protects casein kinase II from self-aggregation and enhances its kinase activity. *The Journal of Biological Chemistry* **267**, 7042–7047.

Miyazawa, T., Shibata, A., Sookwong, P., Kawakami, Y., Eitsuka, T., Asai, A., Oikawa, S. and Nakagawa, K. (2009). Antiangiogenic and anticancer potential of unsaturated vitamin E (tocotrienol). *The Journal of Nutritional Biochemistry* **20**, 79–86.

Mohamed, A., Shoker, A., Bendjelloul, F., Mare, A., Alzrigh, M., Benghuzzi, H. and Desin, T. (2003). Improvement of experimental allergic encephalomyelitis (EAE) by thymoquinone; an oxidative stress inhibitor. *Biomedical Sciences Instrumentation* **39**, 440–445.

Mohsin, S. K., Weiss, H. L., Gutierrez, M. C., Chamness, G. C., Schiff, R., Digiovanna, M. P., Wang, C. X., Hilsenbeck, S. G., Osborne, C. K., Allred, D. C., Elledge, R. and Chang, J. C. (2005). Neoadjuvant trastuzumab induces apoptosis in primary breast cancers. *Journal of Clinical Oncology* **23**, 2460–2468.

Moraes, L. A., Piqueras, L. and Bishop-Bailey, D. (2006). Peroxisome proliferator-activated receptors and inflammation. *Pharmacology & Therapeutics* **110**, 371–385.

Morris, J. P., 4th, Wang, S. C. and Hebrok, M. (2010). KRAS, Hedgehog, Wnt and the twisted developmental biology of pancreatic ductal adenocarcinoma. *Nature Reviews. Cancer* **10**, 683–695.

Mukherjee, P. K., Saha, K., Das, J., Pal, M. and Saha, B. P. (1997). Studies on the anti-inflammatory activity of rhizomes of Nelumbo nucifera. *Planta Medica* **63**, 367–369.

Mukhopadhyay, A., Banerjee, S., Stafford, L. J., Xia, C., Liu, M. and Aggarwal, B. B. (2002). Curcumin-induced suppression of cell proliferation correlates with down-regulation of cyclin D1 expression and CDK4-mediated retinoblastoma protein phosphorylation. *Oncogene* **21**, 8852–8861.

Murakami, A., Ohura, S., Nakamura, Y., Koshimizu, K. and Ohigashi, H. (1996). 1′-Acetoxychavicol acetate, a superoxide anion generation inhibitor, potently inhibits tumor promotion by 12-O-tetradecanoylphorbol-13-acetate in ICR mouse skin. *Oncology* **53**, 386–391.

Murillo, G., Kosmeder, J. W., 2nd, Pezzuto, J. M. and Mehta, R. G. (2003). Deguelin suppresses the formation of carcinogen-induced aberrant crypt foci in the colon of CF-1 mice. *International Journal of Cancer* **104**, 7–11.

Muruganandan, S., Srinivasan, K., Gupta, S., Gupta, P. K. and Lal, J. (2005). Effect of mangiferin on hyperglycemia and atherogenicity in streptozotocin diabetic rats. *Journal of Ethnopharmacology* **97**, 497–501.

Nair, A. S., Shishodia, S., Ahn, K. S., Kunnumakkara, A. B., Sethi, G. and Aggarwal, B. B. (2006). Deguelin, an Akt inhibitor, suppresses IkappaBalpha kinase activation leading to suppression of NF-kappaB-regulated gene expression, potentiation of apoptosis, and inhibition of cellular invasion. *Journal of Immunology* **177**, 5612–5622.

Nakatani, K., Nakahata, N., Arakawa, T., Yasuda, H. and Ohizumi, Y. (2002). Inhibition of cyclooxygenase and prostaglandin E2 synthesis by gamma-mangostin, a xanthone derivative in mangosteen, in C6 rat glioma cells. *Biochemical Pharmacology* **63**, 73–79.

Nakatani, K., Yamakuni, T., Kondo, N., Arakawa, T., Oosawa, K., Shimura, S., Inoue, H. and Ohizumi, Y. (2004). gamma-Mangostin inhibits inhibitor-kappaB kinase activity and decreases lipopolysaccharide-induced cyclooxygenase-2 gene expression in C6 rat glioma cells. *Molecular Pharmacology* **66**, 667–674.

Newman, D. J. and Cragg, G. M. (2007). Natural products as sources of new drugs over the last 25 years. *Journal of Natural Products* **70**, 461–477.

Nikolovska-Coleska, Z., Xu, L., Hu, Z., Tomita, Y., Li, P., Roller, P. P., Wang, R., Fang, X., Guo, R., Zhang, M., Lippman, M. E., Yang, D. *et al.* (2004). Discovery of embelin as a cell-permeable, small-molecular weight inhibitor of XIAP through structure-based computational screening of a traditional herbal medicine three-dimensional structure database. *Journal of Medicinal Chemistry* **47**, 2430–2440.

Nioi, P. and Hayes, J. D. (2004). Contribution of NAD(P)H:quinone oxidoreductase 1 to protection against carcinogenesis, and regulation of its gene by the Nrf2 basic-region leucine zipper and the arylhydrocarbon receptor basic helix-loop-helix transcription factors. *Mutation Research* **555**, 149–171.

Nishino, H., Kitagawa, K., Fujiki, H. and Iwashima, A. (1986). Berberine sulfate inhibits tumor-promoting activity of teleocidin in two-stage carcinogenesis on mouse skin. *Oncology* **43**, 131–134.

Nishisho, I., Nakamura, Y., Miyoshi, Y., Miki, Y., Ando, H., Horii, A., Koyama, K., Utsunomiya, J., Baba, S. and Hedge, P. (1991). Mutations of chromosome 5q21 genes in FAP and colorectal cancer patients. *Science* **253**, 665–669.

Niu, G., Wright, K. L., Huang, M., Song, L., Haura, E., Turkson, J., Zhang, S., Wang, T., Sinibaldi, D., Coppola, D., Heller, R., Ellis, L. M. *et al.* (2002). Constitutive Stat3 activity up-regulates VEGF expression and tumor angiogenesis. *Oncogene* **21**, 2000–2008.

Nozawa, H. (2005). Xanthohumol, the chalcone from beer hops (Humulus lupulus L.), is the ligand for farnesoid X receptor and ameliorates lipid and glucose metabolism in KK-A(y) mice. *Biochemical and Biophysical Research Communications* **336**, 754–761.

Nusslein-Volhard, C. and Wieschaus, E. (1980). Mutations affecting segment number and polarity in Drosophila. *Nature* **287**, 795–801.

Ocete, M. A., Galvez, J., Crespo, M. E., Cruz, T., Gonzalez, M., Torres, M. I. and Zarzuelo, A. (1998). Effects of morin on an experimental model of acute colitis in rats. *Pharmacology* **57**, 261–270.

Odrowaz-Sypniewska, G. (2007). Markers of pro-inflammatory and pro-thrombotic state in the diagnosis of metabolic syndrome. *Advances in Medical Sciences* **52**, 246–250.

Ogasawara, M. and Suzuki, H. (2004). Inhibition by evodiamine of hepatocyte growth factor-induced invasion and migration of tumor cells. *Biological & Pharmaceutical Bulletin* **27**, 578–582.

Ogasawara, M., Matsubara, T. and Suzuki, H. (2001). Inhibitory effects of evodiamine on in vitro invasion and experimental lung metastasis of murine colon cancer cells. *Biological & Pharmaceutical Bulletin* **24**, 917–920.

Oguma, K., Oshima, H., Aoki, M., Uchio, R., Naka, K., Nakamura, S., Hirao, A., Saya, H., Taketo, M. M. and Oshima, M. (2008). Activated macrophages promote Wnt signalling through tumour necrosis factor-alpha in gastric tumour cells. *EMBO Journal* **27**, 1671–1681.

Ohata, T., Fukuda, K., Murakami, A., Ohigashi, H., Sugimura, T. and Wakabayashi, K. (1998). Inhibition by 1′-acetoxychavicol acetate of lipopolysaccharide- and interferon-gamma-induced nitric oxide production through suppression of inducible nitric oxide synthase gene expression in RAW264 cells. *Carcinogenesis* **19**, 1007–1012.

Ohnishi, M., Tanaka, T., Makita, H., Kawamori, T., Mori, H., Satoh, K., Hara, A., Murakami, A., Ohigashi, H. and Koshimizu, K. (1996). Chemopreventive effect of a xanthine oxidase inhibitor, 1′-acetoxychavicol acetate, on rat oral carcinogenesis. *Japanese Journal of Cancer Research* **87**, 349–356.

Opletalova, V. and Sedivy, D. (1999). Chalcones and their heterocyclic analogs as potential antifungal chemotherapeutic agents. *Ceská a Slovenská Farmacie* **48**, 252–255.

Oshima, H., Matsunaga, A., Fujimura, T., Tsukamoto, T., Taketo, M. M. and Oshima, M. (2006). Carcinogenesis in mouse stomach by simultaneous activation of the Wnt signaling and prostaglandin E2 pathway. *Gastroenterology* **131**, 1086–1095.

Packard, R. R. and Libby, P. (2008). Inflammation in atherosclerosis: From vascular biology to biomarker discovery and risk prediction. *Clinical Chemistry* **54**, 24–38.

Palempalli, U. D., Gandhi, U., Kalantari, P., Vunta, H., Arner, R. J., Narayan, V., Ravindran, A. and Prabhu, K. S. (2009). Gambogic acid covalently modifies IkappaB kinase-beta subunit to mediate suppression of lipopolysaccharide-induced activation of NF-kappaB in macrophages. *The Biochemical Journal* **419**, 401–409.

Pan, L. H., Yamauchi, K., Uzuki, M., Nakanishi, T., Takigawa, M., Inoue, H. and Sawai, T. (2001). Type II alveolar epithelial cells and interstitial fibroblasts express connective tissue growth factor in IPF. *The European Respiratory Journal* **17**, 1220–1227.

Pandey, A. and Bani, S. (2010). Hydroxychavicol inhibits immune responses to mitigate cognitive dysfunction in rats. *Journal of Neuroimmunology* **226**, 48–58.

Pandey, M. K., Sandur, S. K., Sung, B., Sethi, G., Kunnumakkara, A. B. and Aggarwal, B. B. (2007a). Butein, a tetrahydroxychalcone, inhibits nuclear factor (NF)-kappaB and NF-kappaB-regulated gene expression through direct inhibition of IkappaBalpha kinase beta on cysteine 179 residue. *The Journal of Biological Chemistry* **282**, 17340–17350.

Pandey, M. K., Sung, B., Ahn, K. S., Kunnumakkara, A. B., Chaturvedi, M. M. and Aggarwal, B. B. (2007b). Gambogic acid, a novel ligand for transferrin receptor, potentiates TNF-induced apoptosis through modulation of the nuclear factor-kappaB signaling pathway. *Blood* **110**, 3517–3525.

Pandey, M. K., Sung, B., Kunnumakkara, A. B., Sethi, G., Chaturvedi, M. M. and Aggarwal, B. B. (2008). Berberine modifies cysteine 179 of IkappaBalpha kinase, suppresses nuclear factor-kappaB-regulated antiapoptotic gene products, and potentiates apoptosis. *Cancer Research* **68**, 5370–5379.

Pandey, M. K., Sung, B., Ahn, K. S. and Aggarwal, B. B. (2009). Butein suppresses constitutive and inducible signal transducer and activator of transcription (STAT) 3 activation and STAT3-regulated gene products through the induction of a protein tyrosine phosphatase SHP-1. *Molecular Pharmacology* **75**, 525–533.

Pandey, A., Bani, S., Dutt, P. and Suri, K. A. (2010a). Modulation of Th1/Th2 cytokines and inflammatory mediators by hydroxychavicol in adjuvant induced arthritic tissues. *Cytokine* **49**, 114–121.

Pandey, M. K., Sung, B. and Aggarwal, B. B. (2010b). Betulinic acid suppresses STAT3 activation pathway through induction of protein tyrosine phosphatase SHP-1 in human multiple myeloma cells. *International Journal of Cancer* **127**, 282–292.

Paris, D., Ganey, N. J., Laporte, V., Patel, N. S., Beaulieu-Abdelahad, D., Bachmeier, C., March, A., Ait-Ghezala, G. and Mullan, M. J. (2010). Reduction of beta-amyloid pathology by celastrol in a transgenic mouse model of Alzheimer's disease. *Journal of Neuroinflammation* **7**, 17.

Park, B., Sung, B., Yadav, V. R., Cho, S. G., Liu, M. and Aggarwal, B. B. (2011a). Acetyl-11-keto-beta-boswellic acid suppresses invasion of pancreatic cancer cells through the downregulation of CXCR4 chemokine receptor expression. *International Journal of Cancer* **129**, 23–33.

Park, M. Y., Kwon, H. J. and Sung, M. K. (2011b). Dietary aloin, aloesin, or aloe-gel exerts anti-inflammatory activity in a rat colitis model. *Life Sciences* **88**, 486–492.

Pathak, A. K., Bhutani, M., Nair, A. S., Ahn, K. S., Chakraborty, A., Kadara, H., Guha, S., Sethi, G. and Aggarwal, B. B. (2007). Ursolic acid inhibits STAT3 activation pathway leading to suppression of proliferation and chemosensitization of human multiple myeloma cells. *Molecular Cancer Research* **5**, 943–955.

Patlolla, J. M., Raju, J., Swamy, M. V. and Rao, C. V. (2006). Beta-escin inhibits colonic aberrant crypt foci formation in rats and regulates the cell cycle growth by inducing p21(waf1/cip1) in colon cancer cells. *Molecular Cancer Therapeutics* **5**, 1459–1466.

Peng, Z. G., Luo, J., Xia, L. H., Chen, Y. and Song, S. J. (2004). CML cell line K562 cell apoptosis induced by mangiferin. *Zhongguo Shi Yan Xue Ye Xue Za Zhi* **12**, 590–594.

Peter, S. and Beglinger, C. (2007). Helicobacter pylori and gastric cancer: The causal relationship. *Digestion* **75**, 25–35.

Petrova, A., Davids, L. M., Rautenbach, F. and Marnewick, J. L. (2011). Photoprotection by honeybush extracts, hesperidin and mangiferin against UVB-induced skin damage in SKH-1 mice. *Journal of Photochemistry and Photobiology. B* **103**, 126–139.

Pinna, G. F., Fiorucci, M., Reimund, J. M., Taquet, N., Arondel, Y. and Muller, C. D. (2004). Celastrol inhibits pro-inflammatory cytokine secretion in Crohn's disease biopsies. *Biochemical and Biophysical Research Communications* **322**, 778–786.

Pinto, M. M., Sousa, M. E. and Nascimento, M. S. (2005). Xanthone derivatives: New insights in biological activities. *Current Medicinal Chemistry* **12**, 2517–2538.

Pisha, E., Chai, H., Lee, I. S., Chagwedera, T. E., Farnsworth, N. R., Cordell, G. A., Beecher, C. W., Fong, H. H., Kinghorn, A. D., Brown, D. M. *et al.* (1995). Discovery of betulinic acid as a selective inhibitor of human melanoma that functions by induction of apoptosis. *Nature Medicine* **1**, 1046–1051.

Poojari, R., Gupta, S., Maru, G., Khade, B. and Bhagwat, S. (2010). Chemopreventive and hepatoprotective effects of embelin on N-nitrosodiethylamine and carbon tetrachloride induced preneoplasia and toxicity in rat liver. *Asian Pacific Journal of Cancer Prevention* **11**, 1015–1020.

Prasad, V. S., Devi, P. U., Rao, B. S. and Kamath, R. (1996). Radiosensitizing effect of plumbagin on mouse melanoma cells grown in vitro. *Indian Journal of Experimental Biology* **34**, 857–858.

Prasad, S., Pandey, M. K., Yadav, V. R. and Aggarwal, B. B. (2011). Gambogic acid inhibits STAT3 phosphorylation through activation of protein tyrosine phosphatase SHP-1: Potential role in proliferation and apoptosis. *Cancer Prevention Research (Philadelphia, Pa.)* **4**, 1084–1094.

Radhakrishnan, N. and Alam, M. (1975). Antifertility activity of embelin in albino rats. *Indian Journal of Experimental Biology* **13**, 70–71.

Rajendran, P., Ekambaram, G., Magesh, V. and Sakthisekaran, D. (2008a). Chemopreventive efficacy of mangiferin against benzo(a)pyrene induced lung carcinogenesis in experimental animals. *Environmental Toxicology and Pharmacology* **26**, 278–282.

Rajendran, P., Ekambaram, G. and Sakthisekaran, D. (2008b). Cytoprotective effect of mangiferin on benzo(a)pyrene-induced lung carcinogenesis in swiss albino mice. *Basic & Clinical Pharmacology & Toxicology* **103**, 137–142.

Rajkamal, G., Suresh, K., Sugunadevi, G., Vijayaanand, M. A. and Rajalingam, K. (2010). Evaluation of chemopreventive effects of Thymoquinone on cell surface glycoconjugates and cytokeratin expression during DMBA induced hamster buccal pouch carcinogenesis. *BMB Reports* **43**, 664–669.

Ramos-Gomez, M., Kwak, M. K., Dolan, P. M., Itoh, K., Yamamoto, M., Talalay, P. and Kensler, T. W. (2001). Sensitivity to carcinogenesis is increased and chemoprotective efficacy of enzyme inducers is lost in nrf2 transcription factor-deficient mice. *Proceedings of the National Academy of Sciences of the United States of America* **98**, 3410–3415.

Ramsey, C. P., Glass, C. A., Montgomery, M. B., Lindl, K. A., Ritson, G. P., Chia, L. A., Hamilton, R. L., Chu, C. T. and Jordan-Sciutto, K. L. (2007). Expression of Nrf2 in neurodegenerative diseases. *Journal of Neuropathology and Experimental Neurology* **66**, 75–85.

Ranelletti, F. O., Ricci, R., Larocca, L. M., Maggiano, N., Capelli, A., Scambia, G., Benedetti-Panici, P., Mancuso, S., Rumi, C. and Piantelli, M. (1992). Growth-inhibitory effect of quercetin and presence of type-II estrogen-binding sites in human colon-cancer cell lines and primary colorectal tumors. *International Journal of Cancer* **50**, 486–492.

Rangasamy, T., Cho, C. Y., Thimmulappa, R. K., Zhen, L., Srisuma, S. S., Kensler, T. W., Yamamoto, M., Petrache, I., Tuder, R. M. and Biswal, S. (2004). Genetic ablation of Nrf2 enhances susceptibility to cigarette smoke-induced emphysema in mice. *The Journal of Clinical Investigation* **114**, 1248–1259.

Rannou, F., Francois, M., Corvol, M. T. and Berenbaum, F. (2006). Cartilage breakdown in rheumatoid arthritis. *Joint, Bone, Spine* **73**, 29–36.

Ravindran, J., Prasad, S. and Aggarwal, B. B. (2009). Curcumin and cancer cells: How many ways can curry kill tumor cells selectively? *The AAPS Journal* **11**, 495–510.

Rayalam, S., Yang, J. Y., Della-Fera, M. A., Park, H. J., Ambati, S. and Baile, C. A. (2009). Anti-obesity effects of xanthohumol plus guggulsterone in 3T3-L1 adipocytes. *Journal of Medicinal Food* **12**, 846–853.

Richmond, A. (2002). Nf-kappa B, chemokine gene transcription and tumour growth. *Nature Reviews Immunology* **2**, 664–674.

Ricote, M. and Glass, C. K. (2007). PPARs and molecular mechanisms of transrepression. *Biochimica et Biophysica Acta* **1771**, 926–935.

Rizzo, G. and Fiorucci, S. (2006). PPARs and other nuclear receptors in inflammation. *Current Opinion in Pharmacology* **6**, 421–427.

Romashkova, J. A. and Makarov, S. S. (1999). NF-kappaB is a target of AKT in anti-apoptotic PDGF signalling. *Nature* **401**, 86–90.

Rothkopf, M. and Vogel, G. (1976). New findings on the efficacy and mode of action of the horse chestnut saponin escin. *Arzneimittel Forschung* **26**, 225–235.

Rzeski, W., Stepulak, A., Szymanski, M., Sifringer, M., Kaczor, J., Wejksza, K., Zdzisinska, B. and Kandefer-Szerszen, M. (2006). Betulinic acid decreases expression of bcl-2 and cyclin D1, inhibits proliferation, migration and induces apoptosis in cancer cells. *Naunyn-Schmiedeberg's Archives of Pharmacology* **374**, 11–20.

Sailer, E. R., Subramanian, L. R., Rall, B., Hoernlein, R. F., Ammon, H. P. and Safayhi, H. (1996). Acetyl-11-keto-beta-boswellic acid (AKBA): Structure requirements for binding and 5-lipoxygenase inhibitory activity. *British Journal of Pharmacology* **117**, 615–618.

Salti, G. I., Kichina, J. V., Das Gupta, T. K., Uddin, S., Bratescu, L., Pezzuto, J. M., Mehta, R. G. and Constantinou, A. I. (2001). Betulinic acid reduces ultraviolet-C-induced DNA breakage in congenital melanocytic naeval cells: Evidence for a potential role as a chemopreventive agent. *Melanoma Research* **11**, 99–104.

Samoszuk, M., Tan, J. and Chorn, G. (2005). The chalcone butein from Rhus verniciflua Stokes inhibits clonogenic growth of human breast cancer cells co-cultured with fibroblasts. *BMC Complementary and Alternative Medicine* **5**, 5.

Sanders, M. M., Liu, A. A., Li, T. K., Wu, H. Y., Desai, S. D., Mao, Y., Rubin, E. H., LaVoie, E. J., Makhey, D. and Liu, L. F. (1998). Selective cytotoxicity of topoisomerase-directed protoberberines against glioblastoma cells. *Biochemical Pharmacology* **56**, 1157–1166.

Sandur, S. K., Ichikawa, H., Sethi, G., Ahn, K. S. and Aggarwal, B. B. (2006). Plumbagin (5-hydroxy-2-methyl-1,4-naphthoquinone) suppresses NF-kappaB activation and NF-kappaB-regulated gene products through modulation of p65 and IkappaBalpha kinase activation, leading to potentiation of apoptosis induced by cytokine and chemotherapeutic agents. *The Journal of Biological Chemistry* **281**, 17023–17033.

Sandur, S. K., Pandey, M. K., Sung, B. and Aggarwal, B. B. (2010). 5-hydroxy-2-methyl-1,4-naphthoquinone, a vitamin K3 analogue, suppresses STAT3 activation pathway through induction of protein tyrosine phosphatase, SHP-1: Potential role in chemosensitization. *Molecular Cancer Research* **8**, 107–118.

Santin, J. R., Lemos, M., Klein-Junior, L. C., Machado, I. D., Costa, P., de Oliveira, A. P., Tilia, C., de Souza, J. P., de Sousa, J. P., Bastos, J. K. and de Andrade, S. F. (2011). Gastroprotective activity of essential oil of the Syzygium aromaticum and its major component eugenol in different animal models. *Naunyn-Schmiedeberg's Archives of Pharmacology* **383**, 149–158.

Sarraf, P., Mueller, E., Smith, W. M., Wright, H. M., Kum, J. B., Aaltonen, L. A., de la Chapelle, A., Spiegelman, B. M. and Eng, C. (1999). Loss-of-function mutations in PPAR gamma associated with human colon cancer. *Molecular Cell* **3**, 799–804.

Sawai, A., Chandarlapaty, S., Greulich, H., Gonen, M., Ye, Q., Arteaga, C. L., Sellers, W., Rosen, N. and Solit, D. B. (2008). Inhibition of Hsp90 down--regulates mutant epidermal growth factor receptor (EGFR) expression and sensitizes EGFR mutant tumors to paclitaxel. *Cancer Research* **68**, 589–596.

Schmidt, M. L., Kuzmanoff, K. L., Ling-Indeck, L. and Pezzuto, J. M. (1997). Betulinic acid induces apoptosis in human neuroblastoma cell lines. *European Journal of Cancer* **33**, 2007–2010.

Schwaiberger, A. V., Heiss, E. H., Cabaravdic, M., Oberan, T., Zaujec, J., Schachner, D., Uhrin, P., Atanasov, A. G., Breuss, J. M., Binder, B. R. and Dirsch, V. M. (2010). Indirubin-3′-monoxime blocks vascular smooth muscle cell proliferation by inhibition of signal transducer and activator of transcription 3 signaling and reduces neointima formation in vivo. *Arteriosclerosis, Thrombosis, and Vascular Biology* **30**, 2475–2481.

Semenza, G. L. (1998). Hypoxia-inducible factor 1: Master regulator of O2 homeostasis. *Current Opinion in Genetics & Development* **8**, 588–594.

Semenza, G. L. (2003). Targeting HIF-1 for cancer therapy. *Nature Reviews. Cancer* **3**, 721–732.

Sen, R. and Baltimore, D. (1986). Multiple nuclear factors interact with the immunoglobulin enhancer sequences. *Cell* **46**, 705–716.

Sen, M., Lauterbach, K., El-Gabalawy, H., Firestein, G. S., Corr, M. and Carson, D. A. (2000). Expression and function of wingless and frizzled homologs in rheumatoid arthritis. *Proceedings of the National Academy of Sciences of the United States of America* **97**, 2791–2796.

Sen, M., Reifert, J., Lauterbach, K., Wolf, V., Rubin, J. S., Corr, M. and Carson, D. A. (2002). Regulation of fibronectin and metalloproteinase expression by Wnt signaling in rheumatoid arthritis synoviocytes. *Arthritis and Rheumatism* **46**, 2867–2877.

Sen, C. K., Khanna, S. and Roy, S. (2004). Tocotrienol: The natural vitamin E to defend the nervous system? *Annals of the New York Academy of Sciences* **1031**, 127–142.

Sen, C. K., Khanna, S. and Roy, S. (2007). Tocotrienols in health and disease: The other half of the natural vitamin E family. *Molecular Aspects of Medicine* **28**, 692–728.

Sengupta, K., Alluri, K. V., Satish, A. R., Mishra, S., Golakoti, T., Sarma, K. V., Dey, D. and Raychaudhuri, S. P. (2008). A double blind, randomized, placebo controlled study of the efficacy and safety of 5-Loxin for treatment of osteoarthritis of the knee. *Arthritis Research & Therapy* **10**, R85.

Sethi, J. M. and Rochester, C. L. (2000). Smoking and chronic obstructive pulmonary disease. *Clinics in Chest Medicine* **21**, 67–86viii.

Sethi, G., Ahn, K. S., Sandur, S. K., Lin, X., Chaturvedi, M. M. and Aggarwal, B. B. (2006). Indirubin enhances tumor necrosis factor-induced apoptosis through modulation of nuclear factor-kappa B signaling pathway. *The Journal of Biological Chemistry* **281**, 23425–23435.

Sethi, G., Sung, B. and Aggarwal, B. B. (2008). Nuclear factor-kappaB activation: From bench to bedside. *Experimental Biology and Medicine (Maywood, N.J.)* **233**, 21–31.

Shah, S. J. and Sylvester, P. W. (2005). Gamma-tocotrienol inhibits neoplastic mammary epithelial cell proliferation by decreasing Akt and nuclear factor kappaB activity. *Experimental Biology and Medicine (Maywood, N.J.)* **230**, 235–241.

Shamma, M. and Guinaudeau, H. (1986). Aporphinoid alkaloids. *Natural Product Reports* **3**, 345–351.

Shaulian, E. and Karin, M. (2001). AP-1 in cell proliferation and survival. *Oncogene* **20**, 2390–2400.

Shaulian, E. and Karin, M. (2002). AP-1 as a regulator of cell life and death. *Nature Cell Biology* **4**, E131–E136.

Sheeja, E., Joshi, S. B. and Jain, D. C. (2010). Bioassay-guided isolation of anti-inflammatory and antinociceptive compound from Plumbago zeylanica leaf. *Pharmaceutical Biology* **48**, 381–387.

Shibata, T., Kokubu, A., Gotoh, M., Ojima, H., Ohta, T., Yamamoto, M. and Hirohashi, S. (2008). Genetic alteration of Keap1 confers constitutive Nrf2 activation and resistance to chemotherapy in gallbladder cancer. *Gastroenterology* **135**, 1358–1368e1351–e1354.

Shieh, D. E., Chen, Y. Y., Yen, M. H., Chiang, L. C. and Lin, C. C. (2004). Emodin-induced apoptosis through p53-dependent pathway in human hepatoma cells. *Life Sciences* **74**, 2279–2290.

Shimizu, K., Mitchell, R. N. and Libby, P. (2006). Inflammation and cellular immune responses in abdominal aortic aneurysms. *Arteriosclerosis, Thrombosis, and Vascular Biology* **26**, 987–994.

Shirogane, T., Fukada, T., Muller, J. M., Shima, D. T., Hibi, M. and Hirano, T. (1999). Synergistic roles for Pim-1 and c-Myc in STAT3-mediated cell cycle progression and antiapoptosis. *Immunity* **11**, 709–719.

Shishodia, S. and Aggarwal, B. B. (2006). Diosgenin inhibits osteoclastogenesis, invasion, and proliferation through the downregulation of Akt, I kappa B kinase activation and NF-kappa B-regulated gene expression. *Oncogene* **25**, 1463–1473.

Shishodia, S., Majumdar, S., Banerjee, S. and Aggarwal, B. B. (2003). Ursolic acid inhibits nuclear factor-kappaB activation induced by carcinogenic agents through suppression of IkappaBalpha kinase and p65 phosphorylation: Correlation with down-regulation of cyclooxygenase 2, matrix metalloproteinase 9, and cyclin D1. *Cancer Research* **63**, 4375–4383.

Shoji, K., Tsubaki, M., Yamazoe, Y., Satou, T., Itoh, T., Kidera, Y., Tanimori, Y., Yanae, M., Matsuda, H., Taga, A., Nakamura, H. and Nishida, S. (2011). Mangiferin induces apoptosis by suppressing Bcl-xL and XIAP expressions and nuclear entry of NF-kappaB in HL-60 cells. *Archives of Pharmacal Research* **34**, 469–475.

Singab, A. N., El-Beshbishy, H. A., Yonekawa, M., Nomura, T. and Fukai, T. (2005). Hypoglycemic effect of Egyptian Morus alba root bark extract: Effect on diabetes and lipid peroxidation of streptozotocin-induced diabetic rats. *Journal of Ethnopharmacology* **100**, 333–338.

Singh, S. (2007). From exotic spice to modern drug? *Cell* **130**, 765–768.

Singh, S., Natarajan, K. and Aggarwal, B. B. (1996). Capsaicin (8-methyl-N-vanillyl-6-nonenamide) is a potent inhibitor of nuclear transcription factor-kappa B activation by diverse agents. *Journal of Immunology* **157**, 4412–4420.

Skardelly, M., Gaber, K., Schwarz, J. and Milosevic, J. (2011). Neuroprotective effects of the beta-catenin stabilization in an oxygen- and glucose-deprived human neural progenitor cell culture system. *International Journal of Developmental Neuroscience* **29**, 543–547.

Son, T. G., Camandola, S., Arumugam, T. V., Cutler, R. G., Telljohann, R. S., Mughal, M. R., Moore, T. A., Luo, W., Yu, Q. S., Johnson, D. A., Johnson, J. A., Greig, N. H. *et al.* (2010). Plumbagin, a novel Nrf2/ARE activator, protects against cerebral ischemia. *Journal of Neurochemistry* **112**, 1316–1326.

Stebbins, C. E., Russo, A. A., Schneider, C., Rosen, N., Hartl, F. U. and Pavletich, N. P. (1997). Crystal structure of an Hsp90-geldanamycin complex: Targeting of a protein chaperone by an antitumor agent. *Cell* **89**, 239–250.

Stewart, G. A., Lowrey, J. A., Wakelin, S. J., Fitch, P. M., Lindey, S., Dallman, M. J., Lamb, J. R. and Howie, S. E. (2002). Sonic hedgehog signaling modulates activation of and cytokine production by human peripheral CD4+ T cells. *Journal of Immunology* **169**, 5451–5457.

Stewart, G. A., Hoyne, G. F., Ahmad, S. A., Jarman, E., Wallace, W. A., Harrison, D. J., Haslett, C., Lamb, J. R. and Howie, S. E. (2003). Expression of the developmental Sonic hedgehog (Shh) signalling pathway is upregulated in chronic lung fibrosis and the Shh receptor patched 1 is present in circulating T lymphocytes. *The Journal of Pathology* **199**, 488–495.

Sugie, S., Okamoto, K., Rahman, K. M., Tanaka, T., Kawai, K., Yamahara, J. and Mori, H. (1998). Inhibitory effects of plumbagin and juglone on azoxymethane-induced intestinal carcinogenesis in rats. *Cancer Letters* **127**, 177–183.

Suiqing, C., Min, Z. and Lirong, C. (2005). Overexpression of phosphorylated-STAT3 correlated with the invasion and metastasis of cutaneous squamous cell carcinoma. *The Journal of Dermatology* **32**, 354–360.

Suksamrarn, S., Komutiban, O., Ratananukul, P., Chimnoi, N., Lartpornmatulee, N. and Suksamrarn, A. (2006). Cytotoxic prenylated xanthones from the young fruit of Garcinia mangostana. *Chemical & Pharmaceutical Bulletin (Tokyo)* **54**, 301–305.

Sun, M., Liu, C., Nadiminty, N., Lou, W., Zhu, Y., Yang, J., Evans, C. P., Zhou, Q. and Gao, A. C. (2011). Inhibition of Stat3 activation by sanguinarine suppresses prostate cancer cell growth and invasion. *The Prostate* **72**, 82–89.

Sung, S., Yang, H., Uryu, K., Lee, E. B., Zhao, L., Shineman, D., Trojanowski, J. Q., Lee, V. M. and Pratico, D. (2004). Modulation of nuclear factor-kappa B activity by indomethacin influences A beta levels but not A beta precursor protein metabolism in a model of Alzheimer's disease. *The American Journal of Pathology* **165**, 2197–2206.

Sung, B., Pandey, M. K. and Aggarwal, B. B. (2007). Fisetin, an inhibitor of cyclin-dependent kinase 6, down-regulates nuclear factor-kappaB-regulated cell proliferation, antiapoptotic and metastatic gene products through the suppression of TAK-1 and receptor-interacting protein-regulated IkappaBalpha kinase activation. *Molecular Pharmacology* **71**, 1703–1714.

Sung, B., Ahn, K. S. and Aggarwal, B. B. (2010). Noscapine, a benzylisoquinoline alkaloid, sensitizes leukemic cells to chemotherapeutic agents and cytokines by modulating the NF-kappaB signaling pathway. *Cancer Research* **70**, 3259–3268.

Tabakoff, B., Rabe, C. S. and Hoffman, P. L. (1991). Selective effects of sedative/hypnotic drugs on excitatory amino acid receptors in brain. *Annals of the New York Academy of Sciences* **625**, 488–495.

Tajouri, L., Mellick, A. S., Ashton, K. J., Tannenberg, A. E., Nagra, R. M., Tourtellotte, W. W. and Griffiths, L. R. (2003). Quantitative and qualitative changes in gene expression patterns characterize the activity of plaques in multiple sclerosis. *Brain Research. Molecular Brain Research* **119**, 170–183.

Takada, Y. and Aggarwal, B. B. (2003). Betulinic acid suppresses carcinogen-induced NF-kappa B activation through inhibition of I kappa B alpha kinase and p65 phosphorylation: Abrogation of cyclooxygenase-2 and matrix metalloprotease-9. *Journal of Immunology* **171**, 3278–3286.

Takada, Y., Ichikawa, H., Badmaev, V. and Aggarwal, B. B. (2006). Acetyl-11-keto-beta-boswellic acid potentiates apoptosis, inhibits invasion, and abolishes osteoclastogenesis by suppressing NF-kappa B and NF-kappa B-regulated gene expression. *Journal of Immunology* **176**, 3127–3140.

Tamatani, M., Che, Y. H., Matsuzaki, H., Ogawa, S., Okado, H., Miyake, S., Mizuno, T. and Tohyama, M. (1999). Tumor necrosis factor induces Bcl-2 and Bcl-x expression through NFkappaB activation in primary hippocampal neurons. *The Journal of Biological Chemistry* **274**, 8531–8538.

Tan, S. M., Li, F., Rajendran, P., Kumar, A. P., Hui, K. M. and Sethi, G. (2010). Identification of beta-escin as a novel inhibitor of signal transducer and activator of transcription 3/Janus-activated kinase 2 signaling pathway that suppresses proliferation and induces apoptosis in human hepatocellular carcinoma cells. *The Journal of Pharmacology and Experimental Therapeutics* **334**, 285–293.

Tanaka, T., Kawabata, K., Kakumoto, M., Makita, H., Matsunaga, K., Mori, H., Satoh, K., Hara, A., Murakami, A., Koshimizu, K. and Ohigashi, H. (1997). Chemoprevention of azoxymethane-induced rat colon carcinogenesis by a xanthine oxidase inhibitor, 1′-acetoxychavicol acetate. *Japanese Journal of Cancer Research* **88**, 821–830.

Tanaka, T., Kawabata, K., Honjo, S., Kohno, H., Murakami, M., Shimada, R., Matsunaga, K., Yamada, Y. and Shimizu, M. (1999). Inhibition of azoxymethane-induced aberrant crypt foci in rats by natural compounds, caffeine, quercetin and morin. *Oncology Reports* **6**, 1333–1340.

Tekeoglu, I., Dogan, A., Ediz, L., Budancamanak, M. and Demirel, A. (2007). Effects of thymoquinone (volatile oil of black cumin) on rheumatoid arthritis in rat models. *Phytotherapy Research* **21**, 895–897.

Terai, K., Matsuo, A. and McGeer, P. L. (1996). Enhancement of immunoreactivity for NF-kappa B in the hippocampal formation and cerebral cortex of Alzheimer's disease. *Brain Research* **735**, 159–168.

Theriault, A., Chao, J. T., Wang, Q., Gapor, A. and Adeli, K. (1999). Tocotrienol: A review of its therapeutic potential. *Clinical Biochemistry* **32**, 309–319.

Thippeswamy, B. S., Mahendran, S., Biradar, M. I., Raj, P., Srivastava, K., Badami, S. and Veerapur, V. P. (2011). Protective effect of embelin against acetic acid induced ulcerative colitis in rats. *European Journal of Pharmacology* **654**, 100–105.

Thuillier, P., Brash, A. R., Kehrer, J. P., Stimmel, J. B., Leesnitzer, L. M., Yang, P., Newman, R. A. and Fischer, S. M. (2002). Inhibition of peroxisome proliferator-activated receptor (PPAR)-mediated keratinocyte differentiation by lipoxygenase inhibitors. *The Biochemical Journal* **366**, 901–910.

Tiedemann, R. E., Schmidt, J., Keats, J. J., Shi, C. X., Zhu, Y. X., Palmer, S. E., Mao, X., Schimmer, A. D. and Stewart, A. K. (2009). Identification of a potent natural triterpenoid inhibitor of proteosome chymotrypsin-like activity and NF-kappaB with antimyeloma activity in vitro and in vivo. *Blood* **113**, 4027–4037.

Tsukamoto, H., Kondo, S., Mukudai, Y., Nagumo, T., Yasuda, A., Kurihara, Y., Kamatani, T. and Shintani, S. (2011). Evaluation of Anticancer Activities of Benzo[c]phenanthridine Alkaloid Sanguinarine in Oral Squamous Cell Carcinoma Cell Line. *Anticancer Research* **31**, 2841–2846.

Udeani, G. O., Gerhauser, C., Thomas, C. F., Moon, R. C., Kosmeder, J. W., Kinghorn, A. D., Moriarty, R. M. and Pezzuto, J. M. (1997). Cancer chemopreventive activity mediated by deguelin, a naturally occurring rotenoid. *Cancer Research* **57**, 3424–3428.

Uemura, T., Hirai, S., Mizoguchi, N., Goto, T., Lee, J. Y., Taketani, K., Nakano, Y., Shono, J., Hoshino, S., Tsuge, N., Narukami, T., Takahashi, N. *et al.* (2010). Diosgenin present in fenugreek improves glucose metabolism by promoting adipocyte differentiation and inhibiting inflammation in adipose tissues. *Molecular Nutrition & Food Research* **54**, 1596–1608.

Venkatesan, T. and Sorimuthu Pillai, S. (2012). Antidiabetic activity of gossypin, a pentahydroxyflavone glucoside in streptozotocin-induced experimental diabetes in rats. *Journal of Diabetes* **4**, 41–46.

Verma, A. K., Bansal, S., Singh, J., Tiwari, R. K., Kasi Sankar, V., Tandon, V. and Chandra, R. (2006). Synthesis and in vitro cytotoxicity of haloderivatives of noscapine. *Bioorganic & Medicinal Chemistry* **14**, 6733–6736.

Vijayan, P., Raghu, C., Ashok, G., Dhanaraj, S. A. and Suresh, B. (2004). Antiviral activity of medicinal plants of Nilgiris. *Indian Journal of Medical Research* **120**, 24–29.

Walterova, D., Ulrichova, J., Valka, I., Vicar, J., Vavreckova, C., Taborska, E., Harjrader, R. J., Meyer, D. L., Cerna, H. and Simanek, V. (1995). Benzo [c]phenanthridine alkaloids sanguinarine and chelerythrine: Biological activities and dental care applications. *Acta Universitatis Palackianae Olomucensis Facultatis Medicae* **139**, 7–16.

Wang, C. Y., Mayo, M. W., Korneluk, R. G., Goeddel, D. V. and Baldwin, A. S., Jr. (1998). NF-kappaB antiapoptosis: Induction of TRAF1 and TRAF2 and c-IAP1 and c-IAP2 to suppress caspase-8 activation. *Science* **281**, 1680–1683.

Wang, W., Abbruzzese, J. L., Evans, D. B. and Chiao, P. J. (1999). Overexpression of urokinase-type plasminogen activator in pancreatic adenocarcinoma is regulated by constitutively activated RelA. *Oncogene* **18**, 4554–4563.

Wang, Z. T., Lau, C. W., Chan, F. L., Yao, X., Chen, Z. Y., He, Z. D. and Huang, Y. (2001). Vasorelaxant effects of cardamonin and alpinetin from Alpinia henryi K. Schum. *Journal of Cardiovascular Pharmacology* **37**, 596–606.

Wang, J., Pham-Mitchell, N., Schindler, C. and Campbell, I. L. (2003). Dysregulated Sonic hedgehog signaling and medulloblastoma consequent to IFN-alpha-stimulated STAT2-independent production of IFN-gamma in the brain. *The Journal of Clinical Investigation* **112**, 535–543.

Wang, J., Huang, H., Liu, P., Tang, F., Qin, J., Huang, W., Chen, F., Guo, F., Liu, W. and Yang, B. (2006a). Inhibition of phosphorylation of p38 MAPK involved in the protection of nephropathy by emodin in diabetic rats. *European Journal of Pharmacology* **553**, 297–303.

Wang, L. H., Choi, Y. L., Hua, X. Y., Shin, Y. K., Song, Y. J., Youn, S. J., Yun, H. Y., Park, S. M., Kim, W. J., Kim, H. J., Choi, J. S. and Kim, S. H. (2006b). Increased expression of sonic hedgehog and altered methylation of its promoter region in gastric cancer and its related lesions. *Modern Pathology* **19**, 675–683.

Wang, G., Sun, B., Gao, Y., Meng, Q. H. and Jiang, H. C. (2007). The effect of emodin-assisted early enteral nutrition on severe acute pancreatitis and secondary hepatic injury. *Mediators of Inflammation* **2007**, 29638.

Wang, C. C., Chiang, Y. M., Sung, S. C., Hsu, Y. L., Chang, J. K. and Kuo, P. L. (2008). Plumbagin induces cell cycle arrest and apoptosis through reactive oxygen species/c-Jun N-terminal kinase pathways in human melanoma A375.S2 cells. *Cancer Letters* **259**, 82–98.

Weerasinghe, P., Hallock, S., Tang, S. C. and Liepins, A. (2001). Sanguinarine induces bimodal cell death in K562 but not in high Bcl-2-expressing JM1 cells. *Pathology Research and Practice* **197**, 717–726.

Wenger, R. H. (2002). Cellular adaptation to hypoxia: O2-sensing protein hydroxylases, hypoxia-inducible transcription factors, and O2-regulated gene expression. *The FASEB Journal* **16**, 1151–1162.

Wetzel, D., Menke, W., Dieter, R., Smasal, V., Giannetti, B. and Bulitta, M. (2002). Escin/diethylammonium salicylate/heparin combination gels for the topical treatment of acute impact injuries: A randomised, double blind, placebo controlled, multicentre study. *British Journal of Sports Medicine* **36**, 183–188.

Wiech, H., Buchner, J., Zimmermann, R. and Jakob, U. (1992). Hsp90 chaperones protein folding in vitro. *Nature* **358**, 169–170.

Williams, S. P., Nowicki, M. O., Liu, F., Press, R., Godlewski, J., Abdel-Rasoul, M., Kaur, B., Fernandez, S. A., Chiocca, E. A. and Lawler, S. E. (2011). Indirubins decrease glioma invasion by blocking migratory phenotypes in both the tumor and stromal endothelial cell compartments. *Cancer Research* **71**, 5374–5380.

Wingo, P. A., Ries, L. A., Giovino, G. A., Miller, D. S., Rosenberg, H. M., Shopland, D. R., Thun, M. J. and Edwards, B. K. (1999). Annual report to the nation on the status of cancer, 1973-1996, with a special section on lung cancer and tobacco smoking. *Journal of the National Cancer Institute* **91**, 675–690.

Wu, T. W., Zeng, L. H., Wu, J. and Fung, K. P. (1994). Morin: A wood pigment that protects three types of human cells in the cardiovascular system against oxyradical damage. *Biochemical Pharmacology* **47**, 1099–1103.

Wu, Z. Q., Guo, Q. L., You, Q. D., Zhao, L. and Gu, H. Y. (2004). Gambogic acid inhibits proliferation of human lung carcinoma SPC-A1 cells in vivo and in vitro and represses telomerase activity and telomerase reverse transcriptase mRNA expression in the cells. *Biological & Pharmaceutical Bulletin* **27**, 1769–1774.

Wu, S. H., Wu, X. H., Lu, C., Dong, L. and Chen, Z. Q. (2006). Lipoxin A4 inhibits proliferation of human lung fibroblasts induced by connective tissue growth factor. *American Journal of Respiratory Cell and Molecular Biology* **34**, 65–72.

Wu, M. Y., Hung, S. K. and Fu, S. L. (2011). Immunosuppressive effects of fisetin in ovalbumin-induced asthma through inhibition of NF-kappaB activity. *Journal of Agricultural and Food Chemistry* **59**, 10496–10504.

Xia, X. M., Wang, F. Y., Wang, Z. K., Wan, H. J., Xu, W. A. and Lu, H. (2010). Emodin enhances alveolar epithelial barrier function in rats with experimental acute pancreatitis. *World Journal of Gastroenterology* **16**, 2994–3001.

Xiong, H., Zhang, Z. G., Tian, X. Q., Sun, D. F., Liang, Q. C., Zhang, Y. J., Lu, R., Chen, Y. X. and Fang, J. Y. (2008). Inhibition of JAK1, 2/STAT3 signaling induces apoptosis, cell cycle arrest, and reduces tumor cell invasion in colorectal cancer cells. *Neoplasia* **10**, 287–297.

Xu, C., Huang, M. T., Shen, G., Yuan, X., Lin, W., Khor, T. O., Conney, A. H. and Kong, A. N. (2006). Inhibition of 7,12-dimethylbenz(a)anthracene-induced skin tumorigenesis in C57BL/6 mice by sulforaphane is mediated by nuclear factor E2-related factor 2. *Cancer Research* **66**, 8293–8296.

Xue, J., Ding, W. and Liu, Y. (2010). Anti-diabetic effects of emodin involved in the activation of PPARgamma on high-fat diet-fed and low dose of streptozotocin-induced diabetic mice. *Fitoterapia* **81**, 173–177.

Yadav, V. R., Prasad, S. and Aggarwal, B. B. (2012a). Cardamonin sensitizes tumor cells to TRAIL through ROS- and CHOP-mediated upregulation of death receptors and downregulation of survival proteins. *British Journal of Pharmacology* **165**, 741–753.

Yadav, V. R., Prasad, S., Sung, B., Gelovani, J. G., Guha, S., Krishnan, S. and Aggarwal, B. B. (2012b). Boswellic acid inhibits growth and metastasis of human colorectal cancer in orthotopic mouse model by downregulating inflammatory, proliferative, invasive and angiogenic biomarkers. *International Journal of Cancer* **130**, 2176–2184.

Yaku, K., Matsui-Yuasa, I., Azuma, H. and Kojima-Yuasa, A. (2011). 1′-Acetoxychavicol acetate enhances the phase II enzyme activities via the increase in intranuclear Nrf2 level and cytosolic p21 level. *The American Journal of Chinese Medicine* **39**, 789–802.

Yamada, M., Katsuma, S., Adachi, T., Hirasawa, A., Shiojima, S., Kadowaki, T., Okuno, Y., Koshimizu, T. A., Fujii, S., Sekiya, Y., Miyamoto, Y., Tamura, M. *et al.* (2005). Inhibition of protein kinase CK2 prevents the progression of glomerulonephritis. *Proceedings of the National Academy of Sciences of the United States of America* **102**, 7736–7741.

Yamai, H., Sawada, N., Yoshida, T., Seike, J., Takizawa, H., Kenzaki, K., Miyoshi, T., Kondo, K., Bando, Y., Ohnishi, Y. and Tangoku, A. (2009). Triterpenes augment the inhibitory effects of anticancer drugs on growth of human esophageal carcinoma cells in vitro and suppress experimental metastasis in vivo. *International Journal of Cancer* **125**, 952–960.

Yamakuni, T., Aoki, K., Nakatani, K., Kondo, N., Oku, H., Ishiguro, K. and Ohizumi, Y. (2006). Garcinone B reduces prostaglandin E2 release and NF-kappaB-mediated transcription in C6 rat glioma cells. *Neuroscience Letters* **394**, 206–210.

Yan, Y., Wang, Y., Tan, Q., Lubet, R. A. and You, M. (2005). Efficacy of deguelin and silibinin on benzo(a)pyrene-induced lung tumorigenesis in A/J mice. *Neoplasia* **7**, 1053–1057.

Yan, Y. Y., Zheng, L. S., Zhang, X., Chen, L. K., Singh, S., Wang, F., Zhang, J. Y., Liang, Y. J., Dai, C. L., Gu, L. Q., Zeng, M. S., Talele, T. T. *et al.* (2011). Blockade of Her2/neu binding to Hsp90 by emodin azide methyl anthraquinone derivative induces proteasomal degradation of Her2/neu. *Molecular Pharmaceutics* **8**, 1687–1697.

Yang, L., Cohn, L., Zhang, D. H., Homer, R., Ray, A. and Ray, P. (1998). Essential role of nuclear factor kappaB in the induction of eosinophilia in allergic airway inflammation. *The Journal of Experimental Medicine* **188**, 1739–1750.

Yang, M. H., Wu, M. Z., Chiou, S. H., Chen, P. M., Chang, S. Y., Liu, C. J., Teng, S. C. and Wu, K. J. (2008). Direct regulation of TWIST by HIF-1alpha promotes metastasis. *Nature Cell Biology* **10**, 295–305.

Yang, L., Xie, G., Fan, Q. and Xie, J. (2010). Activation of the hedgehog-signaling pathway in human cancer and the clinical implications. *Oncogene* **29**, 469–481.

Yao, Y. B., Peng, Z. G., Liu, Z. F., Yang, J. and Luo, J. (2010). Effects of mangiferin on cell cycle status and CDC2/Cyclin B1 expression of HL-60 cells. *Zhong Yao Cai* **33**, 81–85.

Yap, W. N., Zaiden, N., Tan, Y. L., Ngoh, C. P., Zhang, X. W., Wong, Y. C., Ling, M. T. and Yap, Y. L. (2010). Id1, inhibitor of differentiation, is a key protein mediating anti-tumor responses of gamma-tocotrienol in breast cancer cells. *Cancer Letters* **291**, 187–199.

Ye, J. (2009). Emerging role of adipose tissue hypoxia in obesity and insulin resistance. *International Journal of Obesity* **33**, 54–66.

Ye, K., Ke, Y., Keshava, N., Shanks, J., Kapp, J. A., Tekmal, R. R., Petros, J. and Joshi, H. C. (1998). Opium alkaloid noscapine is an antitumor agent that arrests metaphase and induces apoptosis in dividing cells. *Proceedings of the National Academy of Sciences of the United States of America* **95**, 1601–1606.

Yeh, J. L., Hsu, J. H., Hong, Y. S., Wu, J. R., Liang, J. C., Wu, B. N., Chen, I. J. and Liou, S. F. (2011). Eugenolol and glyceryl-isoeugenol suppress LPS-induced iNOS expression by down-regulating NF-kappaB AND AP-1 through inhibition of MAPKS and AKT/IkappaBalpha signaling pathways in macrophages. *International Journal of Immunopathology and Pharmacology* **24**, 345–356.

Yim, E. K., Lee, M. J., Lee, K. H., Um, S. J. and Park, J. S. (2006). Antiproliferative and antiviral mechanisms of ursolic acid and dexamethasone in cervical carcinoma cell lines. *International Journal of Gynecological Cancer* **16**, 2023–2031.

Yin, J., Zhang, H. and Ye, J. (2008). Traditional chinese medicine in treatment of metabolic syndrome. *Endocrine, Metabolic & Immune Disorders Drug Targets* **8**, 99–111.

Yit, C. C. and Das, N. P. (1994). Cytotoxic effect of butein on human colon adenocarcinoma cell proliferation. *Cancer Letters* **82**, 65–72.

Yoshida, T., Hanada, T., Tokuhisa, T., Kosai, K., Sata, M., Kohara, M. and Yoshimura, A. (2002). Activation of STAT3 by the hepatitis C virus core protein leads to cellular transformation. *The Journal of Experimental Medicine* **196**, 641–653.

Yoshimi, N., Matsunaga, K., Katayama, M., Yamada, Y., Kuno, T., Qiao, Z., Hara, A., Yamahara, J. and Mori, H. (2001). The inhibitory effects of mangiferin, a naturally occurring glucosylxanthone, in bowel carcinogenesis of male F344 rats. *Cancer Letters* **163**, 163–170.

Yoshimura, R., Matsuyama, M., Segawa, Y., Hase, T., Mitsuhashi, M., Tsuchida, K., Wada, S., Kawahito, Y., Sano, H. and Nakatani, T. (2003). Expression of peroxisome proliferator-activated receptors (PPARs) in human urinary bladder carcinoma and growth inhibition by its agonists. *International Journal of Cancer* **104**, 597–602.

Yoshitani, S. I., Tanaka, T., Kohno, H. and Takashima, S. (2001). Chemoprevention of azoxymethane-induced rat colon carcinogenesis by dietary capsaicin and rotenone. *International Journal of Oncology* **19**, 929–939.

Young, M. R., Yang, H. S. and Colburn, N. H. (2003). Promising molecular targets for cancer prevention: AP-1, NF-kappa B and Pdcd4. *Trends in Molecular Medicine* **9**, 36–41.

Yu, C. L., Meyer, D. J., Campbell, G. S., Larner, A. C., Carter-Su, C., Schwartz, J. and Jove, R. (1995). Enhanced DNA-binding activity of a Stat3-related protein in cells transformed by the Src oncoprotein. *Science* **269**, 81–83.

Yu, J., Guo, Q. L., You, Q. D., Zhao, L., Gu, H. Y., Yang, Y., Zhang, H. W., Tan, Z. and Wang, X. (2007). Gambogic acid-induced G2/M phase cell-cycle arrest via disturbing CDK7-mediated phosphorylation of CDC2/p34 in human gastric carcinoma BGC-823 cells. *Carcinogenesis* **28**, 632–638.

Yuan, M., Konstantopoulos, N., Lee, J., Hansen, L., Li, Z. W., Karin, M. and Shoelson, S. E. (2001). Reversal of obesity- and diet-induced insulin resistance with salicylates or targeted disruption of Ikkbeta. *Science* **293**, 1673–1677.

Yuan, S. M., Gao, K., Wang, D. M., Quan, X. Z., Liu, J. N., Ma, C. M., Qin, C. and Zhang, L. F. (2011). Evodiamine improves congnitive abilities in SAMP8 and APP(swe)/PS1(DeltaE9) transgenic mouse models of Alzheimer's disease. *Acta Pharmacologica Sinica* **32**, 295–302.

Zakkar, M., Van der Heiden, K., Luong le, A., Chaudhury, H., Cuhlmann, S., Hamdulay, S. S., Krams, R., Edirisinghe, I., Rahman, I., Carlsen, H., Haskard, D. O., Mason, J. C. *et al.* (2009). Activation of Nrf2 in endothelial cells protects arteries from exhibiting a proinflammatory state. *Arteriosclerosis, Thrombosis, and Vascular Biology* **29**, 1851–1857.

Zhang, Y. and Duan, R. D. (2009). Boswellic acid inhibits expression of acid sphingomyelinase in intestinal cells. *Lipids in Health and Disease* **8**, 51.

Zhang, L. and Hung, M. C. (1996). Sensitization of HER-2/neu-overexpressing non-small cell lung cancer cells to chemotherapeutic drugs by tyrosine kinase inhibitor emodin. *Oncogene* **12**, 571–576.

Zhang, L., Chang, C. J., Bacus, S. S. and Hung, M. C. (1995). Suppressed transformation and induced differentiation of HER-2/neu-overexpressing breast cancer cells by emodin. *Cancer Research* **55**, 3890–3896.

Zhang, L., Lau, Y. K., Xi, L., Hong, R. L., Kim, D. S., Chen, C. F., Hortobagyi, G. N., Chang, C. and Hung, M. C. (1998). Tyrosine kinase inhibitors, emodin and its derivative repress HER-2/neu-induced cellular transformation and metastasis-associated properties. *Oncogene* **16**, 2855–2863.

Zhang, Y., Wu, L. J., Tashiro, S., Onodera, S. and Ikejima, T. (2003). Intracellular regulation of evodiamine-induced A375-S2 cell death. *Biological & Pharmaceutical Bulletin* **26**, 1543–1547.

Zhang, H. Z., Kasibhatla, S., Wang, Y., Herich, J., Guastella, J., Tseng, B., Drewe, J. and Cai, S. X. (2004). Discovery, characterization and SAR of gambogic acid as a potent apoptosis inducer by a HTS assay. *Bioorganic & Medicinal Chemistry* **12**, 309–317.

Zhang, L., Chen, W. and Li, X. (2008). A novel anticancer effect of butein: Inhibition of invasion through the ERK1/2 and NF-kappa B signaling pathways in bladder cancer cells. *FEBS Letters* **582**, 1821–1828.

Zhang, L., Yi, Y., Chen, J., Sun, Y., Guo, Q., Zheng, Z. and Song, S. (2010a). Gambogic acid inhibits Hsp90 and deregulates TNF-alpha/NF-kappaB in HeLa cells. *Biochemical and Biophysical Research Communications* **403**, 282–287.

Zhang, Y., Kong, C., Zeng, Y., Wang, L., Li, Z., Wang, H., Xu, C. and Sun, Y. (2010b). Ursolic acid induces PC-3 cell apoptosis via activation of JNK and inhibition of Akt pathways in vitro. *Molecular Carcinogenesis* **49**, 374–385.

Zhang, X., Song, Y., Wu, Y., Dong, Y., Lai, L., Zhang, J., Lu, B., Dai, F., He, L., Liu, M. and Yi, Z. (2011). Indirubin inhibits tumor growth by anti-tumor angiogenesis through blocking VEGFR2 mediated JAK/STAT3 signaling in endothelial cell. *International Journal of Cancer* **129**, 2502–2511.

Zhao, L., Guo, Q. L., You, Q. D., Wu, Z. Q. and Gu, H. Y. (2004). Gambogic acid induces apoptosis and regulates expressions of Bax and Bcl-2 protein in human gastric carcinoma MGC-803 cells. *Biological & Pharmaceutical Bulletin* **27**, 998–1003.

Zhao, S., Venkatasubbarao, K., Lazor, J. W., Sperry, J., Jin, C., Cao, L. and Freeman, J. W. (2008). Inhibition of STAT3 Tyr705 phosphorylation by Smad4 suppresses transforming growth factor beta-mediated invasion and metastasis in pancreatic cancer cells. *Cancer Research* **68**, 4221–4228.

Zhou, J., Gupta, K., Yao, J., Ye, K., Panda, D., Giannakakou, P. and Joshi, H. C. (2002). Paclitaxel-resistant human ovarian cancer cells undergo c-Jun NH2-terminal kinase-mediated apoptosis in response to noscapine. *The Journal of Biological Chemistry* **277**, 39777–39785.

Zhou, J., Zhou, S., Tang, J., Zhang, K., Guang, L., Huang, Y., Xu, Y., Ying, Y., Zhang, L. and Li, D. (2009). Protective effect of berberine on beta cells in streptozotocin- and high-carbohydrate/high-fat diet-induced diabetic rats. *European Journal of Pharmacology* **606**, 262–268.

Zughaier, S., Karna, P., Stephens, D. and Aneja, R. (2010). Potent anti-inflammatory activity of novel microtubule-modulating brominated noscapine analogs. *PLoS One* **5**, e9165.

Zushi, S., Shinomura, Y., Kiyohara, T., Miyazaki, Y., Kondo, S., Sugimachi, M., Higashimoto, Y., Kanayama, S. and Matsuzawa, Y. (1998). STAT3 mediates the survival signal in oncogenic ras-transfected intestinal epithelial cells. *International Journal of Cancer* **78**, 326–330.

# Taxol, Camptothecin and Beyond for Cancer Therapy

WAI-LENG LEE, JENG-YUAN SHIAU AND LIE-FEN SHYUR[1]

*Agricultural Biotechnology Research Center, Academia Sinica, Taipei, Taiwan*



[1]Corresponding author: E-mail: lfshyur@ccvax.sinica.edu.tw

Advances in Botanical Research, Vol. 62

0065-2296/12 $35.00
DOI: 10.1016/B978-0-12-394591-4.00008-8

# ABSTRACT

Plants produce a remarkably diverse array of metabolites, many of which have evolved to confer effectiveness against microbial attack, resist environmental stress or function as signalling molecules in interplant communication. Further, botanical preparations have been used for treating and preventing various human diseases throughout history. More than half of the current anti-cancer drugs in clinical use are natural products or their derivatives, and many are plant-derived agents. Two renowned examples of plant-based drugs used for humans are paclitaxel and camptothecin (CPT). Paclitaxel (taxol), originally identified from Pacific yew *Taxus brevifolia* Nutt, and its semi-synthesized analogues docetaxel (Taxotere®), and CPT, originally identified from *Camptotheca acuminata* Decne, a native tree to China, as well as its more water-soluble derivatives topotecan (Hycamtin) and irinotecan or CPT-11 (Camptosar), are used as chemotherapeutic drugs for various cancers. This review offers a current and integrative account of the two representative chemotherapeutic phytoagents paclitaxel and CPT in terms of their discovery, development of authentic compounds and improvement by synthetic chemistry approaches, as well as their mechanisms of action and clinical applications. We also discuss the bench-side or clinically observed drug resistance and adverse side effects of both drugs, which are the major drawbacks to their use for cancer. Despite intensive efforts and substantial advances in improving treatment, cancer is still a life-threatening and leading cause of death, especially with recurring, metastatic or drug-resistant tumours. The successful experiences with the discovery and development of paclitaxel and CPT as anti-cancer drugs have been significant milestones for the design of alternative strategies and remedies from plant-derived agents in cancer management.

# I. INTRODUCTION

## A. NUTRITIONAL AND PHARMACOLOGICAL VALUES OF PLANT METABOLITES

Much of the world's population relies on plants as a primary source of traditional medication. Medicinal plants have long been an excellent source of pharmaceutical materials: pharmacological effects elicited in humans have been attributable to bioactive compounds in plants (Kaufman *et al.*, 1998). These chemical substances can be considered products of biochemical 'side tracks' in plant cells, traditionally referred to as secondary metabolites, often distributed differentially among different taxonomic groups within the plant kingdom. Plant secondary metabolites may have a dispensable role for daily functioning of the plant, but growing appreciation of their highly diverse biological effects has prompted a re-evaluation of their possible roles in plants. Plant secondary metabolites function as cellular signalling molecules and are involved in plant–plant or plant–environment interactions (Bouwmeester *et al.*, 2003; Buchanan *et al.*, 2002); some have significant pragmatic value as food additives or folk remedies for health management (Andrade-Cetto and Heinrich, 2011; Kaufman *et al.*, 1998).

Among the diverse types of plant secondary metabolites are flavonoids, natural pigments occurring in a wide range of plant families. This group of phytocompounds consists of a central three-ring structure and phenol groups that have a general anti-oxidant effect with potential to prevent several human diseases (Pietta, 2000). *In planta*, flavonoids arrest free radicals generated during photosynthesis. Isoflavones produced by species of Fabaeceae (bean family) are primarily known as phytoestrogens and may have health benefits including reducing the risk of breast cancer, prostate cancer and cardiovascular disease and protection against osteoporosis (bone loss) and menopausal symptoms (Zhao and Mu, 2011).

Terpenoids represent the most diverse category of plant constituents, with more than 25,000 individual compounds identified to date, many of which have been used in herbal remedies. A well-known plant terpenoid-derived drug is paclitaxel. Diterpenoids are non-volatile terpenoids with strong flavours found in several plants such as *Coffea arabica* (coffee). *In planta*, terpenoids may be attractants to pollinators or seed dispersers, or functioned as allelopathic agents that inhibit competing plants. Alkaloids, with less compound numbers identified from plants so far relative to terpenoids and phenolics, are nitrogen-containing compounds with usually a bitter taste, which provides protection against herbivore animals or insect attacks (phytoalexins). Morphine isolated in 1804 from poppy (*Papaver somniferum*) is one of the renowned plant alkaloidal natural products used since ancient times for therapy and recreation (Buchanan *et al.*, 2002; Luch, 2009; Zwenger and Basu, 2008). In addition to morphine, drug discovery from medicinal plants in the nineteenth century led to the identification of several plant secondary metabolites as early drugs, including codeine, digitoxin and quinine, which are still widely used today for different medications (Balunas and Kinghorn, 2005; Kinghorn *et al.*, 2011).

## B. PLANT-BASED DRUG DISCOVERY

Of the approximately 300,000 species of higher plants identified, about 1% of them are used as food and about 10,000 higher plants are used medicinally (McChesney *et al.*, 2007). Recognition of myriad biologically active compounds in plants has fuelled the current research foci in searching for new drugs, antibiotics, insecticides and herbicides from our green flora. The discovery of new drug leads from medicinal plants can be aided by ethnopharmacology, a mode of scientific investigation of particular folk medicines (Andrade-Cetto and Heinrich, 2011). Because cancer remains one of the major causes of human death worldwide (Jemal *et al.*, 2010), exploring therapeutic agents from medicinal plants or other natural resources has become a major topic in anti-cancer drug discovery and requires integrative and multidisciplinary collaboration of multiple technological platforms (Balunas and Kinghorn, 2005; Kinghorn *et al.*, 2011). For about 50 years, substances from terrestrial microorganisms and higher plants have afforded some relief from the scourge of cancer. For instance, cancer chemotherapeutic agents from microbes such as anthracyclines, bleomycin, dactinomycin and mitomycin C, and plant-derived compounds, such as camptothecin (CPT) derivatives, epipodophyllotoxins and taxanes are used therapeutically worldwide (Cragg *et al.*, 2005). Other plant-derived alkaloids, such as colchicines, vinblastine, vincristine, quinine, emetine, (+)-tubocurarine and scopolamine are currently used in clinical trials for various purposes (Raskin *et al.*, 2002). Plant-based drugs include CPT (Wall *et al.*, 1966) from a tree native to China, *Camptotheca acuminata* Decne (Nyssaceae), and paclitaxel (taxol) (Wani *et al.*, 1971) from Pacific yew *Taxus brevifolia* Nutt (Taxaceae) discovered by the group of Drs. Monroe Wall and Mansukh Wani, at Research Triangle Institute, North Carolina. In addition to their pharmaceutical usage, both CPT and paclitaxel possess biological functions *in planta* against viruses, microbes, herbivores and competing plants (Wink, 2003). Currently, two first-generation analogues of CPT, topotecan and irinotecan, are used for treating ovarian, colorectal and small-cell lung cancers (SCLCs), and several second-generation analogues, such as 9-aminocamptothecin, 9-nitrocamptothecin and belotecan are used in clinical trials. Paclitaxel was first approved in 1992 by the U.S. Food and Drug Administration (FDA) for refractory ovarian cancer and today is used against breast and non-small cell lung cancers (NSCLCs) and in Kaposi's sarcoma (Oberlies and Kroll, 2004). Use of these plant-based drugs representing different classes of chemicals with unique mode of action in tumour elimination have led to unveiling of novel drug targets, for a significant milestone in cancer biology and cancer drug development.

History of discovery and development of paclitaxel and CPT has been extensively illustrated in several review articles (Li *et al.*, 2006; Oberlies and Kroll, 2004; Rowinsky and Donehower, 1995; Wall and Wani, 1996). In this chapter, a brief history of both drugs is uncovered based primarily on these preceding literatures. Meanwhile, studies in various disciplines, including compound modification or organic synthesis, pharmacological mechanism of action, and most updated clinical implications that mark important insights and findings in the use of paclitaxel and CPT in cancer therapy are reviewed. PubMed, a National Institutes of Health (NIH) repository for peer-reviewed primary research reports in the life sciences researches was the main tool for literature search and review, while updated clinical studies and data of these two representative plant-based chemotherapeutic agents was reviewed through

IPA (Ingenuity® Systems, www.ingenuity.com), an integrative web-based bioinformatic database, and ClinicalTrials.gov, a registry and results database of federally and privately supported clinical trials conducted in the United States and around the world (www.clinicaltrials.gov).

# II. TAXOL

## A. HISTORY OF INITIAL ACQUISITION

During the mid- to late-1950s, the U.S. National Cancer Institute (NCI) initiated a programme to screen and test plant samples for antineoplastic activity; the randomly collected samples were supplied by the U.S. Department of Agriculture (USDA). In 1962, 650 samples of bark, twigs, leaves and fruits of Pacific yew *T. brevifolia* Nutt. (Taxaceae) from California, Washington and Oregon were collected by USDA botanists led by Arthur S. Barclay (Persinos, 1990). *T. brevifolia* is a scarce and slow-growing evergreen tree localized in the coastal areas of these northwestern United States and had never been chemically investigated until it was assigned to Wall's laboratory at Research Triangle Institute working under contract with NCI by Dr. Hartwell. Owing to previous observations of an excellent correlation between L1210 (mouse leukaemia) *in vivo* activity and 9KB (human oral epidermoid carcinoma) *in vitro* cytotoxicity of CPT, Dr. Wall showed particular interest in an extract of *T. brevifolia* with good activity against 9KB *in vitro*. A 2-year isolation project was thus conducted in Wall's research group (Wall and Wani, 1996). Guided by bioactivity assays of several cancer cell lines and murine tumour models, including 9KB, L1210, P388 and P1534 leukaemias, Walker 256 carcinosarcoma, sarcoma 180 and Lewis lung carcinoma, successive partitioning of the total ethanolic residue between water and chloroform and finally counter current distribution of the plant extracts obtained taxol with a yield of approximately 0.02%. The name 'Taxol' was assigned to the compound purified from *T. brevifolia* because the molecule contained hydroxyl groups and a taxane nucleus. Wall and Wani described this bioactive compound isolated from the stem bark of *T. brevifolia* as a 'novel anti-leukaemic and anti-tumour agent' (Wani *et al.*, 1971).

## B. TAXOL AND ITS BIOSYNTHESIS AND SEMI-SYNTHESIS OF ANALOGUES

Paclitaxel (taxol) (**1**) is a diterpene ester with unique structural feature containing many asymmetric centres; the structure (Fig. 1) was elucidated through a key degradation cleaving the side chain to obtain 10-deacetylbaccatin III (**2**)

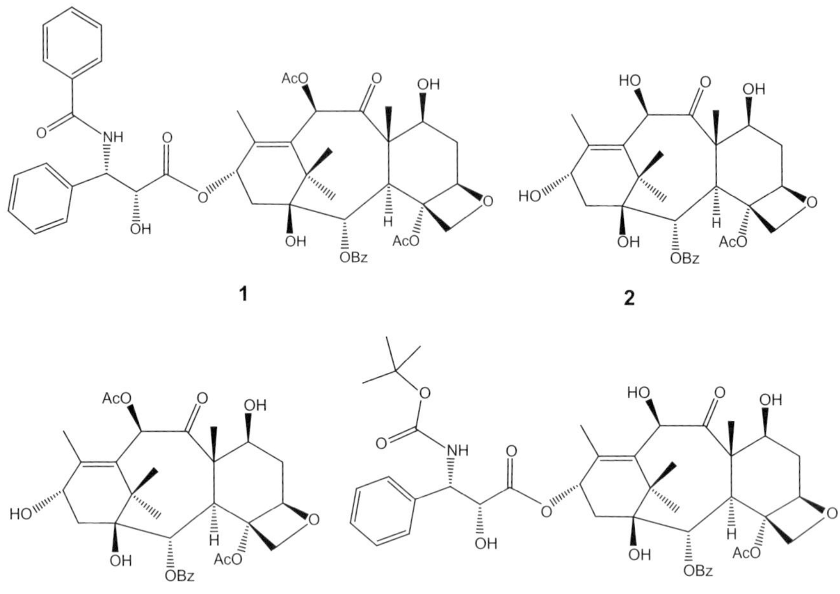


Fig. 1. Chemical structures of paclitaxel (**1**) and respective precursors 10-deacetylbaccatin III (**2**) and baccatin III (**3**) for semi-synthetic drug docetaxel (**4**).

(Fig. 1) and the β-phenylisoserine ester, couple with X-ray studies and $^1$H NMR spectroscopic analysis of the intact molecule and published in 1971 (Wani *et al.*, 1971). The presence of the intact taxane molecule is essential for its anti-tumour activity when ester and tetraol formed by low temperature cleavage of paclitaxel rendered this compound active (Wall and Wani, 1996). The chemistry of paclitaxel has been widely studied by a large number of researchers; Kingston's laboratory is particularly interested in exploring paclitaxel's structure–activity relationship by modifying or removing the hydroxyl groups that decorate the taxane ring system (Kingston, 2007).

Nearly 20 steps are involved in the biosynthesis of paclitaxel, beginning with the prenyl transferase-mediated construction of the universal diterpenoid precursor geranylgeranyl diphosphate. The precursor is cyclized in the committed step to taxa-4(5),11(12)-diene (Jennewein *et al.*, 2004), an olefin structure possessing the unique taxane core, which is then decorated with eight oxygen functional groups (six alcohol functions, one carbonyl and one cyclic ether). The parental olefin is then functionalized by two acetylations, a benzoylation, oxetane ring formation and oxidation at the C9 leading to the late intermediate baccatin III (Croteau *et al.*, 2006), upon which the functionally significant *N*-benzoyl phenylisoserine side chain (Long *et al.*, 2008) is then assembled at the C13-*O*-position to complete the pathway to form paclitaxel. Although significant progress has been made to unveil the biosynthesis of paclitaxel, cloning and expression of all the necessary genes in a suitable host have not been realized (Kingston, 2007). However, total organic synthesis of paclitaxel is formidable because of the large number of asymmetric carbon atoms (Wall and Wani, 1996). The elegant total synthesis of paclitaxel has been devised (Kingston *et al.*, 2002) but is still far from being commercially available because of the cost.

Nevertheless, the destructive isolation of the drug from the original, low-yielding source (the bark of *T. brevifolia*) created a conflict between environmentalists who wished to preserve the old-growth forests of the U.S. Pacific Northwest and dying cancer patients who urgently need the drug for life-saving therapy (Chase, 1991). The short supply of paclitaxel presented a problem for general clinical need. This problem was solved by semi-synthesis of paclitaxel from baccatin III (**3**) or 10-deacetylbaccatin III (**2**) (Fig. 1), by Bristol-Myers Squibb. Semi-synthesis of paclitaxel involved attaching the functionally important *N*-benzoyl phenylisoserine side chain to the C13-*O*-position of 10-deacetylbaccatin III or baccatin III (Denis *et al.*, 1988). Both precursors of paclitaxel are present in much higher quantities than paclitaxel in other *Taxus* species, such as the renewable leaves of the English yew *Taxus baccata* (Wall and Wani, 1996). Docetaxel (Taxotere®) (**4**) (Fig. 1) is a semi-synthetic analogue of paclitaxel, first synthesized in the late 1980s by the research group of Dr. Pierre Potier (Gueritte-Voegelein *et al.*, 1991; Mangatal *et al.*, 1989) and was approved by the U.S. FDA in 1996 and developed by Rhone-Poulenc Rorer (now Aventis). The supply of paclitaxel and its precursors for semi-synthesis still depends on the yew species (Takeya, 2003), and thus resources and the cost of the drug remain a constraint (McCoy, 2004). One productive solution is using a plant tissue culture system, the technique currently used by Bristol-Myers Squibb to produce paclitaxel in Germany (Leistner, 2005). As present, paclitaxel and docetaxel as two clinically available representatives of the taxane drugs share major parts of their structures but differ in pharmacologic and toxicologic profiles (Verweij *et al.*, 1994). Therefore, the parental molecule paclitaxel continues to be the subject of scientific study and therapeutic development.

## C. PHARMACOLOGICAL MECHANISM OF ACTION

### *1. Cell cycle arrest*

Dr. Susan Band Horwitz's research group discovered that by binding to tubulin, paclitaxel had unique anti-tumour activity (Schiff and Horwitz, 1980; Schiff *et al.*, 1979), which was a milestone in clinical application of this natural product. Meanwhile, elucidation of paclitaxel's mode of action unveiled a novel mechanism that was a watershed in our understanding of how tumour cells can be eradicated. Microtubules consist of long, filamentous polymers of tubulin in dynamic equilibrium with heterodimers of alpha and beta protein subunits. The principal function of these protein polymers is the formation of the mitotic spindle during cell division. Microtubules play important roles in cellular activities such as maintenance of cell shape, cellular movement, cell signalling and intracellular transport (Jordan and Wilson, 2004). Thus, microtubules are a clear target for tumour suppression, because cancer cells carry features of uncontrolled cell division for their massive growth (Hanahan and Weinberg, 2000). Paclitaxel was the first known agent to interact specifically with the polymerized form of tubulin, with a manner opposite to that of the *Vinca* alkaloids vinscristine and vinblastine, the microtubule inhibitors already in clinical use that induce disassembly of microtubules (Jordan and Wilson, 2004). Paclitaxel could substitute GTP in promoting the assembly of tubulin into its polymeric form (Schiff *et al.*, 1979), and microtubules formed in the presence of paclitaxel were extraordinarily stable and dysfunctional, which disrupted the normal microtubule dynamics required for cell division and vital interphase processes and subsequently induced mitotic arrest and apoptosis in proliferating cells (Schiff and Horwitz, 1980). X-ray crystallographic solution of the structure of β-tubulin revealed that paclitaxel interacted with the H7 helix of β-tubulin with an optimized *t*-shape structure, also implicated in GTP binding and hydrolysis, to prevent microtubule depolymerization (Snyder *et al.*, 2001). By allowing microtubule attachment but disturbing the tension across the kinetochore in mitosis, paclitaxel arrests the cell cycle at the $G_2$/M phase (Bhalla, 2003; Kelling *et al.*, 2003). Because the $G_2$ and M phases are the most radiosensitive among the cell cycle phases, paclitaxel is also a potent sensitizer to radiotherapy (Tishler *et al.*, 1992). Nevertheless, paclitaxel can also mislead cancer cells passing through the $G_1$/S checkpoint into another cycle of DNA replication (Stewart *et al.*, 1999); consequently, endoreduplication or mitotic slippage occurs when DNA replicates in the absence of cytokinesis, thus leading to the generation of giant cells with 4N, 8N, 16N and even 32N DNA content (Rieder and Maiato, 2004). This paclitaxel-induced mitotic slippage in cells can allow cells to survive and continue dividing or exit the $G_1$ phase and subsequently undergo senescence or apoptosis. Alternatively, paclitaxel-treated cells can escape to the $G_1$, thus overriding the mitotic checkpoint leading to apoptosis in interphase or undergo cell death directly due to mitotic arrest (Blagosklonny, 2007).

*2. Apoptosis*

In 1993, apoptosis was first shown to be a mechanism of paclitaxel-induced cytotoxicity in human lymphoid leukaemia cells (Bhalla *et al.*, 1993). This observation was confirmed by numerous studies of different cancer cell lines and *in vivo* animal model systems (Wang *et al.*, 2000). Although paclitaxel-induced cell cycle arrest has been considered the cause of triggering programmed cell death in cancer cells, the biochemical events occurring after paclitaxel binds to microtubules and leading to apoptotic cell death were not yet fully elucidated in the early 1900s. Accumulated evidence of paclitaxel-mediated activation of several signalling molecules at transcriptional and protein expression levels indicated that paclitaxel initiates apoptosis through multiple mechanisms. For instance, deregulation of factors modulating the checkpoint of mitotic spindle assembly, aberrant activation of cyclin-dependent kinases (Cdks) and c-Jun N-terminal kinases/stress-activated protein kinases (JNK/SAPK) are involved in paclitaxel-induced apoptosis by activating apoptosis signalling-regulating kinase 1 (ASK1) and/or the GTP-binding protein Ras (Wang *et al.*, 1998, 1999, 2000). Breast cancer susceptibility gene 1 (BRCA1) is a tumour suppressor gene functioning by sensing cellular stress and transducing signals to the cell cycle or the apoptotic machinery, thus initiating mitotic arrest and apoptosis (Longley and Johnston, 2005). In response to microtubule inhibitor drugs such as paclitaxel, BRCA1 can activate the $G_2$/M spindle checkpoint (Mullan *et al.*, 2001) or the upstream regulators of both the p38/mitogen-activated protein kinase (MAPK) and JNK/MAPK pathway (Gilmore *et al.*, 2004). p53, a tumour suppressor protein, has been widely studied in cancer biology where it regulates both cell cycle arrest and apoptosis. More than 50% of all cancers harbour p53 mutations (Levine, 1997; Longley and Johnston, 2005). Aberrant activation of cyclin-dependent kinase $p34^{cdc2}$ (also called Cdk1) in p53-mutated cancer cells accounts for greater paclitaxel cytotoxicity, because at the same dose, paclitaxel had a lower effect against cells carrying the wild-type p53 gene. Cdc2 is the ultimate target and checkpoint of pathways that mediate rapid arrest in $G_2$ in response to DNA damage (Stark and Taylor, 2006). The p53-dependent $p21^{WAF1/CIP1}$ inhibiting $p34^{cdc2}$ activity could be one of the mechanisms helping p53(+)-cells survive paclitaxel-induced apoptosis (Stark and Taylor, 2006; Wang *et al.*, 2000).

Paclitaxel-induced apoptosis is also regulated by members of the Bcl-2 family, the key regulators of the mitochondria-mediated intrinsic apoptosis pathway (Adams and Cory, 1998). Bcl-2 is often called the 'guardian of microtubule integrity' (Haldar *et al.*, 1997), and accumulating evidence indicates that the microtubule network integrates mechanistic components of both the cell cycle and apoptosis (Sorger *et al.*, 1997). As a microtubule inhibitor, paclitaxel is of particular interest for its effect on Bcl-2. Activation of $p34^{cdc2}$ is responsible for phosphorylation of Bcl-2 in mitotic cells (Ling *et al.*, 1998). Various signalling pathways regulated by cdc2 (Cdk1), Raf-1, mTOR kinase and JNK are responsive to paclitaxel and lead to inactivation of Bcl-2 and induction of apoptosis (Basu *et al.*, 2006; Haldar *et al.*, 1997; McGrogan *et al.*, 2008). Taken together, the pleiotropic effects of paclitaxel on modulating both the cell cycle and apoptosis, which account for it killing of cancer cells, are sophisticated and remain an area of much research interest and debate. In this regard, numerous reviews have been published (McGrogan *et al.*, 2008; Wang *et al.*, 2000). Nevertheless, elucidation of the mechanisms of paclitaxel-mediated tumour eradication is essential to develop better taxane-based regimens for cancer therapy. Indeed, emerging clinical resistance to taxanes following their extensive use in chemotherapy has attenuated the effectiveness of many of these agents, including paclitaxel, in cancer treatment (Orr *et al.*, 2003).

## D. CLINICAL IMPLICATIONS AND USAGE

### *1. Brief history*

Among many natural compounds identified with taxane ring, paclitaxel is the first shown to have broad preclinical anti-tumour activity before its identification from yew plant extracts (Wani *et al.*, 1971). Although paclitaxel has a novel structure, its development was rather slow because of its moderate effect on experimental tumour models as compared with other agents under development at that time. In addition, large-scale development was arduous because of low yield of the natural product present in authentic Pacific yew tree, which was in limited supply. Fortunately, interest was revived in 1979, when paclitaxel was found to have a unique mode of anti-tumour activity involving binding to tubulin in a novel manner identified by ground-breaking studies of Dr. Susan Band Horwitz's research group (Schiff and Horwitz, 1980; Schiff *et al.*, 1979). Years after its isolation, paclitaxel showed impressive activity on B16 melanoma assay and NCI screening, which includes a panel of human tumour xenografts in nude mice (Wall and Wani, 1996). Because of these important developments, paclitaxel was advanced to clinical trials in the early 1980s.

Rapid progress was made in clinical development of paclitaxel with phase I, II and III clinical trials from 1982 to 1994 in various cancers (Suffness, 1993; Wall, 1993; Wall and Wani, 1994). In 1992, the U.S. FDA approved the use of paclitaxel for refractory ovarian cancer, and Bristol-Myers Squibb (Princeton, NJ, USA) received approval to market paclitaxel with the trade name Taxol®. The effect of paclitaxel was remarkable for recurrent ovarian cancer, for which median survival is usually less than 2 years (Ozols, 2002). This compound was effective in second-line chemotherapy with failure of platinum-based first-line treatment. A combination of carboplatin and paclitaxel increased response rates to 70%, with 3-year survival of 72% (Dizon *et al.*, 2002). The effectiveness of paclitaxel in recurrent ovarian cancer is strongly indicative of its use in cancers that are generally considered refractory to conventional chemotherapy. Subsequently, paclitaxel was approved for treating metastatic breast and lung cancers and Kaposi's sarcoma (Suffness and Wall, 1995).

Besides the scarcity of paclitaxel, solubility was one the factors impeding its development as an anti-tumour agent. The second-generation paclitaxel is a semi-synthetic taxane, docetaxel (Taxotere®), with better solubility. It was originally extracted in 1986 from needles of a smaller shrub-like plant, European yew *T. baccata*, a relatively renewable source (Mangatal *et al.*, 1989), so docetaxel is more readily available. Accumulated research efforts involved comparing the efficacy of paclitaxel and docetaxel against many solid tumours (Suffness, 1993; Wall and Wani, 1994). Docetaxel was approved in 1996 for anthracycline-refractory advanced breast cancer and later also used in lung cancer regimens (Oberlies and Kroll, 2004).

*2. Pharmacokinetics and pharmacodynamics*

Paclitaxel has high binding to plasma protein (88–98%) (Kumar *et al.*, 1993), and the tissue distribution is broad, with a mean distribution half-life of 0.34 h and mean elimination half-life of 5.8 h (Sonnichsen and Relling, 1994). Despite the large steady-state volume of paclitaxel distribution (Kumar *et al.*, 1993), the quantity that crosses the blood–brain barrier is dispensable (Fellner *et al.*, 2002). The clinical dosing schedules commonly used today are 175 mg/m$^2$ three times weekly or 80–100 mg/m$^2$ weekly. The pharmacokinetics of paclitaxel may be non-linear, but doses up to 100 mg/m$^2$ usually maintain concentrations in the linear range of the curve (Gianni *et al.*, 1995; Henningsson *et al.*, 2001). Despite its extensive binding with plasma proteins, the drug is readily cleared from plasma. Renal clearance through urinary excretion accounts for an insignificant proportion of total clearance. Hepatic metabolism by the cytochrome P (CYP) system (isoenzymes CYP2C8 and CYP3A4) and biliary excretion are responsible for systemic elimination of paclitaxel into stools (Spencer and Faulds, 1994). Therefore, patients with renal impairment do not need dose modifications, as compared with those with hepatic dysfunction who are suggested with a reduced dose (Marupudi *et al.*, 2007).

Because of its hydrophobicity, paclitaxel was solubilized in the nonionic surfactant Cremophor EL (CrEL, polyoxyethylated castor oil) and dehydrated ethanol in a 1:1 (v/v) ratio for its first formulation used in clinical practice (Taxol®, Bristol-Myers Squibb) (Dorr, 1994). The non-linear pharmacokinetics of paclitaxel might result from CrEL formulation (Tellingen *et al.*, 1999) because of the micellar entrapment of paclitaxel by CrEL in plasma, which attenuates tumour tissue distribution and clearance of paclitaxel and substantially increases systemic exposure to the drug and increased risk of systemic toxicity (Sparreboom *et al.*, 1999). These factors profoundly affect the therapeutic index and toxicity profile of paclitaxel, which depend on the length of time that sensitive tissues are exposed to a biologically relevant concentration of the drug. When paclitaxel is administered systemically, limited tumour exposure and comparatively high systemic toxicity have led to a low therapeutic index of the drug (Marupudi *et al.*, 2007). Therefore, CrEL-free formulations are needed for alternative delivery strategies with less toxicity during chemotherapy treatment. Different strategies to formulate paclitaxel with albumin-bound, polyglumax, polymeric-micellar, emulsions, nanosphere, nanoparticles and liposomes have been developed. A solvent-free, nanoparticle-based delivery system (ABI-007; Abraxane) has recently demonstrated better efficacy and alleviation of toxic effects as compared with Taxol® during a phase III study of metastatic breast cancer (Gradishar *et al.*, 2005). Clinical use of this formulation was approved by the U.S. FDA in February 2007. A stage III FDA clinical trial is undertaken to examine the clinical use of ABI-007 versus dacarbazine in previously untreated patients with metastatic malignant melanoma (www.clinicaltrials.gov).

### 3. *Drug resistance*

Since paclitaxel was approved for clinical use for advanced ovarian cancer by the U.S. FDA in 1992, paclitaxel or its analogue docetaxel have routinely been used as a single agent or in combination with other anti-cancer agents in the adjuvant and neoadjuvant setting for various solid malignancies, including breast, prostate, ovary, lung, head and neck cancers (Chu *et al.*, 2005; Dombernowsky *et al.*, 1996; Mackler and Pienta, 2005; Wakelee *et al.*, 2005). However, like many cancer therapeutic agents, paclitaxel-induced drug resistance is a serious problem and ultimately leads to relapse and poor prognosis of the tumour. Failure of chemotherapy may due to the inherent resistant trait of the tumour in response to therapeutic agents and/or the acquisition of resistance during therapy. Various mechanisms have been implicated in acquired or secondary taxane resistance; a well-known mechanism is the up-regulation of the multidrug transporter P-glycoprotein (P-gp) and related drug efflux pumps (Gottesman, 2002). Other mechanisms may involve altered expression of tubulin and microtubule-associated proteins (MAPs) (Drukman and Kavallaris, 2002) or functional aberrations in molecular pathways regulating the cell cycle, the spindle assembly checkpoint and cancer cell apoptosis (McGrogan *et al.*, 2008; Patel *et al.*, 2010). The contribution of each of these mechanisms to clinical resistance to paclitaxel or its analogue remains uncertain (Patel *et al.*, 2010). Nevertheless, a number of predictive markers of taxane resistance have been identified, including MAPs, tau and stathmin, which may help identify patients at increased risk of cancer recurrence and those most likely to benefit from taxane treatment (McGrogan *et al.*, 2008).

Altered regulation or expression of the drug efflux transporters or ATP-binding cassette (ABC) transporter family, multidrug resistant protein 1/P-gp, and multiresistance-associated proteins have been shown *in vitro* and *in vivo* to cause intrinsic or acquired resistance to commonly used chemotherapeutic drugs, including taxanes. In this phenotype, known as multidrug resistance (MDR), resistance to one drug can result in cross-resistance to other structurally unrelated drugs (Longley and Johnston, 2005). However, no substantial evidence has shown that both tubulin mutations and isotype composition contribute to taxane resistance (Nicoletti *et al.*, 2001), even though overexpression of MAPs, tau and stathmin (tubulin destabilizing protein) are correlated with paclitaxel resistance (McGrogan *et al.*, 2008). Further, defects in the spindle assembly checkpoint, which is activated in response to taxanes, have led to some promising predictors of taxane resistance, including MAD2, BUBR1, aurora A and synuclein-gamma. Altered expression and activity of regulators in the signalling pathway of cell cycle/apoptosis, such as p53, BRCA1 and Bcl-2, have been related to increased resistance to taxanes, particularly in treatment of breast cancer (McGrogan *et al.*, 2008). As mentioned previously, mutated p53 could sensitize tumours in taxol treatment, but in some cases, the lack of functional p53 increased the resistance to taxanes because of abrogation of p53-mediated apoptosis (Longley and Johnston, 2005). However, BRCA1, a transcriptional regulator of the spindle checkpoint proteins BUBR1 and MAD2, is activated in response to paclitaxel treatment, which implies an interference in the spindle checkpoint (Mullan *et al.*, 2001). In addition, BRCA1 enhances sensitivity to apoptosis induced by paclitaxel (Quinn *et al.*, 2003). Therefore, BRCA1 may be an important predictor of the chemotherapy response to taxanes (Chabalier *et al.*, 2006). Data to predict the chemoresponsiveness of the anti-apoptotic protein Bcl-2 by its expression are inconclusive (Longley and Johnston, 2005; Tothova *et al.*, 2002). Bcl-2 phosphorylation in breast tumours might be associated with increased sensitivity to both paclitaxel and docetaxel (Shitashige *et al.*, 2001).

In many cancers, particular pathways that are essential in cell survival, cell growth and cell cycle progression were found overactive in reducing apoptosis and allowing proliferation. One of these pathways is anti-apoptotic phosphatidylinositol-3-kinase/Akt (PI3K/Akt) (Vara *et al.*, 2004). In several ovarian cancer cell lines, increased activation of Akt conferred resistance to paclitaxel (Page *et al.*, 2000). Hyperactivity of the PI3K/Akt pathway was observed *in vitro* in chemoresistance to paclitaxel, trastuzumab and gemcitabine (Longley and Johnston, 2005). Akt signalling regulates NF-κB, a transcription factor promoting cell proliferation and inhibiting apoptosis (Lin and Karin, 2003). *In vitro* cell-based assay revealed activated NF-κB signalling in treatments with various chemotherapeutic agents resulting in reduced chemotherapy-induced apoptosis (Baldwin, 2001). Suppressed activation of NF-κB increased the efficacy of paclitaxel in *in vitro* and *in vivo* ovarian cancer models (Mabuchi *et al.*, 2004). Thus, resistance to chemotherapeutic agents is a complex phenomenon orchestrated by various pathways in a single cancer cell. In the clinical setting, targeting different oncogenic signalling pathways with combinations of chemotherapeutic agents or with chemotherapeutic drugs combined with inhibitors of a specific signalling pathway or receptor is in general more effective than single-agent therapies. For instance, combining taxanes with trastuzumab, a target therapeutic agent binding to the oncogenic receptor protein HER2, greatly improved overall survival and disease-free survival in breast cancer, which suggests a possible synergistic effect between both agents (Plosker and Keam, 2006).

*4. Adverse effects*

Paclitaxel therapy has a number of adverse effects, such as hypersensitivity reactions, myelosuppression, bradycardia, hypotension, peripheral neuropathy, myalgias, arthralgias, nausea, diarrhoea, mucositis and alopecia (Marupudi *et al.*, 2007). In the early phase I trials, life-threatening type I anaphylactic reaction during paclitaxel administration was one of the obstacles that threatened its further development (Rowinsky and Donehower, 1995), although current paclitaxel-related regimens have reduced the incidence of these side effects to less than 5%. The severe hypersensitivity, including dyspnea with bronchospasm, urticaria and hypotension (Weiss *et al.*, 1990), was suggested to be mediated by histamine. Therefore, the CrEL vehicle of paclitaxel was thought to be responsible for this side effect, because it induced histamine release and similar manifestations in dog (Lorenz *et al.*, 1977; Szebeni *et al.*, 1998). Fortunately, a standard regimen of premedication with corticosteroids such as dexamethasone and histamine $H_1/H_2$ antagonists before paclitaxel infusion was effective in minimizing the risk of hypersensitivity (Weiss *et al.*, 1990).

Paclitaxel's principle haematologic toxic effect is neutropenia (Rowinsky and Donehower, 1995), which results in immunocompromization and increased risk of infectious complications. Despite the dose-dependent effects, myelosuppression is more significantly affected by the schedule of administration; indeed, severe neutropenia ($<500$ cells/mm$^3$) was more frequent with longer duration of infusion (24 h) than shorter duration (3 h) (Eisenhauer *et al.*, 1994). Fortunately, paclitaxel-induced neutropenia is usually brief, which suggests that this drug does not irreversibly damage haematopoietic stem cells. Paclitaxel-induced thrombocytopenia and anaemia are less common than neutropenia (Marupudi *et al.*, 2007). However, combination treatments of paclitaxel and other cytotoxic chemotherapeutic agents seem to enhance myelosuppression. Case reports of early breast cancer patients receiving sequential paclitaxel and doxorubicin-cyclophosphamide showed grade 3/4 haematologic events (neutropenia, thrombocytopenia, anaemia; Simpson and Plosker, 2004). The use of granulocyte colony-stimulating factor (Rowinsky *et al.*, 1993b) with reduced infusion time from 24 to 3 h (Eisenhauer *et al.*, 1994) greatly ameliorated the complications from haematologic toxicity.

Peripheral neuropathy is a cumulative phenomenon (Gerven *et al.*, 1994), representing the most important non-haematological toxic effect associated with paclitaxel administration (Rowinsky *et al.*, 1993a), especially with weekly drug administration. Unfortunately, peripheral neuropathy is encountered in most patients during therapy (Mielke *et al.*, 2003) and is characterized by numbness and paraesthesias in a 'glove-and-stocking' distribution (Lipton *et al.*, 1989). Mechanistically, paclitaxel's effects on microtubule dynamics attenuate normal axonal transport, thus leading to a length-dependent sensorimotor axonal neuropathy, which is a common dose-dependent side effect of paclitaxel treatment (Peltier and Russell, 2002; Sahenk *et al.*, 1994). Overall, the neurotoxic effect of paclitaxel is both dose and time dependent (Scuteri *et al.*, 2006). In contrast to hypersensitivity reactions, peripheral neuropathy was also found severe with CrEL-free paclitaxel formulations (Ibrahim *et al.*, 2002). Neurotoxicity remains the major limitation of the clinical effectiveness of paclitaxel. Therefore, research into palliation of peripheral neuropathy with paclitaxel therapy is important. Cardiac toxic effects were reported in a small number of patients, and most developed total alopecia. Several aspects of paclitaxel use remain to be clarified, including the optimal treatment protocol and infusion time, confirmation of the tolerability profile and efficacy of combination regimens in an expanded range of malignancies. Long-term follow-up of paclitaxel recipients will also allow for monitoring the effects of the drug on survival (Marupudi *et al.*, 2007).

*5. Current clinical trials*

Paclitaxel is a common clinical drug used in standard chemotherapy regimen for breast cancer treatment. Lately in 2011, several stage III FDA clinical trials have just completed to test for the use of bevacizumab, cyclophosphamide, gemcitabine, herceptin and trastuzumab in combined with standard chemotherapy with paclitaxel for treatment of primary or metastatic breast cancer through different therapeutic strategies. In patients with primary epithelial ovarian cancer, primary peritoneal cancer or fallopian tube carcinoma, appropriate doses and sequences of standard treatment involves paclitaxel and cisplatin is still the main concern of current clinical trials. Radiation therapy with or without treatment of paclitaxel and cisplatin following surgery in patients with endometrial cancer is compared in a latest stage III FDA trial (www.clinicaltrials.gov). Besides, combination of paclitaxel, carboplatin and/or radiation therapy has increasingly been considered as a therapeutic approach in treating patients with NSCLC. Intriguingly, natural material such as shark cartilage extract was also been examined for its clinical use in combination therapy with paclitaxel for treatment of irremovable stage III NSCLC, and the stage III clinical trial has just been completed recently in 2011. Meanwhile, in several ongoing trials, combination chemotherapy with paclitaxel are investigated in patients with advanced gastric and cervical cancers (www.clinicaltrials.gov).

# III. CAMPTOTHECIN

## A. HISTORY OF INITIAL ACQUISITION

A joint research effort and program between the U.S. NCI and USDA was established in 1950s, with collection of plant samples for searching for phytosteroids that could be precursors of cortisone. Thousand of plants, particularly of the rare species, were collected and most of their ethanolic extracts were saved (Wall and Wani, 1996). In addition to screening for phytosteroids, the stored plant extracts were also tested for antibiotic, antitumour and antiviral activities. Among the initial 1000 plant extracts tested, that of *C. acuminata* Decne (Nyssaceae), a plant native to China and known as 'happy tree', was the only one with high anti-tumour activity. This finding attracted the intense interest of Dr. Wall's research group at the Research Triangle Institute. From 1963 to 1966, through a research contract with NCI, Wall and co-workers identified CPT (**5**) (Fig. 2) from *C. acuminata* extracts and published its chemical structure; this compound was later described as a 'novel alkaloidal leukaemia and tumour inhibitor' (Wall and Wani, 1996;

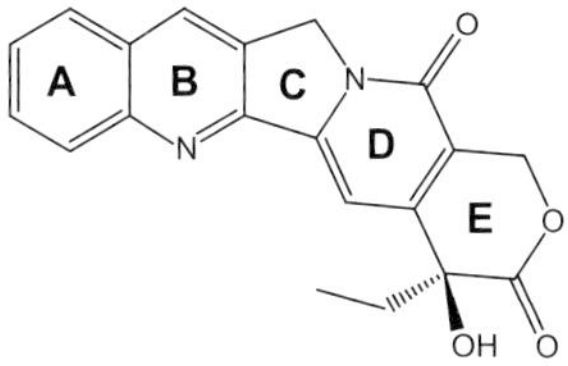

**5**

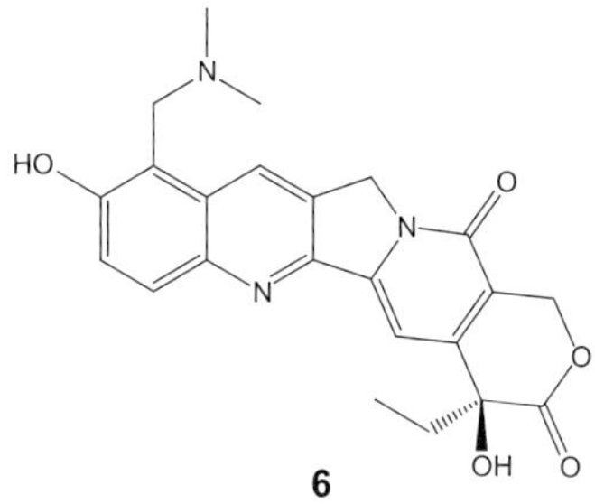

**6**

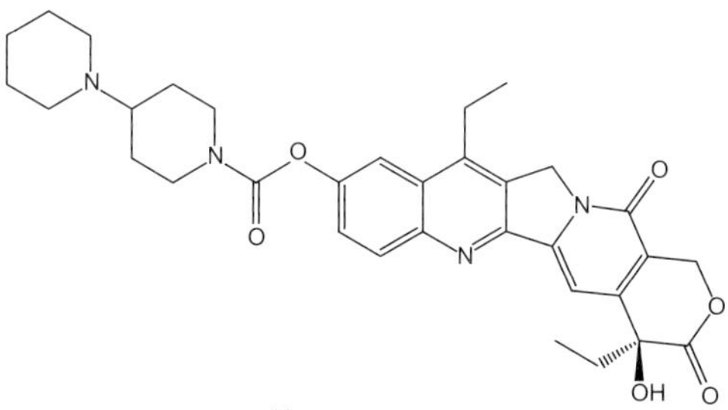

**7**

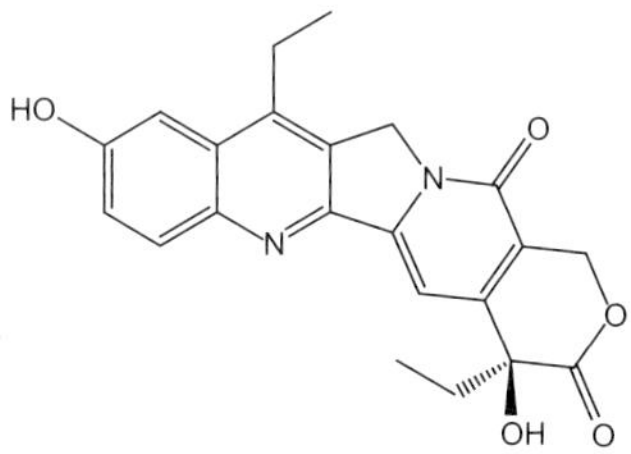

**8**

Fig. 2. Chemical structures of camptothecin (**5**), topotecan (**6**), irinotecan (**7**) and its active metabolite SN-38 (**8**).

Wall *et al.*, 1966). However, CPT had poor stability and solubility, which were observed at the time of its discovery (Wall *et al.*, 1966). Moreover, the unexpected adverse drug responses stalled clinical trials in the early 1970s, with the favourable development of paclitaxel as a broad-spectrum chemotherapeutic agent launched by that time. However, interest was renewed when CPT's unique mode of action was reported in 1985, with the specific molecular target of DNA topoisomerase I (topo I) (Hsiang *et al.*, 1985). To overcome the solubility and stability issues of CPT, various derivatives have been developed; however, only two, irinotecan and topotecan, received FDA approval for clinical use, in 1996. These first-generation analogues of CPT, Hycamtin (topotecan) (**6**, Fig. 2) and Camptosar (irinotecan or CPT-11) (**7**, Fig. 2), were marketed by GlaxoSmithKline and Pharmacia (now Pfizer), respectively, and are currently used for treatment of ovarian and colon cancers (Gore *et al.*, 2001; Saltz *et al.*, 2000).

## B. CAMPTOTHECIN AND ITS BIOSYNTHESIS, SEMI-SYNTHESIS AND ANALOGUES

The structure of CPT (**5**) includes a pyrrolo[3,4-β]-quinoline moiety (rings A, B and C) conjugated with pyridone moiety (ring D) and a chiral centre at position 20 within the alpha-hydroxy lactone ring with an (*S*) configuration (in ring E) (Lu *et al.*, 2007). The 20S chiral carbon is for activity, and a dynamic equilibrium exists between the closed-ring lactone ring moiety and open-ring carboxylic acid forms (Lorence and Nessler, 2004).

Several semi-synthetic approaches for synthesizing CPT analogues have been reported. These organically synthesized CPT analogues share a similar chemical structure with some changes on C-7, C-9, C-10 and C-11 of ring A/B (**5**, Fig. 2; Herben *et al.*, 1998). Tagawa and co-workers were the first to report the asymmetric synthesis of CPT, in which an *N*-tosyl-(*R*)-proline derivative was used as the chiral auxiliary to induce the stereocontrolled assembly of the lone chiral centre (Ejima *et al.*, 1989). Later, the asymmetric synthesis of CPT with the shortest steps was reported (Comins and Nolan, 2001). Two precursors, containing A/B- and D/E-fragments of CPT, are connected by the reaction of *N*-alkylation, then followed by an intramolecular Heck ring closure reaction to reconstitute the CPT. Meanwhile, another strategy involves precursors with A- and D/E-fragments designed to form CPT by a free-radical cascade reaction (Curran and Liu, 1992; Yabu *et al.*, 2001, 2002). The A-ring (2-aminobenzaldehyde) and C/D/E-ring framework can also build the B-ring of CPT by the classical Friedlander reaction (Henegar *et al.*, 1997). Further, Bosch *et al.* described a method involving intramolecular radical cyclization to form the C-ring of CPT, followed by the E-ring of CPT, which was asymmetrically constituted by enolate chemistry (Bennasar *et al.*, 2002). According to Ciufolini and Chavan synthetic strategy, the D-ring of CPT was constituted through intramolecular and intermolecular Michael addition reactions (Chavan and Venkatraman, 1998; Ciufolini and Roschangar, 1996, 1997, 2000). Additionally, a new synthesis strategy of CPT described by Blagg and Boger involved assembly of the CPT based on a Diels-Alder cycloaddition of the electron-rich dienophile with the electron-deficient *N*-sulfonyl-1-aza-1,3-butadiene (Blagg and Boger, 2002).

## C. PHARMACOLOGICAL MECHANISM OF ACTION

### *1. DNA damage*

CPT was found capable of inhibiting DNA synthesis by strand scission, which led to cancer cell death arrested in the S-phase of the cell cycle (Thomas *et al.*, 2004). The discovery of DNA strand scission with CPT treatment that resulted in protein-associated DNA strand break has provided a key sign of the DNA–protein complex as an attack target of CPT (Hsiang *et al.*, 1985). Later, the primary molecular target of CPT was found to be topo I. Topoisomerases are nuclear enzymes involved in DNA transcription and replication via transient breakage of the phosphodiester backbone of DNA to relax torsional strain in supercoiled chromosomal DNA (Osheroff, 1989). Topoisomerases are divided into two types, topo I and II, on the basis of their catalytic mechanism on DNA strand breakage (Redinbo *et al.*, 2000). Topo I involves the cleavage of one strand of a DNA double helix, and topo II cuts both DNA strands. The scission procedure of DNA strand with topo I involves a nucleophilic attack, in which the OH group of Tyr723 of the enzyme interacts with the phosphodiester backbone of DNA to result in breakage of the DNA phosphodiester backbone, and further, a covalent binary DNA–topo I complex is formed (Thomas *et al.*, 2004). CPT is impressively capable of stabilizing the covalent binary complex by decreasing DNA re-ligation catalysed by topo I. The possible interaction mechanism of CPT with the covalent binary DNA–topo I complex was proposed by different groups of researchers using computational docking models. The latest representative models are the Hol base-flipping, Pommier docking and Pilch docking models (Li *et al.*, 2006; Thomas *et al.*, 2004). These structure models may provide some predictive information of the interaction between CPT and its molecular target for the design of new CPT analogues with improved binding efficiency to the DNA–topo I complex. However, the genuine kinetic mechanism of CPT in stabilizing the covalent binary complex has yet to be proved.

*2. Apoptosis*

The major cytotoxicity of CPT and its analogues is S-phase specific and replication related, in which topo I inhibition leads to programmed cell death (Liu *et al.*, 1996). In general, replication-related cell death involves stabilization of p53, activation of checkpoint kinase 1 (Chk1), and phosphorylation of RPA (Shao *et al.*, 1999). Activation of p53, a checkpoint protein in cell cycle control and DNA repair, is known to be involved in the DNA-damaging agent-induced apoptosis (Steller, 1995). CPT-mediated DNA damage through topoisomerase poison predominantly requiring functional p53 for toxicity in tumour cells involves transcriptional activation of pro-apoptotic factors such as FAS, PUMA and BAX (Roos and Kaina, 2006). In the presence of sufficient DNA damage, both p53 and another member of the family, p73, are stabilized and activated (Chen *et al.*, 2001), and these processes appear to strongly depend on Chk1 activity (Urist *et al.*, 2004). Very recently, CPT was found to induce p53-independent apoptosis through activation of caspase-2 in melanoma cells (Rudolf *et al.*, 2011).

CPT activates NF-κB by initial nuclear DNA damage, then degradation of cytoplasmic IkBα by the ubiquitin-proteasome pathway, described as 'a linkage between nuclear DNA damage and cytoplasmic signalling events' (Huang *et al.*, 2000). Inhibition of NF-κB activation augments CPT-induced apoptosis, and protein kinase C was found to be involved in NF-κB activation conferring drug resistance in CPT-treated breast cancer MCF-7 cells (Raveh-Amit *et al.*, 2011). ROS and cellular oxidative stress have long been associated with carcinogenesis, in that cancer cells are more frequently deficient in crucial anti-oxidative enzymes such as catalase, glutathione peroxidase and superoxide dismutase. A high vulnerability of tumour cells to ROS was thus observed. Accordingly, a unique anti-tumour strategy involves delivering excess oxidative stress into tumour cells or targeted disruption of the anti-oxidative defence systems of tumour cells was emerged as 'oxidation therapy' (Fang *et al.*, 2009). Many conventional anti-cancer drugs including CPT activate caspase-3(-like) protease, thus resulting in generation of $H_2O_2$ presumably through the activation of NADPH oxidase, and subsequently inducing apoptosis in cancer cells (Simizu *et al.*, 1998). A more recent finding indicated that CPT sensitizes glioma and medulloblastoma cells, the glioblastomas resistant to Fas-mediated cell death, in drug treatment through ROS-dependent mechanisms (Li *et al.*, 2009; Xia *et al.*, 2005).

## D. CLINICAL IMPLICATIONS AND USAGE

### *1. Brief history*

In early clinical trials, use of unmodified CPT was suspended mainly because of (1) the severe and unpredicted adverse side effects, including diarrhoea, haemorrhagic cystitis and myelotoxicity (Kehrer *et al.*, 2001; Moertel *et al.*, 1972); (2) the extremely negligible water solubility of CPT and (3) the lactone ring of CPT being rapidly inactivated at physiological pH by the hydrolysis reaction but the lactone hydrolysis reversed at acidic pH (Li *et al.*, 2006). Because of poor water solubility, CPT was not convenient for clinical trials, but the unique mode of action of CPT targeting topo I, identified in the mid-1980s, restored interest in this compound (Hsiang *et al.*, 1985). Semi-synthetic and water-soluble forms of CPT derivatives were developed for use in clinical trials during the late 1980s and early 1990s (Lorence and Nessler, 2004). These CPT derivatives, such as Hycamtin (topotecan) (**6**, Fig. 2) and Camptosar (irinotecan or CPT-11) (**7**, Fig. 2), were termed the 'first-generation analogues' and were approved for use in clinical cancer therapy by the U.S. FDA in 1996 (Oberlies and Kroll, 2004); topotecan was approved as a second-line drug to treat ovarian and SCLCs (Herzog, 2002). Compared to other CPT analogues, CPT-11 is a prodrug hydrolyzed by carboxylesterase (CE) to form the active metabolite 7-ethyl-10-hydroxy-CPT (SN-38) (**8**, Fig. 2). It is a water-soluble yellow crystalline powder with slight solubility in aromatic solvents, and has drug potency usually 100–1000-fold higher than that of the precursor CPT compound (Tanizawa *et al.*, 1994). The CPT analogues have also been approved for clinical therapy in Europe (1995) and Japan (1994). In Japan, CPT-11 is used with various cancers, including ovarian, cervical, breast, gastric, skin, and small cell or NSCLCs and melanoma (Herben *et al.*, 1998).

The dosage and administration of CPT-11 are significantly different in different countries. In Europe, the starting dose of CPT-11 was recommended as 350 mg/m$^2$, with intravenous (i.v.) infusion over 30–90 min once every 3 weeks for treating advanced colorectal cancer (CRC). In the United States, the initial dose of CPT-11 was suggested to be 125 mg/m$^2$, with i.v. infusion over 90 min once weekly for 4 weeks, followed by a 2-week rest period. In Japan, the dosage and administration were 100 mg/m$^2$ once a week every 2 weeks (Herben *et al.*, 1998). However, topotecan is used in i.v. or oral formulation, with the initial dose in phase II clinical trials as an oral formulation and i.v. administration of 2.3 and 1.5 mg/m$^2$ per day for 5 days every 21 days, respectively (von Pawel *et al.*, 2001).

*2. CPT analogues in clinical therapy*

Cragg and Newman reviewed the development of the topo I inhibitory agents based on preclinical and clinical trials results in 2004; 26 of the 60 agents investigated (approximately of 43%) belong to CPT derivatives include SN-38, CPT-11 and topotecan (Cragg and Newman, 2004). CPT-11 (irinotecan) has been used for CRC and gastric and NSCLCs. CRC is one of the most commonly occurring cancers worldwide. Local relapse or metastatic disease develops in almost 50% of patients with CRC. CPT-11 is frequently used as first- and second-line chemotherapy for advanced or recurrent CRC and can be combined with 5-fluorouracil/leucovorin (FOLFIRI) as a standard regimen for perhaps prolonging survival time (Ikeguchi *et al.*, 2011). The explicit response rate of advanced CRC is 20–30% for single-use CPT-11, and the response rate can reach 50% with CPT-11 used with other drugs (Douillard *et al.*, 2000). With CPT-11 used as a second-line monotherapy in phase II and III clinical trials for treating advanced CRC, the median overall survival time was 6.6–16.1 and 9.1–10.8 months, respectively (Oostendorp *et al.*, 2010). However, severe side effects have also been described in both phase II and III trials and include diarrhoea (5–39% and 15–36%, respectively), asthenia (0–31% and 4–21%), nausea (1–24% and 5–14%), vomiting (2–22% and 6–14%) and treatment-related mortality (0–2% and 0–5%) (Oostendorp *et al.*, 2010). In several stage III FDA clinical trials just completed in 2011, oxaliplatin, fluorouracil, sunitinib and bevacizumab were included in combined with standard chemotherapy (CPT-11/FOLFIRI) for treatment of CRC (www.clinicaltrials.gov).

Topotecan was used for SCLC and gynaecological and ovarian cancers. Traditionally, the standard first-line chemotherapy regimen for SCLC is the combination of cyclophosphamide, doxorubicin (adriamycin) and vincristine (CAV), but platinum/etoposide (PE) has increasingly been used as chemotherapy (Riemsma *et al.*, 2010). However, compared with CAV, i.v. topotecan had similar response and survival rates as CAV for patients with recurrent SCLC, but topotecan showed less toxicity (von Pawel *et al.*, 1999). Further, i.v. topotecan has been used as second-line chemotherapy for SCLC with 2–6 months recurrence after completion of first-line chemotherapy (Cheng *et al.*, 2007). In recent years, oral and i.v. topotecan showed similar response rates (18% vs. 22%) and 1-year survival rates (33% vs. 29%) as second-line therapy for recurrent SCLC, but the incidence of grade 4 neutropenia was higher for i.v. than oral administration (64% vs. 47%) (Eckardt *et al.*, 2007). Other side effects from topotecan treatment are associated with haematotoxicity, such as neutropenia (61%), thrombocytopenia (38%) and anaemia (25%), with the most common non-haematologic toxicity which is diarrhoea (6%) (Pirker *et al.*, 2010). Combination chemotherapy includes topotecan and other clinical drugs in treating patients with ovarian epithelial/primary peritoneal cancer, neuroblastoma or rhabdomyosarcoma was also examined in a number of FDA clinical studies which were just completed their stage III trials in 2011 (www.clinicaltrials.gov).

*3. Pharmacokinetics and pharmacodynamics*

*a. Topotecan.* A topotecan bioavailability of ~42% after oral administration was reported (Herben *et al.*, 1999). However, both oral and i.v.-delivered topotecan have shown large interpatient variability in exposure (de Jonge, 2000; Montazeri *et al.*, 2000). After oral administration of the drug, a CYP3A system irreversibly catalyzes topotecan into its first-known metabolite *N*-desmethyl topotecan, which is slightly less active but has longer half-life than topotecan (8.8 vs. 2.8 h). Both compounds exist in an active lactone and inactive carboxylate form following a comparable equilibrium (Rosing *et al.*, 1997). As compared with other clinical drugs, topotecan binding to plasma proteins is very low and highly variable, with reported values ranging from 7% to 35% (Kollmannsberger *et al.*, 1999); similarly, only a low level of *N*-desmethyl topotecan has been found in plasma, which suggests minimal CYP3A-related conversion. Both topotecan and *N*-desmethyl topotecan compounds were suggested to undergo further reversible transformation and metabolism into uridine-diphosphate glucuronosyltransferase (UGT)-mediated glucuronide products (i.e., topotecan-*O*-glucuronide and *N*-desmethyl topotecan-*O*-glucuronide). The structural identification and clinical importance of these glucuronide products are unclear (Rosing *et al.*, 1998). The elimination of topotecan is thought to predominantly result from its conversion to carboxylate species, then renal excretion. A total of 76.4% (i.e., 71.3% topotecan and 5.1% *N*-desmethyl topotecan) was traced in excretion products within 9 days after infusion (Kollmannsberger *et al.*, 1999), and the administered dose recovered as unchanged drug in urine ranges from 30% to 50% (Garcia-Carbonero and Supko, 2002).

*b. CPT-11.* CPT-11 (irinotecan) is a prodrug catalysed by liver and intestinal CEs to produce active metabolite 7-ethyl-10-hydroxycamptothecin (SN-38), which is 100 or 1000 times more potent than CPT-11 as a topo I inhibitor (Khanna *et al.*, 2000). Alternatively, a CYP3A-mediated oxidation of CPT-11 may produce the inactive oxidized products, namely 7-ethyl-10-[4-*N*-(5 aminopentanoic acid)-1-piperidino]carbonyloxycamptothecin and NPC (7-ethyl-10-(4-amino-1-piperidino)carbonyloxycamptothecin). Under certain circumstances, CE can transform NPC into SN-38, thereby re-entering the activation pathway of the drug (Santos *et al.*, 2000). In human plasma, CPT-11 presents predominately in its carboxylated form and SN-38 in its lactone form; both reach equilibrium in human plasma (Sparreboom *et al.*, 1998). In contrast to topotecan, about 80% CPT-11 and 99% SN-38 bind with proteins in blood (Combes *et al.*, 2000). Therefore, as compared with other compartments, conversion of CPT-11 into its active form by CEs in tumours may be related more to the chemosensitivity of the patient (Guichard *et al.*, 1999). The terminal half-life of SN-38 is approximately 47 h (Kehrer *et al.*, 2000), and the compound is catalysed by uridine-diphosphate glucuronosyltransferase 1A1 (UGT1A1) to form inactive SN-38 glucuronide (SN-38G) in human liver (Iyer *et al.*, 1998). After the secretion of this metabolized product into the intestinal lumen, a part of the common gut flora is able to reactivate SN-38G by β-glucuronidase (Sperker *et al.*, 1997). Therefore, patient's faeces containing a relatively large amount of active SN-38 suggest the circulation of this toxic agent during excretion, which may contribute substantially to the adverse effects of CPT-11 (i.e., the late form of diarrhoea; Slatter *et al.*, 2000).

*4. Adverse effects*

Despite promising effects in treating a broad spectrum of tumours, CPT and its derivatives have several drawbacks. A major limitation for the clinical use of CPT analogues arises from the chemical equilibrium between CPT lactone form and the E-ring-opened carboxylated form (Basili and Moro, 2009). The carboxylated form has less than 10% the potency of the lactone form as a topo I inhibitor and is inactive in cell culture, probably because of its low permeability through the cell membrane (Adams *et al.*, 2000). Moreover, CPTs are also responsible for severe side effects, as mentioned. For topotecan, neutropenia and thrombocytopenia are the common dose-limiting toxic effects (Schiller *et al.*, 1996), even in combination with other drugs (Herben *et al.*, 1997; Lilenbaum *et al.*, 1995). For CPT-11, neutropenia and diarrhoea are the major dose-limiting toxic effects (Sasaki *et al.*, 1995). Neutropenia is the common haematological side effect with topotecan as well as CPT-11 and is both dose and schedule dependent. Anaemia was frequently observed but rarely clinically significant with CPT-11 or topotecan administration. At higher doses of topotecan, mucositis and (late) diarrhoea are non-haematological side effects (Mathijssen *et al.*, 2002).

CPT-11-related diarrhoea can be divided into an acute and a delayed form. The acute form appears during or within several hours after infusion, as a result of an acute cholinergic-like syndrome (Bleiberg and Cvitkovic, 1996). The cholinergic syndrome is reported only with CPT-11 treatment, not treatment with other CPTs. The structure of CPT-11 may be responsible for this syndrome, given that the piperidino group of this drug shows some similarities to a cholinergic agent, dimethyl-phenylpiperazinium (Gandia *et al.*, 1993). A rapid and reversible inhibition of acetylcholinesterase by the CPT-11 lactone form was proposed as an explanation for the cholinergic syndrome (Dodds and Rivory, 1999). However, cumulative behaviour for delayed diarrhoea with CPT-11 treatment at all dose levels is more frequently reported (Marsh and McLeod, 2004). Owing to the role of secretory mechanism in ailments such as diarrhoea, adverse effects caused by this drug initially was proposed to be related to altered epithelial intestinal cells or abnormal intestinal ion transport (Saliba *et al.*, 1998). UGTs are the principal enzymes converting SN-38 to an inactive metabolite SN-38G, a more water-soluble form excreted into bile and urine. However, the hepatobiliary excreted SN-38G can be deglucuronidated into SN-38 by endogenous bacterial β-glucuronidase in the intestine, thus allowing SN-38 to be reabsorbed from the intestine and enter into the enterohepatic circulation, and subsequently cause delayed-type diarrhoea as a manifestation of local toxic effects of SN-38 on the intestinal mucosa (Iyer *et al.*, 1998; Slatter *et al.*, 2000; Sperker *et al.*, 1997). Abundant evidence suggests that the extensive gene polymorphism of *UGT1A1* caused by polymorphism of a common promoter (*UGT1A1*28*) is strongly associated with differences in CPT-11 toxic effects, which suggests that *UGT1A1* may be a pharmacogenetic marker for patients who would not respond favourably to CPT-containing therapy (Marsh and Hoskins, 2010).

### 5. *Drug resistance*

Previous reports indicated that hydrogen bond likely plays an important role in maintaining the CPT–DNA–enzyme ternary complex. Fujimori *et al.* showed that a human leukaemia cell line resistant to CPT contains a mutant topo I enzyme, in which the Asp533 site might form hydrogen bond interactions with CPT (Fujimori *et al.*, 1995). Further, some other mutated amino acid residues (e.g., Arg364, Asn722 and Lys532) have been demonstrated to play a role in sustained CPT binding and resulting in CPT resistance (Urasaki *et al.*, 2001). On the other hand, significant associations between the activity of CPTs and the expression of epidermal growth factor receptor (EGFR) found in several correlation studies indicated a possible role of EGFR for sensitivity and resistance to CPT (Ling *et al.*, 2001; Vallböhmer *et al.*, 2006). Most recently, Konkimalla and Efferth (2010) reported that the binding of CPT and its derivative 20-*N*,*N*-glycinate with pharmacophore at the EGFR is similar to that of EGFR inhibitor, erlotinib. Further, they demonstrated the preferential activity of CPT towards EGFR-expressing tumour cells (Konkimalla and Efferth, 2010). These results imply that CPT-type drugs may inhibit tumours not only by binding to topo I but also by inhibiting EGFR, a prognostic marker for short survival of patients and therapy resistance of tumours.

Several reports of various cancer cell lines have demonstrated that treatment with CPT analogues can result in marked responses, such as inhibition of DNA synthesis and induction of the expression of DNA damage-response genes such as p53 and p21 (Nelson and Kastan, 1994; Tsao *et al.*, 1992). However, inhibition of topoisomerase activity, induction of topoisomerase mutation and increase of $O^6$-methylguanine-DNA-methyltransferase activity has been found to cause drug resistance of cancer cells with CPT analogue treatment (Fujimori *et al.*, 1995; Yoshikawa *et al.*, 2004). In addition to mutation of topoisomerase-related genes, yeast genetic screening has been used to identify novel mutations that may lead to attenuated cytotoxic effects of CPT. Reid and co-workers demonstrated that mutation of PDR1 (pleiotropic drug resistance protein 1) in yeast significantly affected cell sensitivity to CPT (Reid *et al.*, 1997). Coincidentally, several ABC transporter genes such as PDR5, SNQ2 and YOR1 were found to be the downstream targets of PDR1 (Decottignies *et al.*, 1995; Katzmann *et al.*, 1995); increased level of SNQ2 significantly reduced the intracellular accumulation of CPT (Reid *et al.*, 1997). Multidrug resistance protein 1 (MDR1) has been commonly observed in drug resistance of cancer cells to various anti-cancer drugs such as amsacrine, doxorubicin, etoposide and paclitaxel (Gottesman and Pastan, 1993; Mattern *et al.*, 1993). Similarly, overexpression of MDR1 in cancer cells also caused cell resistance to SN-38, topotecan and 9-aminocamptothecin. Meanwhile, ABCG2, an ABC transporter involved in the regulation of lipid- and sterol-trafficking mechanisms in intestinal mucosa cells, hepatocytes and macrophages (Schmitz *et al.*, 2001), promoted drug resistance to the topo I inhibitors topotecan and SN-38 by increasing drug efflux in

MCF-7 breast cancer cells or S1-M1-80 colon cancer cells (Brangi *et al.*, 1999; Doyle *et al.*, 1998; Nakatomi *et al.*, 2001) because SN-38 is generally accumulated at membrane vesicles. Yoshikawa and co-workers demonstrated that ABCG2 can energetically remove this active compound from PC-6/SN2-5H human lung carcinoma cells in an ATP-dependent manner (Yoshikawa *et al.*, 2004). Investigations of CPT-resistant cancer cells have focused on the synthesis of novel CPT derivatives by modifying the A-ring of CPT to circumvent ABCG2-associated drug resistance (Nakagawa *et al.*, 2006).

# IV. NEW TRENDS IN DISCOVERY OF PLANT-BASED AGENTS FOR CANCER PREVENTION AND TREATMENT

Both paclitaxel and CPT as well as their derivatives have extensively been used in treatments of various cancers with defined molecular action, such as interfering with rapidly dividing cells by destroying microtubules and binding to topoisomerase II (topo II) or DNA directly, and blocking DNA synthesis. Nevertheless, emerging clinical resistance to these drugs with their extensive use as chemotherapy has attenuated their effectiveness. In general, chemotherapeutic drugs can fail to kill cancer cells for various reasons, by mutation of target molecules or more commonly by MDR, which relies on drug efflux from cancer cells mediated by ABC transporters (Gottesman, 2002). As well, conventional chemotherapeutic agents cause serious side effects, including pain, hair loss, diarrhoea, nausea, vomiting and blood-related side effects. Therefore, the discovery of compounds that engage, evade or exploit drug resistance is important to improve chemotherapy efficiency. Further, combining cytotoxic drugs with less toxic biological agents is one way to relieve side effects, especially in adjuvant therapy (Oakman *et al.*, 2009). In this regard, growing evidences have shown that the treatment of anti-cancer drugs can be improved by combined treatment with chemopreventive agents (D'Incalci *et al.*, 2005; Sarkar and Li, 2006). 'Prevention is better than cure' is no doubt the golden rule in health management today. Cancer chemoprevention was first defined as 'a strategy of cancer control by administration of synthetic or natural compounds, single or as a combination, to prevent, reverse or suppress the process of carcinogenesis' (Sporn and Suh, 2002; Sporn *et al.*, 1976). Chemoprevention strategies target three stages of carcinogenesis for anti-initiation (e.g., protecting against DNA damage, detoxification, scavenging of free radicals and metabolism of carcinogens), anti-promotion and anti-progression (e.g., suppression of uncontrolled cancer cell proliferation, induction of differentiation and apoptosis, modulation of inflammation and angiogenesis; Greenwald, 2002; Tsao *et al.*, 2004). Emerging evidence suggests that strategies involving a combination of cancer preventive agents with chemotherapy or radiotherapy by targeting distinct and/or multiple molecular mechanisms are considered promising and alternative approaches for better outcome of cancer treatment, especially for highly metastatic malignancies (Hsan *et al.*, 2010; Lee *et al.*, 2011; Sarkar and Li, 2006).

Recent studies suggest that herbal medicines, botanicals, dietary fruits and vegetables are novel sources of low toxic chemopreventive agents for cancer because of their long history of human consumption (Park and Pezzuto, 2002; Surh, 2003). In fact, an increasing number of plant-derived compounds, such as phenolics genistein, curcumin, (−)-epigallocatechin-3-gallate (EGCG), resveratrol, monogalactosyldiacylglycerols and deoxyelephantopin have shown great potential in chemoprevention of various cancers through their anti-oxidant and anti-inflammation activities or as cell cycle, apoptosis or angiogenesis modulators through regulating important pro-inflammation molecules such as Akt, NF-κB and cyclooxygenase 2 (Bode and Dong, 2009; Hou *et al.*, 2007; Huang *et al.*, 2010; Sarkar and Li, 2006; Sarkar *et al.*, 2009). Owing to the action of multiple-targeting, plant-derived chemopreventive agents are suggested as promising partners for preventing tumour occurrence through increasing drug potency and lowering the doses of toxic anti-cancer drugs (Das *et al.*, 2010). In treatment of paclitaxel, the use of resistance-counteracting chemopreventive compounds may be beneficial in clinical practice with commonly observed chemoresistance of this drug in patients. Aggarwal's laboratory showed that curcumin, a polyphenolic derivative identified from *Curcuma longa* rhizome with anti-inflammatory and anti-angiogenesis activities (Lopez-Lazaro, 2008) can suppress paclitaxel-induced NF-κB activation in breast cancer cells and inhibit lung metastasis in experimental animals (Aggarwal *et al.*, 2005). Curcumin was found to inhibit the growth of MCF-7 breast cancer cell lines with MDR (Labbozzetta *et al.*, 2009). Owing to the promising efficacy of curcumin in suppressing cancer growth and metastasis *in vitro* and *in vivo*, the feasibility and tolerability of the combination of docetaxel and curcumin in patients with advanced and metastatic breast cancer was investigated in a phase I clinical trial. At the recommended dosage of curcumin (6000 mg/day for seven consecutive days every 3 weeks) with a standard dose of docetaxel, most patients showed improved biological and clinical responses (Bayet-Robert *et al.*, 2010).

# V. CONCLUSIONS AND FUTURE PROSPECTS

Taken together, the discovery and development of paclitaxel and CPT provide several lines of valuable research and information on exploring new therapeutic agents and defining novel drug targets. The timeline of discovery and research and development of paclitaxel and CPT since last millennium are summarized in Figs. 3 and 4, respectively. The motivation and persistence of numerous scientists including Drs. Wall and Wani, who overcame obstacles encountered during the 30-plus-year journey of these molecules from bench to bedside, have led to genuine and remarkable achievements in natural products research. From the perspective of application chemistry, both CPT and taxol were discovered using the principles of bioactivity-directed fractionation, especially *in vitro* cytotoxicity as a predictor of *in vivo* efficacy (Wall *et al.*, 1976). Today, these approaches are widely used in laboratories investigating natural product chemistry and phytomedicine research around the world, to discover bioactive compounds or lead compounds from plants, marine organisms or microbes, especially from medicinal plants, which have been used for millennia as main sources of therapeutic or protective remedies. From the bench- or bed-side viewpoint, the discovery of paclitaxel and CPT is a landmark in revealing the power of plant-based natural products to lead to the exploitation and elucidation of novel mechanisms of anti-tumour activity. Emerging 'omics' approaches have been adopted in drug lead discovery and have revealed novel mechanisms of action from phytomedicines and phytocompounds (Huang *et al.*, 2010; Lee *et al.*, 2010; Shyur and Yang, 2008; Wang *et al.*, 2008). Drug lead compounds as small molecule ligands usually can interact with multiple targets via numerous possible mechanisms. The contributions of paclitaxel and CPT research and development are important in providing clues, guidance or experience in biological study of novel anti-cancer compounds and their targets.

As one of the new waves in designing cancer therapy regimens, plant-derived chemotherapeutic agents such as genistein, curcumin, EGCG and resveratrol have great potential with less toxicity and target multiple oncogenic pathways in cancer cells (Bode and Dong, 2009; Das *et al.*, 2010; Sarkar *et al.*, 2009). Hence, plant resources are essential for discovery of new leads for novel drug development as well as useful complementary drugs in chemotherapeutic combination regimens. Only about 10% of higher plants on earth have been investigated for their bioactive compounds, but in most of these plants, a small part of the phytocompounds has been studied in detail. In fact, lower plants, algae, microbes and marine organisms are also natural resources of new natural compounds. In addition, numerous medicinal

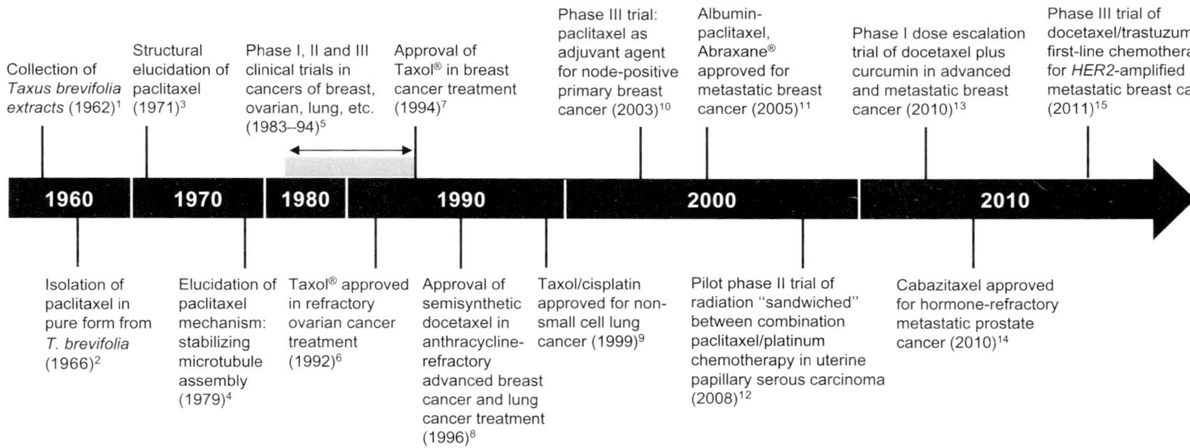


Fig. 3. Timeline of discovery and development of paclitaxel. [1]Persinos (1990), [2]Wall and Wani (1996), [3]Wani *et al.* (1971), [4]Schiff *et al.* (1979), [5]Suffness (1993), Wall (1993), Wall and Wani (1994), [6]Wall and Wani (1996), [7]Suffness and Wall (1995), [8]Oberlies and Kroll (2004), [9]http://www.healthcommunities.com/lung-cancer/chemotherapy, [10]Henderson *et al.* (2003), [11]Gradishar *et al.* (2005), [12]Fields *et al.* (2008), [13]Bayet-Robert *et al.* (2010), [14]www.drugs.com and [15]Valero *et al.* (2011).

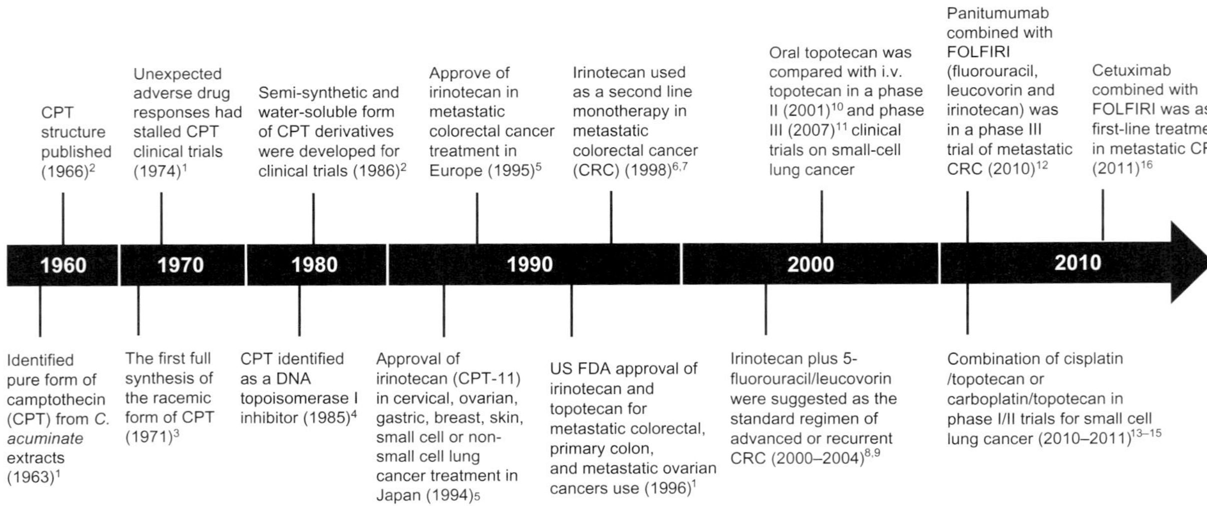


Fig. 4. Timeline of discovery and development of camptothecin (CPT). [1]Oberlies and Kroll (2004), [2]Thomas *et al.* (2004), [3]Li *et al.* (2006), [4]Hsiang *et al.* (1985), [5]Herben *et al.* (1998), [6]Cunningham *et al.* (1998), [7]Rougier *et al.* (1998), [8]de Gramont *et al.* (2000), [9]Tournigand *et al.* (2004), [10]von Pawel *et al.* (2001), [11]Eckardt *et al.* (2007), [12]Peeters *et al.* (2010), [13]Spigel *et al.* (2010), [14]Ichinose *et al.* (2011), [15]Hosomi *et al.* (2011) and [16]Van Cutsem *et al.* (2011).

plants used by traditional healers in Asia, Africa and South America have not been thoroughly explored. Humans can still look forward to a great variety of new compounds useful for better health care in the future (Bernhoft, 2010).

Current treatments for metastatic tumours are still poor in terms of overall response and survival rate. Nevertheless, the clinical efficacy and usefulness of CPT- and paclitaxel-based treatment regimens is still limited because of dose-limiting toxic effects and drug resistance (Marupudi *et al.*, 2007; Pommier, 2006). New treatment options, alone or combined with natural phytocompounds or herbal medicines and current chemotherapeutic drugs may provide breakthrough results in treating life-threatening cancer diseases.

## REFERENCES

Adams, J. M. and Cory, S. (1998). The Bcl-2 protein family: Arbiters of cell survival. *Science* **281**, 1322–1326.

Adams, D. J., Dewhirst, M. W., Flowers, J. L., Gamcsik, M. P., Colvin, O. M., Manikumar, G., Wani, M. C. and Wall, M. E. (2000). Camptothecin analogues with enhanced antitumor activity at acidic pH. *Cancer Chemotherapy and Pharmacology* **46**, 263–271.

Aggarwal, B. B., Shishodia, S., Takada, Y., Banerjee, S., Newman, R. A., Bueso-Ramos, C. E. and Price, J. E. (2005). Curcumin suppresses the paclitaxel-induced nuclear factor-kappa B pathway in breast cancer cells and inhibits lung metastasis of human breast cancer in nude mice. *Clinical Cancer Research* **11**, 7490–7498.

Andrade-Cetto, A. and Heinrich, M. (2011). From the field into the lab: Useful approaches to selecting species based on local knowledge. *Frontiers in Pharmacology* **2**, 1–5.

Baldwin, A. S. (2001). Control of oncogenesis and cancer therapy resistance by the transcriptional factor nuclear factor-kappa B. *Journal of Clinical Investigation* **107**, 241–246.

Balunas, M. J. and Kinghorn, A. D. (2005). Drug discovery from medicinal plants. *Life Sciences* **78**, 431–441.

Basili, S. and Moro, S. (2009). Novel camptothecin derivatives as topoisomerase I inhibitors. *Expert Opinion on Therapeutic Patents* **19**, 555–574.

Basu, A., Dubois, G. and Haldar, S. (2006). Posttranslational modifications of Bcl-2 family members—A potential therapeutic target for human malignancy. *Frontiers in Bioscience* **11**, 1508–1521.

Bayet-Robert, M., Kwiatowski, F., Leheurteur, M., Gachon, F., Planchat, E., Abrial, C., Mouret-Reynier, M. A., Durando, X., Barthomeuf, C. and Chollet, P. (2010). Phase I dose escalation trial of docetaxel plus curcumin in patients with advanced and metastatic breast cancer. *Cancer Biology & Therapy* **9**, 8–14.

Bennasar, M. L., Zulaica, E., Juan, C., Alonso, Y. and Bosch, J. (2002). Addition of ester enolates to *N*-alkyl-2-fluoropyridinium salts: Total synthesis of (+/–)-20-deoxycamptothecin and (+)-camptothecin. *Journal of Organic Chemistry* **67**, 7465–7474.

Bernhoft, A. (2010). Bioactive Compounds in Plants—Benefits and Risks for Man and Animals. The Norwegian Academy of Science and Letters, AIT Otta AS, Oslo, Norway.

Bhalla, K. N. (2003). Microtubule-targeted anticancer agents and apoptosis. *Oncogene* **22**, 9075–9086.

Bhalla, K., Ibrado, A., Tourkina, E., Tang, C., Mahoney, M. and Huang, Y. (1993). Taxol induces internucleosomal DNA fragmentation associated with programmed cell death in human myeloid leukemia cells. *Leukemia* **7**, 563–568.

Blagg, B. S. J. and Boger, D. L. (2002). Total synthesis of (+)-camptothecin. *Tetrahedron* **58**, 6343–6349.

Blagosklonny, M. V. (2007). Mitotic arrest and cell fate: Why and how mitotic inhibition of transcription drives mutually exclusive events. *Cell Cycle* **6**, 70–74.

Bleiberg, H. and Cvitkovic, E. (1996). Characterisation and clinical management of CPT-11 (Irinotecan)-induced adverse events: The European perspective. *European Journal of Cancer* **32A**(Suppl. 3), S18–S23.

Bode, A. M. and Dong, Z. (2009). Cancer prevention research—Then and now. *Nature Reviews. Cancer* **9**, 508–516.

Bouwmeester, H. J., Matusova, R., Sun, Z. and Beale, M. H. (2003). Secondary metabolite signalling in host–parasitic plant interactions. *Current Opinion in Plant Biology* **6**, 358–364.

Brangi, M., Litman, T., Ciotti, M., Nishiyama, K., Kohlhagen, G., Takimoto, C., Robey, R., Pommier, Y., Fojo, T. and Bates, S. E. (1999). Camptothecin resistance: Role of the ATP-binding cassette (ABC), mitoxantrone-resistance half-transporter (MXR), and potential for glucuronidation in MXR-expressing cells. *Cancer Research* **59**, 5938–5946.

Buchanan, B., Gruissem, W. and Jones, R. L. (2002). Biochemistry & Molecular Biology of Plants. American Society of Plant Physiologists, Rockville, MD.

Chabalier, C., Lamare, C., Racca, C., Privat, M., Valette, A. and Larminat, F. (2006). BRCA1 downregulation leads to premature inactivation of spindle checkpoint and confers paclitaxel resistance. *Cell Cycle* **5**, 1001–1007.

Chase, M. (1991). Clashing priorities: A new cancer drug may extend lives. *Wall Street Journal, New York* **9**, A1.

Chavan, S. P. and Venkatraman, M. S. (1998). A practical and efficient synthesis of (+/−)-camptothecin. *Tetrahedron Letters* **39**, 6745–6748.

Chen, X., Zheng, Y., Zhu, J., Jiang, J. and Wang, J. (2001). p73 is transcriptionally regulated by DNA damage. *Oncogene* **20**, 769–774.

Cheng, S., Evans, W. K., Stys-Norman, D., Shepherd, F. A., and Lung Canc Dis Site Grp Canc Care O. (2007). Chemotherapy for relapsed small cell lung cancer: A systematic review and practice guideline. *Journal of Thoracic Oncology* **2**, 348–354.

Chu, Q., Vincent, M., Logan, D., Mackay, J. A. and Evans, W. K. (2005). Taxanes as first-line therapy for advanced non-small cell lung cancer: A systematic review and practice guideline. *Lung Cancer* **50**, 355–374.

Ciufolini, M. A. and Roschangar, F. (1996). Total synthesis of (+)-camptothecin. *Angewandte Chemie International Edition in English* **35**, 1692–1694.

Ciufolini, M. A. and Roschangar, F. (1997). Practical total synthesis of (+)-camptothecin: The full story. *Tetrahedron* **53**, 11049–11060.

Ciufolini, M. A. and Roschangar, F. (2000). Practical synthesis of (20 S)–(+)–camptothecin: The progenitor of a promising group of anticancer agent. *Targets in Heterocyclic Systems* **4**, 25.

Combes, O., Barre, J., Duche, J. C., Vernillet, L., Archimbaud, Y., Marietta, M. P., Tillement, J. P. and Urien, S. (2000). *In vitro* binding and partitioning of

irinotecan (CPT-11) and its metabolite, SN-38, in human blood. *Investigational New Drugs* **18**, 1–5.
Comins, D. L. and Nolan, J. M. (2001). A practical six-step synthesis of (S)-camptothecin. *Organic Letters* **3**, 4255–4257.
Cragg, G. M. and Newman, D. J. (2004). A tale of two tumor targets: Topoisomerase I and tubulin. The Wall and Wani contribution to cancer chemotherapy. *Journal of Natural Products* **67**, 232–244.
Cragg, G. M., Kingston, D. G. I. and Newman, D. J. (2005). Anticancer Agents from Natural Sources. CRC Press/Taylor & Francis, Boca Raton, FL.
Croteau, R., Ketchum, R. E. B., Long, R. M., Kaspera, R. and Wildung, M. R. (2006). Taxol biosynthesis and molecular genetics. *Phytochemistry Reviews* **5**, 75–97.
Cunningham, D., Pyrhonen, S., James, R. D., Punt, C. J. A., Hickish, T. F., Heikkila, R., Johannesen, T. B., Starkhammar, H., Topham, C. A., Awad, L., Jacques, C. and Herait, P. (1998). Randomised trial of irinotecan plus supportive care versus supportive care alone after fluorouracil failure for patients with metastatic colorectal cancer. *Lancet* **352**, 1413–1418.
Curran, D. P. and Liu, H. (1992). New 4+1 radical annulations—A formal total synthesis of (+/−)-camptothecin. *Journal of the American Chemical Society* **114**, 5863–5864.
Das, T., Sa, G., Saha, B. and Das, K. (2010). Multifocal signal modulation therapy of cancer: Ancient weapon. *Molecular and Cellular Biochemistry* **336**, 85–95.
de Gramont, A., Figer, A., Seymour, M., Homerin, M., Hmissi, A., Cassidy, J., Boni, C., Cortes-Funes, H., Cervantes, A., Freyer, G., Papamichael, D., Le Bail, N. *et al.* (2000). Leucovorin and fluorouracil with or without oxaliplatin as first-line treatment in advanced colorectal cancer. *Journal of Clinical Oncology* **18**, 2938–2947.
de Jonge, M. J. A. (2000). Inter- and intrapatient variability in oral topotecan pharmacokinetics: Implications for body-surface area dosage regimens. *Clinical Cancer Research* **6**, 2685–2689.
Decottignies, A., Lambert, L., Catty, P., Degand, H., Epping, E. A., Moyerowley, W. S., Balzi, E. and Goffeau, A. (1995). Identification and characterization of SNQ2, a new multidrug ATP binding cassette transporter of the yeast plasma-membrane. *Journal of Biological Chemistry* **270**, 18150–18157.
Denis, J. N., Greene, A. E., Guenard, D., Gueritte-Voegelein, F., Mangatal, L. and Potier, P. (1988). A highly efficient, practical approach to natural taxol. *Journal of the American Chemical Society* **110**, 5917–5919.
D'Incalci, M., Steward, W. and Gescher, A. (2005). Use of cancer chemopreventive phytochemicals as antineoplastic agents. *The Lancet Oncology* **6**, 899–904.
Dizon, D. S., Hensley, M. L., Poynor, E. A., Sabbatini, P., Aghajanian, C., Hummer, A., Venkatraman, E. and Spriggs, D. R. (2002). Retrospective analysis of carboplatin and paclitaxel as initial second-line therapy for recurrent epithelial ovarian carcinoma: Application toward a dynamic disease state model of ovarian cancer. *Journal of Clinical Oncology* **20**, 1238–1247.
Dodds, H. M. and Rivory, L. P. (1999). The mechanism for the inhibition of acetylcholinesterases by irinotecan (CPT-11). *Molecular Pharmacology* **56**, 1346–1353.
Dombernowsky, P., Gehl, J., Boesgaard, M., Paaske, T. and Jensen, B. V. (1996). Doxorubicin and paclitaxel, a highly active combination in the treatment of metastatic breast cancer. *Seminars in Oncology* **23**, 23–27.
Dorr, R. T. (1994). Pharmacology and toxicology of Cremophor EL diluent. *The Annals of Pharmacotherapy* **28**, S11–S14.

Douillard, J. Y., Cunningham, D., Roth, A. D., Navarro, M., James, R. D., Karasek, P., Jandik, P., Iveson, T., Carmichael, J., Alakl, M., Gruia, G., Awad, L. *et al.* (2000). Irinotecan combined with fluorouracil compared with fluorouracil alone as first-line treatment for metastatic colorectal cancer: A multicentre randomised trial. *Lancet* **355**, 1041–1047.

Doyle, L. A., Yang, W. D., Abruzzo, L. V., Krogmann, T., Gao, Y. M., Rishi, A. K. and Ross, D. D. (1998). A multidrug resistance transporter from human MCF-7 breast cancer cells. *Proceedings of the National Academy of Sciences of the United States of America* **95**, 15665–15670.

Drukman, S. and Kavallaris, M. (2002). Microtubule alterations and resistance to tubulin-binding agents (review). *International Journal of Oncology* **21**, 621–628.

Eckardt, J. R., von Pawel, J., Pujol, J. L., Papai, Z., Quoix, E., Ardizzoni, A., Poulin, R., Preston, A. J., Dane, G. and Ross, G. (2007). Phase III study of oral compared with intravenous topotecan as second-line therapy in small-cell lung cancer. *Journal of Clinical Oncology* **25**, 2086–2092.

Eisenhauer, E. A., Huinink, W. W.t.B., Swenerton, K. D., Gianni, L., Myles, J., Burg, M. E.v.d., Kerr, I., Vermorken, J. B., Buser, K. and Colombo, N. (1994). European-Canadian randomized trial of paclitaxel in relapsed ovarian cancer: High-dose versus low-dose and long versus short infusion. *Journal of Clinical Oncology* **12**, 2654–2666.

Ejima, A., Terasawa, H., Sugimori, M. and Tagawa, H. (1989). Antitumor agents. 1. Asymmetric-synthesis of (S)-camptothecin. *Tetrahedron Letters* **30**, 2639–2640.

Fang, J., Seki, T. and Maeda, H. (2009). Therapeutic strategies by modulating oxygen stress in cancer and inflammation. *Advanced Drug Delivery Reviews* **61**, 290–302.

Fellner, S., Bauer, B., Miller, D. S., Schaffrik, M., Fankhänel, M., Spruß, T., Bernhardt, G., Graeff, C., Färber, L., Gschaidmeier, H., Buschauer, A. and Fricker, G. (2002). Transport of paclitaxel (Taxol) across the blood-brain barrier *in vitro* and *in vivo*. *Journal of Clinical Investigation* **110**, 1309–1318.

Fields, A. L., Einstein, M. H., Novetsky, A. P., Gebb, J. and Goldberg, G. L. (2008). Pilot phase II trial of radiation "sandwiched" between combination paclitaxel/platinum chemotherapy in patients with uterine papillary serous carcinoma (UPSC). *Gynecologic Oncology* **108**, 201–206.

Fujimori, A., Harker, W. G., Kohlhagen, G., Hoki, Y. and Pommier, Y. (1995). Mutation at the catalytic site of topoisomerase I in CEM/C2, a human leukemia cell line resistant to camptothecin. *Cancer Research* **55**, 1339–1346.

Gandia, D., Abigerges, D., Armand, J. P., Chabot, G., Da Costa, L., De Forni, M., Mathieu-Boue, A. and Herait, P. (1993). CPT-11 induced cholinergic effects in cancer patients. *Journal of Clinical Oncology* **11**, 196–197.

Garcia-Carbonero, R. and Supko, J. G. (2002). Current perspectives on the clinical experience, pharmacology, and continued development of the camptothecins. *Clinical Cancer Research* **8**, 641–661.

Gerven, J. M.v., Moll, J. W., Bent, M. J.v.d., Bontenbal, M., Burg, M. E.v.d., Verweij, J. and Vecht, C. J. (1994). Paclitaxel (Taxol) induces cumulative mild neurotoxicity. *European Journal of Cancer* **30A**, 1074–1077.

Gianni, L., Kearns, C., Giani, A., Capri, G., Viganó, L., Lacatelli, A., Bonadonna, G. and Egorin, M. (1995). Nonlinear pharmacokinetics and metabolism of paclitaxel and its pharmacokinetic/pharmacodynamic relationships in humans. *Journal of Clinical Oncology* **13**, 180–190.

Gilmore, P. M., McCabe, N., Quinn, J. E., Kennedy, R. D., Gorski, J. J., Andrews, H. N., McWilliams, S., Carty, M., Mullan, P. B., Duprex, W. P., Liu, E. T., Johnston, P. G. *et al.* (2004). BRCA1 interacts with and is required for paclitaxel-induced activation of mitogen-activated protein kinase kinase kinase 3. *Cancer Research* **64**, 4148–4154.

Gore, M., ten Bokkel Huinink, W., Carmichael, J., Gordon, A., Davidson, N., Coleman, R., Spaczynski, M., Heron, J. F., Bolis, G., Malmstrom, H., Malfetano, J., Scarabelli, C. *et al.* (2001). Clinical evidence for topotecan-paclitaxel non-cross-resistance in ovarian cancer. *Journal of Clinical Oncology* **19**, 1893–1900.

Gottesman, M. M. (2002). Mechanisms of cancer drug resistance. *Annual Review of Medicine* **53**, 615–627.

Gottesman, M. M. and Pastan, I. (1993). Biochemistry of multidrug resistance mediated by the multidrug transporter. *Annual Review of Biochemistry* **62**, 385–427.

Gradishar, W. J., Tjulandin, S., Davidson, N., Shaw, H., Desai, N., Bhar, P., Hawkins, M. and O'Shaughnessy, J. (2005). Phase III trial of nanoparticle albumin-bound paclitaxel compared with polyethylated castor oil-based paclitaxel in women with breast cancer. *Journal of Clinical Oncology* **23**, 7794–7803.

Greenwald, P. (2002). Cancer chemoprevention. *British Medical Journal* **324**, 714–718.

Gueritte-Voegelein, F., Guenard, D., Lavelle, F., Le Goff, M. T., Mangatal, L. and Potier, P. (1991). Relationships between the structure of taxol analogues and their antimitotic activity. *Journal of Medicinal Chemistry* **34**, 992–998.

Guichard, S., Terret, C., Hennebelle, I., Lochon, I., Chevreau, P., Fretigny, E., Selves, J., Chatelut, E., Bugat, R. and Canal, P. (1999). CPT-11 converting carboxylesterase and topoisomerase activities in tumour and normal colon and liver tissues. *British Journal of Cancer* **80**, 364–370.

Haldar, S., Basu, A. and Croce, C. (1997). Bcl2 is the guardian of microtubule integrity. *Cancer Research* **57**, 229–233.

Hanahan, D. and Weinberg, R. A. (2000). The hallmarks of cancer. *Cell* **100**, 57–70.

Henderson, I. C., Berry, D. A., Demetri, G. D., Cirrincione, C. T., Goldstein, L. J., Martino, S., Ingle, J. N., Cooper, M. R., Hayes, D. F., Tkaczuk, K. H., Fleming, G., Holland, J. F. *et al.* (2003). Improved outcomes from adding sequential paclitaxel but not from escalating doxorubicin dose in an adjuvant chemotherapy regimen for patients with node-positive primary breast cancer. *Journal of Clinical Oncology* **21**, 976–983.

Henegar, K. E., Ashford, S. W., Baughman, T. A., Sih, J. C. and Gu, R. L. (1997). Practical asymmetric synthesis of (S)-4-ethyl-7,8-dihydro-4-hydroxy-1H-pyrano 3,4-f indolizine-3,6,10(4H)-trione, a key intermediate for the synthesis of irinotecan and other camptothecin analogs. *Journal of Organic Chemistry* **62**, 6588–6597.

Henningsson, A., Karlsson, M. O., Viganó, L., Gianni, L., Verweij, J. and Sparreboom, A. (2001). Mechanism-based pharmacokinetic model for paclitaxel. *Journal of Clinical Oncology* **19**, 4065–4073.

Herben, V. M. M., ten Bokkel Huinink, W. W., Dubbelman, A. C., Mandjes, I. A., Groot, Y., van Gortel-van Zomeren, D. M. and Beijnen, J. H. (1997). Phase I and pharmacological study of sequential intravenous topotecan and oral etoposide. *British Journal of Cancer* **76**, 1500–1508.

Herben, V. M. M., Huinink, W. W. T., Schellens, J. H. M. and Beijnen, J. H. (1998). Clinical pharmacokinetics of camptothecin topoisomerase I inhibitors. *Pharmacy World & Science* **20**, 161–172.

Herben, V. M. M., Rosing, H., ten Bokkel Huinink, W. W., van Zomeren, D. M., Batchelor, D., Doyle, E., Beusenberg, F. D., Beijnen, J. H. and Schellens, J. H. (1999). Oral topotecan: Bioavailability and effect of food co-administration. *British Journal of Cancer* **80**, 1380–1386.

Herzog, T. J. (2002). Update on the role of topotecan in the treatment of recurrent ovarian cancer. *The Oncologist* **7**, 3–10.

Hosomi, Y., Shibuya, M., Niho, S., Ichinose, Y., Kiura, K., Sakai, H., Takeda, K., Kudo, S., Eguchi, K., Matsui, K., Masuda, N., Ando, M. *et al.* (2011). Phase II study of topotecan with cisplatin in Japanese patients with small cell lung cancer. *Anticancer Research* **31**, 3449–3456.

Hou, C.-C., Chen, Y.-P., Wu, J.-H., Huang, C.-C., Wang, S.-Y., Yang, N.-S. and Shyur, L.-F. (2007). A galactolipid possesses novel cancer chemopreventive effects by suppressing inflammatory mediators and mouse B16 melanoma. *Cancer Research* **67**, 6907–6915.

Hsan, K. M., Chen, C.-C. and Shyur, L.-F. (2010). Current research and development of chemotherapeutic agents for melanoma. *Cancers* **2**, 397–419.

Hsiang, Y. H., Hertzberg, R., Hecht, S. and Liu, L. F. (1985). Camptothecin induces protein-linked DNA breaks via mammalian DNA topoisomerase-I. *Journal of Biological Chemistry* **260**, 4873–4878.

Huang, T. T., Wuerzberger-Davis, S. M., Seufzer, B. J., Shumway, S. D., Kurama, T., Boothman, D. A. and Miyamoto, S. (2000). NF-kB activation by camptothecin: A linkage between nuclear DNA damage and cytoplasmic signaling events. *Journal of Biological Chemistry* **275**, 9501–9509.

Huang, C.-C., Lo, C.-P., Chiu, C.-Y. and Shyur, L.-F. (2010). Deoxyelephantopin, a novel multifunctional agent suppresses mammary tumor growth and lung metastasis and doubles survival time in mice. *British Journal of Pharmacology* **159**, 856–871.

Ibrahim, N. K., Desai, N., Legha, S., Soon-Shiong, P., Theriault, R. L., Rivera, E., Esmaeli, B., Ring, S. E., Bedikian, A., Hortobagyi, G. N. and Ellerhorst, J. A. (2002). Phase I and pharmacokinetic study of ABI-007, a Cremophor-free, protein-stabilized, nanoparticle formulation of paclitaxel. *Clinical Cancer Research* **8**, 1038–1044.

Ichinose, Y., Seto, T., Nishiwaki, Y., Kiura, K., Sakai, H., Yokoyama, A., Segawa, Y., Ando, M. and Watanabe, K. (2011). Phase I study of topotecan and cisplatin in patients with small cell lung cancer. *Japanese Journal of Clinical Oncology* **41**, 197–203.

Ikeguchi, M., Arai, Y., Maeta, Y., Ashida, K., Katano, K. and Wakatsuki, T. (2011). Topoisomerase I expression in tumors as a biological marker for CPT-11 chemosensitivity in patients with colorectal cancer. *Surgery Today* **41**, 1196–1199.

Iyer, L., King, C. D., Whitington, P. F., Green, M. D., Roy, S. K., Tephly, T. R., Coffman, B. L. and Ratain, M. J. (1998). Genetic predisposition to the metabolism of irinotecan (CPT-11). *Journal of Clinical Investigation* **101**, 847–854.

Jemal, A., Siegel, R., Xu, J. and Ward, E. (2010). Cancer statistics, 2010. *CA: A Cancer Journal for Clinicians* **60**, 277–300.

Jennewein, S., Wildung, M. R., Chau, M., Walker, K. and Croteau, R. (2004). Random sequencing of an induced Taxus cell cDNA library for identification of clones involved in Taxol biosynthesis. *Proceedings of the National Academy of Sciences* **101**, 9149–9154.

Jordan, M. A. and Wilson, L. (2004). Microtubules as a target for anticancer drugs. *Nature Reviews. Cancer* **4**, 253–265.

Katzmann, D. J., Hallstrom, T. C., Voet, M., Wysock, W., Golin, J., Volckaert, G. and Moylerowley, W. S. (1995). Expression of an ATP-binding cassette transporter-encoding gene (YOR1) is required for oligomycin resistance in *Saccharomyces cerevisiae*. *Molecular and Cellular Biology* **15**, 6875–6883.

Kaufman, P. B., Pennacchini, L., McKenzie, M. and Hoyt, J. E. (1998). Good and bad uses of these compounds by humans. *In* Natural Products from Plants, (P. B. Kaufman, L. J. Cseke, S. Warber, J. A. Duke and H. L. Brielmann, eds.), pp. 123–155. CRC Press LLC, Boca Raton, FL.

Kehrer, D. F. S., Yamamoto, W., Verweij, J., de Jonge, M. J. A., de Bruijn, P. and Sparreboom, A. (2000). Factors involved in prolongation of the terminal disposition phase of SN-38: Clinical and experimental studies. *Clinical Cancer Research* **6**, 3451–3458.

Kehrer, D. F. S., Soepenberg, O., Loos, W. J., Verweij, J. and Sparreboom, A. (2001). Modulation of camptothecin analogs in the treatment of cancer: A review. *Anticancer Drugs* **12**, 89–105.

Kelling, J., Sullivan, K., Wilson, L. and Jordan, M. A. (2003). Suppression of centromere dynamics by Taxol® in living osteosarcoma cells. *Cancer Research* **63**, 2794–2801.

Khanna, R., Morton, C. L., Danks, M. K. and Potter, P. M. (2000). Proficient metabolism of CPT-11 by a human intestinal carboxylesterase. *Cancer Research* **60**, 4725–4728.

Kinghorn, A. D., Pan, L., Fletcher, J. N. and Chai, H. (2011). The relevance of higher plants in lead compound discovery programs. *Journal of Natural Products* **74**, 1539–1555.

Kingston, D. G. I. (2007). The shape of things to come: Structural and synthetic studies of taxol and related compounds. *Phytochemistry* **68**, 1844–1854.

Kingston, D., Jagtop, P., Yuan, Y. and Samala, L. (2002). The chemistry of Taxol and related taxoids. *Progress in the Chemistry of Organic Natural Products* **84**, 56–225.

Kollmannsberger, C., Mross, K., Jakob, A., Kanz, L. and Bokemeyer, C. (1999). Topotecan—A novel topoisomerase I inhibitor: Pharmacology and clinical experience. *Oncology (Williston Park, N.Y.)* **56**, 1–12.

Konkimalla, V. B. and Efferth, T. (2010). Inhibition of epidermal growth factor receptor-overexpressing cancer cells by camptothecin, 20-(N, N-diethyl) glycinate. *Biochemical Pharmacology* **80**, 39–49.

Kumar, G., Walle, U., Bhalla, K. and Walle, T. (1993). Binding of taxol to human plasma, albumin and alpha 1-acid glycoprotein. *Research Communications in Chemical Pathology and Pharmacology* **80**, 337–344.

Labbozzetta, M., Notarbartolo, M., Poma, P., Maurici, A., Inguglia, L., Marchetti, P., Rizzi, M., Baruchello, R., Simoni, D. and D'Alessandro, N. (2009). Curcumin as a possible lead compound against hormone-independent, multidrug-resistant breast cancer. *Annals of the New York Academy of Sciences* **1155**, 278–283.

Lee, W.-L., Wen, T.-N., Shiau, J.-Y. and Shyur, L.-F. (2010). Differential proteomic profiling identifies novel molecular targets of paclitaxel and phytoagent deoxyelephantopin against mammary adenocarcinoma cells. *Journal of Proteome Research* **9**, 237–253.

Lee, W. K., Bode, A. M. and Dong, Z. (2011). Molecular targets of phytochemicals for cancer prevention. *Nature Reviews. Cancer* **11**, 211–218.

Leistner, E. (2005). Arzneimittel aus der natur. Die biologie der taxane. *Pharmazie in Unserer Zeit* **34**, 98–103.

Levine, A. J. (1997). p53, the cellular gatekeeper review for growth and division. *Cell* **88**, 323–331.

Li, Q. Y., Zu, Y. G., Shi, R. Z. and Yao, L. P. (2006). Review camptothecin: Current perspectives. *Current Medicinal Chemistry* **13**, 2021–2039.
Li, Y., Goodwin, C. R., Sang, Y., Rosen, E., Laterra, J. and Xia, S. (2009). Camptothecin and Fas receptor agonists synergistically induce medulloblastoma cell death: ROS-dependent mechanisms. *Anti-Cancer Drugs* **20**, 770–778.
Lilenbaum, R., Ratain, M., Miller, A., Hargis, J., Hollis, D., Rosner, G., O'Brian, S., Brewster, L., Green, M. and Schilsky, R. (1995). Phase I study of paclitaxel and topotecan in patients with advanced tumors: A cancer and leukemia group B study. *Journal of Clinical Oncology* **13**, 2230–2237.
Lin, A. and Karin, M. (2003). NF-kappaB in cancer: A marked target. *Seminars in Cancer Biology* **13**, 107–114.
Ling, Y. H., Tornos, C. and Perez-Soler, R. (1998). Phosphorylation of bcl-2 is a marker of M phase events and not a determinant of apoptosis. *Journal of Biological Chemistry* **273**, 18984–18991.
Ling, Y. H., Donato, N. J. and Perez-Soler, R. (2001). Sensitivity to topoisomerase I inhibitors and cisplatin is associated with epidermal growth factor receptor expression in human cervical squamous carcinoma ME180 sublines. *Cancer Chemotherapy and Pharmacology* **47**, 473–480.
Lipton, R., Apfel, S., Dutcher, J., Rosenberg, R., Kaplan, J., Berger, A., Einzig, A., Wiernik, P. and Schaumburg, H. (1989). Taxol produces a predominantly sensory neuropathy. *Neurology* **39**, 368–373.
Liu, L. F., Duan, P., Lin, C. T., D'arpa, P. and Wu, J. (1996). Mechanism of action of camptothecin. *Annals of the New York Academy of Sciences* **803**, 44–49.
Long, R. M., Lagisetti, C., Coates, R. M. and Croteau, R. B. (2008). Specificity of the *N*-benzoyl transferase responsible for the last step of Taxol biosynthesis. *Archives of Biochemistry and Biophysics* **477**, 384–389.
Longley, D. and Johnston, P. (2005). Molecular mechanisms of drug resistance. *The Journal of Pathology* **205**, 275–292.
Lopez-Lazaro, M. (2008). Anticancer and carcinogenic properties of curcumin: Considerations for its clinical development as a cancer chemopreventive and chemotherapeutic agent. *Molecular Nutrition & Food Research* **52**, S103–S127.
Lorence, A. and Nessler, C. L. (2004). Molecules of interest—Camptothecin, over four decades of surprising findings. *Phytochemistry* **65**, 2735–2749.
Lorenz, W., Reimann, H., Schmal, A., Dormann, P., Schwarz, B., Neugebauer, E. and Doenicke, A. (1977). Histamine release in dogs by Cremophor El and its derivatives: Oxethylated oleic acid is the most effective constituent. *Agents and Actions* **7**, 63–67.
Lu, A. J., Zhang, Z. S., Zheng, M. Y., Zou, H. J., Luo, X. M. and Jiang, H. L. (2007). 3D-QSAR study of 20 (S)-camptothecin analogs. *Acta Pharmacologica Sinica* **28**, 307–314.
Luch, A. (2009). Molecular, Clinical and Environmental Toxicology. Birkhäuser Verlag AG, Berlin.
Mabuchi, S., Ohmichi, M., Nishio, Y., Hayasaka, T., Kimura, A., Ohta, T., Kawagoe, J., Takahashi, K., Yada-Hashimoto, N., Seino-Noda, H., Sakata, M., Motoyama, T. *et al.* (2004). Inhibition of inhibitor of nuclear factor-kappaB phosphorylation increases the efficacy of paclitaxel in *in vitro* and *in vivo* ovarian cancer models. *Clinical Cancer Research* **10**, 7645–7654.
Mackler, N. J. and Pienta, K. J. (2005). Drug insight: Use of docetaxel in prostate and urothelial cancers. *Nature Clinical Practice Urology* **2**, 92–100.
Mangatal, L., Adeline, M. T., Guenard, D., Gueritte-Voegelein, F. and Potier, P. (1989). Application of the vicinal oxyamination reaction with asymmetric induction to the hemisynthesis of taxol and analogues. *Tetrahedron* **45**, 177–190.

Marsh, S. and Hoskins, J. M. (2010). Irinotecan pharmacogenomics. *Pharmacogenomics* **11**, 1003–1010.

Marsh, S. and McLeod, H. L. (2004). Pharmacogenetics of irinotecan toxicity. *Pharmacogenomics* **5**, 835–843.

Marupudi, N. I., Han, J. E., Li, K. W., Renard, V. M., Tyler, B. M. and Brem, H. (2007). Paclitaxel: A review of adverse toxicities and novel delivery strategies. *Expert Opinion on Drug Safety* **6**, 609–621.

Mathijssen, R. H. J., Loos, W. J., Verweij, J. and Sparreboom, A. (2002). Pharmacology of topoisomerase I inhibitors irinotecan (CPT-11) and topotecan. *Current Cancer Drug Targets* **2**, 103–123.

Mattern, M. R., Hofmann, G. A., Polsky, R. M., Funk, L. R., McCabe, F. L. and Johnson, R. K. (1993). *In vitro* and *in vivo* effects of clinically important camptothecin analogs on multidrug-resistant cells. *Oncology Research* **5**, 467–474.

McChesney, J. D., Venkataraman, S. K. and Henri, J. T. (2007). Plant natural products: Back to the future or into extinction? *Phytochemistry* **68**, 2015–2022.

McCoy, M. (2004). Lining up to make a cancer drug. *Chemical & Engineering News* **82**, 12–14.

McGrogan, B. T., Gilmartin, B., Carney, D. N. and McCann, A. (2008). Taxanes, microtubules and chemoresistant breast cancer. *Biochimica et Biophysica Acta* **1785**, 96–132.

Mielke, S., Mross, K., Gerds, T., Schmidt, A., Wäsch, R., Berger, D., Lange, W. and Behringer, D. (2003). Comparative neurotoxicity of weekly non-break paclitaxel infusions over 1 versus 3 h. *Anti-Cancer Drugs* **14**, 785–792.

Moertel, C. G., Hahn, R. G., Schutt, A. J. and Reitemei, R. (1972). Phase II study of camptothecin (NSC-100880) in treatment of advanced gastrointestinal cancer. *Cancer Chemotherapy Reports* **56**(1), 95–101.

Montazeri, A., Boucaud, M., Lokiec, F., Pinguet, F., Culine, S., Deporte-Fety, R., Albin, N., Laguerre, B., Goupil, A., Bugat, R., Canal, P. and Chatelut, E. (2000). Population pharmacokinetics of topotecan: Intraindividual variability in total drug. *Cancer Chemotherapy and Pharmacology* **46**, 375–381.

Mullan, P. B., Quinn, J. E., Gilmore, P. M., McWilliams, S., Andrews, H., Gervin, C., McCabe, N., McKenna, S., White, P., Song, Y.-H., Maheswaran, S., Liu, E. *et al.* (2001). BRCA1 and GADD45 mediated G2/M cell cycle arrest in response to antimicrotubule agents. *Oncogene* **20**, 6123–6131.

Nakagawa, H., Saito, H., Ikegami, Y., Aida-Hyugaji, S., Seigo, S. and Ishikawa, T. (2006). Molecular modeling of new camptothecin analogues to circumvent ABCG2-mediated drug resistance in cancer. *Cancer Letters* **234**, 81–89.

Nakatomi, K., Yoshikawa, M., Oka, M., Ikegami, Y., Hayasaka, S., Sano, K., Shiozawa, K., Kawabata, S., Soda, H., Ishikawa, T., Tanabe, S. and Kohno, S. (2001). Transport of 7-ethyl-10-hydroxycamptothecin (SN-38) by breast cancer resistance protein ABCG2 in human lung cancer cells. *Biochemical and Biophysical Research Communications* **288**, 827–832.

Nelson, W. G. and Kastan, M. B. (1994). DNA strand breaks: The DNA template alterations that trigger p53-dependent DNA damage response pathways. *Molecular and Cellular Biology* **14**, 1815–1823.

Nicoletti, M. I., Valoti, G., Giannakakou, P., Zhan, Z., Kim, J.-H., Lucchini, V., Landoni, F., Mayo, J. G., Giavazzi, R. and Fojo, T. (2001). Expression of beta-tubulin isotypes in human ovarian carcinoma xenografts and in a subpanel of human cancer cell lines from the NCI-anticancer drug screen: Correlation with sensitivity to microtubule active agents. *Clinical Cancer Research* **7**, 2912–2922.

Oakman, C., Pestrin, M., Cantisani, E., Licitra, S., DeStefanis, M., Biganzoli, L. and Di Leo, A. (2009). Adjuvant chemotherapy—The dark side of clinical trials. Have we learnt more? *Breast* **18**, S18–S24.
Oberlies, N. H. and Kroll, D. J. (2004). Camptothecin and Taxol: Historic achievements in natural products research. *Journal of Natural Products* **67**, 129–135.
Oostendorp, L. J. M., Stalmeier, P. F., Pasker-de Jong, P. C., Van der Graaf, W. T. and Ottevanger, P. B. (2010). Systematic review of benefits and risks of second-line irinotecan monotherapy for advanced colorectal cancer. *Anti-Cancer Drugs* **21**, 749–758.
Orr, G. A., Verdier-Pinard, P., McDaid, H. and Horwitz, S. B. (2003). Mechanisms of Taxol resistance related to microtubules. *Oncogene* **22**, 7280–7295.
Osheroff, N. (1989). Biochemical basis for the interactions of type-I and type-II topoisomerases with DNA. *Pharmacology and Therapeutics* **41**, 223–241.
Ozols, R. F. (2002). Future directions in the treatment of ovarian cancer. *Seminars in Oncology* **29**, 32–42.
Page, C., Lin, H. J., Jin, Y., Castle, V. P., Nunez, G., Hunang, M. and Lin, J. (2000). Overexpression of Akt/AKT can modulate chemotherapy-induced apoptosis. *Anticancer Research* **20**, 407–416.
Park, E. J. and Pezzuto, J. M. (2002). Botanicals in cancer chemoprevention. *Cancer and Metastasis Reviews* **21**, 231–255.
Patel, N., Chatterjee, S., Vrbanac, V., Chung, I., Mu, C., Olsen, R., Waghome, C. and Zetter, B. (2010). Rescue of paclitaxel sensitivity by repression of Prohibitin1 in drug-resistant cancer cells. *Proceedings of the National Academy of Sciences* **107**, 2503–2508.
Peeters, M., Price, T. J., Cervantes, A., Sobrero, A. F., Ducreux, M., Hotko, Y., Andre, T., Chan, E., Lordick, F., Punt, C. J. A., Strickland, A. H., Wilson, G. *et al.* (2010). Randomized phase III study of panitumumab with fluorouracil, leucovorin, and irinotecan (FOLFIRI) compared with FOLFIRI alone as second-line treatment in patients with metastatic colorectal cancer. *Journal of Clinical Oncology* **28**, 4706–4713.
Peltier, A. and Russell, J. (2002). Recent advances in drug-induced neuropathies. *Current Opinion in Neurology* **15**, 633–638.
Persinos, E.E. (1990). Washington Insight, 15 September 1990.
Pietta, P. G. (2000). Flavonoids as antioxidants. *Journal of Natural Products* **63**, 1035–1042.
Pirker, R., Berzinec, P., Brincat, S., Kasan, P., Ostoros, G., Pesek, M., Plate, S., Purkalne, G., Rooneem, R., Skrickova, J., Stanculeanu, D., Timcheva, C. *et al.* (2010). Therapy of small cell lung cancer with emphasis on oral topotecan. *Lung Cancer* **70**, 7–13.
Plosker, G. L. and Keam, S. J. (2006). Trastuzumab: A review of its use in the management of HER2-positive metastatic and early-stage breast cancer. *Drugs* **66**, 449–475.
Pommier, Y. (2006). Topoisomerase I inhibitors: Camptothecins and beyond. *Nature Reviews. Cancer* **6**, 789–802.
Quinn, J. E., Kennedy, R. D., Mullan, P. B., Gilmore, P. M., Carty, M., Johnston, P. G. and Harkin, D. P. (2003). BRCA1 functions as a differential modulator of chemotherapy-induced apoptosis. *Cancer Research* **63**, 6221–6228.
Raskin, I., Ribnicky, D. M., Momarnytsky, S., Ilic, N., Poulev, A., Borisjuk, N., Brinker, A., Moreno, D. A., Ripoll, C., Yakoby, N., O'Neal, J. M., Cornwell, T. *et al.* (2002). Plants and human health in the twenty-first century. *Trends in Biotechnology* **20**, 522–531.
Raveh-Amit, H., Hai, N., Rotem-Dai, N., Shahaf, G., Gopas, J. and Livneh, E. (2011). Protein kinase C eta activates NF-kappaB in response to

camptothecin-induced DNA damage. *Biochemical and Biophysical Research Communications* **412**, 313–317.

Redinbo, M. R., Champoux, J. J. and Hol, W. G. J. (2000). Novel insights into catalytic mechanism from a crystal structure of human topoisomerase I in complex with DNA. *Biochemistry* **39**, 6832–6840.

Reid, R. J. D., Kauh, E. A. and Bjornsti, M. A. (1997). Camptothecin sensitivity is mediated by the pleiotropic drug resistance network in yeast. *Journal of Biological Chemistry* **272**, 12091–12099.

Rieder, C. L. and Maiato, H. (2004). Stuck in division or passing through: What happens when cells cannot satisfy the spindle assembly checkpoint. *Developmental Cell* **7**, 637–651.

Riemsma, R., Simons, J. P., Bashir, Z., Gooch, C. L. and Kleijnen, J. (2010). Systematic review of topotecan (Hycamtin) in relapsed small cell lung cancer. *BMC Cancer* **10**, 436–447.

Roos, W. and Kaina, B. (2006). DNA damage-induced cell death by apoptosis. *Trends in Molecular Medicine* **12**, 440–450.

Rosing, H., Herben, V. M. M., van Gortel-van Zomeren, D. M., Hop, E., Kettenes-van den Bosch, J. J., ten Bokkel Huinink, W. W. and Beijnen, J. H. (1997). Isolation and structural confirmation of *N*-desmethyl topotecan, a metabolite of topotecan. *Cancer Chemotherapy and Pharmacology* **39**, 498–504.

Rosing, H., van Zomeren, D. M., Doyle, E., Bult, A. and Beijnen, J. H. (1998). O-glucuronidation, a newly identified metabolic pathway for topotecan and *N*-desmethyl topotecan. *Anti-Cancer Drugs* **9**, 587–592.

Rougier, P., Van Cutsem, E., Bajetta, E., Niederle, N., Possinger, K., Labianca, R., Navarro, M., Morant, R., Bleiberg, H., Wils, J., Awad, L., Herait, P. *et al.* (1998). Randomised trial of irinotecan versus fluorouracil by continuous infusion after fluorouracil failure in patients with metastatic colorectal cancer. *Lancet* **352**, 1407–1412.

Rowinsky, E. K. and Donehower, R. C. (1995). Paclitaxel (taxol). *The New England Journal of Medicine* **332**, 1004–1014.

Rowinsky, E., Chaudhry, V., Cornblath, D. and Donehower, R. (1993a). Neurotoxicity of taxol. *Journal of the National Cancer Institute. Monographs* **15**, 107–115.

Rowinsky, E., Chaudhry, V., Forastiere, A., Sartorius, S., Ettinger, D., Grochow, L., Lubejko, B., Cornblath, D. and Donehower, R. (1993b). Phase I and pharmacologic study of paclitaxel and cisplatin with granulocyte colony-stimulating factor: Neuromuscular toxicity is dose-limiting. *Journal of Clinical Oncology* **11**, 2010–2020.

Rudolf, E., Rudolf, K. and Cervinka, M. (2011). Camptothecin induces p53-dependent and -independent apoptogenic signaling in melanoma cells. *Apoptosis* **16**, 1165–1176.

Sahenk, Z., Barohn, R., New, P. and Mendell, J. (1994). Taxol neuropathy. Electrodiagnostic and sural nerve biopsy findings. *Archives of Neurology* **51**, 726–729.

Saliba, F., Hagipantelli, R., Misset, J., Bastian, G., Vassal, G., Bonnay, M., Herait, P., Cote, C., Mahjoubi, M., Mignard, D. and Cvitkovic, E. (1998). Pathophysiology and therapy or irinotecan-induced delayed-onset diarrhea in patients with advanced colorectal cancer: A prospective assessment. *Journal of Clinical Oncology* **16**, 2745–2751.

Saltz, L. B., Cox, J. V., Blanke, C., Rosen, L. S., Fehrenbacher, L., Moore, M. J., Maroun, J. A., Ackland, S. P., Locker, P. K., Pirotta, N., Elfring, G. L. and Miller, L. L. (2000). Irinotecan plus fluorouracil and leucovorin for metastatic colorectal cancer. *The New England Journal of Medicine* **343**, 905–914.

Santos, A., Zanetta, S., Cresteil, T., Deroussent, A., Pein, F., Raymond, E., Vernillet, L., Risse, M. L., Boige, V., Gouyette, A. and Vassal, G. (2000). Metabolism of irinotecan (CPT-11) by CYP3A4 and CYP3A5 in humans. *Clinical Cancer Research* **6**, 2012–2020.

Sarkar, F. H. and Li, Y. (2006). Using chemopreventive agents to enhance the efficacy of cancer therapy. *Cancer Research* **66**, 3347–3350.

Sarkar, F., Li, Y., Wang, Z. and Kong, D. (2009). Cellular signaling perturbation by natural products. *Cellular Signalling* **21**, 1541–1547.

Sasaki, Y., Hakusui, H., Mizuno, S., Morita, M., Miya, T., Eguchi, K., Shinkai, T., Tamura, T., Ohe, Y. and Saijo, N. (1995). A pharmacokinetic and pharmacodynamic analysis of CPT-11 and its active metabolite SN-38. *Japanese Journal of Cancer Research* **86**, 101–110.

Schiff, P. B. and Horwitz, S. B. (1980). Taxol stabilizes microtubules in mouse fibroblast cells. *Proceedings of the National Academy of Sciences* **77**, 1561–1565.

Schiff, P. B., Fant, J. and Horwitz, S. B. (1979). Promotion of microtubule assembly *in vitro* by taxol. *Nature* **277**, 665–667.

Schiller, J., Kim, K., Hutson, P., DeVore, R., Glick, J., Stewart, J. and Johnson, D. (1996). Phase II study of topotecan in patients in extensive-stage small-cell carcinoma of the lung: An eastern cooperative oncology group trial. *Journal of Clinical Oncology* **14**, 2345–2352.

Schmitz, G., Langmann, T. and Heimerl, S. (2001). Role of ABCG1 and other ABCG family members in lipid metabolism. *Journal of Lipid Research* **42**, 1513–1520.

Scuteri, A., Nicolini, G., Miloso, M., Bossi, M., Cavaketti, G., Windebank, A. J. and Tredici, G. (2006). Paclitaxel toxicity in post-mitotic dorsal root ganglion (DRG) cells. *Anticancer Research* **26**, 1065–1070.

Shao, R. G., Cao, C. X., Zhang, H., Kohn, K. W., Wold, M. S. and Pommier, Y. (1999). Replication-mediated DNA damage by camptothecin induces phosphorylation of RPA by DNA-dependent protein kinase and dissociates RPA:DNA-PK complexes. *EMBO Journal* **18**, 1397–1406.

Shitashige, M., Toi, M., Yano, T., Shibata, M., Matsuo, T. and Shibasaki, F. (2001). Dissociation of Bax from a Bcl-2/Bax heterodimer triggered by phosphorylation of serine 70 of Bcl-2. *Journal of Biological Chemistry* **130**, 741–748.

Shyur, L.-F. and Yang, N.-S. (2008). Metabolomics for phytomedicine research and drug development. *Current Opinions in Chemical Biology* **12**, 66–71.

Simizu, S., Takada, M., Umezawa, K. and Imoto, M. (1998). Requirement of caspase-3(-like) protease-mediated hydrogen peroxide production for apoptosis induced by various anticancer drugs. *Journal of Biological Chemistry* **273**, 26900–26907.

Simpson, D. and Plosker, G. (2004). Paclitaxel: As adjuvant or neoadjuvant therapy in early breast cancer. *Drugs* **64**, 1839–1847.

Slatter, J. G., Schaaf, L. J., Sams, J. P., Feenstra, K. L., Johnson, M. G., Bombardt, P. A., Cathcart, K. S., Verburg, M. T., Perarson, L. K., Compton, L. D., Miller, L. L., Baker, D. S. *et al.* (2000). Pharmacokinetics, metabolism, and excretion of irinotecan (CPT-11) following i.v. infusion of [14C]CPT-11 in cancer patients. *Drug Metabolism and Disposition* **28**, 423–433.

Snyder, J. P., Nettles, J. H., Cornett, B., Downing, K. H. and Nogales, E. (2001). The binding conformation of Taxol in beta-tubulin: A model based on electron crystallographic density. *Proceedings of the National Academy of Sciences* **98**, 5312–5316.

Sonnichsen, D. and Relling, M. (1994). Clinical pharmacokinetics of paclitaxel. *Clinical Pharmacokinetics* **27**, 256–269.

Sorger, P. K., Dobles, M., Tournebizet, R. and Hymant, A. A. (1997). Coupling cell division and cell death to microtubule dynamics. *Current Opinion in Cell Biology* **9**, 807–814.

Sparreboom, A., de Jonge, M. J. A., de Bruijn, P., Brouwer, E., Nooter, K., Loos, W. J., van Alphen, R. J., Mathijssen, R. H. J., Stoter, G. and Verweij, J. (1998). Irinotecan (CPT-11) metabolism and disposition in cancer patients. *Clinical Cancer Research* **4**, 2747–2754.

Sparreboom, A., Zuylen, L.v., Brouwer, E., Loos, W. J., Bruijn, P.d., Gelderblom, H., Pillay, M., Nooter, K., Stoter, G. and Verweij, J. (1999). Cremophor EL-mediated alteration of paclitaxel distribution in human blood: Clinical pharmacokinetic implications. *Cancer Research* **59**, 1454–1457.

Spencer, C. and Faulds, D. (1994). Paclitaxel. A review of its pharmacodynamic and pharmacokinetic properties and therapeutic potential in the treatment of cancer. *Drugs* **48**, 794–847.

Sperker, B., Backman, J. T. and Kroemer, H. K. (1997). The role of beta-glucuronidase in drug disposition and drug targeting in humans. *Clinical Pharmacokinetics* **33**, 18–31.

Spigel, D. R., Hainsworth, J. D., Gandhi, J. G., Gian, V. G., Peyton, J. D., West-Osterfield, K., Clark, B. L., Vazquez, E. R., Jones, S. F., Burris, H. A. and Greco, F. A. (2010). A phase II trial of carboplatin and weekly topotecan in the first-line treatment of patients with extensive stage small cell lung cancer. *Journal of Thoracic Oncology* **5**, 862–866.

Sporn, M. B. and Suh, N. (2002). Chemoprevention: An essential approach to controlling cancer. *Nature Reviews. Cancer* **2**, 537–543.

Sporn, M. B., Dunlop, N. M., Newton, D. L. and Smith, J. M. (1976). Prevention of chemical carcinogenesis by vitamin A and its synthetic analogs (retinoids). *Federation Proceedings* **35**, 1332–1338.

Stark, G. R. and Taylor, W. R. (2006). Control of the G2/M transition. *Molecular Biotechnology* **32**, 227–248.

Steller, H. (1995). Mechanisms and genes of cellular suicide. *Science* **267**, 1445–1449.

Stewart, Z. A., Mays, D. and Pietenpol, J. A. (1999). Defective G1-S cell cycle checkpoint function sensitizes cells to microtubule inhibitor-induced apoptosis. *Cancer Research* **59**, 3831–3837.

Suffness, M. (1993). Taxol: From discovery to therapeutic use. *In* Annual Reports in Medicinal Chemistry, (J. A. Bristol, ed.), pp. 305–314. Academic Press, San Diego.

Suffness, M. and Wall, M. E. (1995). Discovery and development of taxol. *In* Taxol: Science and Applications, (M. Suffness, ed.), pp. 1–25. CRC Press, Boca Raton.

Surh, Y. J. (2003). Cancer chemoprevention with dietary phytochemicals. *Nature Reviews. Cancer* **3**, 768–780.

Szebeni, J., Muggia, F. M. and Alving, C. R. (1998). Complement activation by cremophor EL as a possible contributor to hypersensitivity to paclitaxel: An in vitro study. *Journal of the National Cancer Institute* **90**, 300–306.

Takeya, Y. (2003). Plant tissue culture of taxoids. *In* Taxus: The Genus Taxus, (H. Itokawa and K. H. Lee, eds.), pp. 134–150. Taylor and Francis, London.

Tanizawa, A., Fujimori, A., Fujimori, Y. and Pommier, Y. (1994). Comparison of topoisomerase-I inhibition, DNA-damage, and cytotoxicity of camptothecin derivatives presently in clinical-trials. *Journal of the National Cancer Institute* **86**, 836–842.

Tellingen, O.v., Huizing, M., Panday, V. N., Schellens, J., Nooijen, W. and Beijnen, J. (1999). Cremophor EL causes (pseudo-) non-linear pharmacokinetics of paclitaxel in patients. *British Journal of Cancer* **81**, 330–335.

Thomas, C. J., Rahier, N. J. and Hecht, S. M. (2004). Camptothecin: Current perspectives. *Bioorganic & Medicinal Chemistry* **12**, 1585–1604.

Tishler, R. B., Schiff, P. B., Geard, C. R. and Hall, E. J. (1992). Taxol: A novel radiation sensitizer. *International Journal of Radiation Oncology, Biology, Physics* **22**, 613–617.

Tothova, E., Fricova, M., Stecova, N., Kafkova, A. and Elbertova, A. (2002). High expression of Bcl-2 protein in acute myeloid leukaemia cells is associated with poor response to chemotherapy. *Neoplasma* **49**, 141–144.

Tournigand, C., Andre, T., Achille, E., Lledo, G., Flesh, M., Mery-Mignard, D., Quinaux, E., Couteau, C., Buyse, M., Ganem, G., Landi, B., Colin, P. *et al.* (2004). FOLFIRI followed by FOLFOX6 or the reverse sequence in advanced colorectal cancer: A randomized GERCOR study. *Journal of Clinical Oncology* **22**, 229–237.

Tsao, Y. P., Darpa, P. and Liu, L. F. (1992). The involvement of active DNA synthesis in camptothecin-induced G2 arrest: Altered regulation of P34CDC2/cyclin B1. *Cancer Research* **52**, 1823–1829.

Tsao, A. S., Kim, E. S. and Hong, W. K. (2004). Chemoprevention of cancer. *CA: A Cancer Journal for Clinicians* **54**, 150–180.

Urasaki, Y., Laco, G. S., Pourquier, P., Takebayashi, Y., Kohlhagen, G., Gioffre, C., Zhang, H. L., Chatterjee, D., Pantazis, P. and Pommier, Y. (2001). Characterization of a novel topoisomerase I mutation from a camptothecin-resistant human prostate cancer cell line. *Cancer Research* **61**, 1964–1969.

Urist, M., Tanaka, T., Poyurovsky, M. and Prives, C. (2004). p73 induction after DNA damage is regulated by checkpoint kinases Chk1 and Chk2. *Genes & Development* **18**, 3041–3054.

Valero, V., Forbes, J., Pegram, M. D., Pienkowski, T., Eiermann, W., Minckwitz, G. v., Roche, H., Martin, M., Crown, J., Mackey, J. R., Fumoleau, P., Rolski, J. *et al.* (2011). Multicenter phase III randomized trial comparing docetaxel and trastuzumab with docetaxel, carboplatin, and trastuzumab as first-line chemotherapy for patients with HER2-gene-amplified metastatic breast cancer (BCIRG 007 study): Two highly active therapeutic regimens. *Journal of Clinical Oncology* **29**, 149–156.

Vallböhmer, D., Iqbal, S., Yang, D. Y., Rhodes, K. E., Zhang, W., Gordon, M., Fazzone, W., Schultheis, A. M., Sherrod, A. E., Danenberg, K. D. and Lenz, H. J. (2006). Molecular determinants of irinotecan efficacy. *International Journal of Cancer* **119**, 2435–2442.

van Cutsem, E., Kohne, C. H., Lang, I., Folprecht, G., Nowacki, M. P., Cascinu, S., Shchepotin, I., Maurel, J., Cunningham, D., Tejpar, S., Schlichting, M., Zubel, A. *et al.* (2011). Cetuximab plus irinotecan, fluorouracil, and leucovorin as first-Line treatment for metastatic colorectal cancer: Updated analysis of overall survival according to tumor KRAS and BRAF mutation status. *Journal of Clinical Oncology* **29**, 2011–2019.

Vara, J. A. F., Casado, E., Castro, J. D., Cejas, P., Belda-Iniesta, C. and Gonzalez-Baron, M. (2004). PI3K/Akt signaling pathway and cancer. *Cancer Treatment Reviews* **30**, 193–204.

Verweij, J., Clavelf, M. and Chevalier, B. (1994). Paclitaxel (Taxol$^{TM}$) and docetaxel (Taxotere$^{TM}$): Not simply two of a kind. *Annals of Oncology* **5**, 495–505.

von Pawel, J., Schiller, J. H., Shepherd, F. A., Fields, S. Z., Kleisbauer, J. P., Chrysson, N. G., Stewart, D. J., Clark, P. I., Palmer, M. C., Depierre, A., Carmichael, J., Krebs, J. B. *et al.* (1999). Topotecan versus cyclophosphamide, doxorubicin, and vincristine for the treatment of recurrent small-cell lung cancer. *Journal of Clinical Oncology* **17**, 658–667.

von Pawel, J., Gatzemeier, U., Pujol, J. L., Moreau, L., Bildat, S., Ranson, M., Richardson, G., Steppert, C., Riviere, A., Camlett, I., Lane, S. and Ross, G. (2001). Phase II comparator study of oral versus intravenous topotecan in patients with chemosensitive small-cell lung cancer. *Journal of Clinical Oncology* **19**, 1743–1749.

Wakelee, H., Ramalingam, S. and Belani, C. (2005). Docetaxel in advanced non-small cell lung cancer. *Expert Review of Anticancer Therapy* **5**, 13–24.

Wall, M. E. (1993). Camptothecin and taxol. *In* Chronicles of Drug Discovery, (D. Lednicer, ed.), pp. 153–165. American Chemical Society, Washington, DC.

Wall, M. E. and Wani, M. C. (1994). Taxol: Discovery to clinic. *In* Economic and Medicinal Plant Research, Vol. 6, (H. Wagner and N. R. Farnsworth, eds.), pp. 299–322. Academic Press, New York.

Wall, M. E. and Wani, M. C. (1996). Camptothecin and taxol: From discovery to clinic. *Journal of Ethnopharmacology* **51**, 239–254.

Wall, M. E., Wani, M. C., Cook, C. E., Palmer, K. H., McPhail, A. T. and Sim, G. A. (1966). Plant antitumor agents. I. The isolation and structure of camptothecin, a novel alkaloidal leukemia and tumor inhibitor from *Camptotheca acuminata*. *Journal of the American Chemical Society* **88**, 3888–3890.

Wall, M., Wani, M. and Taylor, H. (1976). Isolation and chemical characterization of antitumor agents from plants. *Cancer Treatment Report* **60**, 1011–1030.

Wang, T. H., Wang, H. S., Ichijo, H., Giannakakoui, P., Foster, J. S., Fojoi, T. and Wimalasena, J. (1998). Microtubule-interfering agents activate c-Jun N-terminal kinase/stress-activated protein kinase through both Ras and apoptosis signal-regulating kinase pathways. *Journal of Biological Chemistry* **279**, 4928–4936.

Wang, T. H., Popp, D. M., Wang, H. S., Saitohi, M., Mural, J. G., Henley, D. C., Ichijoi, H. and Wimalasena, J. (1999). Microtubule dysfunction induced by paclitaxel initiates apoptosis through both c-Jun N-terminal Kinase (JNK)-dependent and -independent pathways in ovarian cancer cells. *Journal of Biological Chemistry* **274**, 8208–8216.

Wang, T. H., Wang, H. S. and Soong, Y. K. (2000). Paclitaxel-induced cell sheath: Where the cell cycle and apoptosis come together. *Cancer* **88**, 2619–2628.

Wang, C.-Y., Staniforth, V., Chiao, M.-T., Hou, C.-C., Wu, H.-M., Yeh, K.-C., Chen, C.-H., Hwang, P.-I., Wen, T.-N., Shyur, L.-F. and Yang, N.-S. (2008). Genomics and proteomics of immune modulatory effects of a butanol fraction of *Echinacea purpurea* in human dendritic cells. *BMC Genomics* **9**, 479–498.

Wani, M. C., Taylor, H. L., Wall, M. E., Coggon, P. and McPhail, A. T. (1971). Plant antitumor agents. VI. The isolation and structure of Taxol, a novel antileukemic and antitumor agent from *Taxus brevifolia*. *Journal of the American Chemical Society* **93**, 2325–2327.

Weiss, R. B., Donehower, R. C., Wiernik, P. H., Ohnuma, T., Gralla, R. J., Trump, D. L., Baker, J. R. J., Echo, D. A. V., Hoff, D. D. V. and Leyland-Jones, B. (1990). Hypersensitivity reactions from taxol. *Journal of Clinical Oncology* **8**, 1263–1268.

Wink, M. (2003). Evolution of secondary metabolites from an ecological and molecular phylogenetic perspective. *Phytochemistry* **64**, 3–19.

Xia, S., Rosen, E. M. and Laterra, J. (2005). Sensitization of glioma cells to Fas-dependent apoptosis by chemotherapy-induced oxidative stress. *Cancer Research* **65**, 5248–5255.

Yabu, K., Masumoto, S., Yamasaki, S., Hamashima, Y., Kanai, M., Du, W., Curran, D. P. and Shibasaki, M. (2001). Switching enantiofacial selectivities using one chiral source: Catalytic enantioselective synthesis of the key intermediate for (20S)-camptothecin family by (S)-selective cyanosilylation of ketones. *Journal of the American Chemical Society* **123**, 9908–9909.

Yabu, K., Masumoto, S., Kanai, M., Curran, D. P. and Shibasaki, M. (2002). Studies toward practical synthesis of (20S)-camptothecin family through catalytic enantioselective cyanosilylation of ketones: Improved catalyst efficiency by ligand-tuning. *Tetrahedron Letters* **43**, 2923–2926.

Yoshikawa, M., Ikegami, Y., Hayasaka, S., Ishi, K., Ito, A., Sano, K., Suzuki, T., Togawa, T., Yoshida, E., Soda, H., Oka, M., Kohno, S. *et al.* (2004). Novel camptothecin analogues that circumvent ABCG2-associated drug resistance in human tumor cells. *International Journal of Cancer* **110**, 921–927.

Zhao, E. and Mu, Q. (2011). Phytoestrogen biological actions on mammalian reproductive system and cancer growth. *Scientia Pharmaceutica* **79**, 1–20.

Zwenger, S. and Basu, C. (2008). Plant terpenoids: Applications and future potentials. *Biotechnology and Molecular Biology Reviews* **3**, 1–7.

# Effects of Beta-Glucans on Different Immune Cell Populations and Cancers

DANIEL MAN-YUEN SZE*,[1] AND GODFREY CHI-FUNG CHAN†

**Department of Health Technology and Informatics, The Hong Kong Polytechnic University, Kowloon, Hong Kong*
*†Department of Paediatrics & Adolescent Medicine, Li Ka Shing Faculty of Medicine, The University of Hong Kong, Pokfulam, Hong Kong*



[1]Corresponding author: E-mail: daniel.sze@polyu.edu.hk

Advances in Botanical Research, Vol. 62

0065-2296/12 $35.00
DOI: 10.1016/B978-0-12-394591-4.00011-8

# ABSTRACT

β-Glucan is one of the most abundant and widespread form of naturally occurring polysaccharides found inside the cell wall of bacteria, plants and fungi including the medicinal mushrooms. All forms of β-glucans have a basic structure of glucose polymers linked together by a 1→3 linear β-glycosidic chain core. They then differ from each other by their length of the core and the type and extent of branching structures of the two main forms of 1→4 or 1→6 glycosidic chains branching (Stone and Clarke, 1992). These branching assignments tend to be species specific, for instance, 1→6 side branches for fungal β-glucans in contrast to 1→4 side branches for bacterial β-glucans. There are insoluble or particulate forms versus the aqueous soluble forms of β-glucans which act on different target receptors. Various conformations such as triple helix, single helix or random coils also exist. The correlation between β-glucans structure and their biological potency has been elaborated in one of the recent reviews (Barsanti *et al.*, 2011); our review will mainly focus at how β-glucans affect the diverse immunological cells and cytokines, especially those related to anti-cancer action. There are mounting evidences suggesting β-glucans as potent immunomodulators on both innate and adaptive immunity. After our previous review of the immunomodulatory and anti-cancer effects of β-glucans (Chan *et al.*, 2009), there emerged a rich resource of studies on multi-directional immunomodulatory effects of β-glucans. The findings are complex and sometimes conflicting especially to the immune cell subsets. This review therefore aims to provide an up-to-date information on the effects of β-glucan to different immune cell populations and then postulate some possible translational strategies to utilize such knowledge in both anti-cancer research and clinical immunotherapeutic application.


# I. MULTI-DIRECTIONAL IMMUNOMODULATORY ANTI-CANCER EFFECTS

## A. SOURCES OF β-GLUCANS AND THE SPECIFIC RECEPTORS ON IMMUNE CELL SURFACES

The heterogeneous β-glucans are divided primarily into subtypes of plant, bacteria, yeast and medicinal mushrooms for easier categorization (Table I). In brief, the bacterial curdlan has a basic structure of linear β-(1,3) glucan without branching, while the yeast glucan has various numbers of β-(1,6) side chains. Medicinal mushrooms have similar basic structure as yeast with β-(1,3) glucan backbone plus β-(1,6) side chains; but in some mushrooms, other variations have been observed including triple helix form of lentinan, protein components and β-(1,4) linear residues in krestin and heteroglucans in AE2. Plant β-glucans, in contrast, consist of primarily linear β-(1,4) residues with short stretches of β-(1,3) residues. The receptors for these great varieties of β-glucans will be briefly introduced in this section.

Our group found that β-glucans can induce human peripheral blood mononuclear cells proliferation (Chan *et al.*, 2007). It can also enhance phenotypic and functional maturation of monocyte-derived dendritic cells

TABLE I

*Different Sources of β-Glucan and Their Related Immunological Activities*

| Source | Structure | Immunological activities | Reference |
|---|---|---|---|
| **Plant** | | | |
| Oat | Primarily linear with large regions of β-(1,4) linkages separating shorter stretches of β-(1,3) structures | Metastatic spread of injected B16 melanoma cells decrease may be mediated in part by an increase in macrophage cytotoxicity | Murphy *et al.* (2004) |
| Barley | | Tumour suppressed due to anti-bodies, opsonization and classical complement pathway activation glucan | Akramiene *et al.* (2009), Cheung and Modak (2002) |
| **Bacteria** | | | |
| Curdlan | Linear β-(1,3)-glucan without branching | Stimulate Th1, Th17, CTL priming and differentiation | LeibundGut-Landmann *et al.* (2007, 2008) |
| | | Convert Treg into IL-17-producing T cells via Dectin-1 receptor | Osorio *et al.* (2008) |
| | | Curdlan prime Th17 cells, while poly I:C induced a strong IL-12-independent Th1 response | Joffre *et al.* (2010) |
| **Yeast** | | | |
| Zymosan | β-(1,3)-glucan with some β-(1,6)-linked side chains with additional β-(1,3) regions | Increased PGE2 release and TGF-β, PDGF and IL-8 | Lew *et al.* (1991), Noble *et al.* (1993) |
| *Saccharomyces cerevisiae* | | Only the particulate form activate DC and macrophages via Dectin-1 receptor | Qi *et al.* (2011) |
| | | Stimulation of the proliferation and activation of peripheral blood monocytes in patients with advanced breast cancer | Demir *et al.* (2007) |

(*continues*)

Table I (*continued*)

| Source | Structure | Immunological activities | Reference |
|---|---|---|---|
| **Mushroom** | | | |
| *Ganoderma lucidum* | Short β-(1,6)-linked branches coming off of the β-(1,3) backbone | Human PBMC proliferation and phenotypic and functional maturation of DCs with significant IL-12 and IL-10 production | Chan *et al.* (2007) |
| Lentinan | Purified from edible mushroom *Lentinus edodes* Berk. | Synergistic effects of Lentinan and IL-2, two-third spontaneously metastatic mice were completely cured. | Hamuro *et al.* (1994), Murata *et al.* (2002) |
| | | Synergistic effects of Lentinan and IL augmented IL-12, NO and reduction of IL-6 production. | |
| Krestin (PSK) | Purified from *Trametes versicolor* mushroom | PSK induced TNF-α and IL-6 secretion by wild type but not by TLR4-deficient peritoneal macrophages | Price *et al.* (2010) |
| AE2 | Purified from *Astraeus hygrometricus* | Macrophages from treated mice showed higher production of nitric oxide and interleukin-1 and increased phagocytic potential. | Mallick *et al.* (2010) |

(DCs) with significant IL-12 and IL-10 production. The action is, at least in part, through the Dectin-1 receptor. Dectin-1 has been considered as the primary germ-line-encoded pattern recognition receptor (Brown *et al.*, 2003). It is highly expressed on splenic and alveolar macrophages, blood monocytes, neutorphils, DCs and B and T cell subsets (Goodridge *et al.*, 2009). Basically, Dectin-1 receptors are found to have strong affinity for β-(1,3)-D-glucopyronosyl backbone, but not mixed, alternated (1,3)-β and (1,4)-β linkages backbone (Adams *et al.*, 2008). Interestingly, it was further clarified that Dectin-1 does not recognize all β-(1,3)-D-glucans equally, but the specific β-glucan recognition may require a backbone chain length of at least seven glucose subunits and at least one side-chain branch (Adams *et al.*, 2008). The importance of this β-glucan receptor will be further discussed in the next section under different immunological populations.

The second mechanism of β-glucan action is through the activated complement receptor 3 (CR3, CD11b/CD18), which is found on natural killer (NK) cells, neutrophils, monocytes and macrophages (Novak and Vetvicka, 2009). The CR3 pathway is responsible for opsonic recognition of β-glucan leading to phagocytosis and effector cell lysis. β-Glucans bind to the lectin domain of CR3 and prime it for binding to inactivated complement 3b (iC3b) on effector cells surface (Di Renzo *et al.*, 1991; Klein *et al.*, 1990). Thus the CR3 ligation in phagocytes will be triggered because of simultaneous recognition of iC3b and specific microbial β-glucan polysaccharides leading to phagocytosis and degranulation (Hong *et al.*, 2004). In contrast, such lysis or cytotoxicity will not be executed if NK cell or phagocyte CR3 engages to iC3b on red blood cells or tumour cells that lack CR3-binding membrane polysaccharides (Vetvicka *et al.*, 1996). Furthermore, it has been proposed that the orally administered β-(1,3)-D-glucan goes through an intermediate step in which macrophage process and deliver soluble β-(1,3)-D-glucan to the CR3 of granulocytes in the bone marrow (Chan *et al.*, 2009) and the CR3 of marrow monocytes and NK cells (Hong *et al.*, 2004).

Another major route of oral absorption of β-glucans may be through the extensive lymphoid network of gut-associated lymphoid tissue (GALT) that operates via two intertwined dynamic networks (Hong *et al.*, 2004; Rice *et al.*, 2005; Thompson *et al.*, 2010). GALT contains a great variety of immunological populations such as macrophages, DCs and various T and B subsets (Bebawy and Sze, 2008). For instance, the DCs can be further divided into at least five subsets: (i) DCs in the lamina propria of the $CD11b^+$, and $CD172a^+$ phenotype; (ii) DCs in the subepithelial dome of Peyer's patches of the $CD11b^-$, and $CD172a^+$ phenotype; (iii) DCs in the interfollicular region of Peyer's patches of the $CD11b^-$, and $CD172a^-$ phenotype; (iv) DCs in the mesenteric lymph node of the $CD11b^+$, and $CD172a^-$ phenotype and (v) plasmacytoid-type DCs of the CD11c intermediate, and CD4$^+$ phenotype. This diverse DC subset system arguing strongly whether any oversimplified and generalized conclusions about the effects of β-glucans on DCs as if it is a homogenous population is warranted. It further highlights the importance of applying multicolour flow cytometric analysis to define accurately the functional and phenotypic changes in various immune populations and subsets under the β-glucan challenges.

The third receptor of β-glucan has been known as scavenger receptors (Dushkin *et al.*, 1996); these receptors are found on the macrophages surface which can recognize and uptake macromolecules having negative charges such as oxidized low-density lipoprotein, foreign substances and waste materials. The last receptor of β-glucan is lactosylceramide (LacCer; Iwabuchi and Nagaoka, 2002; Zimmerman *et al.*, 1998). The LacCer receptor mainly involves in the initiation of superoxide generation inside the neutrophils. These two receptor types are mainly responsible for innate immune response and phagocytosis.

The relationship between TLR, in particularly TLR2 and TLR4, and β-glucans is an intriguing concept. Dectin-1 amplifies TLR2-dependent induction of cytokines in mouse models (Brown *et al.*, 2003). Furthermore, Dectin-1 has potent synergistic effects with both TLR2 and TLR4 in human PBMCs and macrophages, but such activation seemingly requires the presence of β-(1,6) side chains (Ferwerda *et al.*, 2008). Linear β-(1,3) glucan such as curdlan requires synthetic TLR2 ligand Pam3Cys and the ultrapure TLR4 ligand lipopolysaccharide to exert such action. These data suggest that TLR may not be a direct β-glucan target but has to coordinate with Dectin-1 for the synergistic action.

## II. EFFECTS OF β-GLUCANS ON VARIOUS TYPES OF IMMUNOLOGICAL POPULATIONS

### A. EFFECTS OF β-GLUCANS ON MACROPHAGES

It was found that whole yeast β-glucan particles derived from fungus *Saccharomyces cerevisiae* can stimulate cytokine release from macrophages (Li *et al.*, 2007). Such stimulatory effect is partially dependent on the TLR-related MyD-88 pathway but not affected in CD11b-deficient (CR3 deficient) mice. A recent study used the same yeast β-glucan via oral route to mice (Qi *et al.*, 2011) and they found that macrophage-mediated phagocytosis induced by β-glucan was completely abrogated in Dectin-1-deficient mice but was not affected in CD11b-deficient (CR3 deficient) mice. However, such

Dectin-1-dependent pathway is vital for particulate rather than soluble glucans. These two studies suggested that β-glucans-enhanced macrophage response is mediated by TLR and Dectin-1, but not CR3. The yeast β-glucan can also significantly increase both activated macrophages with high MHC Class II and granulocytes infiltration into the tumour microenvironment (Qi *et al.*, 2011).

Two types of peritoneal macrophages in mouse based on the cells' intracellular thiol redox equilibrium have been described by the Japanese group (Murata *et al.*, 2002). The "Reductive" type is with elevated intracellular glutathione (GSH) in comparison with the oxidative macrophages with reduced amount of GSH. Upon stimulation by the mushroom β-glucan Lentinan, there is a shift to increase in reductive macrophages; such cells demonstrate reduced release of prostaglandins, IL-10 and IL-6, but with increased capability to produce IL-12 and nitric oxide upon *in vitro* triggering. It appears that more evidence is pointing to the fact that macrophage is not a homogenous population. Triggering receptor expressed on myeloid cells-1 expressing intestinal macrophages (Schenk *et al.*, 2007) is one of such examples and it amplifies inflammatory responses by enhancing degranulation and secretion of proinflammatory mediators. $CD14^{+}$ intestinal macrophages that can release proinflammatory cytokines in inflammatory bowel diseases (Kamada *et al.*, 2008) are another example of such macrophage heterogeneity. Swirski (2011)has provided a concise up-to-date overall view of the development map of the macrophage family. Thus, it may be more meaningful if future studies related to the effects of β-glucan on macrophage can look into the individual response of different subsets of macrophages.

Price *et al.* (2010) compared the effects of three distinct mushroom β-glucan preparations on macrophages. These preparations comprise (i) polysaccharide Krestin (PSK), which is a hot-water extract from *Trametes versicolor*; (ii) a Reishi (or Lingzhi) extract from *Ganoderma lucidum* and (iii) scleroglucan, which is a highly purified extract from *Slerotium rolfsii*. They were compared for their potency in inducing cytokine productions from murine macrophages obtained from wild type and TLR4-deficient peritoneal macrophages. They found that both PSK and Reishi induce TNF-α secretion in a TLR4-dependent manner. On the other hand, Dectin-1 was found to mediate TNF-α induction by scleroglucan but does not mediate PSK or Reishi TNF-α induction. Since the scleroglucan has typical (1,3) and (1,6) linkages, the other two types have similar β-glucan composition but with mixed (1,3), (1,4) and (1,6) linkages; it is therefore possible that β-glucan structural differences or the presence of non-glucan constituents may have an impact on receptor binding and effects. This is an area that definitely demands much in-depth study in the future.

### B. EFFECTS OF β-GLUCANS ON NK CELLS

Yoon *et al.* demonstrated that increased NK cytotoxicity effect of yeast β-glucan-stimulated mouse splenoctyes against mouse NK-sensitive cell line YAC-1 cells. The researchers also found that the yeast β-glucan suppresses tumour growth and metastasis. They confirmed that this suppression was related to NK cells activation because pretreatment with anti-asialo GM1 serum that selectively eliminates NK cells enhanced the frequency of lung metastasis (Yoon *et al.*, 2008).

Mallick *et al.* used hot alkaline extract of mushroom β-glucan *Astraeus hygrometricus* (AE2) to enhance the cytotoxicity of NK cells towards the mouse NK-sensitive cell line YAC-1 cells (Mallick *et al.*, 2010). Their observation is consistent with Yoon's finding that increased IL-10 levels seen in the splenocytes may activate NK activity. In addition, the increase of IFN-γ serum levels may also contribute to the increased NK functionality.

Similar to that of the heterogeneity of the macrophage family, it has been suggested that there are at least three distinct NK subsets in human. They are cytokine-producing $CD56^{bright}$ NK; cytotoxic $CD56^{dim}$ as well as the recently found subset of $CD56^{(bright)}CD16^{+}$ cells as an intermediate stage of NK cell maturation (Beziat *et al.*, 2011). Therefore, it is important for future immunomodulatory studies of β-glucan to report the respective effects on each of the distinct NK subsets.

### C. EFFECTS OF β-GLUCANS ON BLOOD DCS, T CELLS AND B CELLS

One of the most significant immunological effects of β-glucans is on the "immunological triad" of blood DCs, T cells and B cells. These three types of cells that are crucial for the development of the adaptive immune response have previously been reviewed including ours (MacLennan *et al.*, 1997).

While it will require an extensive review to describe each of these three types of immunological population, the most single important note could be that these cell populations are not homogenous in nature. For example, there are at least five DC subsets in the circulation and each of them can be identified by a distinct surface marker of CD1b/c, CD16, CD34, CD123 and CD141 as previously reported (Bebawy and Sze, 2008). Similarly, it is clear that CD8 is not a homogenous cytotoxic T cell population, but actually such effector T cells can be divided into at least three subsets of early-, intermediate- and late-stage that based on the surface expression of two markers of CD27 and CD28. Importantly, these three subsets also displayed different cytotoxic capabilities, respectively, as stated in our former review (Sze *et al.*, 2003). Needless to say, there are NKT cells as well as the important suppressive

$CD4^+CD25^+$ regulatory T cells. For B cells, it has also been shown that there is a hierarchical classification of many different types of subsets that may be responsible for different compartments of the body and express either immuno-enhancing or suppressive activities. The summative effects may differ quite significantly from oversimplified *in vitro* setting. In summary, we need to bear in mind when reading the vast amount of β-glucan immunomodulatory literature as many of these studies tend to be conducted in a reductionist approach. In addition, each of the DCs; CD8, CD4 T cells and B cells is a complex population rather than single homogenous entity; therefore, the exact combined response may vary depending on the pre-existed ratio among different subsets.

LeibundGut-Landmann *et al.* demonstrated that the Dectin-1 receptor functions as a pattern recognition receptor for bacterial β-glucans and signals via Syk kinase but independently of the TLR pathway (LeibundGut-Landmann *et al.*, 2007). The activated DCs not only can prime Th1 and Th17 CD4 T cell responses but also can induce antigen-specific expansion of the T cell receptor transgenic $CD8^+$ T cells and their differentiation into cytotoxic T cells *in vitro*.

Qi *et al.* compared the yeast *S. cerevisiae*-derived soluble β-glucan versus particulate β-glucan in a mouse study through oral administration for their respective effects on DCs, T cells and macrophages (Qi *et al.*, 2011). It was found that the particulate but not soluble β-glucan could bind to the C-type lectin receptor Dectin-1. It then activated the increase of $CD8\alpha^+CD11c^+$ DCs infiltration into the tumour. It also promotes Th1 and cytotoxic IFN-γ-producing CD8 T cells but downregulates $CD4^+Foxp3^+$ regulatory T cells tumour infiltration. This tumour infiltration by immune cells is also related to delayed tumour progression. The activation of DC by particulate β-glucan through Dectin-1 pathway is consistent with the observation that human Dectin-1 deficiency associates with mucocutaneous fungal infections (Ferwerda *et al.*, 2009).

Qi *et al.* further elegantly employed both the Dectin-1-deficient and C3-deficient mouse models to verify that the soluble yeast β-glucan counterpart binds to DC independent of Dectin-1 or CR3 receptors, but it does not activate the DCs (Qi *et al.*, 2011). Instead, the soluble yeast β-glucan binds and activates through CR3. Interestingly, the soluble β-glucan alone has no anti-tumour effect but significantly augmented the anti-tumour monoclonal anti-body therapy in that particular study via complementary activation pathway.

Murata *et al.* have shown that mouse $CD4^+$ T cells under the *in vivo* stimulating effects of Lentinan mushroom-derived β-glucan skew towards the Th1 polarity as shown by augmented IFN-γ production (Murata *et al.*, 2002).

There is also reduced IL-4 production upon *in vitro* anti-CD3 stimulation. This Th1 skew pattern and NO increase were also shown by Mallick *et al.* in a mouse study such that the serum IFN-γ levels showed a significant increase on treatment with the hot alkaline extract of mushroom β-glucan *A. hygrometricus* (AE2). While analysing the culture supernatant of the splenocytes obtained from the AE2-treated mice, upon 4-day submitogenic PHA co-culturing, levels of Th1 cytokines of IFN-γ, IL-2 and TNF-α, but not the IL-12, were increased significantly. On the other hand, the Th2 cytokine levels remained similar to that of the control. Furthermore, the level of the immuno-tolerant cytokine IL-10 was found to increase by about threefold after AE2 treatment. Unfortunately, the flow cytometric bead array used in this particular study did not contain another regulatory cytokine TGF-β, and therefore, no data on this key cytokine can be interpreted.

In the same study of Mallick *et al.*, the researchers reported that the mushroom β-glucan *A. hygrometricus* hot alkaline extract AE2 gives no significant change in the B and T cell population as well as their activation status *in vivo* (Mallick *et al.*, 2010). Nevertheless, the CD4/CD8 ratio decreased, and the co-stimulatory molecules of CD80, CD86 and CD71 were found to increase significantly in the whole splenocyte population of AE2-treated groups when compared to the control.

## III. STRUCTURE AND FORM OF β-GLUCANS IN RELATION TO DIFFERENT EFFECTOR FUNCTIONS

With the previous discussion, now we may be able to derive a simplified model trying to explain the observations derived from the previous research studies. Firstly, it is important to note that results derived from purified β-glucans such as those from bacteria, yeast and plant are easier to interpret. In contrary, experimental results from crude glucan extracts need to be cautiously analysed before any implied extrapolation can be deduced. For instance, zymosan consists of only 14% β-glucan (Brown *et al.*, 2003) or Krestin which is a β-glucan–protein complex containing 25–38% protein residues (Ooi and Liu, 2000). Therefore, it is very likely that some responses which were noted in various immune populations may be due to the effects from components other than β-glucan.

Secondly, there is significant difference in immunological response for the soluble form versus the particulate form for a specific purified β-glucan compound. It is not clear if the few known β-glucan receptors of CR3, LacCer scavenger receptors and TLR2 will have similar properties as that of the Dectin-1 which has been shown to bind to particulate but not soluble β-glucans derived from yeast *S. cerevisiae* (Qi *et al.*, 2011).

Thirdly, while the physical form and complexity may affect the potency of the immune response to β-glucans, however at the moment, there is no clear information about the differences in side-chain configuration and complexity will lead to any distinct pattern of specific immune responses. This is definitely a promising and interesting area for future exploration.

# IV. POTENTIAL THERAPEUTIC ANTI-CANCER EFFECTS OF β-GLUCANS

## A. β-GLUCANS AS ANTI-ANGIOGENESIS AGENT

It has been proposed that anti-angiogenesis may be another pathway that β-glucans exert the anti-cancer property. Yamamoto *et al.* (2009) investigated the effect of oral administration of *Sparassis crispa* (SC, or known as Hanabiratake in Japan) on angiogenesis. SC is an edible mushroom with medicinal properties that contains more than 40% β-D-glucan. Using a spontaneous metastatic C57BL/6J mice model, they found that SC can suppress the growth of the primary tumour and also the extent of the metastatic foci in the lung, correlating with the inhibition of tumour induced-angiogenesis. These effects are not a result of direct action on the endothelial cells because cell growth, migration and capillary-like tube formation were not affected in the human umbilical vein endothelial cells under direct exposure of β-glucan. Similarly, using *Coriolus versicolor* mycelia, it was found that there was a reduction of tumour weight and vascular density via suppression of vascular endothelial cell growth factor (VEGF) gene expression (Ho *et al.*, 2004). Anti-angiogenesis effects induced by β-glucans may be acting through an indirect manner and the underlying mechanisms are an emerging area for future studies.

## B. β-GLUCANS AS ADJUVANT TO REDUCE CHEMOTHERAPEUTIC DRUG-INDUCED MYELOSUPPRESSION

Anti-cancer chemotherapeutic drugs have been widely used as standard clinical treatment protocol for cancers. It is also commonly known that these chemotherapeutic agents often cause serious side effects such as nausea, anorexia, nephrotoxicity and myelosuppression (Philpott *et al.*, 1994; Tsujimoto *et al.*, 1996). Masuda *et al.* reported that Maitake mushroom β-glucan may reduce the post-cisplatin-treated bone marrow CFU-GM colony formation units decrement (Masuda *et al.*, 2009). This possible bone marrow protective effect is also reflected by the increase in the number of macrophages, DCs and NK cells but not CD4 or CD8 T cells. Thus, Maitake mushroom β-glucan may reduce cisplatin-induced leucopenia.

The same research group has also reported that the Maitake mushroom-derived β-glucan enhances granulopoiesis and mobilization of granulocytes through increasing the production of granulocyte colony-stimulating factor and downregulates the expression of the chemokine receptor CXCR4 and its ligand stromal cell-derived factor 1 in the bone marrow microenvironment (Ito *et al.*, 2009).

### C. β-GLUCANS AS ADJUVANT TO ANTI-CANCER RADIOTHERAPY

Radiotherapy induces hematopoietic and immunological suppression, in particularly when extensive fields such as total body irradiation or total lymphoid irradiation are applied. β-Glucans have been shown to increase serum levels of cytokines that could be radio-protective. For example, a water-soluble glucan from *Lentinus lepideus* was orally administered everyday for 24 days to 6 Gy irradiated mice (Jin *et al.*, 2003). While the levels of IL-1β, IL-6 and GM-CSF increased for β-glucan-treated animals, the TNF-α level, being originally high after radiotherapy, decreased over time. In addition, the number of colony forming units of the β-glucan-treated mice at day 8 was already close to the level seen in non-irradiated controls.

Cramer *et al.* reported that yeast β-glucans could enhance hematopoietic cell proliferation via promoting a CR3-dependent leukocyte recovery in the setting of allogeneic hematopoietic stem cell transplantation following sublethal irradiation (Cramer *et al.*, 2006).

### D. β-GLUCANS AS ADJUVANT TO ANTI-CANCER MONOCLONAL ANTI BODY AND PEPTIDE TREATMENT

There are basically three mechanisms for anti-cancer monoclonal anti-body cytotoxicity (Akramiene *et al.*, 2007). These three mechanisms are (i) anti-body-dependent cellular cytotoxicity, (ii) complement-dependent cytotoxicity and (iii) CR3-dependent cellular cytotoxicity.

The first mechanism is best demonstrated in an athymic mice with transplanted neuroblastoma model. It was shown that anti-ganglioside monoclonal anti-body 3F8 plus β-glucan produced synergistic anti-cancer effects (Cheung and Modak, 2002). The median survival of the mice with implanted tumours treated with β-glucan and 3F8 combined was significant better than those treated with either β-glucan or 3F8 alone. Furthermore, using anti-asialo GM1 anti-serum to remove NK cells or in SCID-beige mice that lack

NK cells, 3F8 plus β-glucan was significantly more effective than 3F8 alone showing that this synergic anti-cancer effects was, at part, mediated by NK-independent cytotoxicity. The second mechanism involved complement-dependent cytotoxicity which can be demonstrated by the study of Sier *et al.* using the multicellular spheroid model of renal cell carcinoma cells that resembles closely the cancer situation *in vivo*. Addition of β-glucan induced the killing of monoclonal anti-body G250 targeting renal cell carcinoma that are resistant to radiation and chemotherapy and iC3b opsonized spheroids by effector cells (Sier *et al.*, 2004). The last mechanism can be demonstrated in a study using the combination of the complement-activating anti-body Rituximab with Barley β-glucan against $CD20^+$ lymphoma xenografts in SCID mice (Modak *et al.*, 2005). The results showed that the growth of established subcutaneous lymphoma was significantly suppressed in mice with the Rituximab plus β-glucan. Furthermore, survival of mice with disseminated lymphoma was significantly increased in the combination group. This approach of CR3-dependent cytotoxicity has also been shown in targeting human VEGF-positive ovarian tumour by synergistic combination of yeast β-glucan and anti-VEGF Bevacizumab (Salvador *et al.*, 2008).

Other than using monoclonal anti-bodies to target surface antigens on cancer cells, there are reports showing that oral administration of β-glucan with associated prophylactic peptide vaccination could significantly decrease cancer load. Harnack *et al.* used MHC-I restricted peptides that targeted at tumour antigen survivin of B cell lymphoma in mice (Harnack *et al.*, 2009). Another model using HER-2/neu oncogene tranfected mice that can spontaneous develop breast cancer developed by Forni's group, though did not use any of the β-glucan compounds but utilized CpG as the TLR agonist instead, the researchers found that vaccinated mice remained tumour free or were able to control spontaneous tumour growth and exhibited long-lasting cytotoxic T lymphocyte (CTL) responses (Nava-Parada *et al.*, 2007).

## V. CONCLUSIONS

β-Glucans are naturally occurring polysaccharides. They have been shown to exhibit significant biological and functional effects on macrophages, NK cells, DCs, T cells, B cells and many other immunological important cell types. This review discusses about the diverse β-glucan sources and their related biological activities. We also describe the heterogeneity of the important immunological cellular populations such that they are actually made up of many subsets with different functions (Fig. 1).

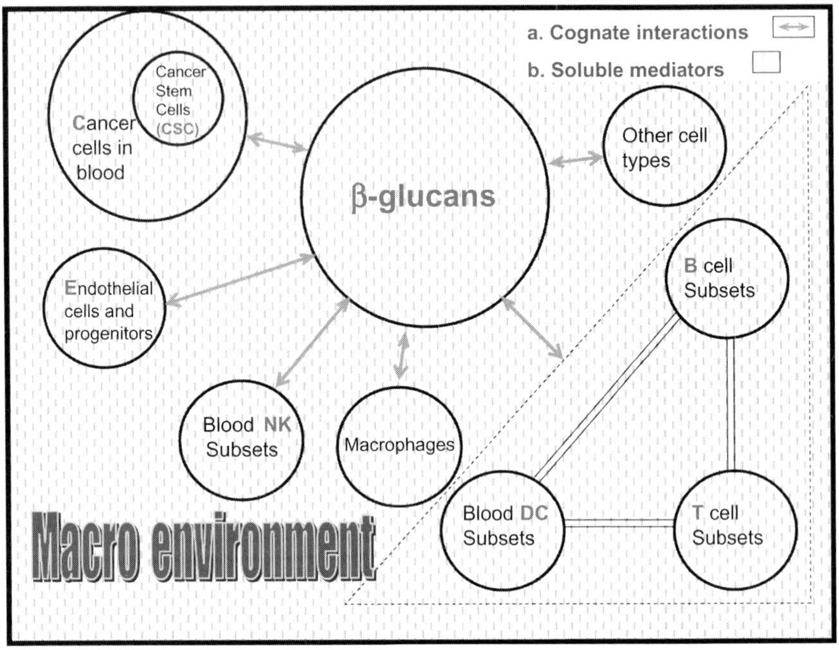


Fig. 1. The effects of β-glucans on the macroenvironment of cancer. A schematic diagram representing the cancer cell macroenvironment. β-glucans have been shown to exhibit multi-directional immunomodulatory activities with many different cells types. These activities can be summarized as either cognate interactions that involve many different membrane surface molecules or secreting cytokines and chemokines as soluble mediators in blood and cellular fluids. (See colour plate section at the back of the book.)

β-Glucans that can be found in bacteria, plant, yeast and also many medicinal mushrooms such as Ganoderma, Coriolus and Cordyceps and all have been shown to produce anti-cancer effect clinically with long track record from ancient China for instance. This review aims to provide a comprehensive review of some currently unfolded mechanisms of β-glucans. They exert their effects in the macroenvironment by inducing the release of cytokines and other soluble factors in blood. It may also act on the microenvironment; orally administered β-glucan has been shown to prevent Listeria-induced myelosuppression by inducing a higher reserve of myeloid progenitors from the bone marrow in a murine model (Torello *et al.*, 2010). This is the consequence of the release of biologically active cytokine (CSFs, IL-1α, IL-6, IFN-γ). In addition, Maitake β-glucan has been shown to enhance stem cells proliferation in both mouse (Lin *et al.*, 2009) and human (Lin *et al.*, 2010). Therefore, it actually affects the stem cell compartment and may impact on the cancer microenvironment as well (Fig. 2). The understanding of the complex immune effects of β-glucans will help us to utilize them as part of our therapeutic strategy against cancer in the future.

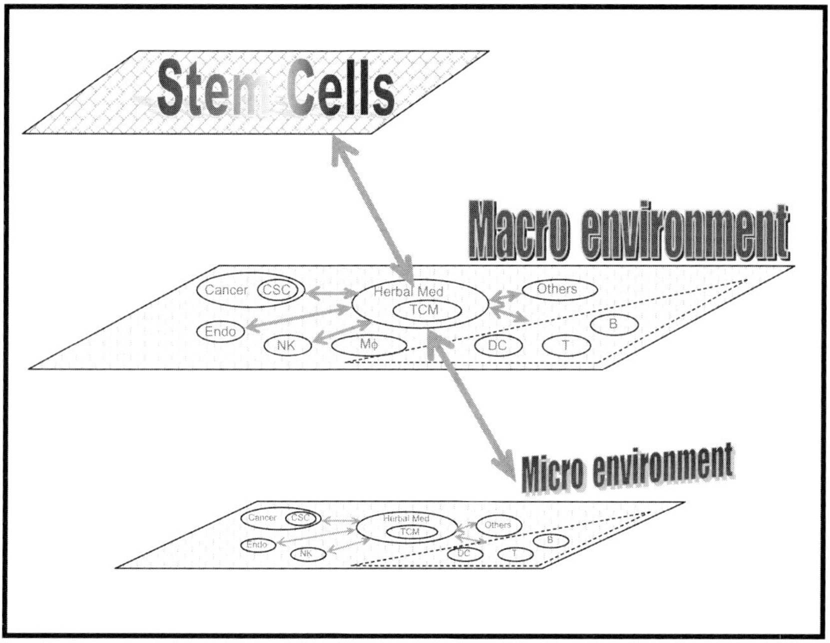


Fig. 2. Proposed view of immunological effects of β-glucans in cancer. A bird-eye view schematic diagram representing the context of cancer cell macroenvironment (Fig. 1) interacting with both the stem cell compartment and the tumour microenvironment. (For colour version of this figure, the reader is referred to the online version of this chapter.)

# REFERENCES


Adams, E. L., Rice, P. J. *et al.* (2008). Differential high-affinity interaction of dectin-1 with natural or synthetic glucans is dependent upon primary structure and is influenced by polymer chain length and side-chain branching. *The Journal of Pharmacology and Experimental Therapeutics* **325**(1), 115–123.

Akramiene, D., Kondrotas, A. *et al.* (2007). Effects of beta-glucans on the immune system. *Medicina (Kaunas, Lithuania)* **43**(8), 597–606.

Akramiene, D., Grazeliene, G. *et al.* (2009). Treatment of Lewis lung carcinoma by photodynamic therapy and glucan from barley. *Medicina (Kaunas)* **45**(6), 480–485.

Barsanti, L., Passarelli, V. *et al.* (2011). Chemistry, physico-chemistry and applications linked to biological activities of beta-glucans. *Natural Product Reports* **28**(3), 457–466.

Bebawy, M. and Sze, D. M. (2008). Targeting P-glycoprotein for effective oral anticancer chemotherapeutics. *Current Cancer Drug Targets* **8**(1), 47–52.

Beziat, V., Duffy, D. *et al.* (2011). CD56brightCD16+ NK cells: A functional intermediate stage of NK cell differentiation. *Journal of Immunology* **186**(12), 6753–6761.

Brown, G. D., Herre, J. *et al.* (2003). Dectin-1 mediates the biological effects of beta-glucans. *The Journal of Experimental Medicine* **197**(9), 1119–1124.

Chan, W. K., Law, H. K. *et al.* (2007). Response of human dendritic cells to different immunomodulatory polysaccharides derived from mushroom and barley. *International Immunology* **19**(7), 891–899.

Chan, G. C., Chan, W. K. *et al.* (2009). The effects of beta-glucan on human immune and cancer cells. *Journal of Hematology & Oncology* **2**, 25.

Cheung, N. K. and Modak, S. (2002). Oral (1 → 3), (1 → 4)-beta-D-glucan synergizes with antiganglioside GD2 monoclonal antibody 3F8 in the therapy of neuroblastoma. *Clinical Cancer Research* **8**(5), 1217–1223.

Cramer, D. E., Allendorf, D. J. *et al.* (2006). Beta-glucan enhances complement-mediated hematopoietic recovery after bone marrow injury. *Blood* **107**(2), 835–840.

Demir, G., Klein, H. O. *et al.* (2007). Beta glucan induces proliferation and activation of monocytes in peripheral blood of patients with advanced breast cancer. *International Imomunopharmacology* **7**(1), 113–116.

Di Renzo, L., Yefenof, E. *et al.* (1991). The function of human NK cells is enhanced by beta-glucan, a ligand of CR3 (CD11b/CD18). *European Journal of Immunology* **21**(7), 1755–1758.

Dushkin, M. I., Safina, A. F. *et al.* (1996). Carboxymethylated beta-1,3-glucan inhibits the binding and degradation of acetylated low density lipoproteins in macrophages in vitro and modulates their plasma clearance in vivo. *Cell Biochemistry and Function* **14**(3), 209–217.

Ferwerda, G., Meyer-Wentrup, F. *et al.* (2008). Dectin-1 synergizes with TLR2 and TLR4 for cytokine production in human primary monocytes and macrophages. *Cellular Microbiology* **10**(10), 2058–2066.

Ferwerda, B., Ferwerda, G. *et al.* (2009). Human dectin-1 deficiency and mucocutaneous fungal infections. *The New England Journal of Medicine* **361**(18), 1760–1767.

Goodridge, H. S., Wolf, A. J. *et al.* (2009). Beta-glucan recognition by the innate immune system. *Immunological Reviews* **230**(1), 38–50.

Hamuro, J., Takatsuki, F. *et al.* (1994). Synergistic antimetastatic effects of lentinan and interleukin 2 with pre- and post-operative treatments. *Japanese Journal of Cancer Research* **85**(12), 1288–1297.

Harnack, U., Eckert, K. *et al.* (2009). Oral administration of a soluble 1-3, 1-6 beta-glucan during prophylactic survivin peptide vaccination diminishes growth of a B cell lymphoma in mice. *International Immunopharmacology* **9**(11), 1298–1303.

Ho, J. C., Konerding, M. A. *et al.* (2004). Fungal polysaccharopeptide inhibits tumor angiogenesis and tumor growth in mice. *Life Sciences* **75**(11), 1343–1356.

Hong, F., Yan, J. *et al.* (2004). Mechanism by which orally administered beta-1,3-glucans enhance the tumoricidal activity of antitumor monoclonal antibodies in murine tumor models. *Journal of Immunology* **173**(2), 797–806.

Ito, K., Masuda, Y. *et al.* (2009). Maitake beta-glucan enhances granulopoiesis and mobilization of granulocytes by increasing G-CSF production and modulating CXCR4/SDF-1 expression. *International Immunopharmacology* **9**(10), 1189–1196.

Iwabuchi, K. and Nagaoka, I. (2002). Lactosylceramide-enriched glycosphingolipid signaling domain mediates superoxide generation from human neutrophils. *Blood* **100**(4), 1454–1464.

Jin, M., Jeon, H. *et al.* (2003). Enhancement of repopulation and hematopoiesis of bone marrow cells in irradiated mice by oral administration of PG101, a water-soluble extract from Lentinus lepideus. *Experimental Biology and Medicine (Maywood, N.J.)* **228**(6), 759–766.

Joffre, O. P., Sancho, D. *et al.* (2010). Efficient and versatile manipulation of the peripheral CD4+ T-cell compartment by antigen targeting to DNGR-1/CLEC9A. *European Journal of Immunology* **40**(5), 1255–1265.

Kamada, N., Hisamatsu, T. *et al.* (2008). Unique CD14 intestinal macrophages contribute to the pathogenesis of Crohn disease via IL-23/IFN-gamma axis. *The Journal of Clinical Investigation* **118**(6), 2269–2280.

Klein, E., Di Renzo, L. *et al.* (1990). Contribution of CR3, CD11b/CD18 to cytolysis by human NK cells. *Molecular Immunology* **27**(12), 1343–1347.

LeibundGut-Landmann, S., Gross, O. *et al.* (2007). Syk- and CARD9-dependent coupling of innate immunity to the induction of T helper cells that produce interleukin 17. *Nature Immunology* **8**(6), 630–638.

Lew, D. B., Leslie, C. C. *et al.* (1991). Role of endogenously derived leukotrienes in the regulation of lysosomal enzyme expression in macrophages exposed to beta 1,3-glucan. *Journal of Leukocyte Biology* **49**(3), 266–276.

Li, B., Cramer, D. *et al.* (2007). Yeast glucan particles activate murine resident macrophages to secrete proinflammatory cytokines via MyD88- and Syk kinase-dependent pathways. *Clinical Immunology* **124**(2), 170–181.

Lin, H., De Stanchina, E. *et al.* (2009). Maitake beta-glucan enhances umbilical cord blood stem cell transplantation in the NOD/SCID mouse. *Experimental Biology and Medicine (Maywood, N.J.)* **234**(3), 342–353.

Lin, H., de Stanchina, E. *et al.* (2010). Maitake beta-glucan promotes recovery of leukocytes and myeloid cell function in peripheral blood from paclitaxel hematotoxicity. *Cancer Immunology, Immunotherapy* **59**(6), 885–897.

MacLennan, I. C., Gulbranson-Judge, A. *et al.* (1997). The changing preference of T and B cells for partners as T-dependent antibody responses develop. *Immunology Reviews* **156**, 53–66.

Mallick, S. K., Maiti, S. *et al.* (2010). Immunostimulatory properties of a polysaccharide isolated from Astraeus hygrometricus. *Journal of Medicinal Food* **13**(3), 665–672.

Masuda, Y., Inoue, M. *et al.* (2009). Maitake beta-glucan enhances therapeutic effect and reduces myelosupression and nephrotoxicity of cisplatin in mice. *International Immunopharmacology* **9**(5), 620–626.

Modak, S., Koehne, G. *et al.* (2005). Rituximab therapy of lymphoma is enhanced by orally administered (1→3), (1→4)-D-beta-glucan. *Leukemia Research* **29** (6), 679–683.

Murata, Y., Shimamura, T. *et al.* (2002). The skewing to Th1 induced by lentinan is directed through the distinctive cytokine production by macrophages with elevated intracellular glutathione content. *International Immunopharmacology* **2**(5), 673–689.

Murphy, E. A., Davis, J. M. *et al.* (2004). Effects of moderate exercise and oat beta-glucan on lung tumor metastases and macrophage antitumor cytotoxicity. *Journal of Applied Physiology* **97**(3), 955–959.

Nava-Parada, P., Forni, G. *et al.* (2007). Peptide vaccine given with a Toll-like receptor agonist is effective for the treatment and prevention of spontaneous breast tumors. *Cancer Research* **67**(3), 1326–1334.

Noble, P. W., Henson, P. M. *et al.* (1993). Transforming growth factor-beta primes macrophages to express inflammatory gene products in response to particulate stimuli by an autocrine/paracrine mechanism. *The Journal of Immunology* **151**(2), 979–989.

Novak, M. and Vetvicka, V. (2009). Glucans as biological response modifiers. *Endocrine, Metabolic & Immune Disorders Drug Targets* **9**(1), 67–75.

Ooi, V. E. and Liu, F. (2000). Immunomodulation and anti-cancer activity of polysaccharide-protein complexes. *Current Medicinal Chemistry* **7**(7), 715–729.

Osorio, F., LeibundGut-Landmann, S. *et al.* (2008). DC activated via dectin-1 convert Treg into IL-17 producers. *European Journal of Immunology* **38**(12), 3274–3281.

Philpott, N., Mehta, J. *et al.* (1994). Idarubicin, high-dose cytarabine and etoposide for remission induction in therapy-related acute myeloid leukemia. *Leukemia & Lymphoma* **15**(1–2), 127–130.

Price, L. A., Wenner, C. A. *et al.* (2010). Role for toll-like receptor 4 in TNF-alpha secretion by murine macrophages in response to polysaccharide Krestin, a Trametes versicolor mushroom extract. *Fitoterapia* **81**(7), 914–919.

Qi, C., Cai, Y. *et al.* (2011). Differential pathways regulating innate and adaptive antitumor immune responses by particulate and soluble yeast-derived beta-glucans. *Blood* **117**(25), 6825–6836.

Rice, P. J., Adams, E. L. *et al.* (2005). Oral delivery and gastrointestinal absorption of soluble glucans stimulate increased resistance to infectious challenge. *The Journal of Pharmacology and Experimental Therapeutics* **314**(3), 1079–1086.

Salvador, C., Li, B. *et al.* (2008). Yeast-derived beta-glucan augments the therapeutic efficacy mediated by anti-vascular endothelial growth factor monoclonal antibody in human carcinoma xenograft models. *Clinical Cancer Research* **14**(4), 1239–1247.

Schenk, M., Bouchon, A. *et al.* (2007). TREM-1-expressing intestinal macrophages crucially amplify chronic inflammation in experimental colitis and inflammatory bowel diseases. *The Journal of Clinical Investigation* **117**(10), 3097–3106.

Sier, C. F., Gelderman, K. A. *et al.* (2004). Beta-glucan enhanced killing of renal cell carcinoma micrometastases by monoclonal antibody G250 directed complement activation. *International Journal of Cancer* **109**(6), 900–908.

Stone, B. A. and Clarke, A. E. (1992). Chemistry and Biology of (1,3)-D-Glucans. La Trobe University Press, Victoria, Australia.

Swirski, F. K. (2011). The spatial and developmental relationships in the macrophage family. *Arteriosclerosis, Thrombosis, and Vascular Biology* **31**(7), 1517–1522.

Sze, D. M., Brown, R. D. *et al.* (2003). Clonal cytotoxic T cells in myeloma. *Leukemia & Lymphoma* **44**(10), 1667–1674.

Thompson, I. J., Oyston, P. C. *et al.* (2010). Potential of the beta-glucans to enhance innate resistance to biological agents. *Expert Review of Anti-Infective Therapy* **8**(3), 339–352.

Torello, C. O., de Souza Queiroz, J. *et al.* (2010). Immunohematopoietic modulation by oral beta-1,3-glucan in mice infected with Listeria monocytogenes. *International Immunopharmacology* **10**(12), 1573–1579.

Tsujimoto, H., Sasaki, S. *et al.* (1996). Cisplatin microcrystals suspended in oil: Pathological study of acute toxicity in mice. *Anti-Cancer Drugs* **7**(3), 240–247.

Vetvicka, V., Thornton, B. P. *et al.* (1996). Soluble beta-glucan polysaccharide binding to the lectin site of neutrophil or natural killer cell complement receptor type 3 (CD11b/CD18) generates a primed state of the receptor capable of mediating cytotoxicity of iC3b-opsonized target cells. *The Journal of Clinical Investigation* **98**(1), 50–61.

Yamamoto, K., Kimura, T. *et al.* (2009). Anti-angiogenic and anti-metastatic effects of beta-1,3-D-glucan purified from Hanabiratake, Sparassis crispa. *Biological and Pharmaceutical Bulletin* **32**(2), 259–263.

Yoon, T. J., Kim, T. J. *et al.* (2008). Anti-tumor metastatic activity of beta-glucan purified from mutated Saccharomyces cerevisiae. *International Immunopharmacology* **8**(1), 36–42.

Zimmerman, J. W., Lindermuth, J. *et al.* (1998). A novel carbohydrate-glycosphingolipid interaction between a beta-(1-3)-glucan immunomodulator, PGG-glucan, and lactosylceramide of human leukocytes. *The Journal of Biological Chemistry* **273**(34), 22014–22020.

# Developing Phytocompounds from Medicinal Plants as Immunomodulators

CHIH-CHUN WEN,* HUI-MING CHEN*,† AND NING-SUN YANG*,[1]

**Agricultural Biotechnology Research Center, Academia Sinica, Taipei, Taiwan*

†*Department and Institute of Pharmacology, National Yang-Ming University, Taipei, Taiwan*



[1]Corresponding author: E-mail: nsyang@gate.sinica.edu.tw

Advances in Botanical Research, Vol. 62

0065-2296/12 $35.00
DOI: 10.1016/B978-0-12-394591-4.00004-0

# ABSTRACT

Imbalance or malfunction of the immune systems is associated with a range of chronic diseases including autoimmune diseases, allergies, cancers and others. Various innate and adaptive immune cells that are integrated in this complex networking system may represent promising targets for developing immunotherapeutics for treating specific immune diseases. A spectrum of phytochemicals have been isolated, characterized and modified for development and use as prevention or treatment of human diseases. Many cytotoxic drugs and antibiotics have been developed from phytocompounds, but the application of traditional or new medicinal plants for use as immunomodulators in treating immune diseases is still relatively limited. In this review, a selected group of medicinal herbs, their derived crude or fractionated phytoextracts and the specific phytochemicals/phytocompounds isolated from them, as well as categorized phytocompound groups with specific chemical structures are discussed in terms of their immunomodulatory bioactivities. We also assess their potential for future development as immunomodulatory or inflammation-regulatory therapeutics or agents. New experimental approaches for evaluating the immunomodulatory activities of candidate phytomedicines are also discussed.

# I. INTRODUCTION

During the past few decades, there has been a paradigm shift in medicine, with interest moving from disease-treatment to disease-prevention health care. Medical care is being evaluated not only according to diagnosis, prevention and treatment of diseases but also according to the enhancement of life quality, maintenance of health and the use of nutritional or medicinal foods. In this context, new strategies for drug discovery with advanced experimental approaches are of importance for the modernization of medicine. Currently, mainstream pharmaceutical research and development still concentrates on single compounds, biochemicals or biologics as candidate drugs or lead compounds, that aim at specific targets associated with a disease. This drug discovery strategy, however, seems to have reached a bottleneck, as it becomes ever more time consuming, labour intensive and costly to develop new drugs (Chen *et al.*, 2008). Such "traditional" Western medical research expects that single compound chemicals confer high potency, low toxicity and high selectivity for targeted molecular/cellular targets and diseases. However, in reality, these ideals are proving hard to achieve.

Therefore, the development of drug candidates from various traditional or alternative and complementary medicines is receiving an increasing worldwide attention (Aravindaram and Yang, 2010; Tu, 2011).

Drug discovery is being transformed from a "game of chance" (mass screening) or overdependence on brute-force new high-throughput technology (Patwardhan and Mashelkar, 2009). A better understanding of approaches, the adaptation of a variety of approaches and cross-disciplinary learning that draws from traditional wisdom are now being considered by many scientists to be critical to make a significant difference (Schmid and Smith, 2004). Drug discovery strategies and development based on systematic and modernized investigation of complementary and alternative systems of medicines (CAM) or/and the traditional systems of medicines are re-emerging as an attractive approach for many pharmacology and pharmaceutical researchers. According to the definition of the National Center for Complementary and Alternative Systems of Medicine (NCCAM), CAM is a group of diverse medical and health-care systems, practices and products that are not generally considered part of "conventional medicine" (Western medicine) (Mansky and Wallerstedt, 2006). Although defining CAM can be arduous, it can generally be categorized into several groups including natural products, mind and body medicine, manipulative and body-based practices and other CAM practices. Speaking in a broad sense, other CAM practices may also include ancient and "self-integrated" medical systems such as the Ayurvedic medicines and traditional Chinese medicines (TCM). It is estimated that more than 70% of the developing world's population still relies primarily on CAM (Azaizeh *et al.*, 2010). Plant materials are the major sources of various therapeutic agents in the CAM categories of natural products, Ayurvedic medicine and TCM. In fact, within TCM, plant-derived medicines have been used for prevention and treatment of various diseases and documented in a systematic way for over five millennia (Lam *et al.*, 2010). Within this context, it can be expected that a spectrum of medicinal plants with a long history of use will be quickly re-recognized as highly valuable for future drug discovery and development, as recently commented (Lam *et al.*, 2010; Tu, 2011).

An imbalance in specific immune systems or their coordination in general is known to be involved in the pathogenesis of various diseases including infection, dermatitis, inflammatory bowel diseases, metabolic syndrome, cancers and a spectrum of inflammation-related diseases (Mantovani *et al.*, 2008; Nestle *et al.*, 2009). Modulation of the immune systems has hence been considered a vital approach for the treatment or control of various immune-related diseases (Cho, 2008; Ouchi *et al.*, 2011). The recent advent and breakthroughs in omics technology and systems biology experimental approaches have created a new era for the investigation and development of novel therapeutics and drugs for diverse disease systems, especially complex immune-related disorders. These technologies and approaches include various genomic, proteomic, metabolomic, cellomic, lipidomic and phenomic approaches, as well as the associated bioinformatics sciences and databases. In general, phytomedicines, including phytoextracts, their subfractions derived from partitioning using organic solvent systems or isolated single phytocompounds or phytochemicals with a long history of medicinal use, are believed to interact with multiple targets to confer pharmacological or physiological effects at the cellular, tissue or organ levels. Experimental uses seem to suggest that they may be relatively safe. An increasing number of studies have shown that traditional phytomedicines can confer a variety of immunomodulatory activities, as recently revealed by others' and our own studies (Hou *et al.*, 2010; Shyur and Yang, 2008). The process of discovering and developing phytocompounds as immunomodulatory agents by evaluation using different experimental approaches and omics platforms is shown in Fig. 1. Here, we review a group of specific medicinal herbs, their derived plant extracts, fractions, and the derived phytochemicals that have been studied for their immunomodulatory bioactivities and assessed for their potential as immunomodulatory and/or inflammation-regulatory therapeutics agents. The functional genomics and proteomics approaches used for characterization of the bioactivities outlined here can be employed as key strategies in future applications. The accompanying findings are discussed in detail and their implications for phytomedicine research are contemplated.

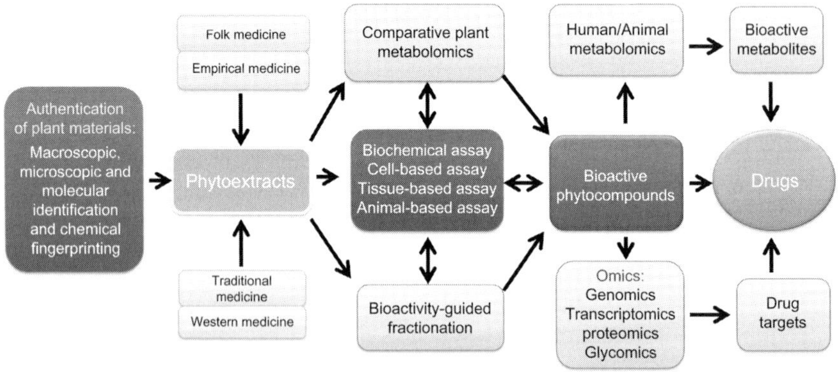


Fig. 1. Schematic representation of technological systems for drug discovery from phytocompounds of medicinal herbs as immunomodulatory agents using various experimental approaches. (For colour version of this figure, the reader is referred to the online version of this chapter.)

# II. MODULATION OF THE IMMUNE SYSTEM

## A. IMMUNE SYSTEMS AND IMMUNE-RELATED DISORDERS

The immune system is, in nature, a uniquely complex network that protects the host body from foreign pathogens, stresses, insults and the resultant illnesses. It can govern the various and interconnecting pathways of inflammation, microbial recognition, microbial clearance, cell and tissue damage and death and wound healing. The homeostatic system requires the well-timed interplay of multiple immune cell types and crosstalk with the specific tissue microenvironment to maintain immune homeostasis. The immune system in vertebrates, at least, is traditionally divided into two types, innate immunity and adaptive immunity (Ullrich, 2010; Vesely *et al.*, 2011), although the distinctions between innate and adaptive immunity have become more intertwined in recent studies (Lanier and Sun, 2009). Both play critical roles in functioning as a defence system against the invasion of microbial pathogens present in our environment. It also provides a regulatory system that controls normal cell turnover and eliminates damaged cells and tumour cells. Typically, innate immunity has been considered to be the first line of defence against pathogens such as bacteria, viruses or fungi. It exhibits characteristic features such as rapid response, infection halting and lack of memory in functions (Schiller *et al.*, 2006). The innate immunity system may include dendritic cells (DCs), macrophages, mast cells, neutrophils, basophils, eosinophils, invariant natural killer cells (NK cells), NKT cells and $\gamma\delta$ T cells (Garg *et al.*, 2010). In comparison, the adaptive immune system or acquired immune response is a relatively slow process mediated by T cells and B cells. It employs diverse antigen receptors that are not encoded in germ line cells but rather *de novo* generated through DNA rearrangement mechanisms in the somatic immune tissues of mammalian organisms (Iwasaki and Medzhitov, 2010; Krogsgaard and Davis, 2005). Characteristics of various immune cell types involved in innate and adaptive immunity are summarized in Fig. 2. A number of recent findings focusing on the specific cellular functions, as well as the complexity and functional specialization of immune cells, have drastically expanded our knowledge of immunity (Medzhitov *et al.*, 2011; Vivier *et al.*, 2011). For example, NK cells were originally defined as effector lymphocytes of innate immunity and were endowed with constitutive cytolytic functions. Recent studies, however, disclosed that NK cells can also mount a form of antigen-specific immunologic memory (Vivier *et al.*, 2011). Therefore, NK cells may also be classified as a new type of immune cell that can exert sophisticated biological functions that contribute to both innate and adaptive immunities. These properties render

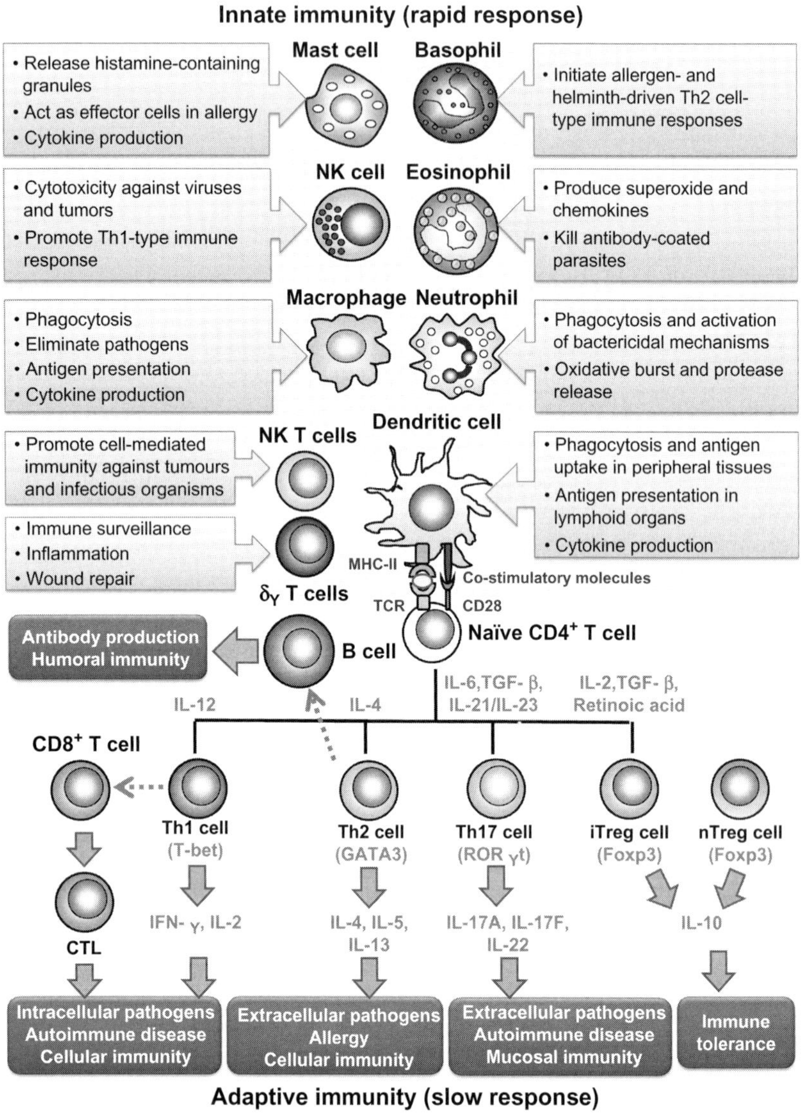


Fig. 2. Characteristics and functions of various innate and adaptive immune cells in the immune system. The immune system can be divided into innate immunity and adaptive immunity. The innate immune system involves the participation of dendritic cells, macrophages, mast cells, granulocytes (neutrophils, eosinophils and basophils), NK cells, NKT cells, δγ T cells and others. The key functions of each cell type are described in the blue grid. The adaptive immune system involves $CD4^+$ T cells, $CD8^+$ T cells, B cells and others. $CD4^+$ T cells can differentiate into Th1, Th2, Th17 and inducible Treg (iTreg) cells under different microenvironments specialized by interactive cytokines and chemokines, and distinct activation of specific transcription

NK cells highly specific and selective in various cellular functions, and thereby able to respond to a broad spectrum of antigens. These new findings and increased appreciation of the importance of the immune systems have led, over the past two decades, to considerable effort being spent on understanding how immune responses against various immune-related diseases are governed and modulated.

Most immune-associated diseases, including viral or bacterial pathogen-mediated infectious diseases, allergic diseases, inflammatory bowel diseases, cancers and a number of chronic diseases, are now known to be correlated with inflammation. Inflammation is probably the most vital immune response induced by noxious stimuli or conditions (Schmid-Schonbein, 2006). Inflammation underlies a wide variety of physiology and pathological processes, enables survival and tissue repair during or after tissue infection or injury and maintains the organ and body homeostasis (Medzhitov, 2008). Inflammation has been known to humans for thousands of years due to observations and experiences with wounded tissue and infections (Medzhitov, 2010). Traditionally, the symptoms of inflammation were characterized by five cardinal signs, redness, swelling, heat, pain as well as the disturbance of functions (Medzhitov, 2010). A typical inflammatory response is composed of four key components: (1) inflammatory inducers, as the signals to initiate inflammation; (2) specialized sensors that detect the inducers; (3) inflammatory mediators induced by the sensors; and (4) the target tissues that are functionally altered by the inflammatory mediators, that is, the effectors of inflammation (Medzhitov, 2008). These components and their relationships are shown in Fig. 3. Each component can be presented in multiple forms and their combinations can result in distinct inflammatory pathways.

The different types of pathways triggered under given physiological and environmental conditions depend on the nature of the inflammatory inducers. Therefore, understanding and characterizing the types of inducers are a key issue that needs to be addressed in studies of inflammation. The inducers of inflammation are, in general, broadly classified into exogenous and

factors. $CD8^+$ T cells are responsible for confirming cytotoxicity against virus-infected cells or tumour cells. Treg cells are generally grouped into two classes, iTreg cells and natural Treg (nTreg) cells. They can regulate specific immune responses, especially immune tolerance, to maintain immune homeostasis. Abbreviations: Th1, T-helper type 1; Th2, T-helper type 2; Th17, T-helper type 17; Treg, T regulatory cells; NK, natural killer cells; NK T cells, natural killer T cells; MHC-II, major histocompatibility complex class II; TCR, T cell receptor; IFN-$\gamma$, interferon-gamma; TGF-$\beta$, transforming growth factor-beta; IL-2, interleukin-2; IL-4, interleukin-4; IL-6, interleukin-6; IL-12, interleukin-12; IL-17, interleukin-17; IL-21, interleukin-21; IL-23, interleukin-23. (See colour plate section at the back of the book.)

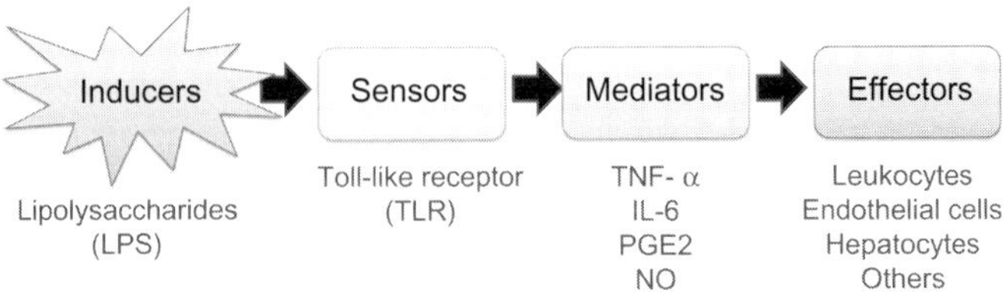


Fig. 3. Schematic representation of the inflammatory pathway. The inflammatory pathway consists of four major components: (1) inducers such as lipopolysaccharides (LPS); (2) sensors such as toll-like receptors (TLRs); (3) mediators such as tumour necrosis factor-alpha (TNF-α), interleukin-6 (IL-6), prostaglandin E2 (PGE2) and nitric oxide (NO); and (4) effectors such as leukocytes, endothelial cells, hepatocytes and others. (For colour version of this figure, the reader is referred to the online version of this chapter.)

TABLE I

*Classification of Inducers of the Inflammatory Pathway (Modified from Medzhitov, 2008)*

| | | | |
|---|---|---|---|
| Inducers | Exogenous | Microbial | 1. PAMPs<br>2. Virulence factors |
| | | Non-microbial | 1. Allergens<br>2. Irritants<br>3. Foreign bodies<br>4. Toxic compounds |
| | Endogenous | Cell derived<br>Tissue derived<br>Plasma derived<br>ECM derived | 1. Inducers released from malfunctioning, stressed or dead cells and from damaged tissues<br>2. Endogenous crystal<br>3. Products of ECM breakdown |

endogenous (Table I) (Medzhitov, 2008, 2010). Exogenous inducers are categorized into two subgroups: microbial and non-microbial. Further, there are two types of microbial inducer: pathogen-associated molecular patterns (PAMPs) and virulence factors. PAMPs are defined as a set of conserved molecular patterns that are characterized by all microorganisms of a given class (Medzhitov and Janeway, 1997). PAMPs are defined in the context that the host has evolved a corresponding set of pattern-recognition receptors (PRRs) that are responsible for detecting the presence of PAMPs. The second class of microbial inducers is composed of various virulence factors and is hence restricted to various pathogens. In contrast to PAMPs, virulence factors are not sensed directly by dedicated or specific receptors. Exogenous inducers of non-microbial origin include various allergens, irritants, foreign materials and toxic compounds (Medzhitov, 2008, 2010). A variety of allergens are sensed because they mimic the virulence activity of parasites. Others may function as irritants on the mucosal epithelia. The inflammatory response induced by both types of inducers is quite similar as the action of the immune system against parasites and environmental irritants depends mainly on expulsion and clearance under the control of the mucosal epithelia. The sensors for allergens, however, are largely unknown.

In essence, endogenous inducers of inflammation are defined as the signals produced by malfunctioning, stressed or damaged cells or tissues and can trigger distinct types of inflammatory responses, suggesting that they can play a vital role in immune response (Medzhitov, 2008). The identity and features of these signals are currently not well defined or understood. However, they apparently belong to different functional classes specified according to the nature and level of the cell or tissue anomalies on which they report. An important and common theme, but not a universal one in detection of acute tissue injuries, is the sensing of the desequestration of cells, organelles or molecules which are normally maintained as separate entities in intact or undamaged cells and tissues. The sequestration activity of these components including various ligands and their receptors or enzymes is provided by various types of compartmentalization that normally and commonly occurs in normal tissues. Among them, some obvious examples are the sequestration bounded by cellular membranes, basement membranes, the surface epithelium and the vascular endothelium (Medzhitov, 2008). For example, for necrotic cell death, the integrity of the plasma membrane is damaged, resulting in release of a number of DAMPs and other constituents, including ATP, uric acid, $K^+$ ions and high-mobility group box 1 protein (HMGB1) (Bianchi, 2007). Another class of endogenous inducers, including crystals of monosodium urate, calcium pyrophosphate dihydrate and advanced glycation end products (AGEs), are correlated with chronic inflammatory conditions. Specific salt crystals can induce the inflammatory conditions gout and pseudogout. AGEs bind to advanced glycation end-product-specific receptor (RAGE, also known as AGER), and this mediates the induction of inflammation. The final class of endogenous inducers is breakdown products of the extracellular matrix (ECM) generated during tissue damage or malfunction (Medzhitov, 2008). One of the best-studied components in such processes is the glycosaminoglycan hyaluronate. Under normal conditions, hyaluronate is present as an inert high-molecular-weight polymer. Tissue injury causes its breakdown into low-molecular-weight fragments, which can induce inflammatory activity via activating toll-like receptor 4 (TLR4), resulting in a tissue-repair response (Jiang *et al.*, 2005; Medzhitov, 2008). This key conversion activity is regulated by reactive oxygen species (ROS)-dependent signalling (Jiang *et al.*, 2007). In fact, a number of endogenous pathways that initiate inflammatory responses are known to be dependent on ROS activity. Therefore, ROS is considered as a promising target for immunomodulation or anti-inflammation.

Recent findings from a spectrum of immunology scientists (Medzhitov, 2008, 2010; Medzhitov *et al.*, 2011) have increased awareness that inflammation comes in distinct forms and modalities, regulated by different molecular and cellular mechanisms of induction, regulation and resolution. Undoubtedly, a well-controlled inflammatory response is beneficial for homeostasis (e.g. in providing protection against tissue injury and pathogen infection); however, it can become very detrimental if dysregulation of the process occurs (e.g. resulting in septic shock). Therefore, it is highly important to govern various dysregulated acute inflammatory disorders with appropriate drugs and specific therapeutics. Interestingly, during the past few decades, the research focus on prevailing inflammatory conditions has shifted from treating acute inflammatory reactions in response to infections or/and tissue wounds to the newly defined chronic inflammatory states that accompany, obesity, type 2 diabetes, atherosclerosis, asthma, cancers and various neurodegenerative diseases (Donath and Shoelson, 2011; Nguyen and Casale, 2011; Ouchi *et al.*, 2011).

## B. IMMUNOMODULATION AND IMMUNOMODULATORS

Therapeutics for immunomodulation can be referred to as a therapeutic approach to intervene or adjust the auto-regulating immune responses to a desired level via immune-stimulation, immune-suppression or induction of immunologic tolerance. An immunomodulator can be defined as a substance or agent that can elicit immunomodulatory activities by altering or affecting immune cell systems to produce the desired immune response through dynamic regulation of the target immune systems (Spelman *et al.*, 2006). Immunomodulators have been traditionally divided into three groups: immunosuppressive agents, immunostimulators and tolerogens. Immunostimulators, also known as immunostimulants, are substances that can stimulate the immune systems by inducing the activation or augmenting the activity of immune system components. They are usually used in the treatment or control of infections, immunodeficiency and cancers. Immunosuppressive agents, also known as immunosuppressants, are substances that can reduce the ability of the immune system by inhibiting activation or decreasing the activity of its components. These types of agents are often used in organ transplantation and/or autoimmune diseases. Tolerogens are recognized to induce immunologic tolerance and make the immune system non-responsive to target antigens. Immunologic adjuvants can be considered as another type of immunomodulator, as they are agents that can stimulate the immune system and increase the response to a vaccine without possessing any specific antigenic effect alone. Various phytomedicines have been found to modulate the components of the inflammatory pathways including the various inducers, sensors, mediators and sensors mentioned above. Based on understanding of various immunomodulation activities and the profound effects of certain traditional medicines on these activities, we suggest that plant-derived secondary metabolites as natural products could be important resources for future development of immunomodulators into immunotherapies.

Taking the treatment of allergic disease as an example, immunomodulation aims at decreasing the pathologic immune response such as inflammation instead of causing an unwanted return to an immunologically naive or unresponsive state (Nguyen and Casale, 2011). On the basis of our knowledge of innate and adaptive immune responses at both the molecular and cellular levels, various immunomodulators for a number of allergic diseases, including asthma, allergic rhinitis and eosinophilic esophagitis, have been developed (Akdis *et al.*, 2005; Chang *et al.*, 2007). As several approaches for exploring immunomodulation activity in mouse models of allergic disease have not been effective to translate into useful results in human clinical trials, the pleiotropic nature of associated or related cytokines/chemokines and the underlying effector mechanisms of the varied phenotypes of these diseases need to be carefully investigated to develop future treatment for such diseases. The common goals for treating these diseases are to decrease the excessive T-helper 2 (Th2) response via various mechanisms such as (1) blocking critical Th2 cytokine activities, (2) inhibiting Th2 cytokine synthesis, (3) blocking critical Th2 effector molecules, (4) inhibiting key cell-type populations involved in Th2 response and (5) stimulating Th1 responses for balance (Nguyen and Casale, 2011). Therapies directed against specific effector molecules, including immunoglobulin E for targeting the IL-4/IL-13 receptor and augmenting the Th1/Th2 balance, are promising targets for immune-modulation therapy of allergic diseases. Herbal extracts from *Ganoderma lucidum*, *Glycyrrhiza radix* and *Sophorae flavescentis* Radix were found to reduce eosinophil infiltration of the lungs and inhibit airway hyperresponsiveness (AHR) in ovalbumin (OVA)-sensitized mice via reducing the levels of IgE and Th2-associated cytokines (IL-5, IL-4 and IL-13) and increasing the level of IFN-$\gamma$ secretion (Busse *et al.*, 2010; Shen *et al.*, 2011).

In addition to the suppression of inflammatory responses, an important approach for immunomodulation is to boost an individual's immune defence systems by giving either physiologic or supraphysiologic dosages of exogenous cytokines or therapeutic substances to treat the associated chronic malignancies and viral infections (Nelson and Ballow, 2003). The most studied approaches consist of pathogen-derived vaccines, tumour cell-based vaccines, DC-based immunotherapy and peptide vaccines (Melief, 2008; Smyth *et al.*, 2001). A number of clinical studies for these approaches have demonstrated the safety, but not necessarily satisfying clinical efficacy of such experimental medicines (Robson *et al.*, 2010). Moreover, there is an emerging consensus that the most efficacious therapies will activate several specific components of the immune system (Whelan *et al.*, 2003). Cancer immunotherapy using cytokines is an important and attractive approach for cancer therapy; however, optimizing the pharmacological doses to avoid cytotoxic reactions remains a very challenging issue (Chada *et al.*, 2003). Several cytotoxic drugs such as paclitaxel have been shown to also confer immunomodulatory effects at relatively low doses and exhibit immunity-dependent curative effects in animal models (Mizumoto *et al.*, 2005; Shin *et al.*, 2003). Combinational therapies using low-dose anti-cancer agents and cytokines together have revealed some benefits in some studies. It has also been shown that inducing T-helper (Th) 1-promoting cytokines using specific adjuvants is vital for enhancing certain anti-tumour immunity, and thereby preventing or reducing tumour growth (Garg *et al.*, 2010; Wen *et al.*, 2011). Therefore, the development of specific phytocompounds from herbal medicines as immunomodulatory agents to be used as either adjuvants or therapeutics for cancer treatment or immunotherapy is an emerging clinical issue. For example, specific phytocompounds from *Dioscorea batatas* (DsCE-I) were shown to increase the promoter activities of nuclear factor kappa B (NF-κB)-inducible ELAM and GM-CSF promoter constructs and protect animals against certain test cancers (Su *et al.*, 2008). Another group of agents have been shown to have the potential to stimulate hematopoietic recovery in patients suffering from cytopenias resulting from disease- or therapy-related bone marrow suppression (Nelson and Ballow, 2003). For example, phytocompounds from *D. batatas* (DsCE-II) extracted using a different fractionation procedure was proposed as adjuvant therapy, to be used alongside chemotherapy (Su *et al.*, 2011b).

## C. VARIOUS IMMUNE CELLS INVOLVED IN THE IMMUNE RESPONSE

### *1. Innate immune cells*

Innate immunity is the first line of host defence against malignant transformation and pathogen infection (Medzhitov and Janeway, 1997). In the inflammatory pathway shown in Fig. 2, the immune cells probably play a vital role in the components of sensors, mediators and effectors. Therefore, exploration and understanding of the roles of the various immune types in the immune system and the underlying mechanisms and interactions in/between the cells related to pathogenesis to the immune-related disorders are critically important for developing immunotherapeutics or immunomodulator agents. As shown in Fig. 2, the innate immune mechanisms known to be involved in immunomodulation are orchestrated by an array of cells, including NK cells, NKT cells, γδ-T cells, macrophages, granulocytes (neutrophils, eosinophils and basophils) and DCs. Adaptive immunity is created by networking among B cells, naïve $CD4^+$ T cells, differentiated $CD4^+$ T cells including helper T cells (including Th1, Th2, Th17 cells), induced regulatory T cells (iTreg cells) and the natural regulatory T cells differentiated from thymus. The immune functions or dysfunction of some of the key immune cell types that play an essential role in various immune diseases are briefly described below.

Among cells involved in innate immunity, DCs not only act as front-line cells to confer phagocytosis and produce cytokines and chemokines against invading pathogens, but they are also the most specialized professional antigen-presentation cells (APCs) with a unique T-cell stimulatory ability that plays a vital role in the follow-up adaptive immune responses in most immune diseases. In general, APCs include DCs, macrophages and B cells, all of which play a crucial role in antigen presentation (Joffre *et al.*, 2009). They mature after encountering various "danger signals" and can initiate subsequent immune processes leading to activation of antigen-specific T-cell response. DCs are well known as key immune cells, as highlighted by the awarding of half the Nobel Prize in Physiology or Medicine 2011 to Ralph M. Steinman for his discovery of DCs and their role in adaptive immunity (Travis, 2011). Steinman demonstrated the presence of this new immune cell type in 1973.

Physiologically, DCs act as sentinels in peripheral tissues where they encounter invading pathogens or other danger signals in the course of an infection. PRRs on the DCs recognize general PAMPs from microbial signatures and enable DCs to detect these molecular species from different pathogens including bacteria and viruses. With binding of PAMPs to PRRs on DCs, this ligand–receptor activity can instigate DC activation and induction of the maturation process (Diebold, 2008; Reis e Sousa, 2001). During their maturation, DCs perform the uptake, processing and presentation of antigen-containing or antigen-expressing materials as epitopes from their environment (Mellman and Steinman, 2001). Immature DCs (iDCs) can usually pick up foreign materials from their environment, but they are inefficient in antigen presentation (Mellman and Steinman, 2001). Activity antigen processing and presentation from the ingested materials are only induced once DCs are activated and undergo maturation (Robson *et al.*, 2010). Particularly, since recycling of these molecules and their passage through the endosomal class II-rich compartments cease upon DC activation, the levels of major histocompatibility complex class II (MHC-II) molecules on the cell surface are elevated (Petersen *et al.*, 2010). Consequently, DCs increase the levels of antigen at the cell surface and impart a snapshot of antigens derived from the target pathogen they encountered during infection. Subsequently, DCs cease to take up and process any new antigenic materials from their environment (West *et al.*, 2004). Further, the DC maturation process entails a change in the upregulation of co-stimulatory molecules such as CD40, CD80 and CD86 molecules on the DC surface which can act as maturation markers and in an increase in chemokine receptor expression level of CCR7 (Scandella *et al.*, 2004). Expression of CCR7 accompanied by inflammatory mediators such as prostaglandin E2 at the site of infection enables DCs to migrate from the inflamed tissue to the draining lymph node (Scandella *et al.*, 2004). Once they arrive in the draining lymph node, the activated DCs interact with naïve T cells. The key determinants of DC-derived signals that induce these interactions and immune response are the levels of antigen presentation (signal 1), the expression level of co-stimulatory molecules (signal 2) and the presence of immunomodulatory factors such as specific cytokines (signal 3) (Diebold, 2008).

Increased levels of antigen presentation and the expression of co-stimulatory molecules on DCs are very important for the expansion of antigen-specific T cells, whereas they are not sufficient for the induction of effector functions (Diebold, 2008). Immunomodulatory factors such as cytokines (signal 3) can determine the differentiation of expanded T cells into effector cells (Sporri and Reis e Sousa, 2005). The ability of DCs to induce differentiated effector functions in T cells enables the immune system to adjust its response to combat diverse classes of pathogens or stimuli. As shown in Fig. 2, the different cytokine expression patterns from DCs can help differentiate distinct forms of effector T cells or regulatory T cells. Dysfunction of DCs is involved in pathogenesis of a variety of immune diseases including type 1 diabetes, rheumatic disease, psoriatic arthritis, inflammation, microbial infection and cancer. Therefore, due to their various unique and multifacet features, DCs are a promising therapeutic target for skewing differentiation of T cells to treat a variety of immune diseases, especially cancers. We believe that a spectrum of phytochemicals, derived from plant secondary metabolites from traditional medicines, may be applicable for use as immunomodulators for regulating various DC functions. For instance, we showed that phytocompound mixtures extracted from the butanol fraction (BF) of a stem and leaf (S + L) extract of *Echinacea purpurea* ([BF/S + L/Ep]) can modulate DC mobility and related cellular physiology in mouse immune systems (Wang *et al.*, 2008a; Yin *et al.*, 2010).

Macrophages are other key players in the innate immunity system (Fig. 2). They are critical effectors and regulators of inflammation and the immediate arm of the immune system; they can, however, also confer antigen-presentation ability. They are the resident cells which perform phagocytosis in lymphoid and non-lymphoid tissues and are involved in steady-state tissue homeostasis via the clearance of cell debris from both apoptotic and necrosis cells, and the production of various growth factors (Geissmann *et al.*, 2010; Qian and Pollard, 2010). Macrophages can use a broad range of pathogen-recognition receptors (PPRs) to become efficient at phagocytosis and induce production of pro-inflammatory cytokines. Timely and efficient production of pro-inflammatory cytokines and nitrogen species as well as extensive production of reactive oxygen from macrophages may serve as protective mechanisms. Different types of macrophages have been recently characterized according to their functional participation in particular immunological responses (Qian and Pollard, 2010). The "activated" macrophages (M1) are defined as cells involved in the responses of type I helper T (Th1) cells to pathogens such as bacteria. This population is activated by IFN-$\gamma$ and engagement of TLRs and has the characteristics of elevated expression level of MHC-II, production of IL-12 and TNF-$\alpha$, generation of nitric oxide (NO) and ROS and the ability to kill pathogens and undesirable or stressed host/endogenous cells. In contrast, "alternatively activated" macrophages (M2) that can differentiate in response to IL-13 and IL-4 play a key role in Th2-type responses, including wound healing and humoral immunity (Qian and Pollard, 2010). The developmental origin and the function of tissue macrophage subsets are very diverse and include microglia, dermal macrophages and liver macrophages (Kupffer cells). These cells remain poorly understood (Qian and Pollard, 2010). Nonetheless, it has been shown that they do play an important role in sepsis, inflammation, liver disease, obesity and cancers (Qian and Pollard, 2010). The use of phytomedicines as immunomodulatory agents for treating macrophage-related immune diseases may be a promising approach for developing new generation of therapeutics. Taking specific phytocompounds as examples, we showed that shikonin selectively inhibits the expression of TNF-$\alpha$ at the mRNA splicing level (Chiu and Yang, 2007) and also significantly inhibits the early mRNA expression of inflammatory cytokines including TNF-$\alpha$, IL-1$\beta$ and IL-4 and chemokines CCL4 and CCL8 (Chiu *et al.*, 2010). Caffeic acid derivatives, ethyl caffeate, a natural phenolic compound isolated from *Bidens pilosa* plant, markedly suppressed lipopolysaccharide (LPS)-induced NO production, mRNA and protein expression of inducible nitric oxide synthase (iNOS), and PGE2 production in RAW 264.7 macrophages and significantly inhibited the TPA-induced COX-2 expression in mouse skin tissues (Chiang *et al.*, 2005).

NK cells are known as effector lymphocytes of the innate immune system and control various types of tumour growth and microbial infection mechanistically by limiting their spread and subsequent tissue damage (Vivier *et al.*, 2008). NK cells have a number of traits in common with $CD8^+$ T cells (Sun *et al.*, 2009b). Recent studies have shed light on a new role for NK cells in different immune responses, suggesting that these innate lymphocytes have the characteristics of both innate and adaptive immunities (Cooper *et al.*, 2009; Vivier *et al.*, 2011). Activation of NK cells through NKG2D can result in cytotoxicity and cytokine production. This activation may be triggered by the disappearance of class I MHC molecules from the cell surface of tumour cells or by exposure to antigens such as MHC class I-related chains A and B (MICA and MICB), which are NKG2D ligands. Their expression is induced by DNA damage in tumour cells (Garg *et al.*, 2010). Phytochemical-derived medicines may prove to be the highly useful resources for developing immunomodulatory agents for controlling NK cell activity. For example, oral administration of the total flavones and polysaccharides of *Epimedium* at doses of 240 mg/kg for 30 days was shown to significantly enhance the activities of NK cells in aged rats (Ma *et al.*, 2011). Another study indicated that the aqueous extract of *Nigella sativa* can significantly enhance NK cytotoxic activity against specific cancer cells, suggesting that the documented anti-tumour effects of *N. sativa* may be in part due to its ability to stimulate NK anti-tumour activity (Majdalawieh *et al.*, 2010).

Natural killer T cells (NKT cells) are lipid antigen-reactive, CD1d-restricted, immunoregulatory T lymphocytes that can enhance cell-mediated immunity against infectious organisms such as bacteria and some self or endogenous antigenic determinants as from tumours (Godfrey *et al.*, 2010). The invariant natural killer T (iNKT) cells are a subset of αβ T-cell receptor $(\alpha\beta TCR)^+$ T cells which are restricted by CD1d molecules. They can modulate the activities of DC cells and B cells and can increase DC-induced B- and T-cell responses. The iNKT cells can amplify TLR-derived signals. It is thought that combinations of specific compounds that can activate iNKT cells may provide a formulation that could serve as a vaccine adjuvant (Cerundolo *et al.*, 2009). In addition, iNKT cells express an invariant T-cell receptor α chain that recognizes glycolipid antigens presented by CD1d molecules present on the surface of tumour cells, allowing receptor/ligand action NKT cells to subsequently elicit their anti-tumour effects primarily via secretion of IFN-γ and directly effect cytotoxicity. iNKT cells are recognized as a unique population of T cells with immunomodulatory properties that can link innate and adaptive immune responses (Cerundolo *et al.*, 2009; Godfrey *et al.*, 2010).

*2. Adaptive immune cells*

The adaptive immune system includes two major types of lymphocytes, T cells and B cells, that are made up of several subsets (Fig. 2). B cells can differentiate into plasma cells that secrete antibodies. T lymphocytes or T cells are further divided into two classes, $CD4^+$ T cells and $CD8^+$ T cells (Janeway, 2005). $CD8^+$ T cells can differentiate into cytotoxic T cells, which kill virus-infected cells and tumour cells, whereas $CD4^+$ T cells differentiate into different subsets of $CD4^+$ T cells which can activate other cell types for execution of specific immune functions.

B lymphocytes are required for the induction of effective antibody-based immunity following pathogen challenge. The antibody response of B cell-mediated humoral immunity can be activated by T-helper cell type-2 (Th2 cells). Currently available vaccines have been mostly developed to explore the specificity of antibodies produced by B lymphocytes, for protection against diseases such as diphtheria, tetanus, hepatitis, measles and pneumococcal and meningococcal infections (Makela, 2000). Phytocompounds, phytochemicals or phytoextracts which can regulate B-lymphocyte effector functions have the potential to be employed as a useful tool for the maintenance of protective immunity; however, the efficacy of specific vaccines is currently limited. For example, Quan *et al.* (2007) reported that the intranasal co-administration of inactivated influenza virus A and *Panax ginseng* on days 0 and 14 significantly increased the levels of influenza virus-specific IgG in the serum as compared to that control in mice, possibly due to high saponin content. In this study, *P. ginseng* was shown to elevate the mouse lung IgA level at 15 days post-challenge with influenza virus, suggesting that phytochemicals from *P. ginseng* can apparently modulate systemic and mucosal immunity and may act as a powerful mucosal adjuvant for vaccination.

$CD8^+$ T cells (cytotoxic T lymphocytes, CTLs) are very potent professional killers, particularly important for protection against virus-infected cells and tumour cells. Some reports suggest that one single activated CTL cell can eliminate hundreds of target tumour cells (Garg *et al.*, 2010). Previous studies also showed that oral administration of *S. cerevisiae*-derived β-glucan in mice elevated the levels of $CD8^+$ intraepithelial lymphocytes (IELs) in comparison with control mice (Tzianabos, 2000).

$CD4^+$ T cells are the major orchestrators and conductors of the adaptive immune response. Upon interaction with antigen-presenting cells such as DCs, naïve CD4+ T cells can differentiate into a variety of effector subsets, including the classic T-helper cells (Th1 and Th2 cells), as well as recently defined Th17 cells and inducible regulatory T (iTreg) cells (Zhou *et al.*, 2009), as described in Fig. 2. Differentiation is determined predominantly by the specific cytokines present in the microenvironment and by the strength of the interaction between the T cell antigen receptor and target antigen (Sakaguchi *et al.*, 2008). Traditionally, Th1 cells produce IFN-$\gamma$ and contribute to cellular immunity against intracellular microorganisms such as bacteria. IL-12 is effectively produced by innate immune cells such as DCs, and the IFN-$\gamma$ produced by both T cells and NK cells can skew the polarization of cells towards Th1 cell differentiation through action of T box transcription factor (T-bet). Th2 cells can produce cytokines IL-4, IL-5 and IL-13, which are essential for humoral immunity in control of infection from helminths and other extracellular pathogens. Th2 cell differentiation attributes to the action of GATA3, which occurs downstream of IL-4 action. Th17 cells can produce IL-17A, IL-17F and IL-22, and they play vital roles in clearance of extracellular fungi and bacteria, especially in mucosal immunity (Medzhitov *et al.*, 2011). Th17 cell differentiation is mediated by retinoid-related orphan receptor (ROR)$\gamma$t, a transcription factor that is activated by TGF-$\beta$ in combination with the pro-inflammatory cytokines such as IL-6, IL-23 and IL-21 (Sakaguchi *et al.*, 2008). Regulatory T (Treg) cells are characterized by the expression of Forkhead box P3 (FOXP3)$^+$ genes and can be classified into two categories: iTreg cells differentiated from naïve CD4$^+$ T cell and natural Treg (nTreg) cells that arise from the thymus. Aberrant control or malfunction of Th1 and Th17 cell responses may contribute to organ-specific autoimmunity, whereas Th2 cells contribute to atopy, allergy and asthma. Treg cells play crucial roles in regulating these effector T cell responses, thereby preventing the body from potential pathogenic effects (Sakaguchi *et al.*, 2010). Various phytochemicals, phytocompounds or phytoextracts from traditional medicines may be of use to maintain or optimize our immune system via the modulation of the different subsets of helper T cells. For instance, *Tripterygium wilfordii* Hook. F (TWHF) has been evaluated for treating autoimmune diseases including rheumatoid arthritis (RA). Triptolide, the diterpene purified from this plant, was shown to inhibit peripheral CD4$^+$ T lymphocytes but increase CD8$^+$ T lymphocyte in Peyer's patches of mice in a collagen-induced mouse arthritis model (Zhou *et al.*, 2006).

## III. MEDICINAL HERBS WITH IMMUNOMODULATORY ACTIVITIES

Five medicinal herbs have been selected in this section for detailed review. All have a long history of human use as traditional or folk medicines. Echinacea was a top-selling herbal remedy in the USA between 1995 and 2005 and has been used as a traditional medicine or nutraceutical in the

USA and Europe for decades or perhaps centuries. The other four medicinal plants have been extensively used in TCM or Taiwanese traditional medicine as single herbs or in formulation with other herbs for specific indications. Evidence accumulated from a series of studies by our group (Chiu *et al.*, 2010; Wang *et al.*, 2006, 2008a,b; Yin *et al.*, 2010) has demonstrated the immunomodulatory activities of *E. purpurea*, *Lithospermum erythrorhizon* and *D. batatas* through *in vitro* and *in vivo* biological assay systems using transgenic and omics research approaches. Artemisinin from *Artemisia annua* has recognized benefit and use in the treatment of malaria, and Dr. Tu recently won the 2011 Lasker Award in medical research for her findings relating to the plant (Tu, 2011). The research revealed the importance of *A. annua* phytochemicals not only for use in malaria but also for its potential application in inflammatory diseases (Tu, 2011). *T. wilfordii* Hook. F has been traditionally used for treating autoimmune diseases including RA (Brinker and Raskin, 2005; Tao *et al.*, 2008). One of its well-known bioactive components, triptolide, has been shown to possess a strong immunosuppressive effect and has the potential to treat a series of autoimmune diseases.

### A. *ECHINACEA PURPUREA*

*Echinacea* is a top-selling herbal remedy in the United States. It has been claimed to confer high immunostimulatory activity by acting as an immunopromoter (Ernst, 2002). It is reputed to alleviate respiratory infections and colds, including sore throats, coughs and other symptoms (S *et al.*, 2011). *Echinacea angustifolia*, *Echinacea pallida* and *E. purpurea* are the three major species used in traditional medicine or nutraceutical applications in the United States and Europe (Borchers *et al.*, 2000). The most common constituents of *Echinacea* are alkamides, caffeic acid derivatives (shown in Fig. 4), polysaccharides and lipoproteins (Pietta *et al.*, 1998). The active components present in *Echinacea* may vary due to differences in plant age and organ portion, agricultural conditions, geographical location and tissue extraction methods (Perry *et al.*, 2001).

An accumulating number of studies have reported the effects of *E. purpurea* from the perspective of immune functions and systems (Brush *et al.*, 2006; Mishima *et al.*, 2004). The most frequently reported pharmacological activities of *Echinacea* are the activation of macrophages and polymorphonuclear neutrophils immune cells (Goel *et al.*, 2005; Sullivan *et al.*, 2008). A recent study has shown that macrophage phagocytosis and NK cell activities can be strongly activated after *ex vivo* exposure of these cells to *E. purpurea* extracts (See *et al.*, 1997). Reports of increased macrophage phagocytic activity from mouse liver and spleen following oral administration of *E. purpurea* extract

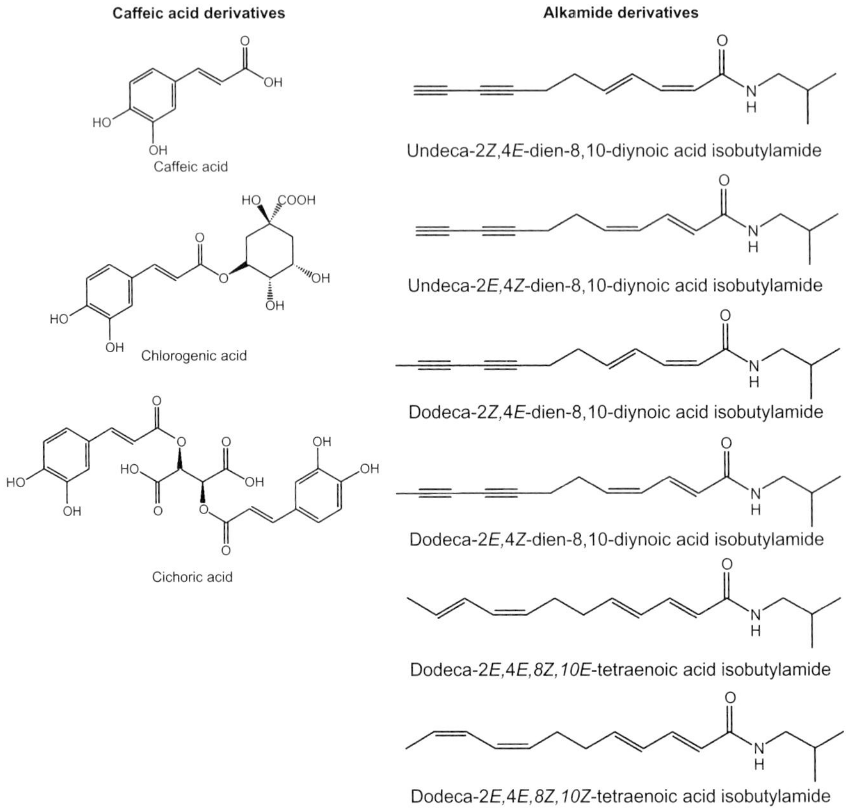


Fig. 4. Chemical structures of caffeic acid derivatives and alkamide derivatives from *Echinacea purpurea*.

have been shown to result in enhanced expression of cytokines including TNF-α, IL-1α, IL-1β, IL-6 and IL-10, and NO production (Rininger *et al.*, 2000). It was also determined that *E. purpurea* extract enhanced phagocytic activity in human peripheral blood mononuclear cells (Rininger *et al.*, 2000). Most recently, Sasagawa *et al.* (2006) found that low concentrations of the ethanolic extracts obtained from aerial portions of *E. purpurea* suppressed the ability of activated T cells to express IL-2, a key cytokine involved in the early phase of Jurkat T-cell activation. Moreover, a recent study reported that differential expression of key accessory molecules was detected in polysaccharide-enriched *E. purpurea* root extract and the ethanolic, alkamides-enriched *E. purpurea* leaf extract (Benson *et al.*, 2010). The *E. purpurea* root extract increased the expression of MHC class II, co-stimulatory markers (CD86 and CD54) and pro-inflammatory cytokines (IL-6 and TNF-α), while the

*E. purpurea* leaf extract decreased the expression of the tested markers and cytokines, suggesting that the root extract and leaf extract from the same *E. purpurea* plant stimulated and inhibited immune activities, respectively (Benson *et al.*, 2010). Similar effects were also observed in our investigation of the immunomodulatory effects of *E. purpurea* extracts on human monocyte-derived DCs and mouse bone marrow-derived DCs (Wang *et al.*, 2006, 2008a). We employed a chemically defined *E. purpurea* extract, termed [BF/S + L/Ep], containing hypoxanthine, chlorogenic acid, caffeic acid, cichoric acid, quercetin-3-*O*-rhamnosyl-(1–6)-galactoside, kaempferol-3-*O*-rhamnosyl-(1–6)-galactoside and rutin as index compounds (Wang *et al.*, 2008a). Our findings suggested that the [BF/S + L/Ep] phytochemical mixture was able to modulate cell adhesion-, cell mobility-, cytokine- and NF-κB signalling-related activities in primary cultures of mouse DCs, and it could also enhance the mobility of DCs to target specific lymphoid tissues in test mice in *in vivo* trafficking experiments (Yin *et al.*, 2010). These studies were performed using a network knowledge-based approach to analyse the genome-wide transcriptome activity *in vitro* and *in vivo*, and to correlate specific proteome activities and special functional genomic phenotypes in test cells (Wang *et al.*, 2006, 2008a). Further, we also showed that the alkamides can play an important role in anti-inflammatory activities of *Echinacea*, as revealed by comparative metabolomics approaches and cell- and gene-based assays (Hou *et al.*, 2010). Further, the possible receptors, cannabinoid (CB1 and CB2), were reported to mediate the bioactivities and pleiotropic effects of *E. purpurea* by manipulating the endocannabinoid system through molecular targeting to receptors, endocannabinoid transport and degradation (Chicca *et al.*, 2009; Hohmann *et al.*, 2011).

The harvest of medicinal herbs from different regions and at different periods of the year has been shown to play a role in the quantity of bioactive components found in plants and their associated pharmacological activities (Jia and Zhao, 2009; Liu *et al.*, 2007b). With *E. purpurea*, for example, plant extracts have been found to display differential profiles and varied amounts of phenolic compounds including caffeic acid, cichoric acid, chlorogenic acid and alkamides, in different seasons and months of plant growth and/or year of harvesting, and sometimes according to different post-harvest treatment (Hou *et al.*, 2010; Liu *et al.*, 2007b). Since some of the phytochemicals of this plant are bioactive components contributing to the immunomodulatory activities of *E. purpurea*, the differential amounts of these components may result in distinguishable biological effects.

Functional and comparative genomics analysis of the cellular and immunological effects of different anti-inflammatory phytoextracts or phytocompounds, especially via microarray analysis, is recognized as a promising approach to distinguish the complex and specific bioactivities of candidate phytomedicines (Chiu *et al.*, 2010; Wang *et al.*, 2008a,b). However, combining this with other sets of data on protein expression such as proteomics or Western blot analyses is critically important for verification of the transcriptome result. For example, Wang *et al.* and her colleagues investigated the specific and differential gene expression in human iDCs in response to treatment with a BF containing defined bioactive phytocompounds extracted from the stems and leaves of *E. purpurea*, denoted as [BF/S + L/Ep] (Wang *et al.*, 2008a,b). The results from Affymetrix DNA microarray showed significant upregulation of specific genes for cytokines (IL-1β, IL-8 and IL-18) and chemokines (CCL-2, CCL-5 and CXCL-2) within 4 h after [BF/S + L/Ep] treatment of iDCs. Bioinformatics analysis of genes expressed in [BF/S + L/Ep]-treated DCs showed a key signalling network involving a number of immunomodulatory molecules, possibly leading to the activation of a downstream molecule, adenylate cyclase 8. Confirmed with proteomic analysis, results also showed upregulation of antioxidant defence enzymes such as Mn-SOD and downregulation of cytoskeletal proteins such as cofilin after treatment with [BF/S + L/Ep] and cichoric acid. These data were further verified by Western blot analyses.

### B. *DIOSCOREA BATATAS*

*Dioscorea* species are widely used plants not only in Eastern traditional medicine but also in modern Western medicine. *D. batatas* (yam), which is widely distributed in East Asia, has long been used as a supplement as a major source of steroid precursors (Li and Ni, 2011) or prescribed to treat poor appetite, chronic diarrhoea, asthma, frequent or uncontrollable urination, diabetes and even emotional instability (Hou *et al.*, 2002). Several active components in tubers of *D. batatas* have been shown to exhibit immunomodulatory activities (Oh and Lim, 2009; Su *et al.*, 2011b). These phytochemicals include mucopolysaccharide, dioscorin, diosgenin (Fig. 5), batatasins and glycoproteins.

Dioscorin, a tuber protein, has been shown to exhibit systemic and mucosal immunomodulatory activities *in vivo* after oral administration (Liu *et al.*, 2009). Dioscorins, the storage protein of *D. batatas* tuber, can enhance the proliferation of $CD4^+$, $CD8^+$ and $CD19^+$ cells in spleen (Lin *et al.*, 2009). Dioscorin also can act as a TLR4 activator and induce macrophage activation via the typical TLR4-signalling pathways via stimulation of multiple signalling molecules (NF-κB, ERK, JNK and p38) and induction of the expression of pro-inflammatory cytokines (TNF-α, IL-1β and IL-6) (Fu *et al.*, 2006). The mucopolysaccharide in *D. batatas* can significantly

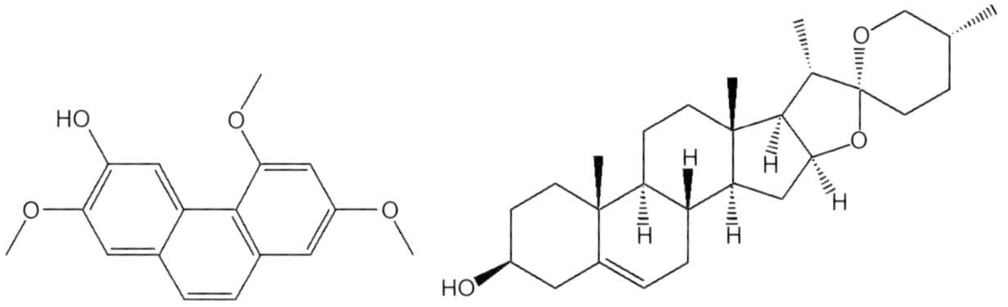

Batatasin I Diosgenin

Fig. 5. Chemical structures of diosgenin and batatasin I from *Dioscorea batata*.

increase IFN-γ production in treated splenocytes, suggesting that it may induce cell-mediated immune responses (Choi *et al.*, 2004). Besides, these mucopolysaccharides (50 μg/ml) were found to increase the uptake capacity and lysosomal phosphatase activity of test peritoneal macrophages (Choi *et al.*, 2004). Batatasin I (Fig. 5), with a well-identified structure of 6-hydroxy-2,4,7-trimethoxyphenanthrene, was shown to inhibit the generation of prostaglandin D2 and leukotriene C4 and degranulation activity in mouse bone marrow-derived mast cells (Lu *et al.*, 2011c). Some glycoproteins in *D. batatas* were shown to inhibit the expressions of IL-4 and IL-10 through modulation of GATA-3, STAT-6, p44/42 MAPK and p38 MAPK in mouse lymphocytes (Lin *et al.*, 2009), leading to the possibility that glycoproteins in such plants may be usefully applied for use as nutraceuticals or health supplements for prevention of Th2-mediated immune disorders (Oh and Lim, 2009). In addition, *Dioscorea* glycoproteins significantly increased the trafficking of macrophages, lymphocytes, neutrophils and monocytes into the peritoneal cavity (Huong *et al.*, 2011). Further, in addition to significant enhancement of proliferation of T cells and B cells in splenocytes of glycoprotein-treated mice, the non-specific cytolytic activity of NK cells and macrophages was significantly increased (Huong *et al.*, 2011). These glycoproteins also can stimulate specific immune system functions, including macrophage activation via increasing the expression levels of iNOS, IL-1β and TNF-α (Huong *et al.*, 2011). In our previous study, we found that a fraction of the *D. batatas* tuber extract significantly increased the GM-CSF promoter activity in normal and inflamed skin tissues (Su *et al.*, 2008). Our previous study reported that a 50–75% ethanol-partitioned fraction of the tuber extract of *D. batatas* (DsCE-II) may confer immunogenic activities (Su *et al.*, 2011b). DsCE-II contained polysaccharides with a high abundance of 1,4-linkage mannose (64%), which can preferentially promote the regeneration of CFU-GM cells in damaged bone marrow tissues in

5-fluorouracil-treated mice fed with DsCE-II (Su *et al.*, 2011b). DsCE-II efficacy level for bone marrow cell restoration was ~85% of that obtained by a subcutaneous administration of recombinant G-CSF proteins (5 μg/kg) in mice tested in parallel, suggesting that the DsCE-II fraction of *D. batatas* extract may be useful for further development as a dietary supplement for use alongside chemotherapy during cancer treatment (Su *et al.*, 2011b). Recently, we have also obtained results indicating that DsCE-I may be employed as an adjuvant for gene-based or protein subunit cancer vaccines (Yang *et al.*, 2011). In addition, the ethanol extract of bark of *D. batatas* was identified to confer anti-inflammatory bioactivity through inhibition of iNOS and COX-2 expression in RAW 264.7 cells, apparently via NF-κB and ERK1/2 inactivation (Jin *et al.*, 2010).

## C. *ARTEMISIA ANNUA*

*A. annua* is an ancient Chinese medicine still in common use today. It has long been utilized to treat malarial and autoimmune diseases, including systemic lupus erythematosus and RA (Christen and Veuthey, 2001). As shown in Fig. 6, artemisinin, also known as qinghaosu, was identified as

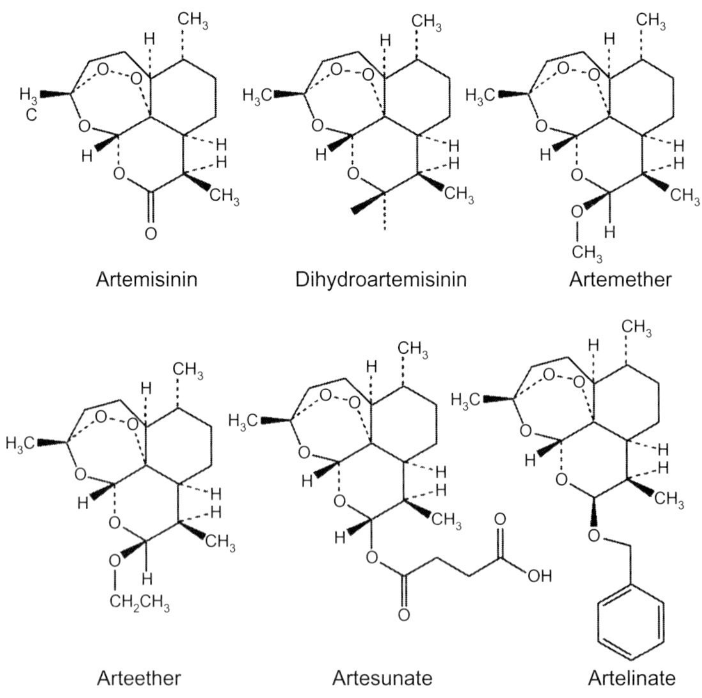


Fig. 6. Chemical structures of artemisinin and its derivatives.

the major active compound isolated from *A. annua* (Christen and Veuthey, 2001). Artemisinin is a sesquiterpene trioxane lactone, and its chemical structure contains a peroxide bridge, considered to be critical for its bioactivity and unique among antimalarial drugs (van Agtmael *et al.*, 1999). From the perspective of drug metabolism, artemisinin is primarily converted to inactive metabolites, while its derivatives, namely, artesunate, artelinate, artemether and arteether, can serve as parent compounds all exhibiting a very short half-life ($<10$ min), and be converted to the highly potent active metabolite, dihydroartemisinin (DHA), which has a much longer half-life ($\sim 1$ h) (Balint, 2001). The antimalarial mechanisms of artemisinin were shown to involve the interference of parasite transport proteins that can disrupt the function of parasite mitochondria, and most important, modulate the host immune response function (Golenser *et al.*, 2006).

Currently, the first-line antimalarial treatment for *Plasmodium falciparum* recommended by the World Health Organization (WHO) is the artemisinin combination therapy (Reyburn, 2010). Several such therapeutic approaches have been developed, for example, the formulation of one of artemisinin-derived phytochemical and one clinically therapeutic antimalarial drug, such as the combination of artemether and lumefantrine (Olliaro and Taylor, 2004). Artemisinin has also been reported to suppress LPS-induced proteolytic degradation of IκB, the translocation of NF-κB, and thus inhibit iNOS transcription, leading to the blockade of NO synthesis (Aldieri *et al.*, 2003). Artemisinin, artesunate and DHA were shown to enhance DNA synthesis by treatment with alloantigens or Con A and increase IL-2 production in mouse splenocytes, indicating that artemisinin and its derivatives may selectively promote T-cell function and accelerate immune reconstitution. These activities may be applicable for future therapy for the restoration of immune function (Yang *et al.*, 1993). Artemisinin has also been reported to inhibit the protein expression of p65 unit of NF-κB, the mRNA expression of NF-κB and TGF-β 1 and the levels of TNF-α and IL-6 in test mice with lupus nephritis, suggesting that artemisinin may be a reliable and effective treatment for lupus nephritis (Wu *et al.*, 2010). Artemisinin can also reduce angiotensin II-induced cardiac hypertrophy via inhibition of the NF-κB binding activity, and the mRNA expression levels of IL-6, TNF-α and MCP-1 (Xiong *et al.*, 2010). Artemisinin was also reported to prevent atherosclerosis via an inhibition of activation of THP-1 monocytes (Wang *et al.*, 2011b). Recently, artemisinin and its derivatives have been found to inhibit generation of NO in the RAW 264.7 mouse macrophage cell line (Konkimalla *et al.*, 2008). Among the compounds studied, artesunate showed the highest NO inhibition activity. Microarray analyses showed that the effects of artesunate in macrophages are associated mainly with NO metabolism and signalling (Konkimalla *et al.*, 2008).

## D. *TRIPTERYGIUM WILFORDII*

TWHF, sometimes named leigongteng (thunder god vine) from the Chinese, is another member of the traditional Chinese pharmacopoeia. The portion of TWHF plant in empirical TCM use is the debarked root, which has been anecdotally used for treating autoimmune diseases including RA, immune complex nephritis, systemic lupus erythematosus, organ transplantation; it has and even been used as an anti-cancer agent (Brinker and Raskin, 2005; Tao *et al.*, 2008). Starting in the 1970s, a series of TWHF-associated products claimed to have high therapeutic value were developed, patented and commercialized. Leigongteng was developed as a multi-glycoside tablet. A number of triterpenes, diterpenes and macrocyclic alkaloids have been identified as secondary metabolites from TWHF plant (Brinker and Raskin, 2005). Triptolide ($C_{20}H_{24}O_6$), a diterpene triepoxide, is the most well-studied component derived from TWHF and was the first recognized diterpenoid triepoxide containing an 18(4+3) abeoabietane skeleton shown in Fig. 7 (Kupchan *et al.*, 1972). Triptolide has been reported to exhibit multiple pharmacological activities including anti-inflammatory (Krakauer *et al.*, 2005), anti-neoplastic, proapoptotic (Antonoff *et al.*, 2009) and anti-angiogenic properties (Zhu *et al.*, 2010). Triptolide can suppress TLR-induced NF-κB activation and downregulate TLR4 and TRIF proteins (Premkumar *et al.*, 2010). Triptolide also can ameliorate Th1-mediated chronic colitis and the disordered immune state in IL-10(−/−) mice (Wei *et al.*, 2008). Triptolide has been shown to suppress the nuclear concentration of NF-κB and the secreted levels of IL-17, IL-21 and IFN-γ in parallel, showing greater potency in Th17 cells from young mice as opposed to older mice (Huang *et al.*, 2008). In addition, the triptolide-mediated inhibition of LPS-induced activation of PI3K/Akt and NF-κB was found to involve the downregulation of COX-2 and CCR7 expression resulting in impaired migration to secondary lymphoid organs of test DCs (Liu *et al.*, 2007a).

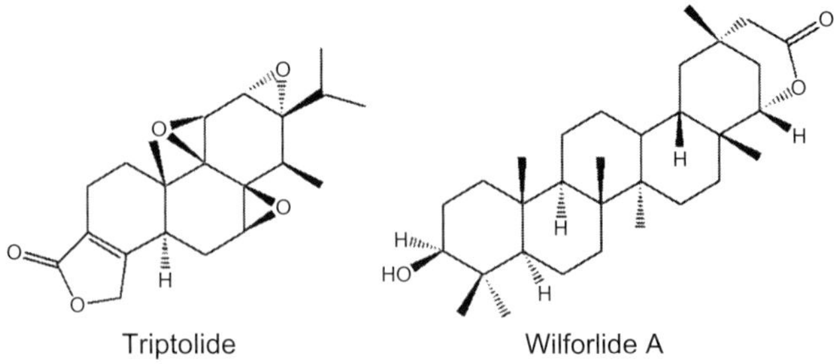


Fig. 7. Chemical structures of triptolide and wilforlide A from *Tripterygium wilfordii.*

Triptolide can inhibit staphylococcal exotoxin-stimulated T-cell proliferation and the expression of IL-1β, IL-6, TNF, IFN-γ, MCP-1, MIP-1α and MIP-1β in human PBMCs (Krakauer *et al.*, 2005). Previous studies showed that triptolide inhibited the secretion of RANTES, TARC and IP-10 from LPS-stimulated DCs, resulting in impaired DC-mediated chemoattraction of neutrophils and T cells under both *in vitro* and *in vivo* test conditions (Liu *et al.*, 2006b). Triptolide, at a high concentration, was also observed to induce apoptosis of DCs through sequential activity in p38 MAP kinase phosphorylation and caspase-3 activation (Liu *et al.*, 2004). It decreased the expression of CD80 and CD86 and the secretion of IL-12p40 and IL-12p70 in THP-1 cells leading to impaired antigen-presenting functions (Liu *et al.*, 2004). In Jurkat T cells, triptolide inhibited phorbol myristate acetate (PMA)/Iono-stimulated IL-2 transcription through regulation of purine-box/antigen receptor response element (ARRE)/nuclear factor of activated T cells (NF-AT) and NF-κB transcriptional activation (Qiu *et al.*, 1999). In a functional genomics study, triptolide treatment affected the expression of 22.5% of 195 immune signalling genes (Premkumar *et al.*, 2010). Shao *et al.* (2004) further provided evidence that triptolide could significantly attenuate TNF-α-induced COX-2, iNOS, PGE2 and NF-κB and suppress the subsequent NO production in human RA synovial fibroblasts (Shao *et al.*, 2004). Triptolide ameliorated the clinical signs of experimental autoimmune encephalomyelitis by induction of heat shock protein 70 and stabilization of NF-κB/IκBα transcriptional complex (Kizelsztein *et al.*, 2009). Triptolide also significantly reduced the inflammatory responses and cartilage damage in the joint tissues in test mice with collagen-induced arthritis (CIA), apparently by interfering with the CIA-induced expression of matrix metalloproteinase (MMP)-13 and -3 and by augmenting tissue inhibitors of metalloproteinases (TIMP) 1 and 2 (Lin *et al.*, 2007). Triptolide significantly inhibited the generation of Th17 cells from murine splenocytes and purified $CD4^+$ T cells in a dose-dependent manner via inhibition of the transcription of IL-17 mRNA and the IL-6-induced phosphorylation of STAT3 (Wang *et al.*, 2008b). Further, triptolide effectively inhibited the expression of IFN-γRα, pJak2, pSTAT1 and ICAM-1 in HaCaT cells (Hongqin *et al.*, 2011). IL-12 and IL-23 produced by antigen-presenting cells are known as key factors for the generation and function of Th1 and Th17 cells, respectively, and they have been strongly implicated in the pathogenesis of a number of autoimmune disorders (Wei *et al.*, 2011). Triptolide was able to inhibit the expression of the *p40* gene at the transcriptional level in part through the activation of CCAAT/enhancer-binding protein-α (C/EBPα), thus inhibiting p40 expression (Zhang and Ma, 2010). Triptolide can activate the transcription of C/EBPα and enhance the phosphorylation of Ser21 and Thr222/226 which are critical for C/EBPα inhibition of p40 (Zhang and Ma, 2010). C/EBPα activation by triptolide is dependent on the upstream kinases ERK1/2 and Akt-GSK3β activities (Zhang and Ma, 2010). Triptolide also inhibited the migration of lymphoma cells to lymph nodes *in vitro*, and blockage of the SDF-1/CXCR4 axis by triptolide may contribute to a potential anti-metastatic effect (Zhang *et al.*, 2006). Triptolide also effectively blocked the induction of miR-155 RNA (Matta *et al.*, 2009). Wilforlide A (Fig. 7), another tripterygium glycoside, has also been found to confer efficacious anti-inflammatory and immune suppressive activities in carrageenan-induced rat pedal swelling and tampon-induced rat granulation models (Xue *et al.*, 2010).

## E. *LITHOSPERMUM ERYTHRORHIZON*

The dried root of *L. erythrorhizon*, known as zicao or purple gromwell and referred to as *shikon* in Japanese, is a commonly used traditional Chinese herbal medicine in China and Taiwan (Novosel'tseva *et al.*, 1979). It has been used for thousands of years for treatment of macular eruptions, measles, smallpox, eczema, carbuncles and burns (Novosel'tseva *et al.*, 1979). Shikonin and its derivatives are the primary active components isolated from root tissues of the traditional Chinese medicinal herb *L. erythrorhizon* and have recently garnered considerable interest for their broad spectrum of anti-inflammatory activities and significant anti-tumour activities (Chen *et al.*, 2002; Staniforth *et al.*, 2004; Su *et al.*, 2008). The chemical structure of shikonin and its derivatives are shown in Fig. 8. In this section, we focus on the primary active compound, shikonin. Our previous study showed that shikonin drastically suppressed the transcriptional activity of GM-CSF promoter by inhibiting the binding of the TFIID protein complex to the TATA box (Su *et al.*, 2008). In addition, shikonin effectively inhibited the promoter/transcriptional activity of the pro-inflammatory cytokine TNF-α (Staniforth *et al.*, 2004). Interestingly, at a relatively low concentration (0.1 μM), shikonin also specifically blocked the splicing of TNF-α pre-mRNA (Chiu and Yang, 2007). Shikonin can further confer a drastic and acute effect in human monocytes at the genomic and proteomic levels (Chiu *et al.*, 2010). We demonstrated that shikonin significantly inhibited the early expression (within 0.5 h) of approximately 50 genes, notably cytokines TNF-α, IL-1β and IL-4, chemokines CCL4 and CCL8 and inflammatory modulators NFATC3 and PTGS2 (Chiu *et al.*, 2010). Previous studies from others have shown that shikonin can possess multiple pharmacological properties such as anti-tumour (Lee *et al.*, 2008; Min *et al.*, 2008), antioxidant (Wang *et al.*, 2010), anti-platelet (Ko *et al.*, 1995) and anti-atherosclerosis

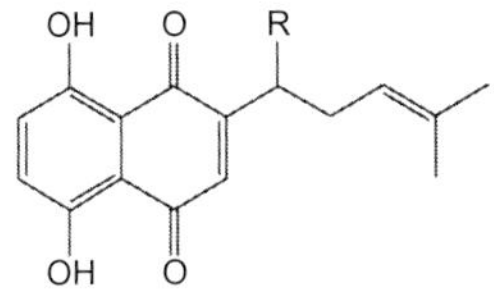


| **Compound** | **R** |
|---|---|
| Deoxyshikonin | H |
| Shikonin (*R*) or Alkanin (*S*) | OH |
| Acetylshikonin | $OCOCH_3$ |
| Isobutylshikonin | $OCOCH(CH_3)_2$ |
| β,β-Dimethylacrylshikonin | $OCOCH{=}C(CH_3)_2$ |
| Isovalerylshikonin | $OCOCH_2CH(CH_3)_2$ |
| α-Methyl-*n*-butylshikonin | $OCOCH(CH_3)CH_2CH_3$ |
| β-Hydroxyisovalerylshikonin | $OCOCH_2C(CH_3)_2(OH)$ |

Fig. 8. Chemical structures of shikonins from *Lithospermum erythrorhizon*.

(An *et al.*, 2007) activities. More recently, it has been reported that the anti-tumour effects of shikonin may be due to its induction of ROS (Chang *et al.*, 2010; Mao *et al.*, 2008), inhibition of proteasome activity (Yang *et al.*, 2009) and the circumvention of cancer drug resistance via induction of necroptosis (Han *et al.*, 2007). Shikonin is also considered as a potential drug for treating allergic diseases by inhibition of PMA+cAMP-induced IL-4 and IL-5 expression through downregulation of the expression of GATA-3 and Maf (Lee *et al.*, 2011). Shikonin significantly prolonged the survival and recovered or increased numbers of $CD3^+$ and $CD19^+$ cells (Long *et al.*, 2011). Other study suggests that the anti-inflammatory effect of shikonin may be due to its proteasome inhibitory activity, resulting in accumulation of IκB-α and ubiquitinated proteins and blockage of p65-NF-κB translocation from the cytoplasm to the nucleus. Further, shikonin was also shown to induce apoptosis and cell death in rat primary macrophage cultures (Lu *et al.*, 2011b). Other reports have revealed that the wound-healing activity of shikonin could result in active proliferation of fibroblasts, an increase in the collagen fibre levels of granuloma tissues and an increase in $CD11b^+$ cell population in granulation tissues (Kaith *et al.*, 1996; Sakaguchi *et al.*, 2001). Accumulating evidence showed that shikonin may serve as a naturally occurring, low-molecular-weight pan-chemokine receptor inhibitor for CCL1, CCL2, CCL3, CCL5, CXCL12 and C5a (Chen *et al.*, 2001, 2002; Chiu *et al.*, 2010). Shikonin downregulated surface expression of CCR5, a primary HIV-1 co-receptor, constituting a basis for the development of novel anti-HIV therapeutic agents (Chen *et al.*, 2003). These findings collectively and strongly suggest that shikonin may confer a spectrum of cellular and molecular activities that can induce specific chemokines and subsequent chemotaxis activities in various and specific immune-responsive cell types.

Recently, it was shown that shikonin may be involved in the inhibition of acetylcholine-induced aorta relaxation response and NO generation in RAW 264.7 cells (Yoshida *et al.*, 2010). Shikonin also downregulated the expression of SREBP-1c and the subsequent expression of PPAR$\gamma$ and C/EBP$\alpha$, resulting in downregulation of lipid metabolizing enzymes and reduced fat accumulation (Lee *et al.*, 2010b). Shikonin effectively suppressed the maturation of OVA and thymic stromal lymphopoietin-induced bone marrow DCs *in vitro* via downregulation of IL-4, IL-5, IL-13 and TNF-$\alpha$, and it inhibited allergic inflammation and AHR in a murine model of asthma (Lee *et al.*, 2010a). In addition, shikonin may also be further evaluated for its potential therapeutic effect on allergic asthma by blocking histamine release from human basophils via suppression of Syk-dependent phosphorylation and inhibition of leukotriene B4 and 5-hydroxyeicosatetraenoic acid (Takano-Ohmuro *et al.*, 2008). Shikonin also significantly inhibited the expression of MMP-1 and upregulated TIMP-1 in mice with CIA, suggesting that shikonin could be developed as a candidate cartilage protective medicine for RA (Dai *et al.*, 2009; Kim *et al.*, 2010).

# IV. CATEGORIZED PHYTOCOMPOUNDS WITH IMMUNOMODULATORY ACTIVITIES

Natural product-derived medicines can be traced back for more than 5000 years, while Western medicine has a relatively short history of a few hundred years (Goldman, 2001). In their review, Balunas *et al.* stated that the medicinal use of more than 85,000 plant species has been documented worldwide (Balunas and Kinghorn, 2005). The WHO also estimated that up to 80% of people in the world, mostly in developing countries, rely on herbal medicines for treatment of various diseases including immune diseases (Licciardi and Underwood, 2011). Moreover, approximately 30% of all FDA-approved drugs are derived from a botanical origin (Licciardi and Underwood, 2011; Onaga, 2001). Based on this evidence, it is important to investigate the chemical structures from traditional phytomedicines to evaluate their usefulness as immunomodulatory agents for immune disorders. Below, we provide examples of phytocompounds whose specific chemical structures and immunomodulating activities have been elucidated. The representative phytocompounds with their chemical structures, molecular targets and associated diseases are summarized.

## A. POLYPHENOL

A "polyphenol" or "phenolic" is defined as a substance that has an aromatic ring with one or more hydroxyl substituents, including functional derivatives (esters, glycosides, etc.) (Shahidi *et al.*, 2005). Polyphenols in foods or natural health products originate from one of the main classes of plant secondary metabolites derived from tyrosine or phenylalanine (Fraga, 2010; Shahidi *et al.*, 2005).

### *1. Stilbene derivatives*

Stilbenes are phenolic compounds that consist of two aromatic rings linked by an ethene bridge (C6–C2–C6) (Lamoral-Theys *et al.*, 2010). Resveratrol (*trans*-3,5,4′-trihydroxystilbene) is well known as a kind of phytoalexin that belongs to the stilbene class (Table II). It is a component of grapes, berries and other TCM such as *Polygonum cuspidatum* and is known to mediate its effects through the modulation of many different pathways (Harikumar and Aggarwal, 2008). Resveratrol has been shown to bind to a wide range of inflammation-related cell-signalling molecules (Harikumar and Aggarwal, 2008; Wood *et al.*, 2010). It has also been shown to regulate various transcription factors (e.g. nuclear factor erythroid-derived 2-like 2 (Nrf-2), NF-κ B, activator protein-1 (AP-1), signal transducer and activator of transcription-3 (STAT3), β-catenin and peroxisome proliferator-activated receptor-gamma (PPAR-γ)), inhibit activation of some protein kinases (e.g. PI3K, JNK and AKT), induce expression of antioxidant enzymes (e.g. catalase, superoxide dismutase (SOD) and hemoxygenase-1(HO-1)), inhibit the expression of inflammatory biomarkers (e.g. cyclooxygenase-2 (COX-2), iNOS, C-reactive protein (CRP) and TNF-α) and inhibit the expression of metastatic and angiogenic genes (e.g. MMPs, vascular endothelial growth factor (VEGF), cathepsin D and intercellular adhesion molecule-1 (ICAM-1)) (Harikumar and Aggarwal, 2008). A number of animal studies have demonstrated that this polyphenol holds promise for use in a variety of age- and inflammation-associated diseases including cancer, diabetes, Alzheimer's disease, cardiovascular and pulmonary diseases as well as in ageing (Richard *et al.*, 2011).

TABLE II

*Chemical Classes, Plant Sources and Molecular Targets of Representative Immunomodulatory Polyphenol Phytocompounds*

| Group/class | Compound/structure | Plant sources | Molecular targets | Targeted diseases | References |
|---|---|---|---|---|---|
| Polyphenol Stilbenes | Resveratrol 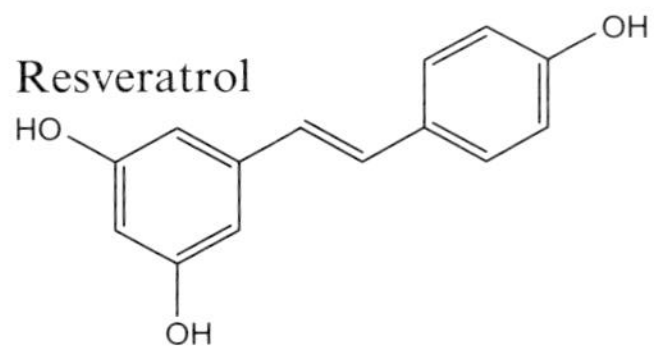  | Grapes (*Vitis vinifera* L.), *Polygonum cuspidatum* | Nrf-2, NF-κB, STAT3, HIF-1α, β-catenin and PPAR-γ, PI3K, JNK and AKT, catalase, SOD, HO-1, MMP2/9, ROS | Inflammation, ageing, cancer, diabetes, Alzheimer's disease, cardiovascular and pulmonary diseases | Harikumar and Aggarwal (2008), El-Mowafy *et al.* (2011), Csiszar (2011), Bereswill *et al.* (2010) |
| Hydroxycinnamic acids | Curcumin 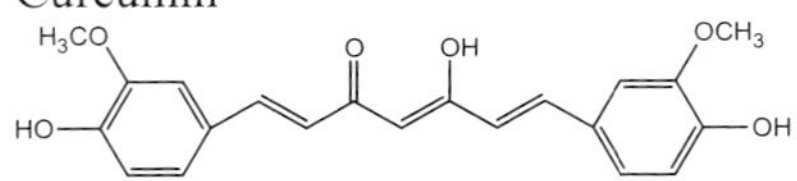  | Turmeric (*Curcuma longa*) | Nrf-2, NF-κB, AP-1, STAT3, PKCα, PI3K, GSK-3β, ERK, JNK, AKT, COX-2, iNOS, IL-6, TNF-α, PGE2, MMP-2/9, VEGF, ROS | Inflammation, arthritis, allergy, asthma, cancer, atherosclerosis, heart disease, Alzheimer's disease, diabetes | Surh (2003), Lamoral-Theys *et al.* (2010), Goel and Aggarwal (2010) |

| | | | | | |
|---|---|---|---|---|---|
| | 6-Gingerol 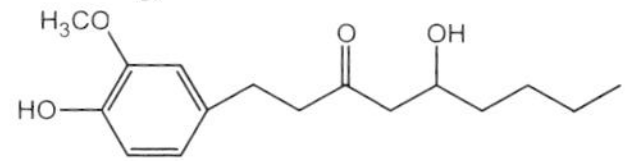 | Ginger (*Zingiber officinale*) | NF-κB, AP-1, PKCα, cyclin D1, COX-2, iNOS, IL-6, TNF-α | Inflammation, hyperlipidaemia, hyperglycaermia, analgesic | Aggarwal and Shishodia (2006), Al-Suhaimi *et al.* (2011), Kim *et al.* (2005) |
| Flavonoids | Quercetin (belongs to flavonol) 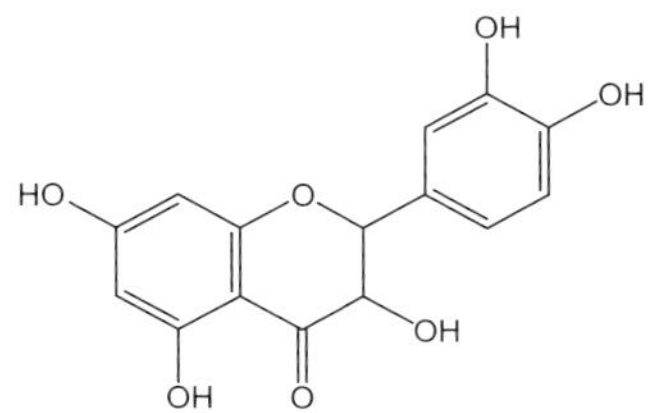 | Grapes, tea, onions, apples, berries | Nrf-2, NF-κB, AKT, iNOS, PGE2, COX-2, TNF-α, IL-1β, IL-6, ROS | Inflammation, ageing, neurodegerative diseases, inflammatory bowel diseases, cancer | Bacon *et al.* (2003), Conforti and Menichini (2011), Dihal *et al.* (2006), Gomes *et al.* (2008), Pan *et al.* (2009) |
| | EGCG (belongs to flavanol) 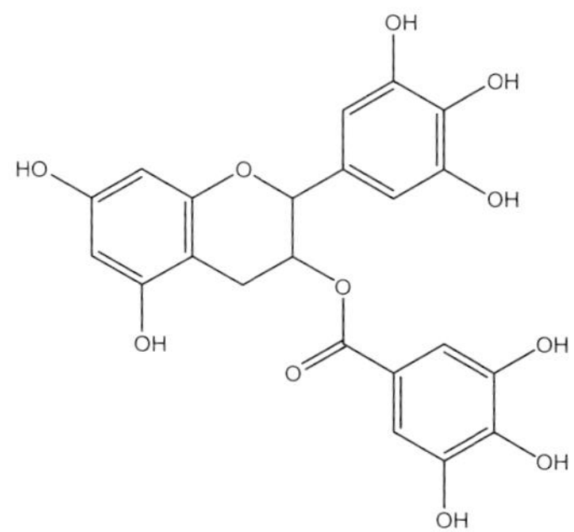 | Tea (*Camellia sinensis*) | EGFR, AKT, NF-κB, AP-1, cyclin D1, VEGF, COX-2, iNOS, MMP-2/-9, IL-12, VCAM-1 | Inflammation, ageing, hepatitis, cancer, upper respiratory tract infections, cardiovascular disease | Babu and Liu (2008), Melgarejo *et al.* (2010), Peairs *et al.* (2010), Shan *et al.* (2008), Shirakami *et al.* (2008) |

*2. Hydrocinnamic acid derivatives*

Curcumin (diferuloylmethane) is a diferuloyl derivative containing 19 carbon atoms (C6–C7–C6) and is a major pigment isolated from *Curcuma longa* (from the Zingiberaceae or Ginger family) (Table II) (Aggarwal, 2010). Curcumin has long been used as part of the daily diet in Asian countries without toxicity (Ammon and Wahl, 1991). It can also be used as a food preservative, drug, a yellow colouring agent and a component in cosmetics. Further, it has probably been most studied as a highly pleiotropic molecule with anti-inflammatory, antioxidant, anti-metabolic, chemopreventive, chemosensitization and radiosensitization activities (Goel and Aggarwal, 2010; Gupta *et al.*, 2011; Lamoral-Theys *et al.*, 2010). The activities of *C. longa* may be due to its modulation of factors at the transcriptional level (e.g. Nrf-2, NF-κB, AP-1 and STAT3), interference with some protein kinases (e.g. PKCα, PI3K, GSK-3, JNK and AKT), enhancement of expression of antioxidant enzymes (e.g. HO-1), suppression of the expression of inflammatory biomarkers (e.g. COX-2, iNOS, IL-6 and TNF-α) and inhibition of metastatic and angiogenic gene expression (e.g. MMP2/9 and VEGF) (Aggarwal, 2010; Goel and Aggarwal, 2010; Yadav and Aggarwal, 2011). The multiple activities of curcumin has meant that it has come to be thought of as somewhat of "a magic bullet" targeted at a broad spectrum of diseases including asthma, allergy, arthritis, atherosclerosis, heart disease, Alzheimer's disease, diabetes and metabolic syndrome (Carroll *et al.*, 2011; Kanai *et al.*, 2011; Sharma *et al.*, 2004). It has already entered clinical trials for cancer treatment at the phase I and II levels in the past 10–15 years (Bayet-Robert *et al.*, 2010; Carroll *et al.*, 2011; Kanai *et al.*, 2011). Another hydrocinnamic acid derivative, 6-gingerol, also shows similar patterns of activity as curcumin (Table II) (Kim *et al.*, 2005; Lee *et al.*, 2009; Park *et al.*, 2008).

*3. Flavonoids*

Flavonoids are one of the most abundant naturally occurring compounds and are ubiquitous in vascular plants (Gomes *et al.*, 2008). Almost all plant tissues can synthesize flavonoids (Pan *et al.*, 2008b), and at least 2000 naturally occurring flavonoids have been found (Pan *et al.*, 2008b). Flavonoids are characterized by a basic backbone of 15 carbon atoms (C6–C3–C6) (Gomes *et al.*, 2008). According to their chemical structures, in general, they are categorized into seven groups: flavones, flavanones, flavonols, flavanonols, isoflavones, flavanols and anthocyanidins (Gomes *et al.*, 2008). They usually exist as a form of aglycone or a form of flavonoid glycoside. Flavonoid glycosides are mainly distributed in the leaves, flowers or fruits, while aglycones appear mainly in woody tissues. Seeds may contain both flavonoid aglycones and glycosides. In addition to their well-known antioxidant activity, flavonoids have long been reported to possess anti-inflammatory, anti-hepatotoxic, anti-atherogenic, anti-osteoporotic, anti-allergic and anti-cancer activities (Gomes *et al.*, 2008). Here, we provide two examples of flavonoids as shown in Table II. Quercetin is a flavonol that is found in grapes, tea, onions, apples and leafy green vegetables. Epigallocatechin-gallate (EGCG) is a potent antioxidant which is the most recognized active component in tea. As shown in Table II, it is not only a potent antioxidant and anti-inflammatory agent that protects human body from the harmful effects induced by free radicals (Conforti and Menichini, 2011) but can also modulate phase I and phase II enzymes (Bacon *et al.*, 2003). The anti-inflammatory mechanisms of action of quercetin and EGCG are believed to be through the inhibition of transcriptional factors (e.g. NF-κB, AP-1) and the enhancement of Nrf-2, resulting in a reduction of pro-inflammatory mediators (Conforti and Menichini, 2011; Fraga, 2010; Shahidi *et al.*, 2005). With these features, these compounds are under evaluation for development as therapies for inflammation-related diseases, ageing, neurodegenerative diseases, inflammatory bowel diseases, cancer and diabetes.

## B. TERPENOIDS

Among natural products, phenolic compounds and terpenoids are the major phytochemicals present in vegetables, fruits and other dietary or medicinal foods (Salminen *et al.*, 2008). Terpenoids are composed of five-carbon isoprene units ($C_5H_8$) which are also often named isoprenoids (de las Heras and Hortelano, 2009). On the basis of biosynthesis and chemical structures, the terpenoids can be divided into five subgroups: (1) monoterpenoids (10 carbons), (2) sesquiterpenoids (15 carbons), (3) diterpenoids (20 carbons), (4) triterpenoids (30 carbons) and (5) carotenoids (40 carbons) (Salminen *et al.*, 2008).

### *1. Monoterpenoids*

The monoterpenoids are, in general, formed from two isoprene units, and have the molecular formula $C_{10}H_{16}$. They are usually present in nature in acyclic, monocyclic or bicyclic forms modified by oxidation, methylation or glycosylation (Bouvier *et al.*, 2005). Most of monoterpenes are volatile in nature. Some monoterpenes have been employed for human used since antiquity. The monoterpene limonene (Fig. 9), originally obtained from citrus fruits, cherries and apricots, was shown to suppress NF-κB activation (Berchtold *et al.*, 2005), and geniposide (Fig. 9), the major ingredient of the fruits of *Gardenia jasminoides*, a traditional herbal medicine used to treat inflammation, fever, headache and hepatic disorders, can inhibit NF-κB and iNOS expression (Koo *et al.*, 2004).

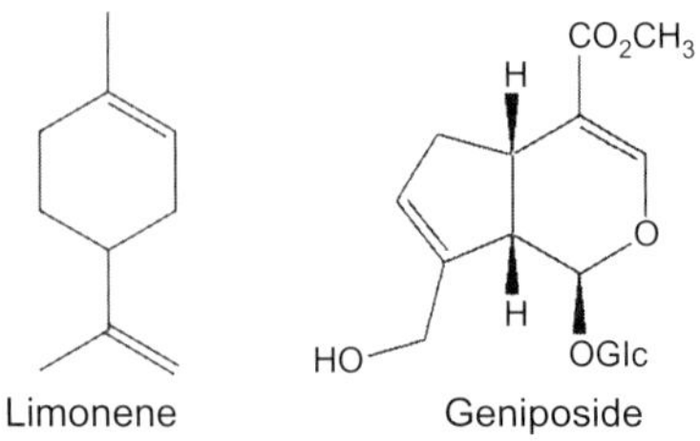


Fig. 9. Chemical structures of representative monoterpenoids.

*2. Sesquiterpenoids*

Sesquiterpenes are generally defined as substances that consist of three isoprene units which can form mono-, bi- or tricyclic compounds (Salminen *et al.*, 2008). Many traditional natural remedies or herbal medicines contain sesquiterpenoids which are modified and structurally rearranged from sesquiterpene structures. Up to the present, more than 7000 sesquiterpene structures have been identified and characterized; however, sesquiterpene lactones are recognized as those most frequently found in nature (Robles *et al.*, 1995). Sesquiterpene lactones are often found to exhibit potential medicinal properties including chemoprevention of certain inflammatory diseases and cancers (Robles *et al.*, 1995; Salminen *et al.*, 2008). Recently, a number of studies suggest that sesquiterpene lactones can be developed into therapeutics for certain diseases (Lee *et al.*, 2010c; Miller and Su, 2011; Shyur *et al.*, 2011). Among them, artemisinin (Fig. 10) is probably the most well known. Artemisinin was isolated from the leaves of *A. annua*, a traditional Chinese medicinal plant (Tu, 2011) (see Section III.C). Artemisinin has been used as an effective antimalarial drug, especially against multidrug-resistant malaria. Artemisinin and its derivatives have also been shown to confer antifungal, anti-cancer, anti-angiogenesis and immunosuppressive properties (Cui and Su, 2009; Miller and Su, 2011). The NF-κB transcription signalling system was suggested to be the target and mode of mechanistic action of artemisinin, resulting in a strong inhibition of inflammation. Further examples of sesquiterpene lactones are the elephantopin derivatives (Fig. 10). They include isodeoxyelephantopin and deoxyelephantopin and are isolated from the *Elephantopus scaber* plant (Ichikawa *et al.*, 2006). Isodeoxyelephantopin and deoxyelephantopin have been shown to not only possess anti-inflammatory activities but also confer anti-cancer activities, again via suppression of NF-κB activation (Huang *et al.*, 2010; Ichikawa *et al.*, 2006; Su *et al.*, 2011a).

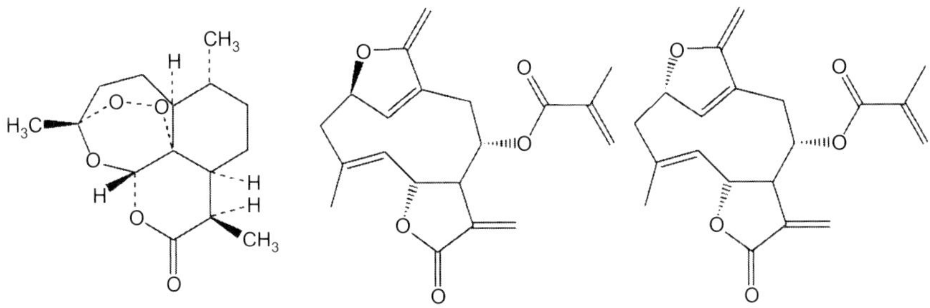

Artemisinin Deoxyelephantopin Isodeoxyelephantopin

Fig. 10. Chemical structures of representative sesquiterpenoids.

*3. Diterpenoids*

Diterpenes consist of four isoprene units and have a basic structure of $C_{20}H_{32}$ (Ajikumar *et al.*, 2008; Robles *et al.*, 1995). Diterpenoids are generally modified and structurally rearranged from diterpene structures. They may be acyclic, but in general, they are present as mono-, bi-, tri-, tetra- or macrocyclic compounds (Ajikumar *et al.*, 2008). Oleoresin from the conifer plant species usually contains a number of diterpenoids (Salminen *et al.*, 2008). Traditionally and clinically, diterpenoid-containing medicines have been applied to a variety of diseases including arthritis, atherosclerosis, cancer and inflammation (Salminen *et al.*, 2008; Thoppil and Bishayee, 2011). Physiologically, these typically active diterpenoids include retinol derivatives, taxanes, phorbols, forskolin and gibberellins (Pan and Ho, 2008).

The retinoids, including all-*trans*-retinoic acid and retinol, are reputed to play essential roles in the function and maintenance of human vision (Pan and Ho, 2008). Another well-known example is taxol, a complex polyoxygenated diterpenoid originating from the bark of the Pacific yew tree, *Taxus brevifolia*. This potent anti-cancer drug is clinically used for treating a number of cancer diseases under the generic name of paclitaxel. Two major bioactive diterpenoids derived from TCM are reputed to be useful for treating various inflammatory diseases: triptolide (Fig. 11), originally isolated from TWHF, and tanshinone IIA (Fig. 11), the major active diterpene quinone from the roots of *Salvia miltiorrhiza*. *S. miltiorrhiza* is a common TCM herb which has been used to treat immunological disorders, cardiovascular diseases, osteoporosis and breast cancer (Gao *et al.*, 2011; Yuan *et al.*, 2003). Studies have shown that tanshinone IIA can inhibit NF-κB signalling and the associated inflammatory mediators (Gao *et al.*, 2011). Another series of diterpenoids containing specific chemical structures of the abietane type have also been found to be potent immunomodulators with potential application to a broad spectrum of diseases. Prevention or blocking of the inducer

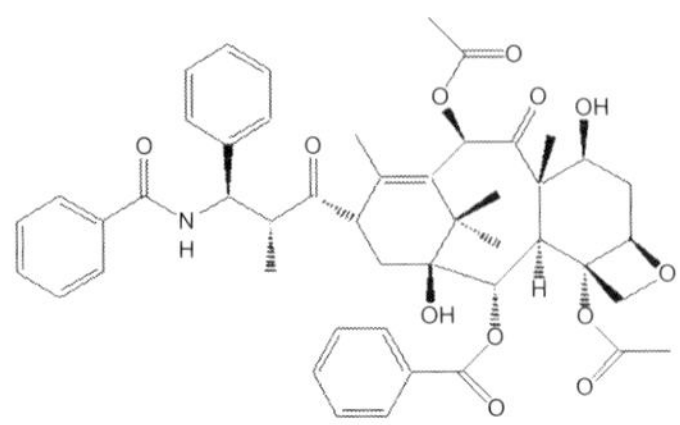

Paclitaxel (Taxol)

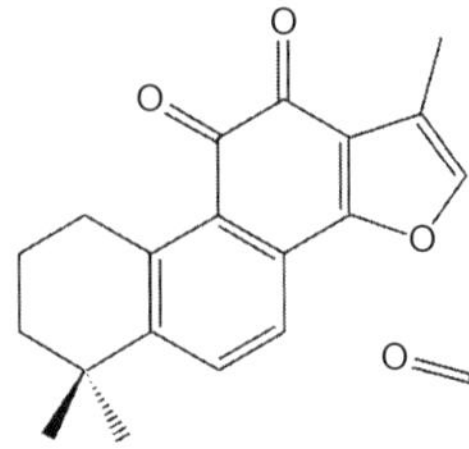

Tashinone IIA

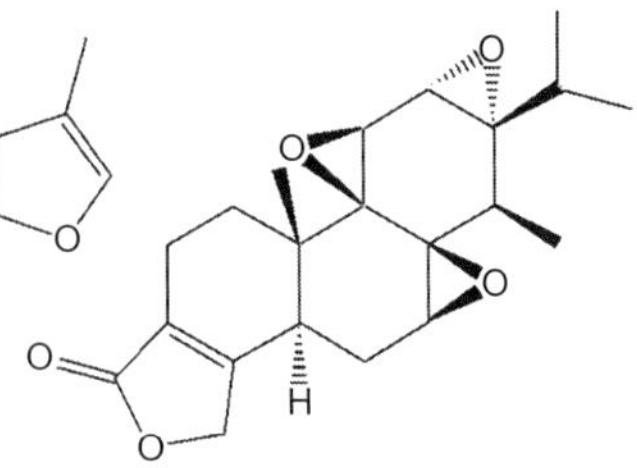

Triptolide

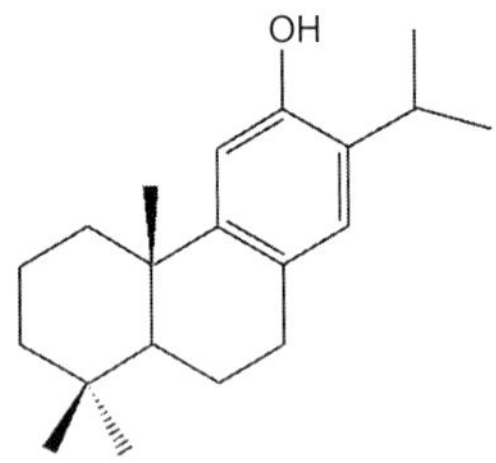

Ferruginol

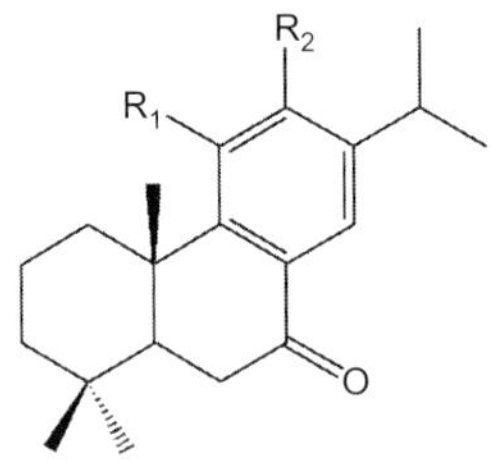

Dehydroabieta-7-one $R_1 = H$ $R_2 = H$
Sugiol $R_1 = H$ $R_2 = OH$
Cryptojaponol $R_1 = OH$ $R_2 = OCH_3$

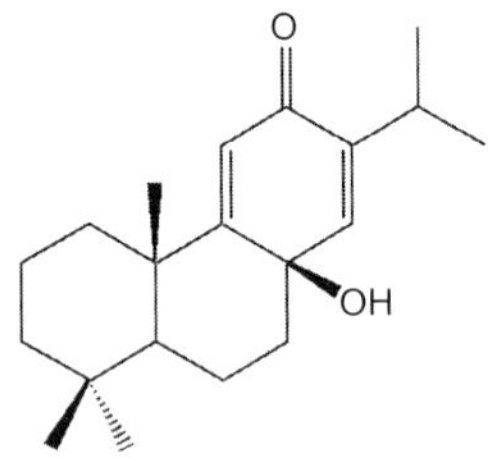

8α-hydroxyabieta-9(11),13-dien-12-one

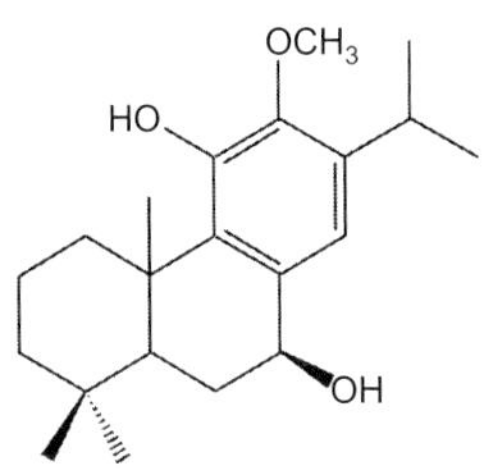

7α-Hydroxydeoxycryptojaponol

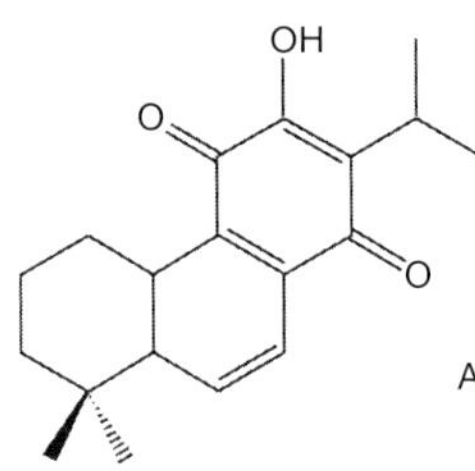

6,7-Dehydroroyleanone

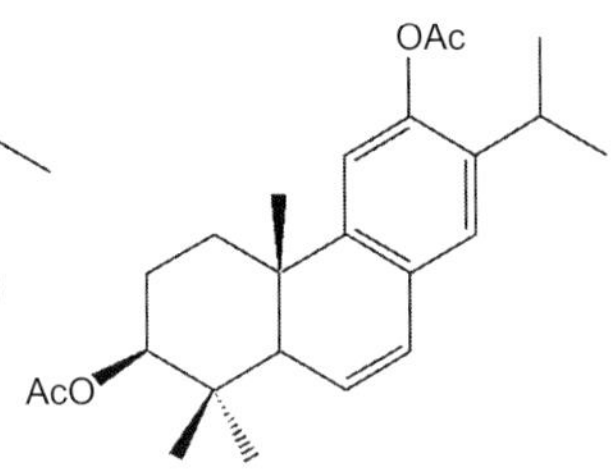

3 α,12-Diacetoxyabieta-6,8,11,13-tetraene

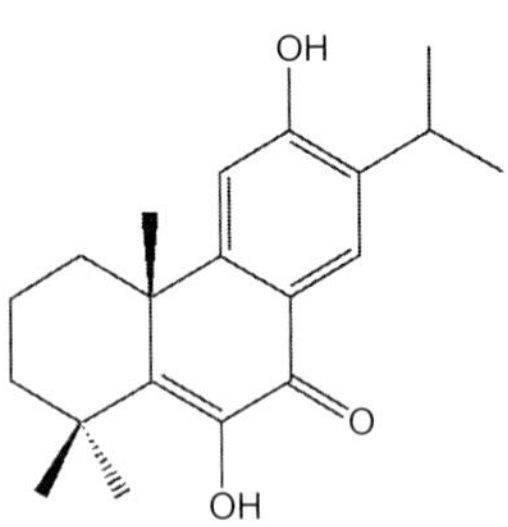

6-Hydroxy-5,6-dehydrosugiol (HDHS)

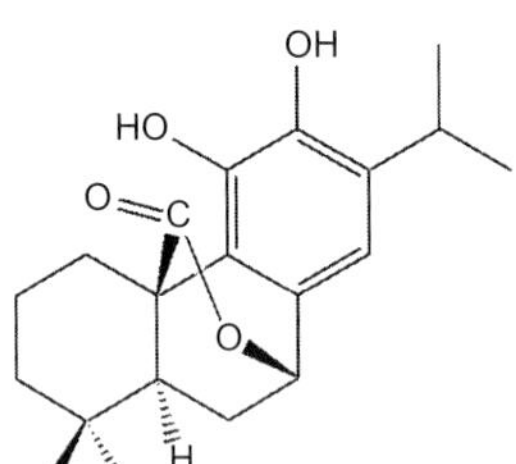

Carnosol

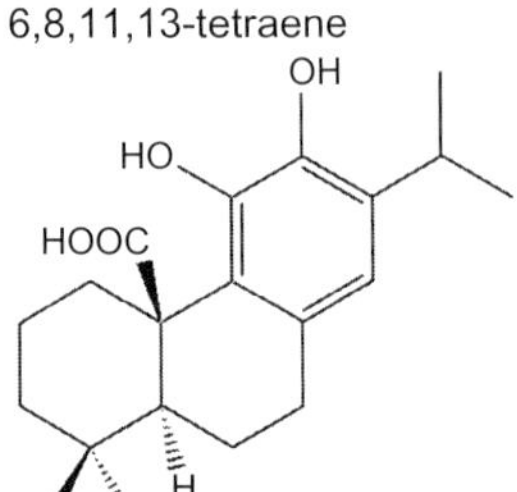

Carnosic acid

Fig. 11. Chemical structures of representative diterpenoids.

from an exogenous pathogen from initiating the inflammatory pathway may be taken as an approach for preventing the specific immune responses. For example, severe acute respiratory syndrome (SARS) is caused by infection with a coronavirus, SARS coronavirus (SARS-CoV), and is characterized by a cytokine storm in the host following infection leading to serious damage to the human body (Skowronski *et al.*, 2005). Our previous study showed that 10 diterpenoids, 8 abietane-type diterpenoids and 2 labdane-type diterpenoids (Fig. 11) can suppress SARS-CoV replication, hence suggesting that they could be further evaluated for use as antiviral agents (Wen *et al.*, 2007). Another two abietane-type diterpenoids, carnosol and carnosic acid have been found in high abundance in Rosemary extracts (*Rosmarinus officinalis*), a frequently used traditional herbal remedy (Salminen *et al.*, 2008). Both of these compounds possess antioxidant and anti-inflammatory activities, probably via the induction of Nrf-2-activated HO-1 expression and inhibition of activation of NF-κB signalling (Pan and Ho, 2008; Salminen *et al.*, 2008). Lin *et al.* (2008) also reported an abietane-type diterpenoid, 6-hydroxy-5,6-dehydrosugiol (HDHS) (Fig. 10) isolated from the stem bark of *Cryptomeria japonica*, can suppress tumour growth in prostate cancer (PCa)-xenografted mice. Based on the various diverse bioactivities of these abietane-type diterpenoids, additional research efforts may need to focus on classifying them into specific subgroups, for example, with regard to whether they suppress or enhance NF-κB signalling directly or indirectly, or serve as an inflammation-modulatory agent or immune-stimulatory agent, depending on their structure/activity relationship. For instance, taxol was reported to activate NF-κB signalling via the TLR4 receptor complex (Li *et al.*, 2004; Tsuda *et al.*, 2007). Further, most of the diterpenoids mentioned above, such as carnosol, possess anti-inflammatory and other therapeutic effects. Taken together, these findings suggest that diterpenoids may serve as a group of promising candidates for drug development.

### 4. *Triterpenoids*

Triterpenes are composed of 6 isoprene units and have 30 carbons. There are more than 20,000 naturally occurring triterpenoids which have cyclic structures (Ajikumar *et al.*, 2008; Liby *et al.*, 2007). Triterpenoids, synthesized in many plants by the cyclization of squalene, are widely used in various traditional and folk medicines (Phillips *et al.*, 2006). Celastrol is a quinone methide pentacyclic triterpenoid and is extracted from the TCM, TWHF (Yang *et al.*, 2006). It has been reported to possess antioxidant, anti-inflammatory and anti-cancer activities (Pinna *et al.*, 2004). Celastrol may act in part through the suppression of NF-κB signalling inhibiting inflammation and tumour growth (Kim *et al.*, 2009; Pinna *et al.*, 2004). Ursolic acid is a different type of pentacyclic triterpene which is the main active ingredient of some traditional herbal remedies, such as rosemary leaves (Liu, 1995). As shown in Fig. 12, ursolic acid is well known to possess a broad spectrum of biological functions that can counteract exogenous and endogenous biological stimuli (Ikeda *et al.*, 2008). In addition, it has been reported to confer various medicinal effects including anti-hyperlipidaemia, anti-cancer and hepatoprotective activities (Ikeda *et al.*, 2008; Pan and Ho, 2008; Salminen *et al.*, 2008). It was reported to inhibit NF-κB activation contributing to the suppression of LPS-induced pro-inflammatory mediators in mouse macrophages and TPA-induced skin tumour promotion (Ikeda *et al.*, 2008). You *et al.* (2001) showed that ursolic acid can also induce NF-κB activation, resulting in release of pro-inflammatory mediators in non-stimulated mouse macrophages. Therefore, it is speculated that, depending on the biological status of test cells and tissues, ursolic acid may exert contrasting anti- and pro-inflammatory activities (Ikeda *et al.*, 2008). Other lupane-type triterpenoids, such as betulinic acid and its derivatives (Fig. 12), have also been considered to have therapeutic potential against pathogen infections (e.g. HIV),

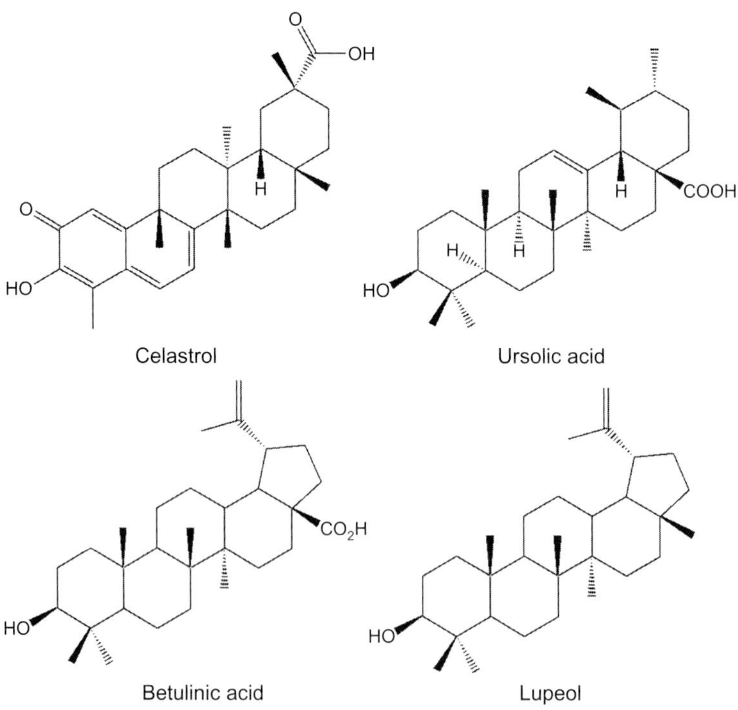


Fig. 12. Chemical structures of representative triterpenoids.

cancers (e.g. melanoma) and different types of inflammation (Fulda, 2009; Takada and Aggarwal, 2003). It was also observed that betulinic acid inhibited the activation of IKKα and NF-κB induced by various NF-κB activators (Takada and Aggarwal, 2003). Our previous study showed that betulinic acid conferred anti-SARS-CoV activities (Wen *et al.*, 2007). Lupeol has a similar chemical structure to betulinic acid and is one of the major constituents of a number of common vegetables, fruits and medicinal herbs (Salminen *et al.*, 2008). It has been studied for possible therapeutic effects for specific cancers (Siddique and Saleem, 2011) and inflammatory disorders (Fernandez *et al.*, 2001; Saleem, 2009). It was shown to inhibit NF-κB signalling via phosphorylation of IκBα protein, NF-κB-dependent reporter gene activity or DNA binding of NF-κB complex (Lee *et al.*, 2004; Saleem *et al.*, 2004). Lupeol can apparently also inhibit other signalling pathways, such as Akt-dependent pathways, and these activities may contribute to its various anti-cancer and anti-inflammatory properties (Fernandez *et al.*, 2001; Salminen *et al.*, 2008).

*5. Carotenoids*

Carotenoids are known as pigmented tetraterpenes typically containing a 40-carbon polyene chain, derived from eight isoprene units with conjugated double bonds, providing strong light absorption and brilliant colour, allowing them to take up excess energy from other molecules through a non-radiative energy transfer mechanism (Pan and Ho, 2008; Salminen *et al.*, 2008). Carotenoids are naturally occurring fat-soluble pigments that give bright colouration to host plants and animals. Plant carotenoids can play an essential role in maintenance of human health (Salminen *et al.*, 2008). They can serve as powerful antioxidants and are reputed to alleviate several chronic diseases, such as cardiovascular disease, osteoporosis and cancer. Some carotenoids such as β-carotene, lutein and lycopene can also offer protection against some inflammatory responses, possibly via modulation of redox-sensitive signalling pathways such as NF-κB and ROS signalling (Chew and Park, 2004; De Stefano *et al.*, 2007; Huang *et al.*, 2007). β-Carotene (Fig. 13) is the most common cyclic tetraterpene and the most potent pro-vitamin A in nature (Pan and Ho, 2008). It is stored in the liver and can be converted to vitamin A. The lipophilic xanthophylls, lutein (Fig. 13), is a dihydroxy derivative of β-carotene and is widely present in a variety of fruits and vegetables as well as in egg yolks. It can protect against oxidative stress and prevent age-related macular degeneration and exhibit a neuroprotective effect in retinal inflammation (Lee *et al.*, 2004; Sasaki *et al.*, 2009). Another acyclic tetraterpene, lycopene (Fig. 13), is the most abundant carotenoid present in the human body (Salminen *et al.*, 2008). It is present mainly in red-colour vegetables and fruits. Lycopene is a powerful

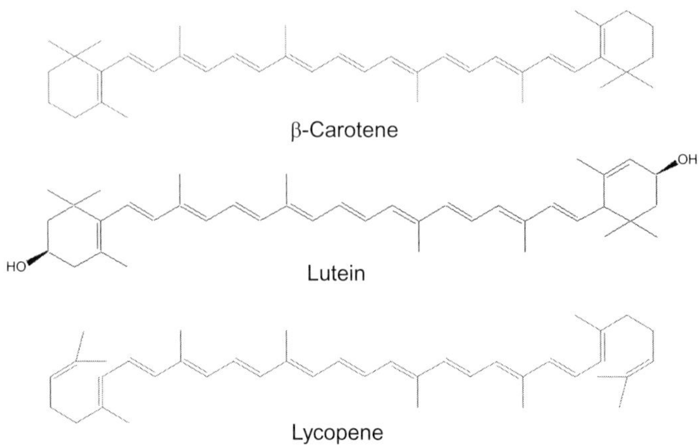


Fig. 13. Chemical structures of representative carotenoids.

antioxidant, more potent than vitamin E, and it can thus prevent cells from free radical attack during oxidative stress. It has also been claimed to reduce the risk for various chronic diseases, such as cardiovascular diseases, RA and atherosclerosis (Pan and Ho, 2008; Salminen *et al.*, 2008). These carotenoids exhibiting antioxidant activities may warrant future development as immunomodulators.

## C. ORGANOSULPHUR-CONTAINING COMPOUNDS

The organosulphur compounds are a special type of phytocompound found in various *Allium* species. The organosulphur compounds in garlic are known to differ slightly from those in onion varieties and consequently may have different health benefits. Two major kinds of organosulphur compounds are present in onion varieties, especially in garlic-γ-glutamyl-*s*-cysteines and cysteine sulphoxides (ca., alliin). When raw garlic cloves are crushed, chopped or chewed, an enzyme known as alliinase is released. Alliinase catalyses the formation of sulphenic acids from cysteine sulphoxides (Fig. 14). Sulphenic acids can spontaneously react with each other to form unstable thiosulphinates compounds. In the case of alliin, the resulting sulphenic acids react with each other to form a thiosulphinate (half-life in crushed garlic at 23 °C is 2.5 days) (Lawson *et al.*, 1998). Thiosulphinate formation is very rapid and can be completed within 10–60 s after crushing a garlic clove. Allicin breaks down and forms a variety of fat-soluble organosulphur compounds, including diallyl trisulphide (DATS), diallyl disulphide (DADS) and diallyl sulphide (DAS), or in the presence of oil or organic solvents, as ajoene and vinyldithiins

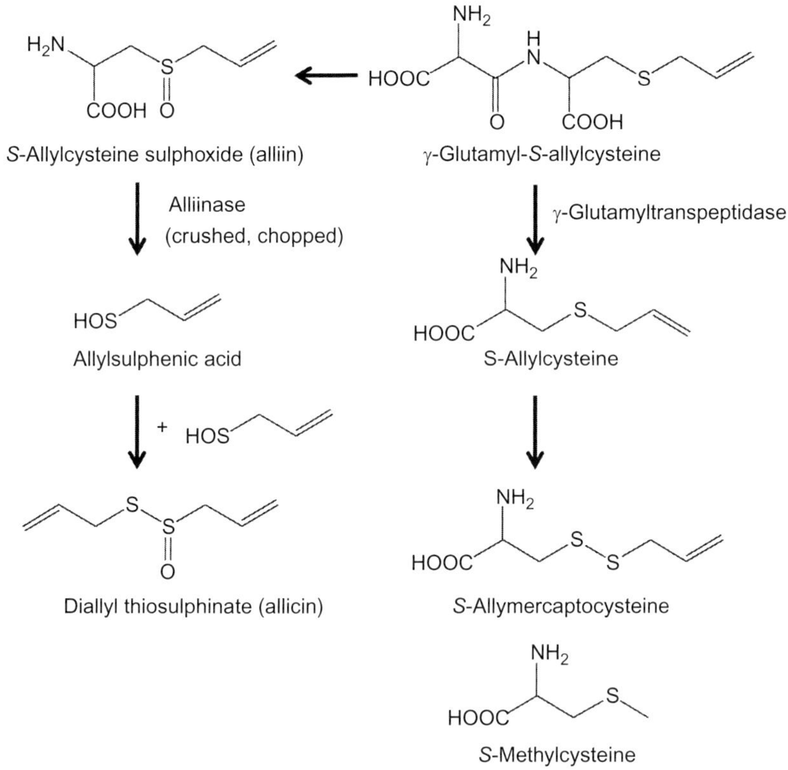


Fig. 14. Biosynthesis and transformation of various organosulphur-containing compounds.

(Block, 1985). Water-soluble organosulphur compounds, such as *S*-allylcysteine (SAC), are formed from γ-glutamylcysteines during long-term incubation of crushed garlic in aqueous solutions, as in the manufacture of mature garlic extracts (Fig. 14).

The oil-soluble organosulphur compound, allicin, is easily transformed into oil-soluble polysulphides, mostly DADS, DAS, DATS and also diallyl tetrasulphide (Fig. 15). Chemical compositions of the various preparations obtained by extraction of oil-soluble garlic fractions also depend on the specific extraction conditions such as temperature, treatment time interval and solvent polarity. Analysis of allicin solution that has been allowed to stand at room temperature for 20 h showed the following bioorganic composition: 66.7% DADS, 14.6% DATS, 13.3% DAS and 5.4% diallyl tetrasulphide (Lee *et al.*, 2003). Various findings suggest that higher poly-sulphides, such as diallyl penta-, hexa- or hepta sulphides, can be formed but

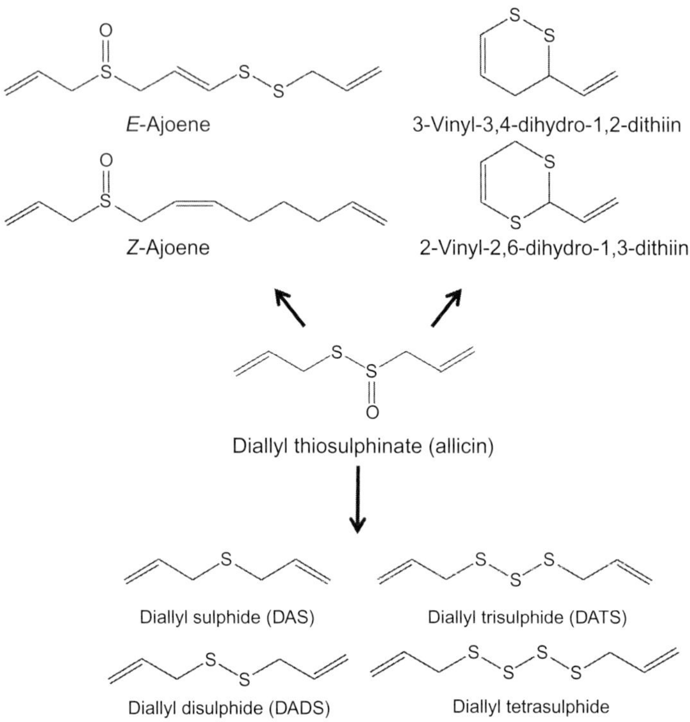


Fig. 15. Biotransformation of various oil-soluble compounds from allicin.

their concentrations are often low (O'Gara *et al.*, 2000). When extraction conditions are optimized, allicin can be transformed into vinyldithiin and structures of the *Z*- or *E*-ajoene type. The vinyldithiin compound was first identified by gas chromatographic analysis as a product of thermal degradation of allicin (Brodnitz *et al.*, 1971; Lee *et al.*, 2003). These structures are formed by dimerization of thioacrolein created via allicin β-elimination. Ajoene (4,5,9-trithiadodeca-1,6,11-triene-9-oxide) was generated via allicin S-thiolation and 2-propenesulphenic acid addition. Originally, ajoene was isolated from an ether fraction of garlic extract as a potential antithrombotic agent (Block *et al.*, 1984).

The reactions of allicin with —SH groups can yield SAC or *S*-allylmercaptocysteine (SAMC), both of which are water-soluble organosulphur compounds (Rabinkov *et al.*, 2000). Unlike oily sulphur compounds, water-soluble compounds are odourless and have a more delicate and less characteristic flavour (Kodera *et al.*, 2002). These phytocompounds are formed during aqueous garlic extraction, when the initial compound γ-glutamyl-*S*-allylcysteine (GSAC) is transformed into SAC. This reaction is catalysed by γ-glutamyltranspeptidase (γ-GT) (Fig. 14). SAC along with its derivatives, *S*-methylcysteine (SMC) and SAMC, are components of aqueous extracts of garlic and possess various biological activities, under both *in vitro* and *in vivo* conditions.

Garlic- and onion-derived organosulphur compounds have been shown to suppress the *in vitro* activities of inflammatory enzymes such as cyclooxygenase and lipoxygenase (Ali *et al.*, 2000) and to inhibit the expression of iNOS in inflammatory white blood cells (macrophages) (Dirsch *et al.*, 1998). Some organosulphur compounds have been shown to inhibit expression of the inflammation signalling molecules in cultured macrophages and human peripheral blood mononuclear cells (Chang *et al.*, 2005). Various findings have demonstrated that garlic extracts and their derived compounds can exhibit anti-inflammatory effects through inhibition of the NF-κB activity induced by various receptor agonists, including TNF-α and LPS (Keiss *et al.*, 2003). Expression of iNOS was also shown to be inhibited by garlic extract in activated macrophages (Dirsch *et al.*, 1998; Liu *et al.*, 2006a). In addition, Youn *et al.* (2008) demonstrated that garlic extracts can modulate inflammatory responses through suppression of TLR activation.

### D. POLYSACCHARIDES

Over 300 types of bioactive polysaccharides have been identified from natural products (Jiang *et al.*, 2010). According to the broad and diverse sources, they can be mainly divided into five categories, including the higher plant, fungal polysaccharides, bacterial, lichen and the algae (Cheung *et al.*, 2009). Polysaccharides, one of main classes of various bioactive substances present in various traditional herbal medicines, have been shown or implicated to confer a spectrum of pharmacological activities, especially on immunomodulatory, anti-tumour effects or cancer chemopreventive effects (Guo *et al.*, 2011). Unfortunately, however, their immunoregulatory activities in terms of molecular and cellular mechanisms are in general not well understood. According to the similarities and differences of their chemical structures, the plant polysaccharides can be roughly categorized into several groups, including the β(1→3)-D-glucans (Fig. 16A), α(1→3)-D-glucans (Fig. 16A), (1→3)-β-linked backbone with (1→6)-β-branches (Fig. 16A), acetylated glucomannans (Fig. 16B), sulphated polysaccharides, arabinans (Fig. 16C), arabinogalactans I, arabinogalactans II (Fig. 16D), rhamnogalacturonan I (RG-I) (Fig. 16E), RG II (Fig. 16F) and pectins (Fig. 16G).

A large volume of studies have reported that various plant polysaccharides can confer potent immunomodulatory activities through regulating the specific functions of various immune cells, including monocytes, macrophages,

NK cells, DCs, T lymphocytes, B lymphocytes and others (Chen *et al.*, 2009a; Thakur *et al.*, 2011; Zhang *et al.*, 2011b). They can be recognized or distinguished by the corresponding receptors on specific immune cells (Table III), and they can activate immune cells to generate a series of specific cellular or molecular events, including the innate immune and acquired immunities. Accumulating evidences have shown that DCs, the potent APCs, are the major immunomodulatory targets of polysaccharides in regulation of innate as well as acquired immunities (Chen *et al.*, 2011; Kim *et al.*, 2009; Li *et al.*, 2010a). Polysaccharides can increase the expression of MHC class II and the co-stimulatory molecules CD80 and CD86. Various polysaccharides can affect the morphological maturity of DCs, upregulate IL-12 and GM-CSF, downregulate phagocytosis and antigen uptake activities of DCs or promote DC differentiation (Jeurissen *et al.*, 2005; Khayrullina *et al.*, 2008). These

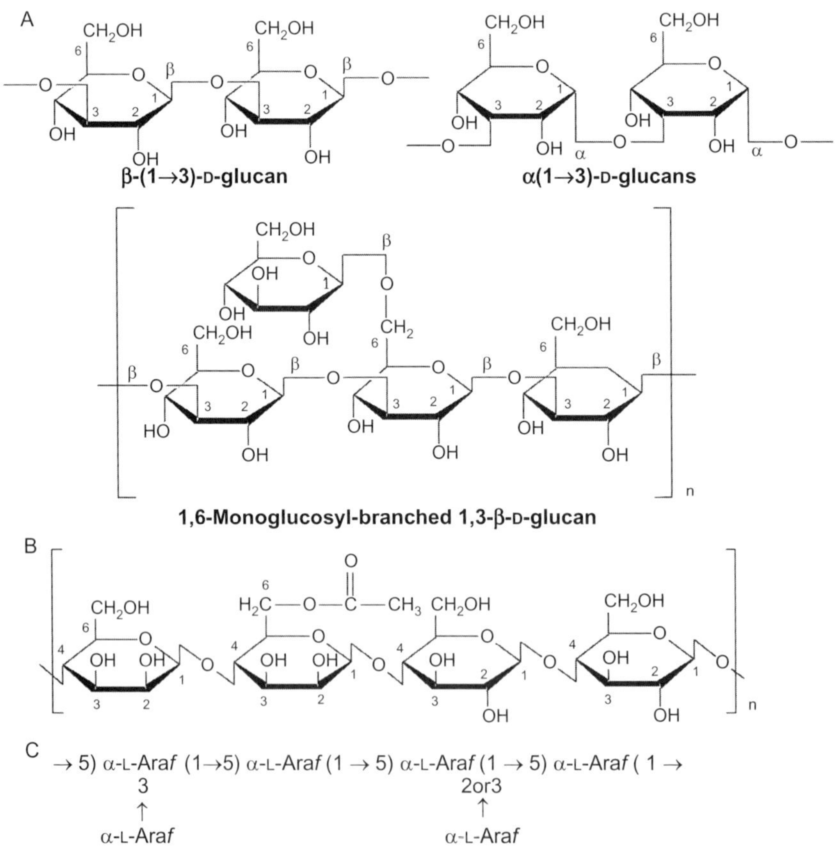


Fig. 16—cont'd

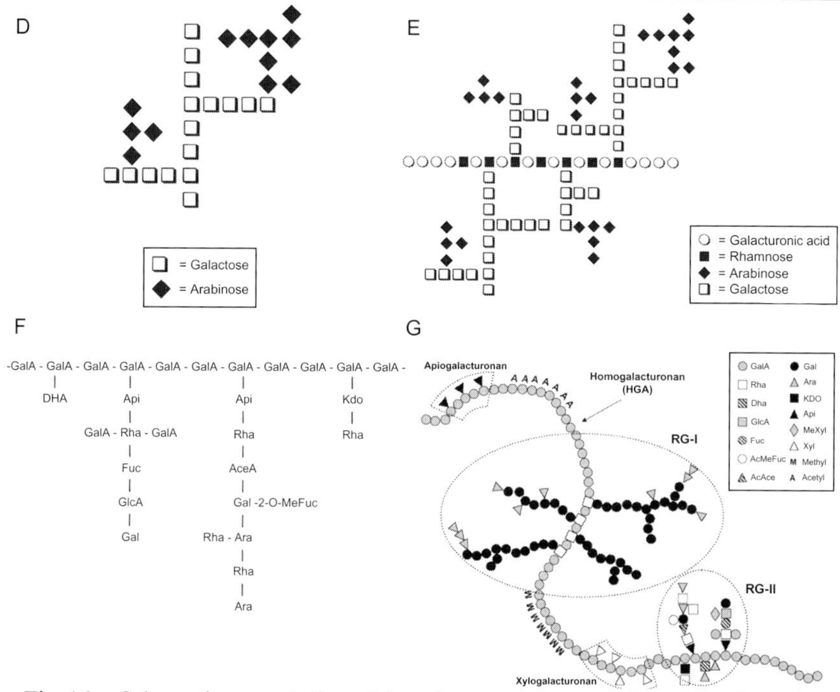


Fig. 16. Schematic presentation of the primary structure of bioactive polysaccharides. (A) and (B) from Moradali *et al.* (2007), (C) from Paulsen and Barsett (2005), (D)–(F) from Paulsen (2002), (G) from Perez *et al.* (2003).

TABLE III

*The Specific Polysaccharides Ligands and Their Target Immune Cells*

| Ligands | Immune cells | Receptors | References |
|---|---|---|---|
| Zymosan, β-1,4-glucan | Myeloid cells (monocytes, macrophages, DCs, epithelial cells, mast cells and neutrophils) | Toll-like receptors | Lu *et al.* (2011a), Graff *et al.* (2009), Han *et al.* (2003), Li *et al.* (2004) |
| Fucoidan, β-glucan | Macrophages, DCs | Scavenger receptors | Wang and Chandawarkar (2010), Means *et al.* (2009) |
| β-1,3-Glucan | Macrophages, DCs, neutrophils, eosinophils, B and T lymphocytes | β-Glucan receptor | Willment *et al.* (2005), Brown *et al.* (2002) |
| Mannan | Macrophages, DCs, hepatic endothelial cells, tracheal smooth muscle and retinal pigment epithelial cells | Mannose receptor | Linehan *et al.* (1999) |
| β-1,3-Glucan | Macrophages, neutrophils, B and T cells and natural killer cells | Complementary receptor type 3 | Hwang *et al.* (2003), Chen *et al.* (2009b) |

findings show that polysaccharides may be employed as a potent adjuvant for design and efficacy of DC-based vaccines. It has been reported that the specific extract isolated from the root of *Echinacea* contain high quantity of polysaccharides and were shown to confer strong immunostimulatory activities for activating DC maturation (Benson *et al.*, 2010). Various plant polysaccharides have also shown to affect another APC type, that is, macrophages. These activities were suggested to involve direct elimination of alien pathogens and the dying/damaged cells and the regulation of various immune effector cells (Wang *et al.*, 1992; Zhang *et al.*, 2011a). Polysaccharides can activate macrophages to secrete pro-inflammatory cytokines (e.g. IL-1, TNF-$\alpha$ and IFN-$\gamma$) (Zheng *et al.*, 2005), increase the production of NO (Xu *et al.*, 2011), ROS (Yang *et al.*, 2004) and myeloperoxide, enhance the activities of cytotoxicity (Choi and Hwang, 2002), phagocytosis (Zheng *et al.*, 2005) and cell proliferation (Su *et al.*, 2011b). In addition to these effects of polysaccharides on the myeloid-lineage immune cells, maintenance or skewing of the Th1/Th2 balance has been reported (Sun *et al.*, 2009a). Polysaccharides can also promote the differentiation of B cells (Han *et al.*, 2003) and the production of IgG and IgM (Nose *et al.*, 1998). We have shown that the specific extract from *D. batatas* can be used as adjuvants for DC-based vaccine (more details are described under the Section III.B (Su *et al.*, 2011b)).

## V. EMERGING APPROACHES FOR MODULATING THE COMPLEX SYSTEMS

### A. EMERGING IMMUNOMODULATORY TARGETS OF MEDICINAL HERBS FOR THERAPEUTIC INTERVENTION

Large volume of evidence shows that the use of complementary and alternative medicines is increasing in supplementing or treating various immune disorders, especially in developed countries (Boon *et al.*, 1999; Ernst and Cassileth, 1998). The use of complementary and alternative medicine has become generally acceptable by the public and becomes more and more popular in cancer patient populations of Western countries (Xu *et al.*, 2006). TCM is one of the complementary and alternative medicines that has a well-documented theoretical framework and a long-established practical history for immune-related diseases, including autoimmune diseases and cancers (Cho, 2010). From the aspect of immunomodulatory characteristics of TCM, they can be generally categorized into two major groups, that is, with pro-inflammatory activity or with anti-inflammatory activity, which are

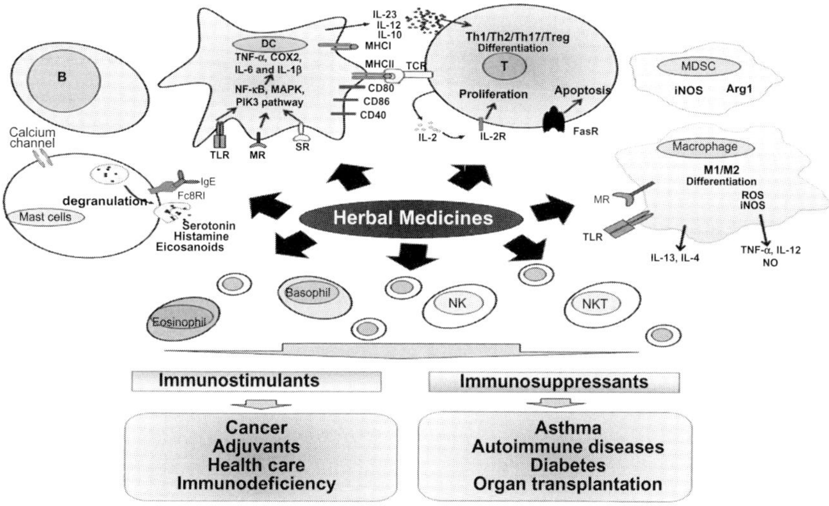


Fig. 17. Possible mechanisms of herbal medicines in immune systems. A number of possible cellular and/or molecular mechanisms of herbal medicines involved in regulation of various immune cells are depicted (black arrow) as follows. The different immunomodulatory activities of specific or defined herbal medicines may be employed for the potential in future/current clinical application of immunotherapies for immune disorders (black box). Abbreviations: NK, natural killer cells; DC, dendritic cells; TLR, toll-like receptor; MR, mannose receptor; SR, scavenger receptor; ROS, reactive oxygen species; iNOS, inducible nitric oxide synthase. (For colour version of this figure, the reader is referred to the online version of this chapter.)

being investigated for potential therapeutic application to adjuvant treatment for cancer or autoimmune diseases, respectively. The possible cellular and/or molecular mechanisms of herbal medicines and their potential applications for future/current clinical immune-therapies of immune disorders are shown in Fig. 17. The following section summarizes in brief published reports, supporting the usage of various TCM in combination with cancer chemotherapy or clinical immune modulators for supportive measures in cancer care and autoimmune disorders.

## B. DEVELOPING MEDICINAL HERBS AS ADJUVANT FOR CANCER THERAPY

As a major global public health problem, cancer has become the major leading cause of death for most developed countries. Chemotherapy is the main stream of current therapeutic approaches; however, it has several drawbacks: (1) serious side effects and complications, (2) poor immune functions for the host and (3) frequent recurrence and poor survival rate

(Qi *et al.*, 2010). The effect of chemotherapy in suppressing the host immune function may in fact worsen the tolerogenic tumour microenvironment orchestrated by tumour-associated macrophages, myeloid-derived suppressor cells (MDSCs) and T regulatory cells (Tregs), which can lead to the escaping of tumour cells form immunosurveillance of host. Therefore, how to effectively break down the invasion and metastasis of tumour microenvironment and to restore the immune functions of cancer patients is one of the most challenging issues we are facing in cancer research today. Growing evidence revealed that TCM may effectively support and enhance the efficacy of cancer chemotherapy via improving certain specific cellular immune functions and diminishing the side effects and complications resulting from conventional cancer therapy (Xu *et al.*, 2006).

MDSCs, identified by the myeloid cell lineage cell antigens-Gr-1 and CD11b (Pan *et al.*, 2008b), is a critical immune cell type involved in maintenance of tolerogenic tumour microenvironment. MDSCs can produce several immunosuppressive factors (e.g. Arginase 1, iNOS and ROS) and specific cytokines (TGF-β and IL-10), leading to the development of Treg (Huang *et al.*, 2006). They can inhibit both the innate and adoptive immunities, subvert immunosurveillance (Pan *et al.*, 2008a) and create a significant impediment in elimination of malignant cells. Curcumin was shown to inhibit tumourigenicity, tumour growth, the expansion of MDSCs and the activation of Stat3 and NF-κB in MDSCs, and to reduce IL-6 levels in a human gastric cancer xenograft model and a mouse colon cancer allograft model (Tu *et al.*, 2011). Curcumin treatment polarized MDSCs towards a M1-like phenotype with an increased expression of CCR7 and decreased expression of dectin-1, *in vivo* and *in vitro* (Tu *et al.*, 2011). The extracts of two Chinese medicinal herbs, namely, *Astragalus membranaceus* and *Ligustrum lucidum*, can exert significant anti-tumour activity via abolition of tumour-associated macrophage suppression (Rittenhouse *et al.*, 1991). Icariin, the major active ingredient of *Herba epimedii*, has been demonstrated to confer anti-inflammatory effect in murine innate immune cells and activated human PBMCs (Zhou *et al.*, 2011). It has been reported that administration of icariin can significantly reduce the percentage of MDSCs with a concomitant activity for differentiation into DCs and macrophages, leading to a downregulation of IL-10, IL-6 and TNF-α production, which may result from decreased expression of S100A8/9 and inhibition of the activation of STAT3 and AKT (Zhou *et al.*, 2011).

In addition to MDSCs, the development of Treg cells in tumour microenvironment is another important determinant for the efficacy of certain cancer immunotherapies. *Radix glycyrrhizae* polysaccharides can reduce Treg population and Foxp3 expression in Treg cells and upregulate Th1/Th2 cytokine ratio (decreased level of IL-10 and TGF-β and increased level of IL-2 and IL-12p70) in sera of H22 hepatocarcinoma-bearing mice (He *et al.*, 2011). Recently, *R. glycyrrhizae* has also been shown to regulate the cellular immunity of tumour-bearing mice by decreasing the proportion of Treg cells and by increasing the spleen lymphocyte transformation ratio (Li *et al.*, 2010b). In addition, glycyrrhizin isolated from *R. glycyrrhizae* was shown to reduce the generation of suppressor macrophages and enhance the efficacy of adoptive transfer therapy of allospecific CTLs (Suzuki *et al.*, 1992). A sulphated polysaccharide-protein complex from Gekko swinhonis Guenther, for a TCM, has been found to confer strong bioactivities for restoring the defective biorheological characteristics of DCs via decreasing the secretion of IL-10 of DCs and thus modifying the tumour microenvironment (Chen *et al.*, 2011). The *Lycium barbarum* polysaccharides was shown to confer anti-tumour activity through increasing the numbers of $CD4^+$ and $CD8^+$ T cells in tumour infiltrated lymphocytes to relieve the immunosuppressive responses and enhance the anti-tumour function of the immune system (He *et al.*, 2005). Bushen Gubiao Recipe, a traditional Chinese herbal medicine, was shown to improve the innate immune function by upregulation of the TLR/NFκB signalling pathway and adjustment of the immune imbalance of T-helper cell (Th) 1/Th2, through reducing the activity of $CD4^+CD25^+Foxp3^+$ Tregs and enhancing the Th1 immune response (Zhou *et al.*, 2010). Radix Astragali (*Astragalus propinquus*, Huangqi) has long been used to modulate the function of the lung and gastrointestinal system, promote healing and reduce fatigue. Currently, a number of immunomodulatory properties of *Astragalus* have been detected, including an increase in expression of interferon and TNF, and the activation of lymphocytes, NK cell and macrophages (Nalbantsoy *et al.*, 2011; Song and Hu, 2009). The polypeptide extract from *Scorpion venom* was shown to inhibit the angiogenesis activity of *Lewis* lung cancer, which may be due to the decreased level of angiogenic factors—factor VIII, α-SMA, Dll4 and Notch1 in test tumour microenvironment (Sun *et al.*, 2011).

In addition to the single herb plants mentioned above, the mixtures of multiple plants and formulated TCM preparations were also shown to confer immunostimulatory activities. Such multiple plant formulations were repeatedly documented in traditional Chinese medicine books. Knowledge and experience presented in such formulations also may provide new and alternative therapy approaches in combination with chemotherapy treatment. A recent interesting example is the successful development of the PHY906 formula for TCM reported by Dr. Y. C. Cheng (Ye *et al.*, 2007). PHY906 is a Chinese medicine formulation composed of four medicinal herbs (Yen *et al.*, 2009; Zhang and Ma, 2010): Huang Qin (dried roots of *Scutellaria baicalensis*

Georgi), Baishao (dried roots of *Paeonia lactiflora* Pall), Gan Cao (dried and honey-fried roots and rhizomes of *Glycyrrhiza uralensis* Fisch, *Glycyrrhiza inflata* Bat or *Glycyrrhiza glabra* L.) and Da Zao (dried fruits of *Ziziphus jujuba* Mill). This formulation was shown as efficacious for use as adjuvant treatment cancer chemotherapy approaches (Ye *et al.*, 2007). It has been found to reduce the chemotherapy-induced gastrointestinal toxicity (Lam *et al.*, 2010) and can be used as an adjuvant therapy for chemotherapy using capecitabine (Yen *et al.*, 2009), irinotecan, 5-fluorouracil and leucovorin (Wang *et al.*, 2011a) in advanced colorectal cancer (Kummar *et al.*, 2011) and pancreatic and other gastrointestinal malignancies (Kummar *et al.*, 2011; Saif *et al.*, 2010; Yen *et al.*, 2009). Further, Juzen-taiho-to (TJ-48) is an extract prepared from a mixture of 10 species of medicinal plants, including *Angelica sinensis*, *P. lactiflora*, *Atractylodes macrocephala*, *Poria cocos*, *Cinnamomum cassia*, *A. membranaceus*, *Liqusticum wallichii*, *G. inflata* and *Rehmannia glutinosa* (Saiki, 2000). This prescription has long been traditionally used against anaemia, anorexia, extreme exhaustion and fatigue (Saiki, 2000). TJ-48 was shown to augment antibody production, the mitogenic activity in splenocytes and B cells, and anti-complementary activity, and to activate macrophages, by oral administration of TJ-48 (Yamada, 1989). TJ-41 (Bu-Zhong-Yi-Qi-Tang) is another traditional herbal formulation, containing *Pinellia tuber*, *S. baicalensis*, *Zingiberis rhizoma*, *Zizyphi fructus*, *Coptidis rhizoma*, *G. radix* and *P. ginseng* (Yang *et al.*, 2010). TJ-41 has been reported to enhance concomitant immunity against tumour development and restore the anti-tumour response of effector T cell in tumour-bearing mice (Li *et al.*, 1999).

Immunogenic chemotherapy has recently emerged as an interesting approach, based on the ability of a cytotoxic compound to induce immunogenic tumour cell death, which are characterized by the changes of danger-associated molecular pattern, including heat shock protein, calreticulin, glucose-related protein and high-mobility group protein box 1 (Garg *et al.*, 2010). This new compelling anti-cancer strategy may offer good therapeutic potential in providing not only a direct tumour-killing effect but also a restoration of tumour-specific immune responses for prevention of tumour recurrence (Ullrich *et al.*, 2008). Unfortunately, there are currently very limited chemotherapeutic drugs that are shown to confer such pharmacological characteristics. Effective reutilization of TCMs as well as its phytochemicals may offer great value in drug discovery, and one of their potentials may be in the area for development of immunogenic chemotherapy. Our laboratory has also identified and tested several phytochemicals, including shikonin and its derivatives and synthetic compounds (Wen *et al.*, 2011), aiming to make use of the immunogenic cell death activity.

### C. DEVELOPING MEDICINAL HERBS FOR USE AGAINST AUTOIMMUNE DISEASES

Autoimmune diseases are a group of illnesses that often involve multiple organs. For clinical applications, autoimmune diseases appear to be either systemic (as in the case of systemic lupus erythematosus) or organ specific (as in the case of type 1 diabetes mellitus). Both the activation and the defective apoptosis of immune effector cells, such as T and B lymphocytes and macrophages, can play critical roles in the pathogenesis of autoimmune disorders (Liu *et al.*, 2011). Current therapy for autoimmune diseases often recommends a combination of several disease-modifying antirheumatic drugs (DMARDs) that are designed to preserve different immunomodulatory mechanisms. Because of the limited success in prevention of RA joint destruction for currently available DMARDs, the development of more effective and less toxic DMARDs is in urgent need.

Two commonly prescribed Chinese antirheumatic herbs, namely, TWHF (as mentioned above in Section III.D) and tetrandrine, were shown to preserve both anti-inflammatory and immunosuppressive effects. Tetrandrine, purified from a creeper *Stephania tetrandra* S Moore, is a bisbenzylisoquinoline alkaloid and has been used as a drug in China for decades (Ho and Lai, 2004; Lai, 2002). The immunosuppressive effect of tetrandrine may be synergistic with current DMARDs, highlighting that tetrandrine is a potential candidate of DMARDs for treatment of autoimmune diseases, especially RA (Lai, 2002). For centuries, *Ganoderma*, a fungus (also named as Ling Zhi in Chinese), has been regarded as a premium remedy for a number of diseases. The extracts of *Ganoderma* have been reported to improve the survival rate of lupus mice, decreased the amount of proteinuria, decreased serum levels of anti-dsDNA autoantibody and showed evidence of decreased perivascular and parenchyma mononuclear cell infiltration in vital organs (Lai *et al.*, 2001). The extract of *Acanthopanax gracilistylus* markedly suppressed the proliferative activities of human peripheral blood lymphocytes stimulated with mitogens concanavalin A and *Staphylococcus aureus* Cowan I. The mechanism of AGE-induced suppression of lymphocytes was shown to involve cell cycle arrest at the G0/G1 stage without a direct cytotoxic effect. AGE also suppressed the alloantigen-specific CTL response (Shan *et al.*, 1999). The ethanol extract of *Celastrus aculeatus Merr.* (*Celastrus*), another Chinese herb, can downmodulate the severity of adjuvant arthritis and reduce the levels of NO (Tong and Moudgil, 2007).

# VI. CHALLENGES, CONCLUSION AND FUTURE PERSPECTIVES

The high value of traditional herbal medicines, specific medicinal plants and the derived phytochemicals for medicinal chemistry study and applications was recently addressed by Dr. Y. Y. Tu in *Nature Medicine* 2011 (Tu, 2011). Her wonderful experience in the discovery of artemisinin from *A. annua* plants and for its use in treatment of malaria very appropriately won her the 2011 Lasker Award in medical research. As elegantly addressed in her article, the wisdom of traditional medicines may need to be re-recognized for the development of future medicines. Within the same context, the recent study on a "multiple formulation" of TCM, consisting of four different medicinal plant species, instead of a single phytochemical, for potential use as a "botanical drug" for cancer treatment was elegantly demonstrated by the group of Y. C. Cheng (Lam *et al.*, 2010) as recently reported in *Science Translational Medicine*. Here, high-quality experimental results were obtained on metabolite profiling, anti-tumour and anti-inflammatory molecular mechanisms and related clinical studies. The above two reports in combination have exemplified the high interest and importance worldwide on research into medicinal plants and phytomedicines.

Our renewed interests in herbal medicines and phytochemistry should not be blindfolded by the complexity, challenge and difficulty in redefining or readdressing the empirical and anecdotal features of a number of traditional medicines, including TCM and Ayurveda. For instance, even though our own laboratory and others have employed the functional genomics, proteomics and limited metabolomics approaches and attempts to define the immune-modulatory activities of *E. purpurea plants extracts* (Hou *et al.*, 2010; Wang *et al.*, 2008a) or other medicinal plants, our new findings, although helpful in exploring possible molecular mechanisms of the action mode on key immune cell type(s), these results are still not able to allow us to demonstrate the "efficacy" or exact function of a spectrum of *E. purpurea* herbal products as the commercial products. Careful and redefined clinical (trail) studies using bioactivity and chemical profiling-defined phytoextracts or phytochemical mixtures may be helpful or required in such future efforts.

The experimental systems and tools for systems biology/omics studies are increasingly available and applicable to research into medicinal plants and their effects on mammalian bioactivities. These research approaches and strategies, however, do not necessarily provide additional or beneficial information on how to improve the use of phytomedicines, due to the complexity of the disease, disorder and our body's normal physiology systems; we therefore should avoid categorically becoming over-optimistic and unrealistic about the future prospect of the science and technology for developing herbal medicines. The same or similar problems are also being recognized for the current development of new chemical drugs from pharmaceutical industry.

Repeated findings on the "readily detectable" antioxidant, anti-inflammation and "anti-tumour" effects of a broad spectrum of herbal medicines may not always serve as a good indication for the effectiveness or "efficacy" of test herbal remedies or phytochemicals. Since a modest level in these bioactivities may simply represent the "reductant" activities of a big spectrum of plant primary and secondary metabolites in common. As a result, there may often be a lack of "true specificity" in detected bioactivities in tested phytoextracts or phytochemicals. Overly simplified or casual claims of "potent anti-inflammatory or anti-tumour activities" may be viewed as hypes and can be very harmful to our research activities in general, and hence they need to be carefully avoided.

The nature or/and appearance of multiple molecular targets for traditional herbal medicines or phytochemicals may be true, but it may not be a unique feature for herbal medicines only, as many single chemical compounds are well known to exhibit their effects via interaction with multiple molecular targets. With the same token, multiple plant formulations, as often prescribed in TCM practice, may not always be accurately viewed as "aiming at multiple cellular/molecular targets". As the key rationales behind the multiple plants in a TCM formulation often reflect the benefit of a king (primary) drug, minister (secondary) drug, the adjuvant and the bioavailability/delivery (carrier) in combination. Therefore, it could be quite specific towards some specific "target(s)".

In conclusion, we are observing a big change in phytomedicine research, with new concepts, tools and approaches becoming increasingly available. What we may need now are systems build-up, networking, integration of collaboration and the data and database sharing at the global level.

## ACKNOWLEDGEMENTS

This work was supported by a grant (99-2324-B-001-003-CC2) from the National Science Council, Taiwan. We thank Ms. Miranda Loney and Ms. Ruth Giodano of Academia Sinica for the professional editing of this chapter.

# REFERENCES


Aggarwal, B. B. (2010). Targeting inflammation-induced obesity and metabolic diseases by curcumin and other nutraceuticals. *Annual Review of Nutrition* **30**, 173–199.

Aggarwal, B. B. and Shishodia, S. (2006). Molecular targets of dietary agents for prevention and therapy of cancer. *Biochemical Pharmacology* **71**, 1397–1421.

Ajikumar, P. K., Tyo, K., Carlsen, S., Mucha, O., Phon, T. H. and Stephanopoulos, G. (2008). Terpenoids: Opportunities for biosynthesis of natural product drugs using engineered microorganisms. *Molecular Pharmacology* **5**, 167–190.

Akdis, M., Blaser, K. and Akdis, C. A. (2005). T regulatory cells in allergy: Novel concepts in the pathogenesis, prevention, and treatment of allergic diseases. *The Journal of Allergy and Clinical Immunology* **116**, 961–968, quiz 969.

Aldieri, E., Atragene, D., Bergandi, L., Riganti, C., Costamagna, C., Bosia, A. and Ghigo, D. (2003). Artemisinin inhibits inducible nitric oxide synthase and nuclear factor NF-kB activation. *FEBS Letters* **552**, 141–144.

Ali, M., Thomson, M. and Afzal, M. (2000). Garlic and onions: Their effect on eicosanoid metabolism and its clinical relevance. *Prostaglandins, Leukotrienes, and Essential Fatty Acids* **62**, 55–73.

Al-Suhaimi, E. A., Al-Riziza, N. A. and Al-Essa, R. A. (2011). Physiological and therapeutical roles of ginger and turmeric on endocrine functions. *The American Journal of Chinese Medicine* **39**, 215–231.

Ammon, H. P. and Wahl, M. A. (1991). Pharmacology of Curcuma longa. *Planta Medica* **57**, 1–7.

An, S., Park, Y. D., Paik, Y. K., Jeong, T. S. and Lee, W. S. (2007). Human ACAT inhibitory effects of shikonin derivatives from Lithospermum erythrorhizon. *Bioorganic & Medicinal Chemistry Letters* **17**, 1112–1116.

Antonoff, M. B., Chugh, R., Borja-Cacho, D., Dudeja, V., Clawson, K. A., Skube, S. J., Sorenson, B. S., Saltzman, D. A., Vickers, S. M. and Saluja, A. K. (2009). Triptolide therapy for neuroblastoma decreases cell viability in vitro and inhibits tumor growth in vivo. *Surgery* **146**, 282–290.

Aravindaram, K. and Yang, N. S. (2010). Anti-inflammatory plant natural products for cancer therapy. *Planta Medica* **76**, 1103–1117.

Azaizeh, H., Saad, B., Cooper, E. and Said, O. (2010). Traditional Arabic and Islamic medicine, a re-emerging health aid. *Evidence-Based Complementary and Alternative Medicine: eCAM* **7**, 419–424.

Babu, P. V. and Liu, D. (2008). Green tea catechins and cardiovascular health: An update. *Current Medicinal Chemistry* **15**, 1840–1850.

Bacon, J. R., Williamson, G., Garner, R. C., Lappin, G., Langouet, S. and Bao, Y. (2003). Sulforaphane and quercetin modulate PhIP-DNA adduct formation in human HepG2 cells and hepatocytes. *Carcinogenesis* **24**, 1903–1911.

Balint, G. A. (2001). Artemisinin and its derivatives: An important new class of antimalarial agents. *Pharmacology & Therapeutics* **90**, 261–265.

Balunas, M. J. and Kinghorn, A. D. (2005). Drug discovery from medicinal plants. *Life Sciences* **78**, 431–441.

Bayet-Robert, M., Kwiatkowski, F., Leheurteur, M., Gachon, F., Planchat, E., Abrial, C., Mouret-Reynier, M. A., Durando, X., Barthomeuf, C. and Chollet, P. (2010). Phase I dose escalation trial of docetaxel plus curcumin in patients with advanced and metastatic breast cancer. *Cancer Biology & Therapy* **9**, 8–14.

Benson, J. M., Pokorny, A. J., Rhule, A., Wenner, C. A., Kandhi, V., Cech, N. B. and Shepherd, D. M. (2010). Echinacea purpurea extracts modulate murine dendritic cell fate and function. *Food and Chemical Toxicology* **48**, 1170–1177.

Berchtold, C. M., Chen, K. S., Miyamoto, S. and Gould, M. N. (2005). Perillyl alcohol inhibits a calcium-dependent constitutive nuclear factor-kappaB pathway. *Cancer Research* **65**, 8558–8566.

Bereswill, S., Munoz, M., Fischer, A., Plickert, R., Haag, L. M., Otto, B., Kuhl, A. A., Loddenkemper, C., Gobel, U. B. and Heimesaat, M. M. (2010). Anti-inflammatory effects of resveratrol, curcumin and simvastatin in acute small intestinal inflammation. *PLoS One* **5**, e15099.

Bianchi, M. E. (2007). DAMPs, PAMPs and alarmins: All we need to know about danger. *Journal of Leukocyte Biology* **81**, 1–5.

Block, E. (1985). The chemistry of garlic and onions. *Scientific American* **252**, 114–119.

Block, E., Ahmad, S., Jain, M. K., Crecely, R. W., Apitz-Castro, R. and Cruz, M. R. (1984). The chemistry of alkyl thiosulfate esters. 8. (E, Z)-Ajoene: A potent antithrombotic agent from garlic. *Journal of the American Chemical Society* **106**, 8295–8296.

Boon, H., Brown, J. B., Gavin, A., Kennard, M. A. and Stewart, M. (1999). Breast cancer survivors' perceptions of complementary/alternative medicine (CAM): Making the decision to use or not to use. *Qualitative Health Research* **9**, 639–653.

Borchers, A. T., Keen, C. L., Stern, J. S. and Gershwin, M. E. (2000). Inflammation and Native American medicine: The role of botanicals. *The American Journal of Clinical Nutrition* **72**, 339–347.

Bouvier, F., Rahier, A. and Camara, B. (2005). Biogenesis, molecular regulation and function of plant isoprenoids. *Progress in Lipid Research* **44**, 357–429.

Brinker, A. M. and Raskin, I. (2005). Determination of triptolide in root extracts of Tripterygium wilfordii by solid-phase extraction and reverse-phase high-performance liquid chromatography. *Journal of Chromatography A* **1070**, 65–70.

Brodnitz, M. H., Pascale, J. V. and Van Derslice, L. (1971). Flavor components of garlic extract. *Journal of Agricultural and Food Chemistry* **19**, 273–275.

Brown, G. D., Taylor, P. R., Reid, D. M., Willment, J. A., Williams, D. L., Martinez-Pomares, L., Wong, S. Y. and Gordon, S. (2002). Dectin-1 is a major beta-glucan receptor on macrophages. *Journal of Experimental Medicine* **196**, 407–412.

Brush, J., Mendenhall, E., Guggenheim, A., Chan, T., Connelly, E., Soumyanath, A., Buresh, R., Barrett, R. and Zwickey, H. (2006). The effect of Echinacea purpurea, Astragalus membranaceus and Glycyrrhiza glabra on CD69 expression and immune cell activation in humans. *Phytotherapy Research* **20**, 687–695.

Busse, P. J., Schofield, B., Birmingham, N., Yang, N., Wen, M. C., Zhang, T., Srivastava, K. and Li, X. M. (2010). The traditional Chinese herbal formula ASHMI inhibits allergic lung inflammation in antigen-sensitized and antigen-challenged aged mice. *Annals of Allergy, Asthma and Immunology* **104**, 236–246.

Carroll, R. E., Benya, R. V., Turgeon, D. K., Vareed, S., Neuman, M., Rodriguez, L., Kakarala, M., Carpenter, P. M., Mclaren, C., Meyskens, F. L., Jr. and Brenner, D. E. (2011). Phase IIa clinical trial of curcumin for the prevention of colorectal neoplasia. *Cancer Prevention Research (Philadelphia, PA)* **4**, 354–364.

Cerundolo, V., Silk, J. D., Masri, S. H. and Salio, M. (2009). Harnessing invariant NKT cells in vaccination strategies. *Nature Reviews. Immunology* **9**, 28–38.

Chada, S., Ramesh, R. and Mhashilkar, A. M. (2003). Cytokine- and chemokine-based gene therapy for cancer. *Current Opinion in Molecular Therapeutics* **5**, 463–474.

Chang, H. P., Huang, S. Y. and Chen, Y. H. (2005). Modulation of cytokine secretion by garlic oil derivatives is associated with suppressed nitric oxide production in stimulated macrophages. *Journal of Agricultural and Food Chemistry* **53**, 2530–2534.

Chang, T. W., Wu, P. C., Hsu, C. L. and Hung, A. F. (2007). Anti-IgE antibodies for the treatment of IgE-mediated allergic diseases. *Advances in Immunology* **93**, 63–119.

Chang, I. C., Huang, Y. J., Chiang, T. I., Yeh, C. W. and Hsu, L. S. (2010). Shikonin induces apoptosis through reactive oxygen species/extracellular signal-regulated kinase pathway in osteosarcoma cells. *Biological and Pharmaceutical Bulletin* **33**, 816–824.

Chen, X., Oppenheim, J. and Howard, O. M. (2001). Shikonin, a component of antiinflammatory Chinese herbal medicine, selectively blocks chemokine binding to CC chemokine receptor-1. *International Immunopharmacology* **1**, 229–236.

Chen, X., Yang, L., Zhang, N., Turpin, J. A., Buckheit, R. W., Osterling, C., Oppenheim, J. J. and Howard, O. M. (2003). Shikonin, a component of Chinese herbal medicine, inhibits chemokine receptor function and suppresses human immunodeficiency virus type 1. *Antimicrobial Agents and Chemotherapy* **47**, 2810–2816.

Chen, S. T., Dou, J., Temple, R., Agarwal, R., Wu, K. M. and Walker, S. (2008). New therapies from old medicines. *Nature Biotechnology* **26**, 1077–1083.

Chen, Q., Liu, Z. and He, J. H. (2009a). Achyranthes bidentata polysaccharide enhances immune response in weaned piglets. *Immunopharmacology and Immunotoxicology* **31**, 253–260.

Chen, S., Yin, D. K., Yao, W. B., Wang, Y. D., Zhang, Y. R. and Gao, X. D. (2009b). Macrophage receptors of polysaccharide isolated from a marine filamentous fungus Phoma herbarum YS4108. *Acta Pharmacologica Sinica* **30**, 1008–1014.

Chen, D., Zhang, X., Du, Y., Jia, B., Ka, W., Sun, D., Yao, W. and Wen, Z. (2011). Effects of Gekko sulfated polysaccharide-protein complex on the defective biorheological characters of dendritic cells under tumor microenvironment. *Cell Biochemistry and Biophysics* **62**, 193–201.

Chen, X., Yang, L., Oppenheim, J. J. and Howard, M. Z. (2002). Cellular pharmacology studies of shikonin derivatives. *Phytotherapy Research* **16**, 199–209.

Cheung, J. K., Li, J., Cheung, A. W., Zhu, Y., Zheng, K. Y., Bi, C. W., Duan, R., Choi, R. C., Lau, D. T., Dong, T. T., Lau, B. W. and Tsim, K. W. (2009). Cordysinocan, a polysaccharide isolated from cultured Cordyceps, activates immune responses in cultured T-lymphocytes and macrophages: Signaling cascade and induction of cytokines. *Journal of Ethnopharmacology* **124**, 61–68.

Chew, B. P. and Park, J. S. (2004). Carotenoid action on the immune response. *The Journal of Nutrition* **134**, 257S–261S.

Chiang, Y. M., Lo, C. P., Chen, Y. P., Wang, S. Y., Yang, N. S., Kuo, Y. H. and Shyur, L. F. (2005). Ethyl caffeate suppresses NF-kappaB activation and its downstream inflammatory mediators, iNOS, COX-2, and PGE2 in vitro or in mouse skin. *British Journal of Pharmacology* **146**, 352–363.

Chicca, A., Raduner, S., Pellati, F., Strompen, T., Altmann, K. H., Schoop, R. and Gertsch, J. (2009). Synergistic immunomopharmacological effects of N-alkylamides in Echinacea purpurea herbal extracts. *International Immunopharmacology* **9**, 850–858.

Chiu, S. C. and Yang, N. S. (2007). Inhibition of tumor necrosis factor-alpha through selective blockade of Pre-mRNA splicing by shikonin. *Molecular Pharmacology* **71**, 1640–1645.

Chiu, S. C., Tsao, S. W., Hwang, P. I., Vanisree, S., Chen, Y. A. and Yang, N. S. (2010). Differential functional genomic effects of anti-inflammatory phytocompounds on immune signaling. *BMC Genomics* **11**, 513.

Cho, J. H. (2008). The genetics and immunopathogenesis of inflammatory bowel disease. *Nature Reviews. Immunology* **8**, 458–466.

Cho, W. C. (2010). Scientific evidence on the supportive cancer care with Chinese medicine. *Zhongguo Fei Ai Za Zhi* **13**, 190–194.

Choi, E. M. and Hwang, J. K. (2002). Enhancement of oxidative response and cytokine production by yam mucopolysaccharide in murine peritoneal macrophage. *Fitoterapia* **73**, 629–637.

Choi, E. M., Koo, S. J. and Hwang, J. K. (2004). Immune cell stimulating activity of mucopolysaccharide isolated from yam (Dioscorea batatas). *Journal of Ethnopharmacology* **91**, 1–6.

Christen, P. and Veuthey, J. L. (2001). New trends in extraction, identification and quantification of artemisinin and its derivatives. *Current Medicinal Chemistry* **8**, 1827–1839.

Conforti, F. and Menichini, F. (2011). Phenolic compounds from plants as nitric oxide production inhibitors. *Current Medicinal Chemistry* **18**, 1137–1145.

Cooper, M. A., Colonna, M. and Yokoyama, W. M. (2009). Hidden talents of natural killers: NK cells in innate and adaptive immunity. *EMBO Reports* **10**, 1103–1110.

Cui, L. and Su, X. Z. (2009). Discovery, mechanisms of action and combination therapy of artemisinin. *Expert Review of Anti-Infective Therapy* **7**, 999–1013.

Csiszar, A. (2011). Anti-inflammatory effects of resveratrol: possible role in prevention of age-related cardiovascular disease. *Annals of the New York Academy of Sciences* **1215**, 117–122.

Dai, Q., Fang, J. and Zhang, F. S. (2009). Dual role of shikonin in early and late stages of collagen type II arthritis. *Molecular Biology Reports* **36**, 1597–1604.

De Las Heras, B. and Hortelano, S. (2009). Molecular basis of the anti-inflammatory effects of terpenoids. *Inflammation and Allergy Drug Targets* **8**, 28–39.

De Stefano, D., Maiuri, M. C., Simeon, V., Grassia, G., Soscia, A., Cinelli, M. P. and Carnuccio, R. (2007). Lycopene, quercetin and tyrosol prevent macrophage activation induced by gliadin and IFN-gamma. *European Journal of Pharmacology* **566**, 192–199.

Diebold, S. S. (2008). Determination of T-cell fate by dendritic cells. *Immunology and Cell Biology* **86**, 389–397.

Dihal, A. A., De Boer, V. C., Van Der Woude, H., Tilburgs, C., Bruijntjes, J. P., Alink, G. M., Rietjens, I. M., Woutersen, R. A. and Stierum, R. H. (2006). Quercetin, but not its glycosidated conjugate rutin, inhibits azoxymethane-induced colorectal carcinogenesis in F344 rats. *The Journal of Nutrition* **136**, 2862–2867.

Dirsch, V. M., Kiemer, A. K., Wagner, H. and Vollmar, A. M. (1998). Effect of allicin and ajoene, two compounds of garlic, on inducible nitric oxide synthase. *Atherosclerosis* **139**, 333–339.

Donath, M. Y. and Shoelson, S. E. (2011). Type 2 diabetes as an inflammatory disease. *Nature Reviews. Immunology* **11**, 98–107.

El-Mowafy, A. M., Salem, H. A., Al-Gayyar, M. M., El-Mesery, M. E. and El-Azab, M. F. (2011). Evaluation of renal protective effects of the green-tea (EGCG) and red grape resveratrol: Role of oxidative stress and inflammatory cytokines. *Natural Product Research* **25**, 850–856.

Ernst, E. (2002). The risk-benefit profile of commonly used herbal therapies: Ginkgo, St. John's Wort, Ginseng, Echinacea, Saw Palmetto, and Kava. *Annals of Internal Medicine* **136**, 42–53.

Ernst, E. and Cassileth, B. R. (1998). The prevalence of complementary/alternative medicine in cancer: A systematic review. *Cancer* **83**, 777–782.

Fernandez, M. A., De Las Heras, B., Garcia, M. D., Saenz, M. T. and Villar, A. (2001). New insights into the mechanism of action of the anti-inflammatory triterpene lupeol. *The Journal of Pharmacy and Pharmacology* **53**, 1533–1539.

Fraga, C. G. (2010). Plant Phenolics and Human Health: Biochemistry, Nutrition, and Pharmacology. Wiley, Hoboken, NJ.

Fu, S. L., Hsu, Y. H., Lee, P. Y., Hou, W. C., Hung, L. C., Lin, C. H., Chen, C. M. and Huang, Y. J. (2006). Dioscorin isolated from Dioscorea alata activates TLR4-signaling pathways and induces cytokine expression in macrophages. *Biochemical and Biophysical Research Communications* **339**, 137–144.

Fulda, S. (2009). Betulinic acid: A natural product with anticancer activity. *Molecular Nutrition and Food Research* **53**, 140–146.

Gao, S., Liu, Z., Li, H., Little, P. J., Liu, P. and Xu, S. (2012). Cardiovascular actions and therapeutic potential of tanshinone IIA. *Atherosclerosis* **220**, 3–10.

Garg, A. D., Nowis, D., Golab, J., Vandenabeele, P., Krysko, D. V. and Agostinis, P. (2010). Immunogenic cell death, DAMPs and anticancer therapeutics: An emerging amalgamation. *Biochimica et Biophysica Acta* **1805**, 53–71.

Geissmann, F., Manz, M. G., Jung, S., Sieweke, M. H., Merad, M. and Ley, K. (2010). Development of monocytes, macrophages, and dendritic cells. *Science* **327**, 656–661.

Godfrey, D. I., Stankovic, S. and Baxter, A. G. (2010). Raising the NKT cell family. *Nature Immunology* **11**, 197–206.

Goel, A. and Aggarwal, B. B. (2010). Curcumin, the golden spice from Indian saffron, is a chemosensitizer and radiosensitizer for tumors and chemoprotector and radioprotector for normal organs. *Nutrition and Cancer* **62**, 919–930.

Goel, V., Lovlin, R., Chang, C., Slama, J. V., Barton, R., Gahler, R., Bauer, R., Goonewardene, L. and Basu, T. K. (2005). A proprietary extract from the echinacea plant (Echinacea purpurea) enhances systemic immune response during a common cold. *Phytotherapy Research* **19**, 689–694.

Goldman, P. (2001). Herbal medicines today and the roots of modern pharmacology. *Annals of Internal Medicine* **135**, 594–600.

Golenser, J., Waknine, J. H., Krugliak, M., Hunt, N. H. and Grau, G. E. (2006). Current perspectives on the mechanism of action of artemisinins. *International Journal for Parasitology* **36**, 1427–1441.

Gomes, A., Fernandes, E., Lima, J. L., Mira, L. and Corvo, M. L. (2008). Molecular mechanisms of anti-inflammatory activity mediated by flavonoids. *Current Medicinal Chemistry* **15**, 1586–1605.

Graff, J. C., Kimmel, E. M., Freedman, B., Schepetkin, I. A., Holderness, J., Quinn, M. T., Jutila, M. A. and Hedges, J. F. (2009). Polysaccharides derived from Yamoa (Funtumia elastica) prime gammadelta T cells in vitro and enhance innate immune responses in vivo. *International Immunopharmacology* **9**, 1313–1322.

Guo, L., Bai, S. P., Zhao, L. and Wang, X. H. (2011). Astragalus polysaccharide injection integrated with vinorelbine and cisplatin for patients with advanced non-small cell lung cancer: Effects on quality of life and survival. *Medical Oncology*.

Gupta, S. C., Prasad, S., Kim, J. H., Patchva, S., Webb, L. J., Priyadarsini, I. K. and Aggarwal, B. B. (2011). Multitargeting by curcumin as revealed by molecular interaction studies. *Natural Product Reports* **28**, 1937–1955.

Han, S. B., Yoon, Y. D., Ahn, H. J., Lee, H. S., Lee, C. W., Yoon, W. K., Park, S. K. and Kim, H. M. (2003). Toll-like receptor-mediated activation of B cells and macrophages by polysaccharide isolated from cell culture of Acanthopanax senticosus. *International Immunopharmacology* **3**, 1301–1312.

Han, W., Li, L., Qiu, S., Lu, Q., Pan, Q., Gu, Y., Luo, J. and Hu, X. (2007). Shikonin circumvents cancer drug resistance by induction of a necroptotic death. *Molecular Cancer Therapeutics* **6**, 1641–1649.

Harikumar, K. B. and Aggarwal, B. B. (2008). Resveratrol: A multitargeted agent for age-associated chronic diseases. *Cell Cycle* **7**, 1020–1035.

He, Y. L., Ying, Y., Xu, Y. L., Su, J. F., Luo, H. and Wang, H. F. (2005). Effects of Lycium barbarum polysaccharide on tumor microenvironment T-lymphocyte subsets and dendritic cells in H22-bearing mice. *Zhong Xi Yi Jie He Xue Bao* **3**, 374–377.

He, X., Li, X., Liu, B., Xu, L., Zhao, H. and Lu, A. (2011). Down-regulation of Treg cells and up-regulation of TH1/TH2 cytokine ratio were induced by polysaccharide from Radix Glycyrrhizae in H22 hepatocarcinoma bearing mice. *Molecules* **16**, 8343–8352.

Ho, L. J. and Lai, J. H. (2004). Chinese herbs as immunomodulators and potential disease-modifying antirheumatic drugs in autoimmune disorders. *Current Drug Metabolism* **5**, 181–192.

Hohmann, J., Redei, D., Forgo, P., Szabo, P., Freund, T. F., Haller, J., Bojnik, E. and Benyhe, S. (2011). Alkamides and a neolignan from Echinacea purpurea roots and the interaction of alkamides with G-protein-coupled cannabinoid receptors. *Phytochemistry* **72**, 1848–1853.

Hongqin, T., Xinyu, L., Heng, G., Lanfang, X., Yongfang, W. and Shasha, S. (2011). Triptolide inhibits IFN-gamma signaling via the Jak/STAT pathway in HaCaT keratinocytes. *Phytotherapy Research* **25**, 1678–1685.

Hou, W. C., Hsu, F. L. and Lee, M. H. (2002). Yam (Dioscorea batatas) tuber mucilage exhibited antioxidant activities in vitro. *Planta Medica* **68**, 1072–1076.

Hou, C. C., Chen, C. H., Yang, N. S., Chen, Y. P., Lo, C. P., Wang, S. Y., Tien, Y. J., Tsai, P. W. and Shyur, L. F. (2010). Comparative metabolomics approach coupled with cell- and gene-based assays for species classification and anti-inflammatory bioactivity validation of Echinacea plants. *The Journal of Nutritional Biochemistry* **21**, 1045–1059.

Huang, B., Pan, P. Y., Li, Q., Sato, A. I., Levy, D. E., Bromberg, J., Divino, C. M. and Chen, S. H. (2006). Gr-1+CD115+ immature myeloid suppressor cells mediate the development of tumor-induced T regulatory cells and T-cell anergy in tumor-bearing host. *Cancer Research* **66**, 1123–1131.

Huang, C. S., Fan, Y. E., Lin, C. Y. and Hu, M. L. (2007). Lycopene inhibits matrix metalloproteinase-9 expression and down-regulates the binding activity of nuclear factor-kappa B and stimulatory protein-1. *The Journal of Nutritional Biochemistry* **18**, 449–456.

Huang, M. C., Liao, J. J., Bonasera, S., Longo, D. L. and Goetzl, E. J. (2008). Nuclear factor-kappaB-dependent reversal of aging-induced alterations in T cell cytokines. *The FASEB Journal* **22**, 2142–2150.

Huang, C. C., Lo, C. P., Chiu, C. Y. and Shyur, L. F. (2010). Deoxyelephantopin, a novel multifunctional agent, suppresses mammary tumour growth and lung metastasis and doubles survival time in mice. *British Journal of Pharmacology* **159**, 856–871.

Huong, P. T., Lee, C. H., Li, M. H., Lee, M. Y., Kim, J. K., Lee, S. M., Seon, J. H., Lee, D. C. and Jeon, Y. J. (2011). Characterization and immunopotentiating effects of the glycoprotein isolated from dioscorea batatas. *Korean Journal of Physiology and Pharmacology* **15**, 101–106.

Hwang, J. S., Chung, H. K., Bae, E. K., Lee, A. Y., Ji, H. J., Park, D. W., Jung, H. J., Cho, C. W., Choi, H. J., Lee, D. S., Lee, K. R. and Youn, H. J. (2003). The polysaccharide fraction AIP1 from Artemisia iwayomogi suppresses apoptotic death of the mouse spleen cells in culture. *Archives of Pharmacal Research* **26**, 294–300.

Ichikawa, H., Nair, M. S., Takada, Y., Sheeja, D. B., Kumar, M. A., Oommen, O. V. and Aggarwal, B. B. (2006). Isodeoxyelephantopin, a novel sesquiterpene lactone, potentiates apoptosis, inhibits invasion, and abolishes osteoclastogenesis through suppression of nuclear factor-kappaB (nf-kappaB) activation and nf-kappaB-regulated gene expression. *Clinical Cancer Research* **12**, 5910–5918.

Ikeda, Y., Murakami, A. and Ohigashi, H. (2008). Ursolic acid: An anti- and proinflammatory triterpenoid. *Molecular Nutrition and Food Research* **52**, 26–42.

Iwasaki, A. and Medzhitov, R. (2010). Regulation of adaptive immunity by the innate immune system. *Science* **327**, 291–295.

Janeway, C. (2005). Immunobiology: The Immune System in Health and Disease. Garland Science, New York.

Jeurissen, A., Van Etten, E., Overbergh, L., Wuyts, G., Heremans, H., Matthys, P., Ceuppens, J. L., Mathieu, C. and Bossuyt, X. (2005). 1alpha,25-Dihydroxyvitamin D3 modulates the murine antibody response to pneumococcal capsular polysaccharide serotype 3 through IL-12. *European Journal of Immunology* **35**, 1841–1848.

Jia, L. and Zhao, Y. (2009). Current evaluation of the millennium phytomedicine—Ginseng (I): Etymology, pharmacognosy, phytochemistry, market and regulations. *Current Medicinal Chemistry* **16**, 2475–2484.

Jiang, D., Liang, J., Fan, J., Yu, S., Chen, S., Luo, Y., Prestwich, G. D., Mascarenhas, M. M., Garg, H. G., Quinn, D. A., Homer, R. J., Goldstein, D. R. *et al.* (2005). Regulation of lung injury and repair by Toll-like receptors and hyaluronan. *Nature Medicine* **11**, 1173–1179.

Jiang, D., Liang, J. and Noble, P. W. (2007). Hyaluronan in tissue injury and repair. *Annual Review of Cell and Developmental Biology* **23**, 435–461.

Jiang, M. H., Zhu, L. and Jiang, J. G. (2010). Immunoregulatory actions of polysaccharides from Chinese herbal medicine. *Expert Opinion on Therapeutic Targets* **14**, 1367–1402.

Jin, M., Suh, S. J., Yang, J. H., Lu, Y., Kim, S. J., Kwon, S., Jo, T. H., Kim, J. W., Park, Y. I., Ahn, G. W., Lee, C. K., Kim, C. H. *et al.* (2010). Anti-inflammatory activity of bark of Dioscorea batatas DECNE through the inhibition of iNOS and COX-2 expressions in RAW264.7 cells via NF-kappaB and ERK1/2 inactivation. *Food and Chemical Toxicology* **48**, 3073–3079.

Joffre, O., Nolte, M. A., Sporri, R. and Reis e Sousa, C. (2009). Inflammatory signals in dendritic cell activation and the induction of adaptive immunity. *Immunological Reviews* **227**, 234–247.

Kaith, B. S., Kaith, N. S. and Chauhan, N. S. (1996). Anti-inflammatory effect of Arnebia euchroma root extracts in rats. *Journal of Ethnopharmacology* **55**, 77–80.

Kanai, M., Yoshimura, K., Asada, M., Imaizumi, A., Suzuki, C., Matsumoto, S., Nishimura, T., Mori, Y., Masui, T., Kawaguchi, Y., Yanagihara, K.,

Yazumi, S. *et al.* (2011). A phase I/II study of gemcitabine-based chemotherapy plus curcumin for patients with gemcitabine-resistant pancreatic cancer. *Cancer Chemotherapy and Pharmacology* **68**, 157–164.

Keiss, H. P., Dirsch, V. M., Hartung, T., Haffner, T., Trueman, L., Auger, J., Kahane, R. and Vollmar, A. M. (2003). Garlic (Allium sativum L.) modulates cytokine expression in lipopolysaccharide-activated human blood thereby inhibiting NF-kappaB activity. *The Journal of Nutrition* **133**, 2171–2175.

Khayrullina, T., Yen, J. H., Jing, H. and Ganea, D. (2008). In vitro differentiation of dendritic cells in the presence of prostaglandin E2 alters the IL-12/IL-23 balance and promotes differentiation of Th17 cells. *Journal of Immunology* **181**, 721–735.

Kim, E. C., Min, J. K., Kim, T. Y., Lee, S. J., Yang, H. O., Han, S., Kim, Y. M. and Kwon, Y. G. (2005). [6]-Gingerol, a pungent ingredient of ginger, inhibits angiogenesis in vitro and in vivo. *Biochemical and Biophysical Research Communications* **335**, 300–308.

Kim, D. H., Shin, E. K., Kim, Y. H., Lee, B. W., Jun, J. G., Park, J. H. and Kim, J. K. (2009). Suppression of inflammatory responses by celastrol, a quinone methide triterpenoid isolated from Celastrus regelii. *European Journal of Clinical Investigation* **39**, 819–827.

Kim, Y. O., Hong, S. J. and Yim, S. V. (2010). The efficacy of shikonin on cartilage protection in a mouse model of rheumatoid arthritis. *Korean Journal of Physiology and Pharmacology* **14**, 199–204.

Kizelsztein, P., Komarnytsky, S. and Raskin, I. (2009). Oral administration of triptolide ameliorates the clinical signs of experimental autoimmune encephalomyelitis (EAE) by induction of HSP70 and stabilization of NF-kappaB/IkappaBalpha transcriptional complex. *Journal of Neuroimmunology* **217**, 28–37.

Ko, F. N., Lee, Y. S., Kuo, S. C., Chang, Y. S. and Teng, C. M. (1995). Inhibition on platelet activation by shikonin derivatives isolated from Arnebia euchroma. *Biochimica et Biophysica Acta* **1268**, 329–334.

Kodera, Y., Suzuki, A., Imada, O., Kasuga, S., Sumioka, I., Kanezawa, A., Taru, N., Fujikawa, M., Nagae, S., Masamoto, K., Maeshige, K. and Ono, K. (2002). Physical, chemical, and biological properties of s-allylcysteine, an amino acid derived from garlic. *Journal of Agricultural and Food Chemistry* **50**, 622–632.

Konkimalla, V. B., Blunder, M., Korn, B., Soomro, S. A., Jansen, H., Chang, W., Posner, G. H., Bauer, R. and Efferth, T. (2008). Effect of artemisinins and other endoperoxides on nitric oxide-related signaling pathway in RAW 264.7 mouse macrophage cells. *Nitric Oxide* **19**, 184–191.

Koo, H. J., Song, Y. S., Kim, H. J., Lee, Y. H., Hong, S. M., Kim, S. J., Kim, B. C., Jin, C., Lim, C. J. and Park, E. H. (2004). Antiinflammatory effects of genipin, an active principle of gardenia. *European Journal of Pharmacology* **495**, 201–208.

Krakauer, T., Chen, X., Howard, O. M. and Young, H. A. (2005). Triptolide attenuates endotoxin- and staphylococcal exotoxin-induced T-cell proliferation and production of cytokines and chemokines. *Immunopharmacology and Immunotoxicology* **27**, 53–66.

Krogsgaard, M. and Davis, M. M. (2005). How T cells 'see' antigen. *Nature Immunology* **6**, 239–245.

Kummar, S., Copur, M. S., Rose, M., Wadler, S., Stephenson, J., O'rourke, M., Brenckman, W., Tilton, R., Liu, S. H., Jiang, Z., Su, T., Cheng, Y. C. *et al.* (2011). A phase I study of the Chinese herbal medicine PHY906 as a modulator of irinotecan-based chemotherapy in patients with advanced colorectal cancer. *Clinical Colorectal Cancer* **10**, 85–96.

Kupchan, S. M., Court, W. A., Dailey, R. G., Jr., Gilmore, C. J. and Bryan, R. F. (1972). Triptolide and tripdiolide, novel antileukemic diterpenoid triepoxides from Tripterygium wilfordii. *Journal of the American Chemical Society* **94**, 7194–7195.

Lai, J. H. (2002). Immunomodulatory effects and mechanisms of plant alkaloid tetrandrine in autoimmune diseases. *Acta Pharmacologica Sinica* **23**, 1093–1101.

Lai, N. S., Lin, R. H., Lai, R. S., Kun, U. C. and Leu, S. C. (2001). Prevention of autoantibody formation and prolonged survival in New Zealand Black/New Zealand White F1 mice with an ancient Chinese herb, Ganoderma tsugae. *Lupus* **10**, 461–465.

Lam, W., Bussom, S., Guan, F., Jiang, Z., Zhang, W., Gullen, E. A., Liu, S. H. and Cheng, Y. C. (2010). The four-herb Chinese medicine PHY906 reduces chemotherapy-induced gastrointestinal toxicity. *Science Translational Medicine* **2**, 45ra59.

Lamoral-Theys, D., Pottier, L., Dufrasne, F., Neve, J., Dubois, J., Kornienko, A., Kiss, R. and Ingrassia, L. (2010). Natural polyphenols that display anticancer properties through inhibition of kinase activity. *Current Medicinal Chemistry* **17**, 812–825.

Lanier, L. L. and Sun, J. C. (2009). Do the terms innate and adaptive immunity create conceptual barriers? *Nature Reviews. Immunology* **9**, 302–303.

Lawson, L. D., Bauer, R. and American Chemical Society, Division of Agricultural and Food Chemistry & American Chemical Society (1998).Meeting 1998. Phytomedicines of Europe: Chemistry and Biological Activity. American Chemical Society, Washington, DC.

Lee, S. N., Kim, N. S. and Lee, D. S. (2003). Comparative study of extraction techniques for determination of garlic flavor components by gas chromatography-mass spectrometry. *Analytical and Bioanalytical Chemistry* **377**, 749–756.

Lee, E. H., Faulhaber, D., Hanson, K. M., Ding, W., Peters, S., Kodali, S. and Granstein, R. D. (2004). Dietary lutein reduces ultraviolet radiation-induced inflammation and immunosuppression. *The Journal of Investigative Dermatology* **122**, 510–517.

Lee, H. J., Magesh, V., Nam, D., Lee, E. O., Ahn, K. S., Jung, M. H., Kim, D. K., Kim, J. Y. and Kim, S. H. (2008). Shikonin, acetylshikonin, and isobutyroylshikonin inhibit VEGF-induced angiogenesis and suppress tumor growth in lewis lung carcinoma-bearing mice. *Yakugaku Zasshi: Journal of the Pharmaceutical Society of Japan* **128**, 1681–1688.

Lee, T. Y., Lee, K. C., Chen, S. Y. and Chang, H. H. (2009). 6-Gingerol inhibits ROS and iNOS through the suppression of PKC-alpha and NF-kappaB pathways in lipopolysaccharide-stimulated mouse macrophages. *Biochemical and Biophysical Research Communications* **382**, 134–139.

Lee, C. C., Wang, C. N., Lai, Y. T., Kang, J. J., Liao, J. W., Chiang, B. L., Chen, H. C. and Cheng, Y. W. (2010a). Shikonin inhibits maturation of bone marrow-derived dendritic cells and suppresses allergic airway inflammation in a murine model of asthma. *British Journal of Pharmacology* **161**, 1496–1511.

Lee, H., Kang, R. and Yoon, Y. (2010b). Shikonin inhibits fat accumulation in 3T3-L1 adipocytes. *Phytotherapy Research* **24**, 344–351.

Lee, W. L., Wen, T. N., Shiau, J. Y. and Shyur, L. F. (2010c). Differential proteomic profiling identifies novel molecular targets of paclitaxel and phytoagent deoxyelephantopin against mammary adenocarcinoma cells. *Journal of Proteome Research* **9**, 237–253.

Lee, C. C., Kang, J. J., Chiang, B. L., Wang, C. N. and Cheng, Y. W. (2011). Shikonin inhibited mitogen-activated IL-4 and IL-5 production on EL-4 cells through downregulation of GATA-3 and c-Maf induction. *Life Sciences* **89**, 364–370.

Li, H. and Ni, J. (2011). Treatment of wastewater from Dioscorea zingiberensis tubers used for producing steroid hormones in a microbial fuel cell. *Bioresource Technology* **102**, 2731–2735.

Li, T., Tamada, K., Abe, K., Tada, H., Onoe, Y., Tatsugami, K., Harada, M., Kubo, C. and Nomoto, K. (1999). The restoration of the antitumor T cell response from stress-induced suppression using a traditional Chinese herbal medicine Hochu-ekki-to (TJ-41:Bu-Zhong-Yi-Qi-Tang). *Immunopharmacology* **43**, 11–21.

Li, W., Yajima, T., Saito, K., Nishimura, H., Fushimi, T., Ohshima, Y., Tsukamoto, Y. and Yoshikai, Y. (2004). Immunostimulating properties of intragastrically administered Acetobacter-derived soluble branched (1,4)-beta-D-glucans decrease murine susceptibility to Listeria monocytogenes. *Infection and Immunity* **72**, 7005–7011.

Li, X., Xu, W. and Chen, J. (2010a). Polysaccharide purified from Polyporus umbellatus (Per) Fr induces the activation and maturation of murine bone-derived dendritic cells via toll-like receptor 4. *Cellular Immunology* **265**, 50–56.

Li, X. B., He, X. J., Liu, B., Xu, L., Ju, D. H., Jiang, M. and Lu, A. P. (2010b). Immunoregulatory function of Radix Glycyrrhizae polysaccharide in tumor-bearing mice. *Zhong Xi Yi Jie He Xue Bao* **8**, 363–367.

Liby, K. T., Yore, M. M. and Sporn, M. B. (2007). Triterpenoids and rexinoids as multifunctional agents for the prevention and treatment of cancer. *Nature Reviews. Cancer* **7**, 357–369.

Licciardi, P. V. and Underwood, J. R. (2011). Plant-derived medicines: A novel class of immunological adjuvants. *International Immunopharmacology* **11**, 390–398.

Lin, N., Liu, C., Xiao, C., Jia, H., Imada, K., Wu, H. and Ito, A. (2007). Triptolide, a diterpenoid triepoxide, suppresses inflammation and cartilage destruction in collagen-induced arthritis mice. *Biochemical Pharmacology* **73**, 136–146.

Lin, F. M., Tsai, C. H., Yang, Y. C., Tu, W. C., Chen, L. R., Liang, Y. S., Wang, S. Y., Shyur, L. F., Chien, S. C., Cha, T. L. and Hsiao, P. W. (2008). A novel diterpene suppresses CWR22Rv1 tumor growth in vivo through antiproliferation and proapoptosis. *Cancer Research* **68**, 6634–6642.

Lin, P. L., Lin, K. W., Weng, C. F. and Lin, K. C. (2009). Yam storage protein dioscorins from Dioscorea alata and Dioscorea japonica exhibit distinct immunomodulatory activities in mice. *Journal of Agricultural and Food Chemistry* **57**, 4606–4613.

Linehan, S. A., Martinez-Pomares, L., Stahl, P. D. and Gordon, S. (1999). Mannose receptor and its putative ligands in normal murine lymphoid and nonlymphoid organs: In situ expression of mannose receptor by selected macrophages, endothelial cells, perivascular microglia, and mesangial cells, but not dendritic cells. *The Journal of Experimental Medicine* **189**, 1961–1972.

Liu, J. (1995). Pharmacology of oleanolic acid and ursolic acid. *Journal of Ethnopharmacology* **49**, 57–68.

Liu, Q., Chen, T., Chen, H., Zhang, M., Li, N., Lu, Z., Ma, P. and Cao, X. (2004). Triptolide (PG-490) induces apoptosis of dendritic cells through sequential p38 MAP kinase phosphorylation and caspase 3 activation. *Biochemical and Biophysical Research Communications* **319**, 980–986.

Liu, K. L., Chen, H. W., Wang, R. Y., Lei, Y. P., Sheen, L. Y. and Lii, C. K. (2006a). DATS reduces LPS-induced iNOS expression, NO production, oxidative stress, and NF-kappaB activation in RAW 264.7 macrophages. *Journal of Agricultural and Food Chemistry* **54**, 3472–3478.

Liu, Q., Chen, T., Chen, G., Li, N., Wang, J., Ma, P. and Cao, X. (2006b). Immunosuppressant triptolide inhibits dendritic cell-mediated chemoattraction of neutrophils and T cells through inhibiting Stat3 phosphorylation and NF-kappaB activation. *Biochemical and Biophysical Research Communications* **345**, 1122–1130.

Liu, Q., Chen, T., Chen, G., Shu, X., Sun, A., Ma, P., Lu, L. and Cao, X. (2007a). Triptolide impairs dendritic cell migration by inhibiting CCR7 and COX-2 expression through PI3-K/Akt and NF-kappaB pathways. *Molecular Immunology* **44**, 2686–2696.

Liu, Y. C., Zeng, J. G., Chen, B. and Yao, S. Z. (2007b). Investigation of phenolic constituents in Echinacea purpurea grown in China. *Planta Medica* **73**, 1600–1605.

Liu, Y. W., Liu, J. C., Huang, C. Y., Wang, C. K., Shang, H. F. and Hou, W. C. (2009). Effects of oral administration of yam tuber storage protein, dioscorin, to BALB/c mice for 21-days on immune responses. *Journal of Agricultural and Food Chemistry* **57**, 9274–9279.

Liu, J., Wan, L., Sheng, C. J. and Xie, X. L. (2011). The correlative study on pulmonary function changes and Th1/Th2 cells & regulatory T cells in adjuvant arthritis rats. *Xi Bao Yu Fen Zi Mian Yi Xue Za Zhi* **27**, 56–60.

Long, S., Guangzhi, Y., Baojie, G., Wei, X., Yanyong, H., Yingli, W., Yang, Z. and Lihua, L. (2011). Shikonin derivatives protect immune organs from damage and promote immune responses in vivo in tumour-bearing mice. *Phytotherapy Research* **26**, 26–33.

Lu, H., Yang, Y., Gad, E., Wenner, C. A., Chang, A., Larson, E. R., Dang, Y., Martzen, M., Standish, L. J. and Disis, M. L. (2011a). Polysaccharide krestin is a novel TLR2 agonist that mediates inhibition of tumor growth via stimulation of CD8 T cells and NK cells. *Clinical Cancer Research* **17**, 67–76.

Lu, L., Qin, A., Huang, H., Zhou, P., Zhang, C., Liu, N., Li, S., Wen, G., Dong, W., Wang, X., Dou, Q. P. and Liu, J. (2011b). Shikonin extracted from medicinal Chinese herbs exerts anti-inflammatory effect via proteasome inhibition. *European Journal of Pharmacology* **658**, 242–247.

Lu, Y., Jin, M., Park, S. J., Son, K. H., Son, J. K. and Chang, H. W. (2011c). Batatasin I, a naturally occurring phenanthrene derivative, isolated from tuberous roots of Dioscorea batatas suppresses eicosanoids generation and degranulation in bone marrow derived-mast cells. *Biological and Pharmaceutical Bulletin* **34**, 1021–1025.

Ma, H., He, X., Yang, Y., Li, M., Hao, D. and Jia, Z. (2011). The genus Epimedium: An ethnopharmacological and phytochemical review. *Journal of Ethnopharmacology* **134**, 519–541.

Majdalawieh, A. F., Hmaidan, R. and Carr, R. I. (2010). Nigella sativa modulates splenocyte proliferation, Th1/Th2 cytokine profile, macrophage function and NK anti-tumor activity. *Journal of Ethnopharmacology* **131**, 268–275.

Makela, P. H. (2000). Vaccines, coming of age after 200 years. *FEMS Microbiology Reviews* **24**, 9–20.

Mansky, P. J. and Wallerstedt, D. B. (2006). Complementary medicine in palliative care and cancer symptom management. *Cancer Journal* **12**, 425–431.

Mantovani, A., Allavena, P., Sica, A. and Balkwill, F. (2008). Cancer-related inflammation. *Nature* **454**, 436–444.

Mao, X., Yu, C. R., Li, W. H. and Li, W. X. (2008). Induction of apoptosis by shikonin through a ROS/JNK-mediated process in Bcr/Abl-positive chronic myelogenous leukemia (CML) cells. *Cell Research* **18**, 879–888.

Matta, R., Wang, X., Ge, H., Ray, W., Nelin, L. D. and Liu, Y. (2009). Triptolide induces anti-inflammatory cellular responses. *American Journal of Translational Research* **1**, 267–282.

Means, T. K., Mylonakis, E., Tampakakis, E., Colvin, R. A., Seung, E., Puckett, L., Tai, M. F., Stewart, C. R., Pukkila-Worley, R., Hickman, S. E., Moore, K. J., Calderwood, S. B. *et al.* (2009). Evolutionarily conserved recognition and innate immunity to fungal pathogens by the scavenger receptors SCARF1 and CD36. *The Journal of Experimental Medicine* **206**, 637–653.

Medzhitov, R. (2008). Origin and physiological roles of inflammation. *Nature* **454**, 428–435.

Medzhitov, R. (2010). Inflammation 2010: New adventures of an old flame. *Cell* **140**, 771–776.

Medzhitov, R. and Janeway, C. A., Jr. (1997). Innate immunity: The virtues of a nonclonal system of recognition. *Cell* **91**, 295–298.

Medzhitov, R., Shevach, E. M., Trinchieri, G., Mellor, A. L., Munn, D. H., Gordon, S., Libby, P., Hansson, G. K., Shortman, K., Dong, C., Gabrilovich, D., Gabrysova, L. *et al.* (2011). Highlights of 10 years of immunology in Nature Reviews Immunology. *Nature Reviews. Immunology* **11**, 693–702.

Melgarejo, E., Medina, M. A., Sanchez-Jimenez, F. and Urdiales, J. L. (2010). Targeting of histamine producing cells by EGCG: A green dart against inflammation? *Journal of Physiology and Biochemistry* **66**, 265–270.

Melief, C. J. (2008). Cancer immunotherapy by dendritic cells. *Immunity* **29**, 372–383.

Mellman, I. and Steinman, R. M. (2001). Dendritic cells: Specialized and regulated antigen processing machines. *Cell* **106**, 255–258.

Miller, L. H. and Su, X. (2011). Artemisinin: Discovery from the Chinese herbal garden. *Cell* **146**, 855–858.

Min, R., Tong, J., Wenjun, Y., Wenhu, D., Xiaojian, Z., Jiacai, H., Jian, Z., Wantao, C. and Chenping, Z. (2008). Growth inhibition and induction of apoptosis in human oral squamous cell carcinoma Tca-8113 cell lines by Shikonin was partly through the inactivation of NF-kappaB pathway. *Phytotherapy Research* **22**, 407–415.

Mishima, S., Saito, K., Maruyama, H., Inoue, M., Yamashita, T., Ishida, T. and Gu, Y. (2004). Antioxidant and immuno-enhancing effects of Echinacea purpurea. *Biological and Pharmaceutical Bulletin* **27**, 1004–1009.

Mizumoto, N., Gao, J., Matsushima, H., Ogawa, Y., Tanaka, H. and Takashima, A. (2005). Discovery of novel immunostimulants by dendritic-cell-based functional screening. *Blood* **106**, 3082–3089.

Moradali, M. F., Mostafavi, H., Ghods, S. and Hedjaroude, G. A. (2007). Immunomodulating and anticancer agents in the realm of macromycetes fungi (macrofungi). *International Immunopharmacology* **7**, 701–724.

Nalbantsoy, A., Nesil, T., Erden, S., Calis, I. and Bedir, E. (2011). Adjuvant effects of Astragalus saponins macrophyllosaponin B and astragaloside VII. *Journal of Ethnopharmacology* **134**, 897–903.

Nelson, R. P., Jr. and Ballow, M. (2003). 26. Immunomodulation and immunotherapy: Drugs, cytokines, cytokine receptors, and antibodies. *The Journal of Allergy and Clinical Immunology* **111**, S720–S743.

Nestle, F. O., Di Meglio, P., Qin, J. Z. and Nickoloff, B. J. (2009). Skin immune sentinels in health and disease. *Nature Reviews. Immunology* **9**, 679–691.

Nguyen, T. H. and Casale, T. B. (2011). Immune modulation for treatment of allergic disease. *Immunological Reviews* **242**, 258–271.

Nose, M., Terawaki, K., Oguri, K., Ogihara, Y., Yoshimatsu, K. and Shimomura, K. (1998). Activation of macrophages by crude polysaccharide fractions obtained from shoots of Glycyrrhiza glabra and hairy roots of Glycyrrhiza uralensis in vitro. *Biological and Pharmaceutical Bulletin* **21**, 1110–1112.

Novosel'tseva, N. P., Rabinovich, A. M. and Tareeva, N. V. (1979). Biological characteristics of the gromwell (Lithospermum erythrorhizon Sieb. et Zucc.) and its content of shikonin when grown in Moscow Province. *Farmatsiia* **28**, 28–31.

O'gara, E. A., Hill, D. J. and Maslin, D. J. (2000). Activities of garlic oil, garlic powder, and their diallyl constituents against Helicobacter pylori. *Applied and Environmental Microbiology* **66**, 2269–2273.

Oh, P. S. and Lim, K. T. (2009). Glycoprotein isolated from Dioscorea batatas Decne modulates expressions of IL-4 and IL-10 in primary-cultured mouse lymphocytes. *Cell Biochemistry and Function* **27**, 316–322.

Olliaro, P. L. and Taylor, W. R. (2004). Developing artemisinin based drug combinations for the treatment of drug resistant falciparum malaria: A review. *Journal of Postgraduate Medicine* **50**, 40–44.

Onaga, L. (2001). Cashing in on nature's pharmacy: Bioprospecting and protection of biodiversity could go hand in hand. *EMBO Reports* **2**, 263–265.

Ouchi, N., Parker, J. L., Lugus, J. J. and Walsh, K. (2011). Adipokines in inflammation and metabolic disease. *Nature Reviews. Immunology* **11**, 85–97.

Pan, M. H. and Ho, C. T. (2008). Chemopreventive effects of natural dietary compounds on cancer development. *Chemical Society Reviews* **37**, 2558–2574.

Pan, P. Y., Ozao, J., Zhou, Z. and Chen, S. H. (2008a). Advancements in immune tolerance. *Advanced Drug Delivery Reviews* **60**, 91–105.

Pan, P. Y., Wang, G. X., Yin, B., Ozao, J., Ku, T., Divino, C. M. and Chen, S. H. (2008b). Reversion of immune tolerance in advanced malignancy: Modulation of myeloid-derived suppressor cell development by blockade of stem-cell factor function. *Blood* **111**, 219–228.

Pan, M. H., Lai, C. S., Dushenkov, S. and Ho, C. T. (2009). Modulation of inflammatory genes by natural dietary bioactive compounds. *Journal of Agricultural and Food Chemistry* **57**, 4467–4477.

Park, M., Bae, J. and Lee, D. S. (2008). Antibacterial activity of [10]-gingerol and [12]-gingerol isolated from ginger rhizome against periodontal bacteria. *Phytotherapy Research* **22**, 1446–1449.

Patwardhan, B. and Mashelkar, R. A. (2009). Traditional medicine-inspired approaches to drug discovery: Can Ayurveda show the way forward? *Drug Discovery Today* **14**, 804–811.

Paulsen, B. S. (2002). Biologically active polysaccharides as possible lead compounds. *Phytochemistry Reviews* **1**, 379–387.

Paulsen, B. S. and Barsett, H. (2005). Bioactive pectic polysaccharides. *Advances in Polymer Science* **186**, 69–101.

Peairs, A., Dai, R., Gan, L., Shimp, S., Rylander, M. N., Li, L. and Reilly, C. M. (2010). Epigallocatechin-3-gallate (EGCG) attenuates inflammation in MRL/lpr mouse mesangial cells. *Cellular and Molecular Immunology* **7**, 123–132.

Perez, S., Rodriguez-Carvajal, M. A. and Doco, T. (2003). A complex plant cell wall polysaccharide: Rhamnogalacturonan II. A structure in quest of a function. *Biochimie* **85**, 109–121.

Perry, N. B., Burgess, E. J. and Glennie, V. L. (2001). Echinacea standardization: Analytical methods for phenolic compounds and typical levels in medicinal species. *Journal of Agricultural and Food Chemistry* **49**, 1702–1706.

Petersen, T. R., Dickgreber, N. and Hermans, I. F. (2010). Tumor antigen presentation by dendritic cells. *Critical Reviews in Immunology* **30**, 345–386.

Phillips, D. R., Rasbery, J. M., Bartel, B. and Matsuda, S. P. (2006). Biosynthetic diversity in plant triterpene cyclization. *Current Opinion in Plant Biology* **9**, 305–314.

Pietta, P., Mauri, P. and Bauer, R. (1998). MEKC analysis of different Echinacea species. *Planta Medica* **64**, 649–652.

Pinna, G. F., Fiorucci, M., Reimund, J. M., Taquet, N., Arondel, Y. and Muller, C. D. (2004). Celastrol inhibits pro-inflammatory cytokine secretion in Crohn's disease biopsies. *Biochemical and Biophysical Research Communications* **322**, 778–786.

Premkumar, V., Dey, M., Dorn, R. and Raskin, I. (2010). MyD88-dependent and independent pathways of Toll-Like Receptors are engaged in biological activity of Triptolide in ligand-stimulated macrophages. *BMC Chemical Biology* **10**, 3.

Qi, F., Li, A., Inagaki, Y., Gao, J., Li, J., Kokudo, N., Li, X. K. and Tang, W. (2010). Chinese herbal medicines as adjuvant treatment during chemo- or radiotherapy for cancer. *Bioscience Trends* **4**, 297–307.

Qian, B. Z. and Pollard, J. W. (2010). Macrophage diversity enhances tumor progression and metastasis. *Cell* **141**, 39–51.

Qiu, D., Zhao, G., Aoki, Y., Shi, L., Uyei, A., Nazarian, S., Ng, J. C. and Kao, P. N. (1999). Immunosuppressant PG490 (triptolide) inhibits T-cell interleukin-2 expression at the level of purine-box/nuclear factor of activated T-cells and NF-kappaB transcriptional activation. *The Journal of Biological Chemistry* **274**, 13443–13450.

Quan, F. S., Compans, R. W., Cho, Y. K. and Kang, S. M. (2007). Ginseng and Salviae herbs play a role as immune activators and modulate immune responses during influenza virus infection. *Vaccine* **25**, 272–282.

Rabinkov, A., Miron, T., Mirelman, D., Wilchek, M., Glozman, S., Yavin, E. and Weiner, L. (2000). S-Allylmercaptoglutathione: The reaction product of allicin with glutathione possesses SH-modifying and antioxidant properties. *Biochimica et Biophysica Acta* **1499**, 144–153.

Reis e Sousa, C. (2001). Dendritic cells as sensors of infection. *Immunity* **14**, 495–498.

Reyburn, H. (2010). New WHO guidelines for the treatment of malaria. *BMJ* **340**, c2637.

Richard, T., Pawlus, A. D., Iglesias, M. L., Pedrot, E., Waffo-Teguo, P., Merillon, J. M. and Monti, J. P. (2011). Neuroprotective properties of resveratrol and derivatives. *Annals of the New York Academy of Sciences* **1215**, 103–108.

Rininger, J. A., Kickner, S., Chigurupati, P., Mclean, A. and Franck, Z. (2000). Immunopharmacological activity of Echinacea preparations following simulated digestion on murine macrophages and human peripheral blood mononuclear cells. *Journal of Leukocyte Biology* **68**, 503–510.

Rittenhouse, J. R., Lui, P. D. and Lau, B. H. (1991). Chinese medicinal herbs reverse macrophage suppression induced by urological tumors. *The Journal of Urology* **146**, 486–490.

Robles, M., Aregullin, M., West, J. and Rodriguez, E. (1995). Recent studies on the zoopharmacognosy, pharmacology and neurotoxicology of sesquiterpene lactones. *Planta Medica* **61**, 199–203.

Robson, N. C., Hoves, S., Maraskovsky, E. and Schnurr, M. (2010). Presentation of tumour antigens by dendritic cells and challenges faced. *Current Opinion in Immunology* **22**, 137–144.

S, P., S, C., C, C., E, B., Ml, T. and G, B. (2011). Efficacy of Arnica Echinacea Powder in umbilical cord care in a large cohort study. *Journal of Maternal-Fetal and Neonatal Medicine* **1**, 1–3.

Saif, M. W., Lansigan, F., Ruta, S., Lamb, L., Mezes, M., Elligers, K., Grant, N., Jiang, Z. L., Liu, S. H. and Cheng, Y. C. (2010). Phase I study of the botanical formulation PHY906 with capecitabine in advanced pancreatic and other gastrointestinal malignancies. *Phytomedicine* **17**, 161–169.

Saiki, I. (2000). A Kampo medicine "Juzen-taiho-to"—Prevention of malignant progression and metastasis of tumor cells and the mechanism of action. *Biological and Pharmaceutical Bulletin* **23**, 677–688.

Sakaguchi, I., Tsujimura, M., Ikeda, N., Minamino, M., Kato, Y., Watabe, K., Yano, I. and Kaneda, K. (2001). Granulomatous tissue formation of shikon and shikonin by air pouch method. *Biological and Pharmaceutical Bulletin* **24**, 650–655.

Sakaguchi, S., Yamaguchi, T., Nomura, T. and Ono, M. (2008). Regulatory T cells and immune tolerance. *Cell* **133**, 775–787.

Sakaguchi, S., Miyara, M., Costantino, C. M. and Hafler, D. A. (2010). FOXP3+ regulatory T cells in the human immune system. *Nature Reviews. Immunology* **10**, 490–500.

Saleem, M. (2009). Lupeol, a novel anti-inflammatory and anti-cancer dietary triterpene. *Cancer Letters* **285**, 109–115.

Saleem, M., Afaq, F., Adhami, V. M. and Mukhtar, H. (2004). Lupeol modulates NF-kappaB and PI3K/Akt pathways and inhibits skin cancer in CD-1 mice. *Oncogene* **23**, 5203–5214.

Salminen, A., Lehtonen, M., Suuronen, T., Kaarniranta, K. and Huuskonen, J. (2008). Terpenoids: Natural inhibitors of NF-kappaB signaling with anti-inflammatory and anticancer potential. *Cellular and Molecular Life Sciences* **65**, 2979–2999.

Sasagawa, M., Cech, N. B., Gray, D. E., Elmer, G. W. and Wenner, C. A. (2006). Echinacea alkylamides inhibit interleukin-2 production by Jurkat T cells. *International Immunopharmacology* **6**, 1214–1221.

Sasaki, M., Ozawa, Y., Kurihara, T., Noda, K., Imamura, Y., Kobayashi, S., Ishida, S. and Tsubota, K. (2009). Neuroprotective effect of an antioxidant, lutein, during retinal inflammation. *Investigative Ophthalmology and Visual Science* **50**, 1433–1439.

Scandella, E., Men, Y., Legler, D. F., Gillessen, S., Prikler, L., Ludewig, B. and Groettrup, M. (2004). CCL19/CCL21-triggered signal transduction and migration of dendritic cells requires prostaglandin E2. *Blood* **103**, 1595–1601.

Schiller, M., Metze, D., Luger, T. A., Grabbe, S. and Gunzer, M. (2006). Immune response modifiers—Mode of action. *Experimental Dermatology* **15**, 331–341.

Schmid, E. F. and Smith, D. A. (2004). Is pharmaceutical R&D just a game of chance or can strategy make a difference? *Drug Discovery Today* **9**, 18–26.

Schmid-Schonbein, G. W. (2006). Analysis of inflammation. *Annual Review of Biomedical Engineering* **8**, 93–131.

See, D. M., Broumand, N., Sahl, L. and Tilles, J. G. (1997). In vitro effects of echinacea and ginseng on natural killer and antibody-dependent cell cytotoxicity in healthy subjects and chronic fatigue syndrome or acquired immunodeficiency syndrome patients. *Immunopharmacology* **35**, 229–235.

Shahidi, F., Ho, C.-T. and American Chemical Society. Division of Agricultural and Food Chemistry (2005). Phenolic Compounds in Foods and Natural Health Products. American Chemical Society: Distributed by Oxford University Press, Washington, DC.

Shan, B. E., Yoshita, Y., Sugiura, T. and Yamashita, U. (1999). Suppressive effect of Chinese medicinal herb, Acanthopanax gracilistylus, extract on human lymphocytes in vitro. *Clinical and Experimental Immunology* **118**, 41–48.

Shan, D., Fang, Y., Ye, Y. and Liu, J. (2008). EGCG reducing the susceptibility to cholesterol gallstone formation through the regulation of inflammation. *Biomedicine and Pharmacotherapy* **62**, 677–683.

Shao, X. T., Feng, L., Yao, H. P., Sun, W. J. and Zhang, L. H. (2004). Effect of Triptolide on TNFalpha-induced activation of NF-kappaB and expression of COX-2 and iNOS in human rheumatoid arthritis synovial fibroblasts. *Zhejiang Da Xue Xue Bao. Yi Xue Ban* **33**, 160–165.

Sharma, R. A., Euden, S. A., Platton, S. L., Cooke, D. N., Shafayat, A., Hewitt, H. R., Marczylo, T. H., Morgan, B., Hemingway, D., Plummer, S. M., Pirmohamed, M., Gescher, A. J. *et al.* (2004). Phase I clinical trial of oral curcumin: Biomarkers of systemic activity and compliance. *Clinical Cancer Research* **10**, 6847–6854.

Shen, J. J., Chiang, M. S., Kuo, M. L., Leu, Y. L., Hwang, T. L., Liou, C. J. and Huang, W. C. (2011). Partially purified extract and viscolin from Viscum coloratum attenuate airway inflammation and eosinophil infiltration in ovalbumin-sensitized mice. *Journal of Ethnopharmacology* **135**, 646–653.

Shin, J. Y., Lee, S. K., Kang, C. D., Chung, J. S., Lee, E. Y., Seo, S. Y., Lee, S. Y., Baek, S. Y., Kim, B. S., Kim, J. B. and Yoon, S. (2003). Antitumor effect of intratumoral administration of dendritic cell combination with vincristine chemotherapy in a murine fibrosarcoma model. *Histology and Histopathology* **18**, 435–447.

Shirakami, Y., Shimizu, M., Tsurumi, H., Hara, Y., Tanaka, T. and Moriwaki, H. (2008). EGCG and Polyphenon E attenuate inflammation-related mouse colon carcinogenesis induced by AOM plus DDS. *Molecular Medicine Reports* **1**, 355–361.

Shyur, L. F. and Yang, N. S. (2008). Metabolomics for phytomedicine research and drug development. *Current Opinion in Chemical Biology* **12**, 66–71.

Shyur, L. F., Huang, C. C., Hsu, Y. Y., Cheng, Y. W. and Yang, S. D. (2011). A sesquiterpenol extract potently suppresses inflammation in macrophages and mice skin and prevents chronic liver damage in mice through JNK-dependent HO-1 expression. *Phytochemistry* **72**, 391–399.

Siddique, H. R. and Saleem, M. (2011). Beneficial health effects of lupeol triterpene: A review of preclinical studies. *Life Sciences* **88**, 285–293.

Skowronski, D. M., Astell, C., Brunham, R. C., Low, D. E., Petric, M., Roper, R. L., Talbot, P. J., Tam, T. and Babiuk, L. (2005). Severe acute respiratory syndrome (SARS): A year in review. *Annual Review of Medicine* **56**, 357–381.

Smyth, M. J., Godfrey, D. I. and Trapani, J. A. (2001). A fresh look at tumor immunosurveillance and immunotherapy. *Nature Immunology* **2**, 293–299.

Song, X. and Hu, S. (2009). Adjuvant activities of saponins from traditional Chinese medicinal herbs. *Vaccine* **27**, 4883–4890.

Spelman, K., Burns, J., Nichols, D., Winters, N., Ottersberg, S. and Tenborg, M. (2006). Modulation of cytokine expression by traditional medicines: A review of herbal immunomodulators. *Alternative Medicine Review* **11**, 128–150.

Sporri, R. and Reis e Sousa, C. (2005). Inflammatory mediators are insufficient for full dendritic cell activation and promote expansion of CD4+ T cell populations lacking helper function. *Nature Immunology* **6**, 163–170.

Staniforth, V., Wang, S. Y., Shyur, L. F. and Yang, N. S. (2004). Shikonins, phytocompounds from Lithospermum erythrorhizon, inhibit the transcriptional activation of human tumor necrosis factor alpha promoter in vivo. *The Journal of Biological Chemistry* **279**, 5877–5885.

Su, P. F., Staniforth, V., Li, C. J., Wang, C. Y., Chiao, M. T., Wang, S. Y., Shyur, L. F. and Yang, N. S. (2008). Immunomodulatory effects of phytocompounds characterized by in vivo transgenic human GM-CSF promoter activity in skin tissues. *Journal of Biomedical Science* **15**, 813–822.

Su, M., Chung, H. Y. and Li, Y. (2011a). Deoxyelephantopin from Elephantopus scaber L. induces cell-cycle arrest and apoptosis in the human nasopharyngeal cancer CNE cells. *Biochemical and Biophysical Research Communications* **411**, 342–347.

Su, P. F., Li, C. J., Hsu, C. C., Benson, S., Wang, S. Y., Aravindaram, K., Chan, S. I., Wu, S. H., Yang, F. L., Huang, W. C., Shyur, L. F. and Yang, N. S. (2011b). Dioscorea phytocompounds enhance murine splenocyte proliferation ex vivo and improve regeneration of bone marrow cells in vivo. *Evidence-Based Complementary and Alternative Medicine: eCAM* **2011**, 731308.

Sullivan, A. M., Laba, J. G., Moore, J. A. and Lee, T. D. (2008). Echinacea-induced macrophage activation. *Immunopharmacology and Immunotoxicology* **30**, 553–574.

Sun, H. X., Wang, H., Xu, H. S. and Ni, Y. (2009a). Novel polysaccharide adjuvant from the roots of Actinidia eriantha with dual Th1 and Th2 potentiating activity. *Vaccine* **27**, 3984–3991.

Sun, J. C., Beilke, J. N. and Lanier, L. L. (2009b). Adaptive immune features of natural killer cells. *Nature* **457**, 557–561.

Sun, X., Zhang, Y., Jia, Q., Wang, Z. and Zhang, W. (2011). Effect of polypeptide extract from scorpion venom (PESV) with chemotherapy inhibited angiogenesis of Lewis lung carcinomas. *Zhongguo Zhong Yao Za Zhi* **36**, 1644–1649.

Surh, Y. J. (2003). Cancer chemoprevention with dietary phytochemicals. *Nature Reviews. Cancer* **3**, 768–780.

Suzuki, F., Schmitt, D. A., Utsunomiya, T. and Pollard, R. B. (1992). Stimulation of host resistance against tumors by glycyrrhizin, an active component of licorice roots. *In Vivo* **6**, 589–596.

Takada, Y. and Aggarwal, B. B. (2003). Betulinic acid suppresses carcinogen-induced NF-kappa B activation through inhibition of I kappa B alpha kinase and p65 phosphorylation: Abrogation of cyclooxygenase-2 and matrix metalloprotease-9. *Journal of Immunology* **171**, 3278–3286.

Takano-Ohmuro, H., Yoshida, L. S., Yuda, Y., Morioka, K. and Kitani, S. (2008). Shikonin inhibits IgE-mediated histamine release by human basophils and Syk kinase activity. *Inflammation Research* **57**, 484–488.

Tao, X., Fan, F., Hoffmann, V., Gao, C. Y., Longo, N. S., Zerfas, P. and Lipsky, P. E. (2008). Effective therapy for nephritis in (NZB x NZW)F1 mice with triptolide and tripdiolide, the principal active components of the Chinese herbal remedy Tripterygium wilfordii Hook F. *Arthritis and Rheumatism* **58**, 1774–1783.

Thakur, M., Connellan, P., Deseo, M. A., Morris, C. and Dixit, V. K. (2011). Immunomodulatory polysaccharide from Chlorophytum borivilianum roots. *Evidence-Based Complementary and Alternative Medicine: eCAM* **2011**, 598521.

Thoppil, R. J. and Bishayee, A. (2011). Terpenoids as potential chemopreventive and therapeutic agents in liver cancer. *World Journal of Hepatology* **3**, 228–249.

Tong, L. and Moudgil, K. D. (2007). Celastrus aculeatus Merr. suppresses the induction and progression of autoimmune arthritis by modulating immune response to heat-shock protein 65. *Arthritis Research and Therapy* **9**, 70.

Travis, J. (2011). Nobel Prize in physiology or medicine. Immunology prize overshadowed by untimely death of awardee. *Science* **334**, 31.

Tsuda, N., Chang, D. Z., Mine, T., Efferson, C., Garcia-Sastre, A., Wang, X., Ferrone, S. and Ioannides, C. G. (2007). Taxol increases the amount and T cell activating ability of self-immune stimulatory multimolecular complexes found in ovarian cancer cells. *Cancer Research* **67**, 8378–8387.

Tu, Y. (2011). The discovery of artemisinin (qinghaosu) and gifts from Chinese medicine. *Nature Medicine* **17**, 1217–1220.

Tu, S. P., Jin, H., Shi, J. D., Zhu, L. M., Suo, Y., Lu, G., Liu, B., Wang, T. and Yang, C. S. (2011). Curcumin induces the differentiation of myeloid-derived suppressor cells and inhibits their interaction with cancer cells and related tumor growth. *Cancer Prevention Research (Philadelphia, PA)* **5**, 205–215.

Tzianabos, A. O. (2000). Polysaccharide immunomodulators as therapeutic agents: Structural aspects and biologic function. *Clinical Microbiology Reviews* **13**, 523–533.

Ullrich, S. E. (2010). Two-way traffic on the bridge from innate to adaptive immunity. *The Journal of Investigative Dermatology* **130**, 1773–1775.

Ullrich, E., Bonmort, M., Mignot, G., Kroemer, G. and Zitvogel, L. (2008). Tumor stress, cell death and the ensuing immune response. *Cell Death and Differentiation* **15**, 21–28.

Van Agtmael, M. A., Eggelte, T. A. and Van Boxtel, C. J. (1999). Artemisinin drugs in the treatment of malaria: From medicinal herb to registered medication. *Trends in Pharmacological Sciences* **20**, 199–205.

Vesely, M. D., Kershaw, M. H., Schreiber, R. D. and Smyth, M. J. (2011). Natural innate and adaptive immunity to cancer. *Annual Review of Immunology* **29**, 235–271.

Vivier, E., Tomasello, E., Baratin, M., Walzer, T. and Ugolini, S. (2008). Functions of natural killer cells. *Nature Immunology* **9**, 503–510.

Vivier, E., Raulet, D. H., Moretta, A., Caligiuri, M. A., Zitvogel, L., Lanier, L. L., Yokoyama, W. M. and Ugolini, S. (2011). Innate or adaptive immunity? The example of natural killer cells. *Science* **331**, 44–49.

Wang, R. and Chandawarkar, R. Y. (2010). Phagocytosis of fungal agents and yeast via macrophage cell surface scavenger receptors. *The Journal of Surgical Research* **164**, e273–e279.

Wang, J. Z., Tsumura, H., Shimura, K. and Ito, H. (1992). Antitumor activity of polysaccharide from a Chinese medicinal herb, Acanthopanax giraldii Harms. *Cancer Letters* **65**, 79–84.

Wang, C. Y., Chiao, M. T., Yen, P. J., Huang, W. C., Hou, C. C., Chien, S. C., Yeh, K. C., Yang, W. C., Shyur, L. F. and Yang, N. S. (2006). Modulatory effects of Echinacea purpurea extracts on human dendritic cells: A cell- and gene-based study. *Genomics* **88**, 801–808.

Wang, C. Y., Staniforth, V., Chiao, M. T., Hou, C. C., Wu, H. M., Yeh, K. C., Chen, C. H., Hwang, P. I., Wen, T. N., Shyur, L. F. and Yang, N. S. (2008a). Genomics and proteomics of immune modulatory effects of a butanol fraction of echinacea purpurea in human dendritic cells. *BMC Genomics* **9**, 479.

Wang, Y., Jia, L. and Wu, C. Y. (2008b). Triptolide inhibits the differentiation of Th17 cells and suppresses collagen-induced arthritis. *Scandinavian Journal of Immunology* **68**, 383–390.

Wang, Z., Liu, T., Gan, L., Wang, T., Yuan, X., Zhang, B., Chen, H. and Zheng, Q. (2010). Shikonin protects mouse brain against cerebral ischemia/reperfusion injury through its antioxidant activity. *European Journal of Pharmacology* **643**, 211–217.

Wang, E., Bussom, S., Chen, J., Quinn, C., Bedognetti, D., Lam, W., Guan, F., Jiang, Z., Mark, Y., Zhao, Y., Stroncek, D. F., White, J. *et al.* (2011a). Interaction of a traditional Chinese Medicine (PHY906) and CPT-11 on the inflammatory process in the tumor microenvironment. *BMC Medical Genomics* **4**, 38.

Wang, Y., Huang, Z., Wang, L., Meng, S., Fan, Y., Chen, T., Cao, J., Jiang, R. and Wang, C. (2011b). The anti-malarial artemisinin inhibits pro-inflammatory cytokines via the NF-kappaB canonical signaling pathway in PMA-induced THP-1 monocytes. *International Journal of Molecular Medicine* **27**, 233–241.

Wei, X., Gong, J., Zhu, J., Wang, P., Li, N., Zhu, W. and Li, J. (2008). The suppressive effect of triptolide on chronic colitis and TNF-alpha/TNFR2 signal pathway in interleukin-10 deficient mice. *Clinical Immunology* **129**, 211–218.

Wei, W. C., Su, Y. H., Chen, S. S., Sheu, J. H. and Yang, N. S. (2011). GM-CSF plays a key role in zymosan-stimulated human dendritic cells for activation of Th1 and Th17 cells. *Cytokine* **55**, 79–89.

Wen, C. C., Kuo, Y. H., Jan, J. T., Liang, P. H., Wang, S. Y., Liu, H. G., Lee, C. K., Chang, S. T., Kuo, C. J., Lee, S. S., Hou, C. C., Hsiao, P. W. *et al.* (2007). Specific plant terpenoids and lignoids possess potent antiviral activities against severe acute respiratory syndrome coronavirus. *Journal of Medicinal Chemistry* **50**, 4087–4095.

Wen, C. C., Chen, H. M., Chen, S. S., Huang, L. T., Chang, W. T., Wei, W. C., Chou, L. C., Arulselvan, P., Wu, J. B., Kuo, S. C. and Yang, N. S. (2011). Specific microtubule-depolymerizing agents augment efficacy of dendritic cell-based cancer vaccines. *Journal of Biomedical Science* **18**, 44.

West, M. A., Wallin, R. P., Matthews, S. P., Svensson, H. G., Zaru, R., Ljunggren, H. G., Prescott, A. R. and Watts, C. (2004). Enhanced dendritic cell antigen capture via toll-like receptor-induced actin remodeling. *Science* **305**, 1153–1157.

Whelan, M., Whelan, J., Russell, N. and Dalgleish, A. (2003). Cancer immunotherapy: An embarrassment of riches? *Drug Discovery Today* **8**, 253–258.

Willment, J. A., Marshall, A. S., Reid, D. M., Williams, D. L., Wong, S. Y., Gordon, S. and Brown, G. D. (2005). The human beta-glucan receptor is widely expressed and functionally equivalent to murine Dectin-1 on primary cells. *European Journal of Immunology* **35**, 1539–1547.

Wood, L. G., Wark, P. A. and Garg, M. L. (2010). Antioxidant and anti-inflammatory effects of resveratrol in airway disease. *Antioxidants and Redox Signaling* **13**, 1535–1548.

Wu, X., Zhang, W., Shi, X., An, P., Sun, W. and Wang, Z. (2010). Therapeutic effect of artemisinin on lupus nephritis mice and its mechanisms. *Acta Biochimica et Biophysica Sinica* **42**, 916–923.

Xiong, Z., Sun, G., Zhu, C., Cheng, B., Zhang, C., Ma, Y. and Dong, Y. (2010). Artemisinin, an anti-malarial agent, inhibits rat cardiac hypertrophy via inhibition of NF-kappaB signaling. *European Journal of Pharmacology* **649**, 277–284.

Xu, W., Towers, A. D., Li, P. and Collet, J. P. (2006). Traditional Chinese medicine in cancer care: Perspectives and experiences of patients and professionals in China. *European Journal of Cancer Care* **15**, 397–403.

Xu, X., Pan, C., Zhang, L. and Ashida, H. (2011). Immunomodulatory beta-glucan from Lentinus edodes activates mitogen-activated protein kinases and nuclear factor-kappaB in murine RAW 264.7 macrophages. *The Journal of Biological Chemistry* **286**, 31194–31198.

Xue, M., Jiang, Z. Z., Liu, J. P., Zhang, L. Y., Wang, T., Wang, H., Liu, L. and Zhou, Z. X. (2010). Comparative study on the anti-inflammatory and immune suppressive effect of Wilforlide A. *Fitoterapia* **81**, 1109–1112.

Yadav, V. R. and Aggarwal, B. B. (2011). Curcumin: A component of the golden spice, targets multiple angiogenic pathways. *Cancer Biology and Therapy* **11**, 236–241.

Yamada, H. (1989). Chemical characterization and biological activity of the immunologically active substances in Juzen-taiho-to (Japanese kampo prescription). *Gan to Kagaku Ryoho* **16**, 1500–1505.

Yang, S. X., Xie, S. S., Gao, H. L. and Long, Z. Z. (1993). Artemisinin and its derivatives enhance T lymphocyte-mediated immune responses in normal mice and accelerate immunoreconstitution of mice with syngeneic bone marrow transplantation. *Clinical Immunology and Immunopathology* **69**, 143–148.

Yang, X. B., Mei, Q. B., Zhou, S. Y., Teng, Z. H. and Wang, H. F. (2004). The role of Angelica polysaccharides in inducing effector molecule release by peritoneal macrophages. *Xi Bao Yu Fen Zi Mian Yi Xue Za Zhi* **20**, 747–749.

Yang, H., Chen, D., Cui, Q. C., Yuan, X. and Dou, Q. P. (2006). Celastrol, a triterpene extracted from the Chinese "Thunder of God Vine," is a potent proteasome inhibitor and suppresses human prostate cancer growth in nude mice. *Cancer Research* **66**, 4758–4765.

Yang, H., Zhou, P., Huang, H., Chen, D., Ma, N., Cui, Q. C., Shen, S., Dong, W., Zhang, X., Lian, W., Wang, X., Dou, Q. P. *et al.* (2009). Shikonin exerts antitumor activity via proteasome inhibition and cell death induction in vitro and in vivo. *International Journal of Cancer* **124**, 2450–2459.

Yang, A. K., He, S. M., Liu, L., Liu, J. P., Wei, M. Q. and Zhou, S. F. (2010). Herbal interactions with anticancer drugs: Mechanistic and clinical considerations. *Current Medicinal Chemistry* **17**, 1635–1678.

Yang, N. S., Wang, J. H. and Wei, W. C. (2011). Dioscorea extracts for enhancing immune system United States patent application 12/202,528.

Ye, M., Liu, S. H., Jiang, Z., Lee, Y., Tilton, R. and Cheng, Y. C. (2007). Liquid chromatography/mass spectrometry analysis of PHY906, a Chinese medicine formulation for cancer therapy. *Rapid Communications in Mass Spectrometry* **21**, 3593–3607.

Yen, Y., So, S., Rose, M., Saif, M. W., Chu, E., Liu, S. H., Foo, A., Jiang, Z., Su, T. and Cheng, Y. C. (2009). Phase I/II study of PHY906/capecitabine in advanced hepatocellular carcinoma. *Anticancer Research* **29**, 4083–4092.

Yin, S. Y., Wang, W. H., Wang, B. X., Aravindaram, K., Hwang, P. I., Wu, H. M. and Yang, N. S. (2010). Stimulatory effect of Echinacea purpurea extract on the trafficking activity of mouse dendritic cells: Revealed by genomic and proteomic analyses. *BMC Genomics* **11**, 612.

Yoshida, L. S., Kawada, T., Irie, K., Yuda, Y., Himi, T., Ikemoto, F. and Takano-Ohmuro, H. (2010). Shikonin directly inhibits nitric oxide synthases: Possible targets that affect thoracic aorta relaxation response and nitric oxide release from RAW 264.7 macrophages. *Journal of Pharmaceutical Sciences* **112**, 343–351.

You, H. J., Choi, C. Y., Kim, J. Y., Park, S. J., Hahm, K. S. and Jeong, H. G. (2001). Ursolic acid enhances nitric oxide and tumor necrosis factor-alpha production via nuclear factor-kappaB activation in the resting macrophages. *FEBS Letters* **509**, 156–160.

Youn, H. S., Lim, H. J., Lee, H. J., Hwang, D., Yang, M., Jeon, R. and Ryu, J. H. (2008). Garlic (Allium sativum) extract inhibits lipopolysaccharide-induced Toll-like receptor 4 dimerization. *Bioscience, Biotechnology, and Biochemistry* **72**, 368–375.

Yuan, S. L., Wang, X. J. and Wei, Y. Q. (2003). Anticancer effect of tanshinone and its mechanisms. *Chinese Journal of Cancer* **22**, 1363–1366.

Zhang, Y. and Ma, X. (2010). Triptolide inhibits IL-12/IL-23 expression in APCs via CCAAT/enhancer-binding protein alpha. *Journal of Immunology* **184**, 3866–3877.

Zhang, C., Cui, G. H., Liu, F., Wu, Q. L. and Chen, Y. (2006). Inhibitory effect of triptolide on lymph node metastasis in patients with non-Hodgkin lymphoma by regulating SDF-1/CXCR4 axis in vitro. *Acta Pharmacologica Sinica* **27**, 1438–1446.

Zhang, L. J., Liu, H. K., Hsiao, P. C., Kuo, L. M., Lee, I. J., Wu, T. S., Chiou, W. F. and Kuo, Y. H. (2011a). New isoflavonoid glycosides and related constituents from astragali radix (Astragalus membranaceus) and their inhibitory activity on nitric oxide production. *Journal of Agricultural and Food Chemistry* **59**, 1131–1137.

Zhang, X. R., Zhou, W. X., Zhang, Y. X., Qi, C. H., Yan, H., Wang, Z. F. and Wang, B. (2011b). Macrophages, rather than T and B cells are principal immunostimulatory target cells of Lycium barbarum L. polysaccharide LBPF4-OL. *Journal of Ethnopharmacology* **136**, 465–472.

Zheng, R., Jie, S., Hanchuan, D. and Moucheng, W. (2005). Characterization and immunomodulating activities of polysaccharide from Lentinus edodes. *International Immunopharmacology* **5**, 811–820.

Zhou, J., Xiao, C., Zhao, L., Jia, H., Zhao, N., Lu, C., Yang, D., Tang, J. C., Chan, A. S. and Lu, A. P. (2006). The effect of triptolide on CD4+ and CD8+ cells in Peyer's patch of SD rats with collagen induced arthritis. *International Immunopharmacology* **6**, 198–203.

Zhou, L., Chong, M. M. and Littman, D. R. (2009). Plasticity of CD4+ T cell lineage differentiation. *Immunity* **30**, 646–655.

Zhou, Y. B., Yu, J. E., Wu, J., Bai, L., Huo, L. L., Zhang, X. G. and Li, L. Q. (2010). Effects of Chinese herbal medicine Bushen Gubiao Recipe on toll-like receptor 4 and CD4(+)CD25(+)foxp3(+) regulatory T cells in mice with recurrent respiratory tract infections. *Zhong Xi Yi Jie He Xue Bao* **8**, 1053–1059.

Zhou, J., Wu, J., Chen, X., Fortenbery, N., Eksioglu, E., Kodumudi, K. N., Pk, E. B., Dong, J., Djeu, J. Y. and Wei, S. (2011). Icariin and its derivative, ICT, exert anti-inflammatory, anti-tumor effects, and modulate myeloid derived suppressive cells (MDSCs) functions. *International Immunopharmacology* **11**, 890–898.

Zhu, B., Wang, Y. J., Zhu, C. F., Lin, Y., Zhu, X. L., Wei, S., Lu, Y. and Cheng, X. X. (2010). Triptolide inhibits extracellular matrix protein synthesis by suppressing the Smad2 but not the MAPK pathway in TGF-beta1-stimulated NRK-49F cells. *Nephrology, Dialysis, Transplantation* **25**, 3180–3191.

# Chinese Herbal Medicines for Rheumatoid Arthritis

CHI ZHANG,[*] MIAO JIANG,[*] HONGTAO GUO,[*,†]
GUANG ZHENG,[*,‡] JING YANG[*] AND AIPING LU[*,1]

**Institute of Basic Research in Clinical Medicine, China Academy of Chinese Medical Sciences, Beijing, China*
*†Shanghai University of Traditional Chinese Medicine, Shanghai, China*
*‡Information Science and Engineering School, Lanzhou University, Lanzhou, China*



[1]Corresponding author: E-mail: lap64067611@126.com

Advances in Botanical Research, Vol. 62 0065-2296/12 $35.00

DOI: 10.1016/B978-0-12-394591-4.00005-2

# ABSTRACT

Herbal medicines have become a major source and main stream for future drug development and for human health care. This chapter provides a brief overview and perspective of the research topic related to Chinese herbal medicines (CHMs) used for the management of rheumatoid arthritis (RA). Herbal medicine-related research is important to integrate the medical and pharmacological sciences with the plant sciences including agriculture.

RA is a chronic, inflammatory, systemic autoimmune disease characterized by pain, swelling and stiffness. Increased attention is being given to the use of CHMs for the treatment of RA due to the limitations of existing drugs. In this review, an attempt has been made to highlight research on herbs with anti-arthritic potential. This chapter gathers valuable information from a wide variety of sources and supplies it to readers in figures and convenient tables. This chapter also considers various single herbs and herbal mixtures (products) in the context of clinical studies. In addition, experimental investigations that focus on the mechanisms of anti-arthritic activity are discussed.

Our objective is to stimulate the researchers to perform further research on the potential use of medicinal plants with anti-arthritic properties.


# I. INTRODUCTION

Rheumatoid arthritis (RA) is a chronic, inflammatory, systemic autoimmune disease that affects about 1% of the general population in Western countries (Hegen *et al.*, 2008). RA has its roots in the nineteenth century and a pedigree that goes back to the twentieth century. Its name was introduced in the 1850s (Arya *et al.*, 2011). RA is characterized by persistent synovitis, systemic inflammation and autoantibodies (Scott *et al.*, 2008). Conventional treatments for RA, including non-steroidal anti-inflammatory drugs (NSAIDs), disease-modifying anti-rheumatoid drugs (DMARDs) and corticosteroids, aim to reduce the patient's pain and joint inflammation, minimize loss of function and decrease the progression of joint damage. However, such treatments are rarely totally effective and some pharmacological therapies have the potential to cause side effects (Agarwal, 2010).

Herbal medicines are popular, self-prescribed treatments for rheumatic conditions. A recent U.S. survey suggested that approximately 90% of arthritic patients use alternative therapies such as herbal medicines (Herman *et al.*, 2004). Many patients try multiple alternative treatments (Efthimiou *et al.*, 2010; Ernst *et al.*, 2007). As more patients start using alternative medicine, conventional health care practitioners increasingly feel the need to acquaint themselves with herbal medicines (Manek *et al.*, 2010; Osborn *et al.*, 2004).

Chinese herbal medicines (CHMs) originated over thousands of years ago and were developed by trial and error. In Chinese mythology, the emperors

Huangdi and Shennong wrote medical manuals (Huang Di Nei Jing, The Yellow Emperor's Manual on Corporeal Medicine and Shen Nong Ben Cao Jing, Classical Pharmacopoeia of the Heavenly Husbandman) about 5000 years ago (Wiseman and Ye, 1998). Traditional Chinese medicine (TCM) is largely based on experience and is guided by a holistic concept.

RA is mainly categorized as Bi Zheng in TCM, which means 'impediment condition'. Bodily vacuity with external invasion of pathogenic evils; long living in a damp environment and contraction of wind; cold, damp, and/or heat evils are many recognized causes of RA in TCM. For the treatment of RA, a prescription of herbs based on these theories may comprise a single herb or more commonly a mixture of herbs in differing amounts. The prescribed herbs are classified according to four properties: Jun, Chen, Zuo and Shi. Jun is the major or chief herb. Chen is an adjuvant herb. Zhou is the assistant or helper-like herb and Shi are guider herbs (Cheng, 2000). There are different systems of classification that are based on the physical properties of the herbs. Another traditional classification involves four properties of herbs (cold, warm, heat and cool) so that hot symptoms are treated with cold herbs (Cheng, 2000).

Classification of the Chinese herbs according to their chemical composition can be a complicated process because of the large and diverse groups of compounds present in each herb and is further complicated by the mixtures of herbs used.

There are more and more studies providing evidence of the efficacy of TCM prescriptions in RA (Cameron *et al.*, 2009; Little *et al.*, 2000). The number of randomized controlled trials (RCTs) is increasing both in the Chinese and the English literature (Liu *et al.*, 2006; Shi *et al.*, 1999). Many studies of the efficacy of TCM include comparative studies of CHM with conventional Western drugs (Zhang *et al.*, 2011a). However, a review of some of these studies indicated that the quality of Chinese trials was poor and that many had serious methodological deficiencies (Zhang *et al.*, 2011b). In any case, CHM is a valuable foundation for modern drug development. More recently, there has been research and development in the areas of drugs for the treatment of arthritis.

## II. THE HISTORY OF CHMS FOR RA

The earliest evidence of herbalism used for RA in China is from two graves from the Han Era (202 B.C.E. to 220 C.E.) (Lesley, 2011). Ancient physicians identified the connection between a specific pattern and a corresponding formula and recorded the 'trial', which was then tested by the successors repeatedly for hundreds or even thousands of years. Thus, this is a unique clinical trial.

Now in China, many TCM practitioners are still using various plants for the treatment of different types of arthritic conditions (Izzo and Ernst, 2009; Zhang *et al.*, 2010). Although the application of these medicaments has a sound tradition and a rational background according to the TCM system of medicine, to be accepted for medical use worldwide, it is essential to investigate their use in modern scientific terms. The scientific studies to work out the actual efficacy and other limitations, including safety of these TCM drugs, would definitely widen their scope for future use if they come out to be really effective.

Over the past 30 years, an increasing number of TCM and biomedicine researchers are focusing their attention on identifying the efficacy and mechanisms of CHMs for RA treatment. Many interventions have been tested in numerous experimental and clinical studies. At the same time, more papers were published in English and indexed in PubMed. The data presented in Fig. 1 are from the Chinese BioMedical Literature Database (CBM) and PubMed. CBM is produced by the Chinese Academy of Medical Sciences, Institute of Medical Information. This database indexes articles published in biomedical journals, reviews and conference proceedings in China from 1978 to the present (accessed 15th Aug 2011; http://sinomed.imicams.ac.cn/). The number of papers in these two references increased rapidly in nearly every 11-year period shown in Fig. 1. Although the numbers

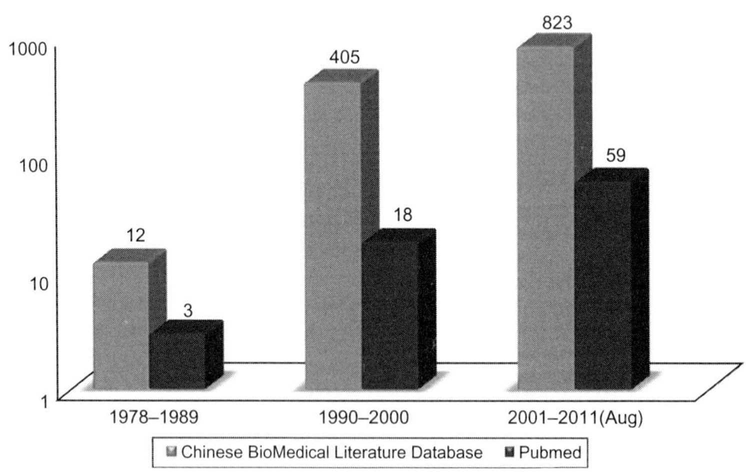


Fig. 1. Comparison by a number of studies on CHMs for RA published from 1978 to 2011, indexed in CBM and PubMed. (For colour version of this figure, the reader is referred to the online version of this chapter.)

of papers are not the most valuable indication that TCM is effective, it does show that Chinese and Western researchers are looking for CHMs as a therapeutic option for RA management.

Amongst the large number of published papers, non-randomized controlled clinical trials, observational studies and case reports have indicated an efficacy of TCM for treatment of RA, whereas there is still a deficiency of RCTs in this area. With the introduction and practice of evidence-based medicine (EBM), numerous RCTs have been conducted in China to evaluate the effectiveness of CHMs for RA. As evidence-gathering tools, systematic reviews (SRs) and meta-analysis provide synthesis of available research. The number of SRs and meta-analysis on CHM has increased rapidly since EBM was introduced into China in the 1990s. Figure 2 is an evidence pyramid figure which illustrates the large number of studies in the field of CHMs for treatment of RA in China, with the corresponding levels of evidence.

For CHM treatment of RA, individually prescribed decoctions have their own particular indications and uses in a large outpatient population. In recent decades, proprietary Chinese medicines (PCMs) for the treatment of RA generated from well-known Chinese medicine formulae or from currently effective practice formulas are easy to adapt to modern life. These interventions are tested in some part by RCTs. Besides formulas, new forms of the effective ingredients such as tablets and capsules have become available in the market and are now the mainstream for RA treatment in China (Fig. 3).

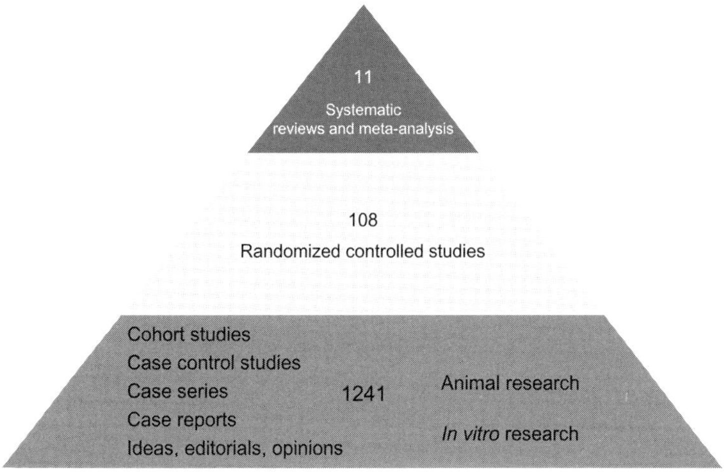


Fig. 2. A diagrammatic representation of the levels of studies on CHMs for RA published from 1978 to 2011, indexed in CBM. (For colour version of this figure, the reader is referred to the online version of this chapter.)

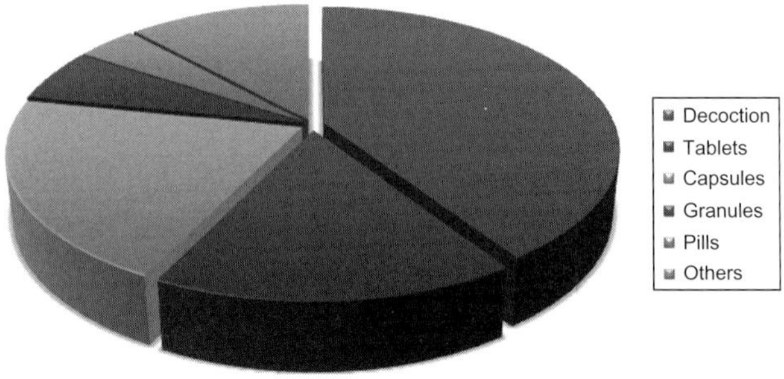


Fig. 3. Dosage forms of interventions in 108 RCTs on CHMs for RA published from 1978 to 2011, indexed in CBM and PubMed. (For colour version of this figure, the reader is referred to the online version of this chapter.)

There are well over 300 herbs and formulas for RA treatment in China. Can we, then, list the most important? There exists a need to search for these interlinking data from reliable databases. Data mining can help us. Presently in our leading research database, the amount of biomedical data is growing rapidly, so it is possible to get relevant and meaningful information through the techniques developed in the field of data mining. Our previous study proposed a hierarchical analysis algorithm called a discrete derivative, which is based on the frequencies of MeSH (The Medical Subject Headings) terms (Zheng *et al.*, 2011). Relevant literature was identified through the databases of both PubMed and SinoMed with the same method. These data sets were downloaded, structured and then stored in Microsoft® SQL® for further treatment. As to the visualization of the treating results, they were done in Cytoscape so as to depict the semantic relationship among objects (TCM herbs, herbal formulas or PCMs). After data mining, we got some well-structured networks (Figs. 4–6). Figure 4 illustrates herb–herb association network. There are a total of 29 herbs used for RA treatment. These lines actually represent connections among herbs in clinical practice. Figure 5 shows commonly used herbal formulas, and Fig. 6 shows commonly used PCMs. The related herbal compositions of the formulas listed in Fig. 5 and PCMs listed in Fig. 6 were shown in Tables I and II, respectively. In both Tables I and II, the pinyin name for the CHMs is given in the first column, followed by the pinyin name and the botanical name (in parentheses) in the second column.

In China, integrative medicine blends the best of Western conventional approaches and CHMs to develop a diagnostic and therapeutic program. DMARDs, NSAIDs and corticosteroids are commonly used with CHMs for RA in practice (Fig. 7), and conventional Western monotherapy medicines are added to Chinese herbal products for RA in drug design for better efficacy (Table III).

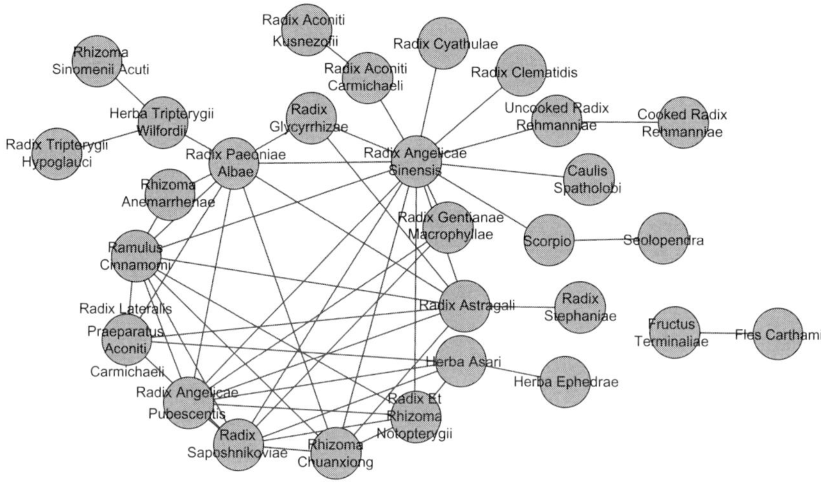


Fig. 4. The MeSH terms were obtained from Chinese BioMedical Literature Database (CBM) (till 1st Jul 2011). The results indicate 29 commonly used herbs and their networks that together use for rheumatoid arthritis (RA) in practice. (For colour version of this figure, the reader is referred to the online version of this chapter.)

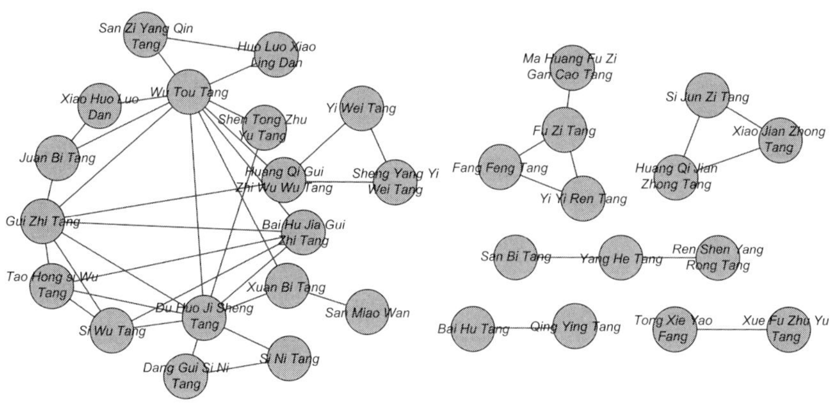


Fig. 5. The MeSH terms were obtained from Chinese BioMedical Literature Database (CBM) (till 1st Jul 2011). The results indicate 32 commonly used Chinese herbal decoctions and their networks that together use for rheumatoid arthritis (RA) in practice. (For colour version of this figure, the reader is referred to the online version of this chapter.)

In summary, the challenge for CHMs is to develop repeatable and provable outcomes, standardization of ingredients, and safety and quality assurance. The scientific bases of herbs in therapy are rapidly being established.

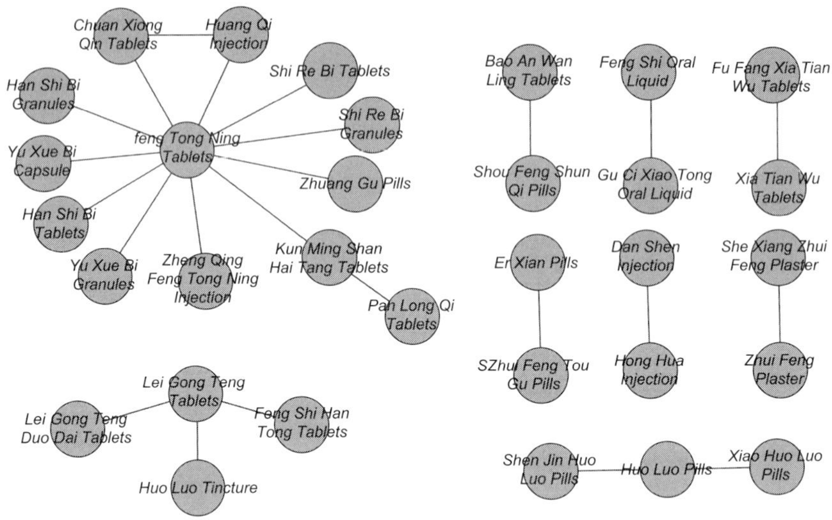


Fig. 6. The MeSH terms were obtained from Chinese BioMedical Literature Database (CBM) (till 1st July 2011). The results indicate 32 commonly used proprietary Chinese medicines (PCMs) and their networks that together use for rheumatoid arthritis (RA) in practice. (For colour version of this figure, the reader is referred to the online version of this chapter.)

# III. EXPERIMENTAL INVESTIGATIONS ON CHMS FOR RA

At present, many CHMs shown to be effective in the therapy of RA have been analyzed and the related mechanisms have been further evaluated. In experimental studies on CHMs, a variety of approaches have been used. Some studies have focused on chemical analysis, others on biological activities, safety, bioavailability and mode of action using either cell or animal models. In the following section, we will describe some anti-rheumatic herbs.

## A. EXPERIMENTAL STUDIES ON BIOACTIVE EXTRACTS OR SINGLE HERBS

Within the Chinese traditional medicine system, Thunder God Vine (Tripterygium wilfordii Hook F, TwHF) occupies the foremost place for the treatment of autoimmune and inflammatory diseases such as RA (Lipsky and Tao, 1997; Ma *et al.*, 2007; Tao and Lipsky, 2000; Tao *et al.*, 2001). It is a perennial vine-like plant growing in southern China. Triptolide is the main active component which imparts the anti-inflammatory and free radical scavenging potential. Dose-dependent studies of an ethyl acetate extract of TwHF roots have resulted in symptomatic improvement in 50% of RA cases

TABLE I

*List of Chinese Herbal Formulas in Fig. 5*

| Chinese herbal formulas | Ingredients | CM treatment principles |
|---|---|---|
| *Bai Hu Jia Gui Zhi Tang* | *Shi Gao* (Gypsum Fibrosum), *Geng Mi* (Fructus Oryzae), *Zhi Mu* (Rhizoma Anemarrhenae), *Gui Zhi* (Ramulus Cinnamomi), *Huang Bai* (Cortex Phellodendri), *Cang Zhu* (Rhizoma Atractylodis), *Nan Xing* (Rhizoma Arisaematis), mix-fried *Gan Cao* (Radix Glycyrrhizae) | Clear heat and disinherit dampness, dispel wind and free the flow of the network vessels |
| *Bai Hu Tang* | Uncooked *Shi Gao* (Gypsum Fibrosum), *Bai Mao Gen* (Rhizoma Imperatae), *Ban Lan Gen* (Radix Isatidis/ Baphicacanthi), *Sheng Di* (uncooked Radix Rehmanniae), *Da Qing Ye* (Folium Isatidis), *Zhi Mu* (Rhizoma Anemarrhenae), *Dan Zhu Ye* (Herba Lopthatheri), *Gan Cao* (Radix Glycyrrhizae) | Greatly clear heat from the qi aspect or division |
| *Dang Gui Si Ni Tang* | *Dang Gui* (Radix Angelicae Sinensis), *Bai Shao* (Radix Paeoniae Albae), *Gui Zhi* (Ramulus Cinnamomi), *Xi Xin* (Herba Asari), *Mu Tong* (Caulis Akebiae), *Gan Cao* (Radix Glycyrrhizae), *Da Zao* (Fructus Jujubae) | Nourish the blood, warm the channels and free the flow of impediment |
| *Du Huo Ji Sheng Tang* | *Sang Ji Sheng* (Herba Taxilli), *Shu Di* (cooked Radix Rehmanniae), *Dang Gui* (Radix Angelicae Sinensis), *Dang Shen* (Radix Codonopsitis), *Fu Ling* (Poria), *Qin Jiao* (Gentianae Macrophyllae), *Du Zhong* (Cortex Eucommiae), *Du Huo* (Radix Angelicae Pubescentis), *Fang Feng* (Radix Saposhnikoviae), *Chuan Xiong* (Rhizoma Chuanxiong), *Bai Shao* (Radix Paeoniae Albae), *Niu Xi* (Radix Achyranthis Bidentatae), *Xi Xin* (Herba Asari), *Rou Gui* (Cortex Cinnamomi), *Gan Cao* (Radix Glycyrrhizae) | Dispel wind and eliminate dampness, diffuse impediment and free the flow of the network vessels |

(*continues*)

TABLE I (*continued*)

| Chinese herbal formulas | Ingredients | CM treatment principles |
|---|---|---|
| *Fang Feng Tang* | *Fang Feng* (Radix Saposhnikoviae), *Fu Ling* (Poria), *Si Gua Luo* (Fasciculus Luffae), *Qin Jiao* (Radix Gentianae Macrophyllae), *Ge Gen* (Radix Puerariae), *Sang Zhi* (Ramulus Mori), *Qiang Huo* (Radix Et Rhizoma Notopterygii), *Gui Zhi* (Ramulus Cinnamomi), *Di Long* (Pheretima) | Dispel wind and eliminate dampness, supplement the qi and nourish the blood |
| *Fu Zi Tang* | *Fu Zi* (Radix Lateralis Praeparatus Aconiti Carmichaeli), *Gan Cao* (Radix Glycyrrhizae), *Sheng Jiang* (uncooked Rhizoma Zingiberis), *Da Zao* (Fructus Jujubae), *Hui Xiang* (Fructus Foeniculi), *Fu Ling* (Poria), *Shan Yao* (Radix Dioscoreae) | Scatter cold and stop pain assisted by dispelling wind and eliminating dampness |
| *Gui Zhi Tang* | *Gui Zhi* (Ramulus Cinnamomi), *Bai Shao* (Radix Paeoniae Albae), *Sheng Jiang* (uncooked Rhizoma Zingiberis), *Da Zao* (Fructus Jujubae), *Gan Cao* (Radix Glycyrrhizae) | Dispel wind and scatter cold, free the flow of the network vessels and diffuse impediment |
| *Huang Qi Gui Zhi Wu Wu Tang* | *Huang Qi* (Radix Astragali), *Ji Xue Teng* (Caulis Spatholobi), *Chi Shao* (Radix Paeoniae Rubrae), *Bai Shao* (Radix Paeoniae), *Gui Zhi* (Ramulus Cinnamomi), *Ge Gen* (Radix Puerariae), *Sheng Jiang* (uncooked Rhizoma Zingiberis), *Da Zao* (Fructus Jujubae) | Supplement the qi and nourish the blood, free the flow of the network vessels and move the impediment |
| *Huang Qi Jian Zhong Tang* | *Yi Tang* (Maltose), *Gui Zhi* (Ramulus Cinnamomi), *Bai Shao* (Radix Paeoniae Albae), *Da Zao* (Fructus Jujubae), *Sheng Jiang* (uncooked Rhizoma Zingiberis), *Huang Qi* (Radix Astragali), *Gan Cao* (Radix Glycyrrhizae) | Boost the qi and nourish the blood, warm the channels and free the flow of impediment |
| *Huo Luo Xiao Ling Dan* | *Dang Gui* (Radix Angelicae Sinensis), *Du Huo* (Radix Angelicae Pubescentis), *Niu Xi* (Radix Achyranthis Bidentatae), *Wu Shao She* (Zaocys), *Ru Xiang* (Olibanum), *Mo Yao* (Myrrha) | Quicken the blood and dispel stasis, free the flow of the network vessels and stop pain |

| | | |
|---|---|---|
| *Juan Bi Tang* | *Huang Qi* (Radix Astragali), *Dang Gui* (Radix Angelicae Sinensis), *Chuan Xiong* (Rhizoma Chuanxiong), *Luo Shi Teng* (Caulis Trachelospermi), *Qiang Huo* (Radix Et Rhizoma Notopterygii), *Du Huo* (Radix Angelicae Pubescentis), *Gui Zhi* (Ramulus Cinnamomi), *Qin Jiao* (Radix Gentianae Macrophyllae), *Hai Feng Teng* (Caulis Piperis Kadsurae), *Sang Zhi* (Ramulus Mori), *Ru Xiang* (Olibanum), *Jiang Huang* (Rhizoma Curcumae Longae) | Eliminate wind, overcome dampness, scatter cold, quicken the blood, supplement vacuity, and free the flow of the network vessels |
| *Ma Huang Fu Zi Gan Cao Tang* | *Fu Zi* (Radix Lateralis Praeparatus Aconiti Carmichaeli), *Ma Huang* (Herba Ephedrae), *Xi Xin* (Herba Asari) | Scatter cold and dispel dampness, warm both the spleen and kidneys |
| *Qing Ying Tang* | *Sheng Di* (uncooked Radix Rehmanniae), *Dan Pi* (Cortex Moutan), *Zi Cao* (Radix Arnebiae/Lithospermi), *Dan Shen* (Radix Salviae Miltiorrhizae), *Bai Mao Gen* (Rhizoma Imperatae), *Xuan Shen* (Radix Scrophulariae), *Chi Shao* (Radix Paeoniae Rubrae), *Jin Yin Hua* (Flos Lonicerae), *Xia Ku Cao* (Spica Prunellae), *Mai Men Dong* (Tuber Ophiopogonis) | Clear heat and resolve toxins, cool the blood and nourish yin |
| *Ren Shen Yang Rong Tang* | *Huang Qi* (Radix Astragali), *Shu Di* (cooked Radix Rehmanniae), *Dang Gui* (Radix Angelicae Sinensis), *Bai Shao* (Radix Paeoniae Albae), *Dang Shen* (Radix Codonopsitis), *Bai Zhu* (Rhizoma Atractylodis Macrocephalaex), *Fu Ling* (Poria), *Wu Wei Zi* (Fructus Schisandrae), *Niu Xi* (Radix Achyranthis Bidentatae), *Huang Jing* (Rhizoma Polygonati), *Rou Gui* (Cortex Cinnamomi) | Fortify the spleen and supplement the kidneys |

(*continues*)

TABLE I (*continued*)

| Chinese herbal formulas | Ingredients | CM treatment principles |
|---|---|---|
| *San Bi Tang* | *Huang Qi* (Radix Astragali), *Fang Feng* (Radix Saposhnikoviae), *Qin Jiao* (Radix Gentianae Macrophyllae), *Qiang Huo* (Radix Et Rhizoma Notopterygii), *Xu Duan* (Radix Dipsaci), *Du Zhong* (Cortex Eucommiae), *Fu Ling* (Poria), *Chuan Xiong* (Rhizoma Chuanxiong), *Jiang Huang* (Rhizoma Curcumae Longae), *Niu Xi* (Radix Achyranthis Bidentatae), *Xi Xin* (Herba Asari), *Dang Shen* (Radix Codonopsitis), *Dang Gui* (Radix Angelicae Sinensis), *Bai Shao* (Radix Paeoniae Albae), *Gan Cao* (Radix Glycyrrhizae), *Rou Gui* (Cortex Cinnamomi) | Dispel wind, scatter cold, and free the flow of the network vessels |
| *San Miao Wan* | *Huang Bai* (Cortex Phellodendri), *Cang Zhu* (Rhizoma Atractylodis), *Niu Xi* (Radix Achyranthis Bidentatae) | Clear heat and disinherit dampness |
| *San Zi Yang Qin Tang* | *Ban Xia* (Rhizoma Pinelliae), *Chen Pi* (Pericarpium Citri Reticulatae), *Fu Ling* (Poria), *Su Zi* (Fructus Perillae), *Bai Jie Zi* (Semen Sinapis), *Lai Fu Zi* (Semen Raphani), *Hou Po* (Cortex Magnoliae Officinalis), *Xing Ren* (Semen Armeniacae), mix-fried *Gan Cao* (Radix Glycyrrhizae), *Wu Mei* (Fructus Mume) | Transform phlegm, down bear the qi and stop panting |
| *Shen Tong Zhu Yu Tang* | *Tao Ren* (Semen Persicae), *Hong Hua* (Flos Carthami), *Dang Gui* (Radix Angelicae Sinensis), *Chuan Xiong* (Rhizoma Chuanxiong), *Qiang Huo* (Radix Et Rhizoma Notopterygii), *Di Long* (Pheretima), *Mo Yao* (Myrrha), *Wu Ling Zhi* (Feces Trogopterori), *Gan Cao* (Radix Glycyrrhizae), *Xiang Fu* (Rhizoma Cyperi) | Quicken the blood and dispel stasis, free the flow of the network vessels and stop pain |

| | | |
|---|---|---|
| *Sheng Yang Yi Wei Tang* | *Dang Shen* (Radix Codonopsitis), *Huang Qi* (Radix Astragali), *Bai Zhu* (Rhizoma Atractylodis Macrocephalae), *Fu Ling* (Poria), *Ze Xie* (Rhizoma Alismatis), *Bai Shao* (Radix Paeoniae Albae), *Huang Lian* (Rhizoma Coptidis), *Ban Xia* (Rhizoma Pinelliae), *Qiang Huo* (Radix Et Rhizoma Notopterygii), *Du Huo* (Radix Angelicae Pubescentis), *Chen Pi* (Pericarpium Citri Reticulatae), *Fang Feng* (Radix Saposhnikoviae), *Chai Hu* (Radix Bupleuri), *Gan Cao* (Radix Glycyrrhizae) | Warm the viscera and drain evils |
| *Si Jun Zi Tang* | Processed *Ban Xia* (Rhizoma Pinelliae), *Fu Ling* (Poria), *Dang Shen* (Radix Codonopsitis), *Bai Zhu* (Radix Atractylodis Macrocephalae), mix-fried *Gan Cao* (Radix Glycyrrhizae), *Chen Pi* (Pericarpium Citri Reticulatae), *Sheng Jiang* (uncooked Rhizoma Zingiberis) | Fortify the spleen and eliminate dampness, soften the hard, transform phlegm, and scatter nodulation |
| *Si Ni Tang* | *Fu Zi* (Radix Lateralis Praeparatus Aconiti Carmichaeli), *Gan Jiang* (dry Rhizoma Zingiberis), *Gan Cao* (Radix Glycyrrhizae) | Course the liver and resolve depression, supplement the qi |
| *Si Wu Tang* | *Shu Di* (cooked Radix Rehmanniae), *Ji Xue Teng* (Caulis Spatholobi), *Dang Gui* (Radix Angelicae Sinensis), *Bai Shao* (Radix Paeoniae Albae), *Chuan Xiong* (Rhizoma Chuanxiong), *Dan Shen* (Radix Salviae Miltiorrhizae) | Nourish the blood and harmonise the constructive |
| *Tao Hong Si Wu Tang* | *Sheng Di* (uncooked Radix Rehmanniae), *Chi Shao* (Radix Paeoniae Rubrae), *Dang Gui* (Radix Angelicae Sinensis), *Chuan Xiong* (Rhizoma Chuanxiong), *Tao Ren* (Semen Persicae), *Hong Hua* (Flos Carthami) | Quicken the blood and dispel stasis |
| *Tong Xie Yao Fang* | *Bai Zhu* (Rhizoma Atractylodis Macrocephalae), *Bai Shao* (Radix Paeoniae Albae), *Fang Feng* (Radix Saposhnikoviae), *Chen Pi* (Pericarpium Citri Reticulatae) | Course the liver and rectify the qi, fortify the spleen and boost the qi |

(*continues*)

TABLE I (*continued*)

| Chinese herbal formulas | Ingredients | CM treatment principles |
|---|---|---|
| *Wu Tou Tang* | *Feng Mi* (Mel), *Ji Xue Teng* (Caulis Spatholobi), *Huang Qi* (Radix Astragali), *Bai Shao* (Radix Paeoniae Albae), processed *Chuan Wu* (Radix Aconiti Carmichaeli), mix-fried *Gan Cao* (Radix Glycyrrhizae), *Bai Zhu* (Rhizoma Atractylodis Macrocephalae), *Ma Huang* (Herba Ephedrae) | Scatter cold and stop pain assisted by dispelling wind and eliminating dampness |
| *Xiao Huo Luo Dan* | Processed *Chuan Wu* (Radix Aconiti Carmichaeli), processed *Cao Wu* (Radix Aconiti Kusnezoffi), *Dang Gui* (Radix Angelicae Sinensis), *Chuan Xiong* (Rhizoma Chuanxiong), *Bai Shao* (Radix Paeoniae Albae). *Ru Xiang* (Olibanum), *Mo Yao* (Myrrha), *Di Long* (Pheretima), *Xiang Fu* (Rhizoma Cyperi), *Dan Xing* (Rhizoma Arisaematis) | Clear wind and quicken the blood, free the flow of the network vessels and stop pain |
| *Xiao Jian Zhong Tang* | *Yi Tang* (Maltose), *Gui Zhi* (Ramulus Cinnamomi), *Bai Shao* (Radix Paeoniae Albae), *Da Zao* (Fructus Jujubae), *Sheng Jiang* (uncooked Rhizoma Zingiberis), *Gan Cao* (Radix Glycyrrhizae) | Warm the channels and free the flow of impediment |
| *Xuan Bi Tang* | *Ban Xia* (Rhizoma Pinelliae), *Han Fang Ji* (Radix Stephaniae), *Xing Ren* (Semen Armeniacae), *Hua Shi* (Talcum), *Yi Yi Ren* (Semen Coicis), *Zhi Zi* (Fructus Gardeniae), *Can Sha* (Feces Bombycis), *Lian Qiao* (Fructus Forsythiae), *Chi Xiao Dou* (Semen Phaseoli) | Dispel wind and clear heat, flush phlegm and free the flow of the network vessels |
| *Xue Fu Zhu Yu Tang* | *Tao Ren* (Semen Persicae), *Sheng Di* (uncooked Radix Rehmanniae), *Hong Hua* (Flos Carthami), *Dang Gui* (Radix Angelicae Sinensis), *Chi Shao* (Radix Paeoniae Rubrae), *Chuan Niu Xi* (Radix Cyathulae), *Zhi Ke* (Fructus Aurantii), *Chuan Xiong* (Rhizoma Chuanxiong), *Chai Hu* (Radix Bupleuri), *Jie Geng* (Radix Platycodi), *Gan Cao* (Radix Glycyrrhizae) | Quicken the blood and free the flow of the network vessels |

| | | |
|---|---|---|
| *Yang He Tang* | *Shu Di* (cooked Radix Rehmanniae), *Huang Qi* (Radix Astragali), *Dan Shen* (Radix Salviae Miltiorrhizae), *Dang Gui* (Radix Angelicae Sinensis), *Ji Xue Teng* (Caulis Spatholobi), *Rou Gui* (Cortex Cinnamomi), *Gan Jiang* (dry Rhizoma Zingiberis), *Di Long* (Pheretima), *Ma Huang* (Herba Ephedrae), *Gan Cao* (Radix Glycyrrhizae) | Warm yang and scatter cold, quicken the blood and free the flow of the vessels |
| *Yi Wei Tang* | *Bie Sha Shen* (Radix Glehniae), *Shi Hu* (Herba Dendrobii), *Ge Gen* (Radix Puerariae), *Huang Qi* (Radix Astragali), *Gou Qi Zi* (Fructus Lycii), *Mai Men Dong* (Tuber Ophiopogonis), *Dang Shen* (Radix Codonopsitis), *Huang Jing* (Rhizoma Polygonati), *Bai Zhu* (Rhizoma Atractylodis Macrocephalae), *Gan Cao* (Radix Glycyrrhizae) | Supplement the spleen and boost the qi, nourish yin and engender fluids |
| *Yi Yi Ren Tang* | *Yi Yi Ren* (Semen Coicis), *Ji Xue Teng* (Caulis Spatholobi), *Hai Tong Pi* (Cortex Erythinae), *Han Fang Ji* (Radix Stephaniae), *Xi Xian Cao* (Herba Siegesbeckiae), *Dang Gui* (Radix Angelicae Sinensis), *Du Huo* (Radix Angelicae Pubescentis), *Qiang Huo* (Radix Et Rhizoma Notopterygii), *Gui Zhi* (Ramulus Cinnamomi), *Bai Zhu* (Rhizoma Atractylodis Macrocephalae), *Cang Zhu* (Rhizoma Atractylodis), *Chuan Xiong* (Rhizoma Chuanxiong), *Fang Feng* (Radix Saposhnikoviae) | Eliminate dampness and free the flow of impediment assisted by dispelling wind and scattering cold |

TABLE II

*List of Proprietary Chinese Medicines in Fig. 6*

| Proprietary Chinese medicines | Ingredients |
| --- | --- |
| *Bao An Wan Ling Tablets* | *Jing Jie* (Herba Schizonepetae), *Fang Feng* (Radix Saposhnikoviae), *Qiang Huo* (Radix Et Rhizoma Notopterygii), *Ma Huang* (Herba Ephedrae), *Xi Xin* (Herba Asari), *Chuan Xiong* (Rhizoma Chuanxiong), *Chuan Wu* (Radix Aconiti Carmichaeli), *Cao Wu* (Radix Aconiti Kusnezoffii), *Tian Ma* (Rhizoma Gastrodiae), *Dang Gui* (Radix Angelicae Sinensis), *Cang Zhu* (Rhizoma Atractylodis), *Gan Cao* (Radix Glycyrrhizae), *Shi Hu* (Herba Dendrobii), *He Shou Wu* (Radix Polygoni Multiflori), *Quan Xie* (Scorpio), *Xiong Huang* (Realgar), *Zhu Sha* (Cinnabar) |
| *Chuan Xiong Qin Tablets* | Ligustrazine |
| *Dan Shen Injection* | *Dan Shen* (Radix Salviae Miltiorrhizae) |
| *Er Xian Pills* | *Hei Mu Er* (Auricularia polytricha), *Cang Zhu* (Rhizoma Atractylodis), *Chuan Wu* (Radix Aconiti Carmichaeli), *Cao Wu* (Radix Aconiti Kusnezoffi), *Du Zhong* (Cortex Eucommiae), *Niu Xi* (Radix Achyranthis Bidentatae), *Sheng Ma* (Rhizoma Cimicifugae), *Shen Qu* (Massa Medica Fermentata) |
| *Feng Shi Han Tong Tablets* | *Qing Feng Teng* (Rhizoma Sinomenii Acuti), *Gui Zhi* (Ramulus Cinnamomi), *Du Huo* (Radix Angelicae Pubescentis), *Qiang Huo* (Radix Et Rhizoma Notopterygii), *Niu Xi* (Radix Achyranthis Bidentatae), *Sang Ji Sheng* (Herba Taxilli), *Fu Ling* (Poria), *Lu Rong* (Cornu Parvum Cervi), *Wei Ling Xian* (Radix Clematidis), *Yi Yi Ren* (Semen Coicis), *Dang Shen* (Radix Codonopsitis), *Huang Qi* (Radix Astragali), *Gou Qi Zi* (Fructus Lycii), *Bai Zhu* (Rhizoma Atractylodis Macrocephalae), *Qin Jiao* (Radix Gentianae Macrophyllae), *Zhi Fu Zi* (Radix Lateralis Praeparatus Aconiti Carmichaeli), *Chi Shao* (Radix Paeoniae Rubrae), *Dang Gui* (Radix Angelicae Sinensis), *Mu Xiang* (Radix Auklandiae), *Yan Hu Suo* (Rhizoma Corydalis), *Huang Qin* (Radix Scutellariae) |
| *Feng Shi Oral Liquid* | *Du Huo* (Radix Angelicae Pubescentis), *Qiang Huo* (Radix Et Rhizoma Notopterygii), *Lu Jiao Jiao* (Gelatinum Cornu Cervi), *Bie Jia* (Carapax Trionycis), *Chuan Xiong* (Rhizoma Chuanxiong), *Sang Ji Sheng* (Herba Taxilli) |
| *Feng Tong Ning Tablets* | Processed *Chuan Wu* (Radix Aconiti Carmichaeli), processed *Cao Wu* (Radix Aconiti Kusnezoffi), *Qiang Huo* (Radix Et Rhizoma Notopterygii), *Du Huo* (Radix Angelicae Pubescentis), *Ru Xiang* (Olibanum), *Mo Yao* (Myrrha), *Dang Gui* (Radix Angelicae Sinensis), *Chuan Niu Xi* (Radix Cyathulae), *Mu Gua* (Fructus Chaenomelis), *Ma Huang* (Herba Ephedrae), *Gui Zhi* |

(*continues*)

TABLE II (*continued*)

| Proprietary Chinese medicines | Ingredients |
|---|---|
| | (Ramulus Cinnamomi), *Wu Gong* (Scolopendra), *Chuan Xiong* (Rhizoma Chuanxiong), *Ma Qian Zi* (Semen Strychnotis) |
| *Fu Fang Xia Tian Wu Tablets* | Rhizoma corydalis decumbentis and other 31 ingredients |
| *Gu Ci Xiao Tong Oral Liquid* | *Chuan Wu* (Radix Aconiti Carmichaeli), *Cao Wu* (Radix Aconiti Kusnezofii), *Ma Huang* (Herba Ephedrae), *Niu Xi* (Radix Cyathulae), *Du Huo* (Radix Angelicae Pubescentis), *Wei Ling Xian* (Radix Clematidis), *Dang Gui* (Radix Angelicae Sinensis), *Chuan Xiong* (Rhizoma Chuanxiong), *Wu Mei* (Fructus Mume), *Niu Xi* (Radix Achyranthis Bidentatae), *Mu Gua* (Fructus Chaenomelis) |
| *Han Shi Bi Granules* | *Bai Shao* (Radix Paeoniae Albae), *Bai Zhu* (Rhizoma Atractylodis Macrocephalae), *Dang Gui* (Radix Angelicae Sinensis), *Fu Zi* (Radix Lateralis Praeparatus Aconiti Carmichaeli), *Gui Zhi* (Ramulus Cinnamomi), *Huang Qi* (Radix Astragali), *Ma Huang* (Herba Ephedrae), *Mu Gua* (Fructus Chaenomelis), *Wei Ling Xian* (Radix Clematidis), *Xi Xin* (Herba Asari), processed *Chuan Wu* (Radix Aconiti Carmichaeli) |
| *Hong Hua Injection* | *Hong Hua* (Flos Carthami) |
| *Huang Qi Injection* | *Huang Qi* (Radix Astragali) |
| *Huo Luo Pills* | Processed *Chuan Wu* (Radix Aconiti Carmichaeli), processed *Cao Wu* (Radix Aconiti Kusnezoffi), *Ru Xiang* (Olibanum), *Mo Yao* (Myrrha), *Dan Nan Xing* (bile-processed Rhizoma Arisaematis), *Di Long* (Pheretima) |
| *Huo Luo Tincture* | *Chuan Wu* (Radix Aconiti Carmichaeli), *Fang Ji* (Radix Stephaniae), *Shan Nai* (Rhizoma Kaempferiae), Lian Qian Cao (*Radix* Cycleae Rachemosae), *Yan Hu Suo* (Rhizoma Corydalis), *Gan Jiang* (dry Rhizoma Zingiberis), *Bing Pian* (Borneolum), *La Jiao* (cayenne pepper), *Chan Su* (Venenum Bufonis) |
| *Kun Ming Shan Hai Tang Tablets* | Tripterygii hypoglauci |
| *Lei Gong Teng Duo Dai Tablets* | Tripterygium glycosides |
| *Lei Gong Teng Tablets* | *Lei Gong Teng* (Herba Tripterygii Wilfordii) |
| *Pan Long Qi Tablets* | Rhizome of Scopulosus Bergenia and 28 herbs |
| *She Xiang Zhui Feng Plaster* | *She Xiang* (Moschus), *Du Huo* (Radix Angelicae Pubescentis), *Wu Jia Pi* (Cortex Acanthopanacis), *Hai Feng Teng* (Caulis Piperis Kadsurae), *Su Mu* (Lignum Sappan), *Hai Tong Pi* (Cortex Erythinae), *Yan Hu Suo* (Rhizoma Corydalis), *Chuan Wu* (Radix Aconiti Carmichaeli), *Cao Wu* (Radix Aconiti Kusnezofii), *Wei Ling Xian* (Radix Clematidis), *Xue Jie* |

(*continues*)

TABLE II (*continued*)

| Proprietary Chinese medicines | Ingredients |
|---|---|
| | (Sanguis Draconis), *Mu Xiang* (Radix Auklandiae), *Ru Xiang* (Olibanum), *Mo Yao* (Myrrha), *Wu Yao* (Radix Linderae), *Hong Hua* (Flos Carthami), *Dang Gui* (Radix Angelicae Sinensis), *Shu Di* (cooked Radix Rehmanniae), *Sheng Di* (uncooked Radix Rehmanniae), *Ma Huang* (Herba Ephedrae), *Niu Xi* (Radix Cyathulae), *Bing Pian* (Borneolum), Camphor, Oleum Eucalypti, Oleum Cinnamomi, Oleum Ocimi Gratissimi, Methyl salicylate |
| *Shen Jin Huo Luo Pills* | *Ma Qian Zi* (Semen Strychnotis), *Chuan Wu* (Radix Aconiti Carmichaeli), *Cao Wu* (Radix Aconiti Kusnezofii), *Mu Gua* (Fructus Chaenomelis), *Dang Gui* (Radix Angelicae Sinensis), *Niu Xi* (Radix Achyranthis Bidentatae), *Du Zhong* (Cortex Eucommiae), *Mu Xiang* (Radix Auklandiae), *Quan Xie* (Scorpio) |
| *Shi Re Bi Granules/ Tablets* | *Cang Zhu* (Rhizoma Atractylodis), *Ren Dong Teng* (Caulis Lonicerae), *Di Long* (Pheretima), *Lian Qiao* (Fructus Forsythiae), *Huang Bai* (Cortex Phellodendri), *Yi Yi Ren* (Semen Coicis), *Fang Feng* (Radix Saposhnikoviae), *Chuan Niu Xi* (Radix Cyathulae), *Bi Xie* (Rhizoma Dioscoreae Hypoglaucae), *Sang Zhi* (Ramulus Mori), *Wei Ling Xian* (Radix Clematidis) |
| *Shou Feng Shun Qi Pills* | *Da Huang* (Radix Et Rhizoma Rhei), *Huo Ma Ren* (Semen Cannabis), *Yu Li Ren* (Semen Pruni), *Shan Yao* (Radix Dioscoreae), *Shan Yu Rou* (Fructus Corni), *Che Qian Zi* (Semen Plantaginis), *Niu Xi* (Radix Cyathulae), *Tu Si Zi* (Semen Cuscutae), *Du Huo* (Radix Angelicae Pubescentis), *Fang Feng* (Radix Saposhnikoviae), *Bing Lang* (Semen Arecae), *Zhi Ke* (Fructus Aurantii) |
| *Xia Tian Wu Tablets* | Rhizoma corydalis decumbentis |
| *Xiao Huo Luo Pills* | Processed *Chuan Wu* (Radix Aconiti Carmichaeli), processed *Cao Wu* (Radix Aconiti Kusnezoffi), *Dang Gui* (Radix Angelicae Sinensis), *Chuan Xiong* (Rhizoma Chuanxiong), *Bai Shao* (Radix Paeoniae Albae), *Ru Xiang* (Olibanum), *Mo Yao* (Myrrha), *Di Long* (Pheretima), *Xiang Fu* (Rhizoma Cyperi), *Dan Xing* (Rhizoma Arisaematis) |
| *Yu Xue Bi Capsule/ Granules* | *Ru Xiang* (Olibanum), *Wei Ling Xian* (Radix Clematidis), *Hong Hua* (Flos Carthami), *Dan Shen* (Radix Salviae Miltiorrhizae), *Mo Yao* (Myrrha), *Chuan Niu Xi* (Radix Cyathulae), *Dang Gui* (Radix Angelicae Sinensis), *Jiang Huang* (Rhizoma Curcumae Longae), *Xiang Fu* (Rhizoma Cyperi), *Huang Qi* (Radix Astragali), *Chuan Cao* (Podostemales) |
| *Zheng Qing Feng Tong Ning Injection* | *Qing Feng Teng* (Rhizoma Sinomenii Acuti) Sinomenone Hydrochloride |

(*continues*)

TABLE II (*continued*)

| Proprietary Chinese medicines | Ingredients |
|---|---|
| *Zhuang Gu Pills* | Panthera tigris L., *Qian Shi* (Semen Euryalis), *Suan Zao Ren* (Semen Zizyphi Spinosae), *Gou Qi Zi* (Fructus Lycii), *Wu Wei Zi* (Fructus Schisandrae), *Chuan Xiong* (Rhizoma Chuanxiong), *Dang Gui* (Radix Angelicae Sinensis), *Shi Hu* (Herba Dendrobii), *Carapaxet Plastrum Testudinis, Fu Zi* (Radix Lateralis Praeparatus Aconiti Carmichaeli), *Du Zhong* (Cortex Eucommiae), *Huang Bai* (Cortex Phellodendri), *Du Huo* (Radix Angelicae Pubescentis), *Qiang Huo* (Radix Et Rhizoma Notopterygii), *Fang Feng* (Radix Saposhnikoviae), *Niu Xi* (Radix Achyranthis Bidentatae), *Mu Gua* (Fructus Chaenomelis), *Bai Shao* (Radix Paeoniae Albae), *Bai Zhu* (Rhizoma Atractylodis Macrocephalae), *Cang Zhu* (Rhizoma Atractylodis), *Dang Shen* (Radix Codonopsitis), *Sha Shen* (Radix Glehniae), *Huang Qi* (Radix Astragali), *Bu Gu Zhi* (Fructus Psoraleae), *Shan Yao* (Radix Dioscoreae), *Yi Yi Ren* (Semen Coicis), *Fang Ji* (Radix Stephaniae), *Gui Zhi* (Ramulus Cinnamomi), *Di Yu* (Radix Sanguisorbae), *Shu Di* (cooked Radix Rehmanniae), *Tu Si Zi* (Semen Cuscutae), *Fu Ling* (Poria), *Gan Jiang* (dry Rhizoma Zingiberis), Zhen Zhu Cao (*Herba* Phyllanthi Urinariae) |
| *Zhui Feng Plaster* | *Ma Huang* (Herba Ephedrae), *Du Huo* (Radix Angelicae Pubescentis), *Qiang Huo* (Radix Et Rhizoma Notopterygii), *Ru Xiang* (Olibanum), *Mo Yao* (Myrrha), *Gao Ben* (Rhizoma Ligustici), *Mu Gua* (Fructus Chaenomelis), *Chuan Wu* (Radix Aconiti Carmichaeli), *Cao Wu* (Radix Aconiti Kusnezoffi), *Fang Feng* (Radix Saposhnikoviae), *Bai Zhi* (Radix Angelicae Dahuricae), *Jing Jie* (Herba Schizonepetae), *Dang Gui* (Radix Angelicae Sinensis), *Chuan Xiong* (Rhizoma Chuanxiong), *Wu Jia Pi* (Cortex Acanthopanacis), *Chi Shao* (Radix Paeoniae Rubrae), *Chai Hu* (Radix Bupleuri), *Niu Xi* (Radix Achyranthis Bidentatae), *Du Zhong* (Cortex Eucommiae), *Zhi Ke* (Fructus Aurantii), *Xiang Fu* (Rhizoma Cyperi), *Gui Zhi* (Ramulus Cinnamomi), *Gao Liang Jiang* (Rhizoma Alpiniae Officinari), *Lian Qiao* (Fructus Forsythiae), *Chen Pi* (Pericarpium Citri Reticulatae), *Di Huang* (Radix Rehmanniae), *Da Huang* (Radix Et Rhizoma Rhei), *Xiao Hui Xiang* (Fructus Foeniculi), *Rou Gui* (Cortex Cinnamomi), *Mu Xiang* (Radix Auklandiae) |
| *Zhui Feng Tou Gu Pills* | *Zhi Chuan Wu* (Radix Praeparatus Aconiti Carmichael), *Bai Zhi* (Radix Angelicae Dahuricae), processed *Cao Wu* (Radix Aconiti Kusnezofii), *Xiang Fu* (Rhizoma Cyperi), *Gan Cao* (Radix Glycyrrhizae), |

(*continues*)

TABLE II (*continued*)

| Proprietary Chinese medicines | Ingredients |
|---|---|
| | *Bai Zhu* (Radix Atractylodis Macrocephalae), *Mo Yao* (Myrrha), *Ru Xiang* (Olibanum), *Dang Gui* (Radix Angelicae Sinensis). *Chuan Xiong* (Radix Chuanxiong), *Qin Jiao* (Radix Gentianae Macrophyllae), *Gui Zhi* (Ramulus Cinnamomi), *Di Long* (Pheretima), *Qiang Huo* (Radix Et Rhizoma Notopterygii), *Tian Nan Xing* (Rhizoma Arisaematis), *Fu Ling* (Poria), *Chi Xiao Dou* (Semen Phaseoli), *Tian Ma* (Rhizoma Gastrodiae), *Chi Shao* (Radix Paeoniae Rubrae), *Zhu Sha* (Cinnabar), *Gan Song* (Radix Et Rhizoma Nardostachyos) |

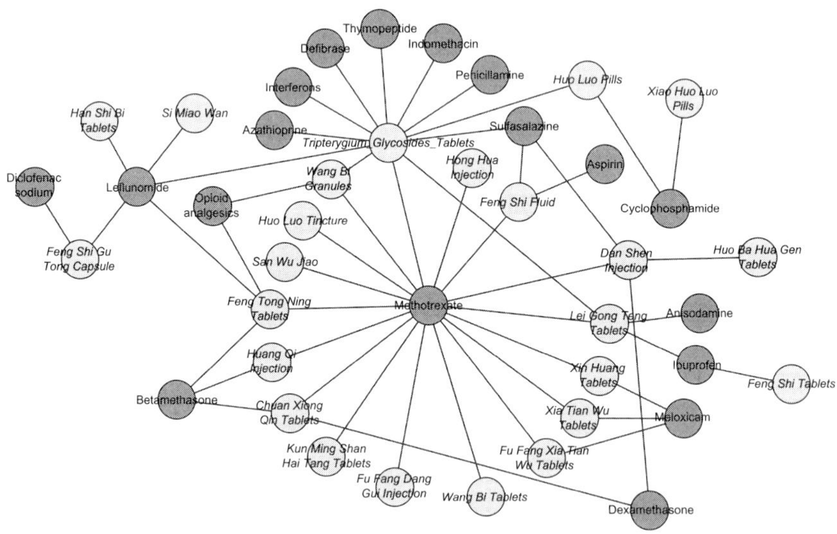


Fig. 7. The MeSH terms were obtained from Chinese BioMedical Literature Database (CBM) (till 1st July 2011). The results indicate 24 commonly used Chinese herbs (yellow) and 18 conventional Western medicines (green) and their combination used for RA management. (For interpretation of the references to colour in this figure legend, the reader is referred to the online version of this chapter.)

TABLE III

*Conventional Western Medicines Added to Chinese Herbal Products for RA*

| Chinese herbal product | Conventional drugs detected |
|---|---|
| *Ba Bao Feng Shi Huo Luo Dan* | Diclofenac |
| *Da Huo Luo Dan* | Berberine |
| *She Xiang Zhui Feng Tou Gu Wan* | Diclofenac |
| *Xin Huang Pian* | Indomethacin |

during the first 4 weeks of treatment. Suppression of adhesion molecules such as E-selectin, intracellular adhesion molecule-1 (ICAM-1) and vascular cell adhesion molecule-1 (VCAM-1) contribute to the anti-inflammatory action of the TwHF extracts (Hiramitsu *et al.*, 2006). A 2001 study (Sylvester *et al.*, 2001) demonstrated that pro-inflammatory cytokines, IL-1, IL-17 and TNF-a, induce matrix metalloproteinase-3 (MMP-3) and MMP-13 RNA and protein in primary human and bovine chondrocytes. The extracts of TWHF, possibly triptolide, reproducibly suppress this induction in part by interfering with AP-1 and NF-kB DNA binding activities. This agent could serve as a source and template for novel anti-arthritic drugs.

Another Chinese folk medicine Qing Feng Teng (Rhizoma Sinomenii) has long been utilized in the treatment of RA (Zhang *et al.*, 2009). Sinomenine, an alkaloid isolated from the stem, is reported to have clinically good curative effects on RA. It has been shown that sinomenine inhibits both the humoral and cellular immunity of mice (Zhang *et al.*, 2009). A new study (Ou *et al.*, 2011) suggests a possible mechanism of sinomenine on treatment of RA is the inhibitory effect of sinomenine on cell invasion and migration ability, which strongly correlates with repressing the expression of CD147, MMP-2 and MMP-9.

The Chinese herbal medicine Qin Jiao (Gentiana macrophylla, Gentianaceae) is also reported to be used in the treatment of rheumatism, and its main active constituents comprise four iridoid glycosides, that is, loganic acid, swertiamarin, gentiopicroside and sweroside (Zhou *et al.*, 2010). It has been reported that roots of Gentiana have an anti-inflammatory effect in RA patients, which is quite comparable to Prednisone, a conventional drug used to treat RA (Yu *et al.*, 2004).

## B. EXPERIMENTAL STUDIES ON MEDICINAL FORMULAE AND CHINESE MATERIAL MEDICINE

Chinese researchers have attempted to elucidate how the herbs used in traditional arthritis formulas alleviate the symptoms from the modern viewpoint, by analysing blood constituents of patients in numerous studies. According to these investigations (Li, 1996; Xu, 1996; Zhou *et al.*, 1999), the mechanism of action that may be dominant in those situations with good therapeutic results is a reduction in the levels of pro-inflammatory cytokines, such as interleukin-1 (IL-1). The biological effect is then to alter the levels of T-cells and the production of activated antibodies and other components. In addition, or as a result, the properties of the blood and its circulation also change, with lowered sedimentation rate and improved circulation to the extremities.

It is important to note that the most recent medical literature on experimental studies, including English publications as found in the *Journal of Traditional Chinese Medicine* and the *Journal of Integrated Western and Traditional Medicine*, reveals that there is much experimentation still going on with CHMs for RA in China for efficacy on formulas still not accepted widely by Western researchers.

### C. SUMMARY

In general, the CHMs discussed in this review have been examined for their anti-inflammatory and anti-arthritic activity. The *in vivo* testing was performed using well-established experimental models of human RA (Funk *et al.*, 2006a,b, 2009), whereas the *in vitro* testing was based on cultures of defined cell types (Matta *et al.*, 2009). For the *in vivo* studies, the plant products were tested either as an extract (Tong and Moudgil, 2007; Yu *et al.*, 2004) or as a purified bioactive compound and acetyl-11-keto-beta-boswellic acid (AKBA) (Matta *et al.*, 2009). Oral feeding and intraperitoneal injection represent the two major routes of administration employed for the *in vivo* testing. The readout for the efficacy of CHMs in the arthritis models included assessment of the severity of arthritis using clinical criteria for grading or objective parameters such as paw volume, histopathological evaluation of tissue damage in the joints and bone mineral density (Zhu *et al.*, 2005).

For studies with *in vitro* models, specific compounds purified from the natural product were added to the cell culture in the presence of inflammatory stimuli and lipopolysaccharide (Matta *et al.*, 2009). CHMs mentioned in bold font were studied in the RA model. The cells tested were derived either from mice or rats or from cell lines. The readouts of these cellular assays comprise various biochemical and molecular mediators of inflammation as discussed below in detail (Table IV). The active compound identified in each herbal extract is listed.

## IV. CLINICAL STUDY OF CHMS FOR RA

TCM has over 3000 years of continuous practice and refinement through observation. TCM practitioners and researchers are always interested in clinical studies of CHMs. More and more patients are seeking help from CHMs for its clinical effect. To more accurately assess the efficacy of CHMs treatment, innovative approaches and different methods of testing need to be developed. This section focuses on clinical studies on medicinal formulae,

TABLE IV
*Some Examples of Chinese Herbs That Are Biochemical and Molecular Mediators in Rheumatoid Arthritis*

| Herbs | Active compound(s) | Reference |
|---|---|---|
| *Biochemical mediators* | | |
| Herbs targeting $PLA_2$, COX-2, LOX, $PGE_2$ and/or $LTB_4$ | | |
| *Jiang Huang (Curcuma longa)* | Curcumin | Funk *et al.* (2006a,b) |
| *Jiang (Zingiber officinale)* | Gingerol and Zingerone | Funk *et al.* (2009) |
| *Lei Gong Teng (Tripterygium wilfordii)* | Triptolide, triptonide and celastrol | Wang *et al.* (2008), Ramgolam *et al.* (2000), Setty and Sigal (2005) |
| *Qin Jiao (Gentiana macrophylla)* | – | Yu *et al.* (2004) |
| *Qing Feng Teng (Sinomenium acutum)* | Sinomenine | Yang *et al.* (2005) |
| Herbs targeting MMPs and/or TIMPs | | |
| *Bai Shao (Paeonia lactiflora)* | Paeoniflorin | Zhu *et al.* (2005) |
| *Gou Ji (Cibotium barametz)* | Cibotinoside, cyathenosin A | Li *et al.* (2008) |
| *Niu Xi (Achyranthes bidentata)* | Oleanolic acid | Sun *et al.* (2008) |
| Herbs targeting NO, iNOS and/or SOD | | |
| *Nan She Teng (Celastrus aculeatus)* | Celastrol | Tong and Moudgil (2007) |
| *Qing Feng Teng (Sinomenium acutum)* | Sinomenine | Yang *et al.* (2005) |
| *Molecular targets* | | |
| Herbs targeting nuclear factors (NF-kB and/or AP-1) | | |
| *Bai Shao (Paeonia lactiflora)* | Paeoniflorin | Liu *et al.* (2010) |
| *Qing Feng Teng (Sinomenium acutum)* | Sinomenine | Yang *et al.* (2005) |
| *Lei Gong Teng (Tripterygium wilfordii)* | Triptolide, triptonide and celastrol | Matta *et al.* (2009), Li *et al.* (1997) |

clinical studies on Chinese material medicine, clinical studies on integrated therapy, incorporation of TCM pattern classification in CHM clinical trials, SRs on CHMs and adverse effects (AEs) (i.e. safety) of CHMs.

## A. CLINICAL STUDIES ON MEDICINAL FORMULAE

Individually prescribed formulas are the professional standard of care in China (Flaws and Sionneau, 2005). Chinese researchers are very interested in clinical trials of CHM formulae. Unfortunately, most research articles are published in Chinese, are inaccessible to Western doctors or researchers, are often not included in SRs of Western researchers and thus have failed to promote CHM research. Due to concerns about the quality of design, conduct and reporting of clinical trials for Chinese herbs, rather than reviewing the broad range of reports that can be found by searching the literature, examples will be provided here to illustrate the claimed results. This study was selected for it is the first Chinese herbal formulae research published in an SCI-indexed journal.

A study of integrated therapy was reported by Xi Bao *et al.* (2006). In this study, Ling Zhi and San Miao San were used as CHMs. Ling Zhi belongs to the Ganoderma family of mushrooms and has been a valuable medicine for longevity in ancient China. San Miao San is a CHM that contains Cang Zhu (Rhizoma atractylodis), Huang Bai (Cortex phellodendri) and Niu Xi (Radix achyranthes bidentatae). San Miao San exhibits analgesic and anti-inflammatory activities and has been used for RA treatment in China. In the study, 65 RA patients were recruited and randomized to receive either Ling Zhi plus San Miao San or placebo for 24 weeks. The CHMs were added on top of their current Western medications. After 24 weeks of treatment, patients who took Ling Zhi and San Miao San experienced significantly lower inflammatory responses as measured by inflammatory cytokine IL-18. Most important, RA patients who took the TCM had significantly greater pain relief than patients in the placebo group ($p < 0.05$). The study suggested integrating Chinese herbal formulae with Western medications can have beneficial anti-inflammatory effects on patients with RA.

## B. CLINICAL STUDIES ON CHINESE MATERIAL MEDICINE

More Chinese material medicines can be used for RA treatment; however, single herb use is not so common in clinical practice. The widely used single herb for RA management includes TWHF and total glucosides of peony (TGP).

### *1. Tripterygium wilfordii Hook F*

TWHF, a CHM used as an anti-inflammatory, was recently published in an SCI-indexed journal (Goldbach *et al.*, 2009; Marks 2011). Although undertaken by the same research group, data from two studies of orally administered TWHF could not be pooled because the extracts, interventions and measures differed between the two trials (Tao *et al.*, 1989, 2002). The 1989 study is the first prospective, controlled, double-blind, crossover study of TWHF; a 60 mg daily extract dose of TWHF was compared with placebo in a crossover trial, the first arm of which included joint tenderness (0–3; none to severe), joint swelling of 60 joints only, morning stiffness in hours, grip strength (mmHg) and 15 m walking time in seconds. Improvements were reported in favour of the extract with statistically significant decreases in joint tenderness (MD −14.00, CI −19.02 to −8.98, $p<0.01$) and swollen joint count (MD −3.10, CI −5.53 to −0.67, $p=0.01$). Non-significant decreases were reported in morning stiffness (MD −1.40, CI −4.18 to 1.38, $p=0.32$) and walking time (MD −10.40, CI −22.07 to 1.27, $p=0.08$). An increase in grip strength was also demonstrated (MD 3.20, CI −20.01 to 26.41). In the later study (Tao *et al.*, 2002) relative to a 1989 study (Tao *et al.*, 1989), larger extract doses of TWHF were used, 180 mg ($n=10$) and 360 mg ($n=10$) per day. Clinical outcomes were assessed using the ACR 20%, 50% and 70% improvement levels. Eight high-dose and four low-dose patients satisfied the ACR20 improvement criteria at the end of the intervention period compared to none of the patients in the placebo group. These dichotomous data convert to an odds ratio of 17.31 (CI 0.80 to 373.45) and 85.00 (CI 3.61 to 2001.33) of satisfying ACR20 improvement criteria when patients consume the extract in a dose of 180 or 360 mg TWHF, respectively, compared with placebo. Graphic presentation of between-group comparisons of each of the disease activity component measures (tender joint count, swollen joint count, pain, physical function, patient global, physician global, erythrocyte sedimentation rate, C-reactive protein) indicated that all outcomes were improved in the high-dose group over both the low-dose and placebo groups and that improvements were observed in all measures in the low-dose group over the placebo group. These data were not reported in re-analysis.

Another randomized double-blind placebo controlled trial involved 61 patients with RA who participated in a 6-week trial of a topically applied tincture of Tripterygium wilfordii (Cibere *et al.*, 2003). It is very important to note that the placebo was prepared and packaged to be indistinguishable from the intervention. The tincture was applied up to six times per day to the swollen or tender joints, and results were reported using the core set of ACR response criteria, aggregated into a slightly modified form of the 20% improvement level. At the end of the intervention period, statistically significant differences were identified between the two groups on most clinical outcomes: 42 tender joint count (MD 1.50, CI 0.58 to 2.42, $p=0.001$), 40 swollen joint count (MD 4.40, CI 2.76 to 6.04, $p<0.001$), grip strength in kiloPascals (MD 39, CI 25.70 to 52.30, $p < 0.001$) and duration of morning stiffness in hours (MD 0.80, CI 0.54 to 1.06, $p < 0.001$). A significant difference was noted between the groups for the number of ACR20 responders; 18 of 31 patients in the Tripterygium group and 6 of 30 patients in the placebo group. These data represent an odds ratio of 5.54 (CI 1.76 to 17.39, $p = 0.003$) for the Tripterygium group in satisfying ACR20 improvement criteria. These results may be considered as evidence that Tripterygium application is superior to placebo for reducing RA disease activity.

*2. Total glucosides of peony*

The first multi-centre double-blind trial of TGP was conducted in 1993, when 450 patients with RA were randomized to receive TGP (1.8 g/day) or methotrexate (MTX) (10 mg/week) for 12 weeks. Therapeutic response was achieved in 71.7% of TGP-treated patients and 81.7% of MTX-treated patients, respectively. The adverse events of TGP occurred in 13.3% patients, and the adverse events of MTX occurred in 56.7% patients. And most adverse events of TGP were gastrointestinal tract disturbances, mostly mild diarrhoea. No adverse events following hepatic, renal or hematological tests were found. Similar results were confirmed in a phase III clinical trial of 1016 patients with RA in 1997 (He and Dai, 2011; Min *et al.*, 2005; Zhou and Li, 2003).

### C. CLINICAL STUDIES ON INTEGRATED THERAPY

In a study (Liu *et al.*, 2007) that included 120 patients, one group was treated orally with Qing Bi Tablet, a patented Chinese herbal preparation, and the control group was treated with MTX, oral intake of voltaren 75 mg and hydroxychloroquine 0.2 g once a day. Prednisone was given to all patients; the initiating dosage used in the treated group (15 mg once a day) was less than that in the control group (20 mg once a day). The clinical index, dosage and adverse reaction of prednisone were recorded every 2 weeks. The improvement in resting pain, patient's self-evaluation and doctor's evaluation in the treated group was better than those in the control group. The 20-week total amount of prednisone used in the treated group was less than that of the control group (32,935 mg vs. 51,170 mg), while the dosage of prednisone used in various observation time points between the two groups was also significantly different, respectively ($p < 0.05$), the former was less than that of the latter.

Another trial to assess the clinical effectiveness and safety of Kampo used in combination with oral MTX in an attempt to control patients whose disease remains active despite treatment with MTX (Kogure *et al.*, 2010).

Patients ($n = 13$) with RA who achieved only a sub-optimal response to MTX therapy ($\geq 6$ mg/week and $\geq 6$ months) were enrolled in this assessment. All patients were treated with Keishinieppiittokaryojutsubu decoction. Every 3 months, joint symptoms were examined, and routine blood analysis and general serological tests including anticyclic citrullinated peptide antibody (aCCP) were performed, and then the disease activity score of 28 joints (DAS28) was calculated. The results showed that combination of Kampo with MTX is safe and well tolerated and provides clinical and economic benefits. This is a positive demonstration of the complementary nature of MTX and Kampo for RA treatment.

Dr P. J. G. Forster first described a patient with RA who showed astonishing improvement while taking a Chinese 'herbal' medicine (Forster *et al.*, 1979). Within months, the authors heard that public interest in Holland had led to analysis of the tablets by the Dutch Government Institute of Medicines. The results of the analysis were published in the Dutch press (McConkey, 2003) under the headline 'Hong Kong-pill Chinese roulette'. The pills had been found to contain dexamethasone and indomethacin. Conventional Western medicines added to Chinese herbal products for RA as integrated therapy is commonly used (Table III). It is therefore very important to ensure product labels provide an accurate list of all ingredients.

## D. INCORPORATION OF TCM PATTERN CLASSIFICATION IN CLINICAL TRIALS

TCM pattern classification is the hallmark of the TCM therapy. The TCM pattern is a summary of the body's health condition at a certain stage in the disease process and includes all signs and symptoms and many other significant aspects of the disease process. The indication of a Chinese herbal medicine requires specification with TCM pattern differentiation, which means that a CHM should show its best effectiveness only in part of a population of patients with a specific disease. The RCT design has distinct limitations when applied to TCM, because TCM is holistic and conceptual, and it identifies and treats 'syndromes' rather than diseases. Most of these projects employed 'very standard' randomized controlled designs without TCM theory; thus, they were not very suitable for TCM trials design.

It is important to incorporate the concepts of both biomedicine and TCM into research protocols, and this necessitates collaboration between biomedicine and TCM practitioners (Tsang 2007). It is necessary to assess the TCM clinical efficacy with better, more innovative approaches (Tsang *et al.*, 2010). Thus, incorporation of pattern classification has become more acceptable to the researchers. Increasing numbers of medical researchers recognize that the combination of disease diagnosis in biomedicine and pattern classification in TCM is essential for clinical practice when TCM is included as part of the treatment alongside biomedicine diagnostic approaches. This has become a common practice model in China, as it focuses on a broader or better assessment of clinical effects (Lu *et al.*, 2008; Manheimer *et al.*, 2009).

The classification has been partially proven in RA clinical trials. The predictive roles of symptom combination traditionally evaluated in TCM in the treatment of RA were explored in a multi-centre, randomized, controlled trial. Factor analysis classifies the variances (symptoms) into certain groups according to the correlation among the variance. All 18 symptoms observed before randomization were used for factor analysis and four symptom combinations (factors) were obtained. According to the factor loadings, four factors are correlated with different symptoms, respectively, and the symptoms in each factor were called symptom combination. The symptom combination in factor 1 (symptom combination 1) includes joint tenderness, joint swelling and joint stiffness, which are related to joint inflammation; the symptom combination in factor 2 (symptom combination 2) includes cold joint, cold limbs and intolerance to cold, which are related to cold pattern in TCM; the symptom combination in factor 3 (symptom combination 3) includes weakness in waist, fatigue, dizziness, heavy limbs, and nocturia and numb limbs, which are related to the deficiency pattern in TCM; the symptom combination in factor 4 (symptom combination 4) includes joint heat, vexation, thirst, turbid urine with yellow colour and fever, which are related to the hot pattern in TCM. And the symptom combinations are pretty much similar to RA pattern in TCM theory. TCM would be more effective in patients with weak-symptom combination 3 (deficiency pattern in TCM), and MTX and Sulfasalazine (SSZ) would be more effective in patients with symptom combination 2 (cold pattern in TCM) (He *et al.*, 2008). It has also been reported that the TCM pattern classification can specify the indications for the combination biomedical therapy in the treatment of existing RA patients (He *et al.*, 2007; Lu *et al.*, 2009).

Therefore, TCM pattern classification helps to identify individuals who are at high response of therapy for RA and should receive further therapy. Predictive pattern classifications help to identify those drugs to which patients are most responsive (or unresponsive). A two-stage multi-centre trial of Chinese herbal medicine for the management of RA is being conducted (Zhang *et al.*, 2011c).The stage-one trial is an open-label trial that aims to explore what groups of TCM information (such as symptoms) correlates with better efficacy, and the stage-two trial is a randomized, controlled, double-blind, double-dummy clinical trial that incorporates the efficacy-related information identified in the stage-one trial into the inclusion criteria. The indication of a Chinese herbal formula is a specific Chinese medicine pattern and not a single disease, and stratifying a disease into several patterns with a group of symptoms is a feasible procedure in clinical trials. This study is the first to investigate whether this approach in the design of Chinese herbal medicine trials can improve responses.

Another trial was designed with incorporation of TCM pattern classification during patient recruitment (Shen *et al.*, 2011). Eighty patients with RA belonging to syndrome of damp-heat obstruction were randomly divided into a treatment group and a control group. Forty cases in the control group took Loxoprofen Sodium Tablet and Leifumite Tablet orally and the other 40 cases in the treatment group orally took a Chinese medicine decoction. After 12 weeks of treatment, the improvement of morning stiffness time, number of swelling joints, index of swelling joints, grip strength, HAQ and patient's assessment in the treatment group was much better than that in the control groups with statistical difference ($p < 0.05$). ACR20, ACR50 and ACR70 were 27.5% (11/40), 37.5% (15/40) and 22.5% (9/40), respectively, in the treatment group and 40% (16/40), 27.5% (11/42) and 10.0% (4/40), respectively, in the control group. The incidence of adverse reactions in the control group (25.00%) was higher than that in the treatment group (6.67%) ($p < 0.05$). Definite therapeutic effect and high safety can be achieved in using the therapy combining disease with syndrome to treat RA.

## E. SRS ON CHINESE MATERIAL MEDICINE

Here are recently published findings of SRs that explore a range of evidence on CHM for RA. There are 12 SRs/meta-analysis of CHMs that have been published online (Table IV). Three SRs (Jiang *et al.*, 2009a,b; Xu *et al.*, 2004) evaluate the effect and safety of TWHF and its extracts. Authors presented conclusions that although limited evidence suggested some favourable results in clinical trials, these findings should be carefully interpreted due to the low methodological quality and limited number of trials. TGP has been recognized as valuable traditional herbs used for the treatment of RA. A SR (Shang *et al.*, 2010) evaluates the efficacy of TGP when combined with MTX, and a second evaluates TGP with immunosuppressant for RA (Zhong *et al.*, 2010). Results from these reviews show that these combination or complementary therapies are better than biomedicine monotherapy. High-quality and large-scale RCTs are needed to further validate the results because of the limited quality of the included studies. Chinese herbal medicine, as a whole intervention in SR, was tested. Two authors gave details described in their master's theses (Yang, 2008; Yu, 2009), but these data are still limited because of the limited quality of the included trials. Further,

Compositae, a species of snow lotus from China used for treating RA, was tested. This poor quality report did not appear valid and convincing (Huan *et al.*, 2008). The low quality of clinical design in many of these RCTs prevented meaningful meta-analysis from providing evidence to prove the efficacy of TCM in the treatment of RA patients (Zhang *et al.*, 2011b). And many SRs currently on CHMs for RA showed that there is no solid clinical evidence to support the efficacy of the TCM intervention (Zhang *et al.*, 2011a). Several reasons could be identified to account for the clinical research of TCM for RA. There was a failure to consider fundamental concepts of TCM pattern (Table V).

## F. ADVERSE EFFECTS OF CHMS

TCM doctors may utilize ingredients that are toxic and cannot be widely used in Western practice, such as tripterygium or the powerful muscle stimulant strychnos (which contains curare). Such ingredients may have a more profound effect than the commonly used and gentle components of the formulas described above. Since Western patients are usually seeking treatments to substitute for (alternative therapy) or complement (complementary therapy) modern monotherapy drugs that cause unacceptable adverse side effects in some cases, the herbal prescriptions used must be essentially non-toxic. The safety profile for preparations from TWHF is particularly concerning. This herb is no longer included in the English edition of the Chinese Pharmacopoeia, probably due to the toxicity of the coactive triptolides. AEs include dysmenorrhoea, decrease of male fertility, renal insufficiency, haematotoxicity, embryotoxicity and immune suppression, demonstrated by increased rate of infections (Canter *et al.*, 2006). In both clinical studies (Tao *et al.*, 1989, 2002), more AEs were seen among patients receiving TWHF than among patients receiving placebo. In the earlier study, AEs resulted in four withdrawals, and a severe reaction (fever and aplastic anaemia) occurred in one patient following an overdose of TWHF. One death occurred also, although not thought to be related to the intervention. In the later study, four patients in the placebo group reported AEs, as did six patients in the high-dose and five patients in the low-dose group. In this study, none of the AEs was reported as severe (did not require hospitalization). Commonly reported AEs included diarrhoea, headache and hair loss.

Sinomenine treatment is also a major concern. In a review (Xu *et al.*, 2008), AEs occurred less frequently in the digestive system during sinomenine treatment than during NSAID treatment (OR 0.04, 95% CI 0.01 to 0.23; three studies) but occurred more frequently in the dermatomucosal system

TABLE V

*Systematic Reviews/Meta-analysis of Chinese Herbal Medicine for RA*

| Title | Authors | Year published | Trials | Participants | Author's comments |
|---|---|---|---|---|---|
| A systematic review of randomised clinical trials of Tripterygium wilfordii for rheumatoid arthritis[a] | Canter *et al.* (2006) | 2006 | 2[b] | 105 | Limited evidence suggested that T. wilfordii seems to have beneficial effects on the symptoms of RA; however, serious adverse events associated with the herb mean that this treatment cannot be recommended. |
| A systematic review on clinical efficacy of Xuelian for treatment of rheumatoid arthritis | Huan *et al.* (2008) | 2008 | 2[c] | 198 | The authors cannot draw a conclusion that Xuelian may increase the cure rate and total effective rate of RA. Much higher quality trials are required. |
| Compound Tripterygium wilfordii for rheumatoid arthritis: a systematic review | Jiang *et al.* (2009a) | 2009 | 8[b] | 470 | Topical or oral administration with compound Tripterygium wilfordii for RA is effective. No serious adverse event is reported. |
| Tripterygium wilfordii extract for treating rheumatoid arthritis: systematic review | Jiang *et al.* (2009b) | 2009 | 7[b] | 393 | Tripterygium wilfordii extract (TWE) can be used as an alternative medicine in treating RA. But owing to lack of data from high-quality RCTs, the efficacy and safety of TWE need to be further studied. |
| Systematics evaluation on traditional Chinese medicine in treating rheumatoid arthritis contrasts to methotrexate | Qi and Liu (2010) | 2010 | 7[b] | 767 | Because of the different treatment methods of RA, the few literature which can be merged to carry out meta-analysis and the low methodology quality of the literature which is internalised and may have the publication bias, the availability and safety of RA treated by TCM are pending to confirm with large sample, multi-centre double-blind RCT, which must be designed strictly. |

(*continues*)

TABLE V (*continued*)

| Title | Authors | Year published | Trials | Participants | Author's comments |
|---|---|---|---|---|---|
| Meta-analysis on total glucosides of paeony combined with methotrexate in treatment for rheumatoid arthritis | Shang *et al.* (2010) | 2010 | 7[b] | 849 | The total effective rate of total glucosides of paeony combined with MTX is much higher than that of MTX in treating the patients with RA. At the same time, it can reduce adverse effects, such as reducing the incidence rate of impaired liver function significantly. |
| Sinomenine versus NSAIDs for the treatment of rheumatoid arthritis: a systematic review and meta-analysis[a] | Xu *et al.* (2008) | 2008 | 10 | 1185 | Sinomenine (SIN) may be a valuable remedy to treat RA clinically, although current evidence needs to be further verified by more high-quality trials. |
| Chinese herbal medicines versus non-steroidal anti-inflammatory drugs for rheumatoid arthritis: a systematic review | Xu *et al.* (2004) | 2004 | 10[d] | 1475 | Chinese herbal medicines showed additional better effects than non-steroidal anti-inflammatory drugs for RA. Due to the low methodological quality of trials. Large-sample, multi-centre, randomized, double-blind, controlled trials with limiting clinical heterogeneity are warranted. |
| Traditional Chinese medicine for treating rheumatoid arthritis: a systematic review | Yang (2008) | 2008 | 110[b] | None indicated | The clinical effects of traditional Chinese medicine or integrated Western medicine against RA have not been confirmed yet. The quality of the RCTs' reports of traditional Chinese medicine against RA was low. |
| The systematic review of the recent efficacy for the treatment of RA with Chinese herb alone or Chinese herb with western medicine | Yu (2009) | 2009 | 112[b] | 10,703 | The evidence is not strong enough because of low quality of the trials. |

| | | | | | |
|---|---|---|---|---|---|
| Chinese herbal medicines versus disease modifying antirhumatic drugs for management of rheumatoid arthritis: a systematic review[a] | *Zhang* et al. (2011a,b,c) | 2011 | 20[b] | 1849 | Despite some favourable results in clinical trials, these findings should be carefully interpreted due to the low methodological quality. Further clinical trials with optimal CHM treatment protocols, rigorous and well-controlled randomized design are warranted. |
| Meta-analysis of the efficacy and safety of Total Glucosides of Paeony combined with immunosuppressant in the treatment of rheumatoid arthritis | Zhong *et al.* (2010) | 2010 | 9[b] | 707 | The therapy using the total glucosides of paeony combined with immunosuppressant might be more effective than immunosuppressant alone for RA in the early stage of treatment. High-quality and large-scale RCTs are needed to further prove the results of study because of low quality of the included studies. |

[a]In English.
[b]Randomized controlled trials.
[c]Quasi-randomized trials.
[d]Randomized controlled trials and quasi-randomized trials.

with sinomenine treatment (OR 3.07, 95% CI 1.14 to 8.22; four studies). There was no significant heterogeneity for these comparisons. Adverse events of the nervous system were similar for both treatments.

In general, AEs of CHM include those related to the herbs themselves and the toxic effects due to the contaminants. Problems arise when herbs are adulterated with toxic herbs or toxic chemicals. Interactions between herbs should be carefully observed and documented in CHMs due to the current limited evidence.

### G. COMMENTARY

To assess accurately any potential benefits and risks in the treatment of RA, treatment with Chinese herbs must be scientifically evaluated. Standard treatment and outcome indices need to be developed based on the principles used by the Chinese herb practitioners. The studies that have been conducted in China always varied in outcome index. Another important issue is the 'one size fits all' hypothesis, which is not consistent with TCM practice, where a remedy is prescribed individually on the basis of each RA patient's particular symptoms and not on the disease category they may or may not fall into. Treatments to be tested should be selected and standardized so that potentially effective and important treatments are evaluated first. In general, adverse reactions due to CHM treatment are low and mild although these have not been monitored in all studies. AE monitoring should definitely be a part of future studies, so the incidence and severity of AEs due to CHM treatment are well known. The rates of AEs are suspected to be lower in well-managed CHM practices, but this remains to be proven. The best evidence should be systematically reviewed, summarized and disseminated, providing the basis for evidence-based decision making in therapy for RA with CHMs. Though the scientific bases of treatment with herbs are rapidly being established, more well-designed, controlled clinical studies are required if these treatments are ever to be used for first-line therapy worldwide.

## V. FUTURE PERSPECTIVES

A number of important changes are necessary in the regulation of CHMs that could improve the safety and appropriate use of these products. These include (1) requiring the incorporation of TCM pattern classification in experimental and clinical studies, (2) requiring all clinical trials to be approved by the Clinical Trials Registry, (3) mandating safety tests and (4) ensuring that product labels provide an accurate list of all ingredients. While these changes will almost certainly improve the safety of therapy with herbal products, additional changes are also required to improve and promote high-quality research. The most critical element will be to define specific standards, especially syndrome(s) standards for CHM products to ensure consistency between studies.

Even recent studies that used high-quality methodology were open to criticism about the formulation of the herbal product used because no clear, well-established standard of the chemical fingerprint exists for most commonly used herbs. The effort to create standards for herbal products must be encouraged. Research would also proceed more rapidly if some form of patent protection existed, so that manufacturers who invest in expensive studies and document the efficacy of their products could be rewarded financially.

Finally, techniques to speed the progress of clinical research would help deal with the backlog of herbal products that have limited evidence of safety and efficacy. In a recent study conducted in 10 centres of China, investigators were able to try a new study design (Zhang *et al.*, 2011c). Although this technique is still under investigation, it has the potential to demonstrate that therapy with more herbal preparations is efficacious for existing, complex syndromes.

# VI. CONCLUSIONS

It is interesting to note that therapy with a large number of plant products has been pursued up to the stage of clinical trials. But many of the said herbs and formulas have not been tested in experimental models of inflammation. This is very different from chemical or biological drug discovery and development. Usually these investigations suggest selective anti-inflammatory activity of selected plant constituents, and then the entire plant product goes to the stage of trials. This may be due to the fact that most of the clinical research work done in China is academically oriented, pursued by researchers with experience in clinical practice.

CHMs are rapidly gaining attention in the West as sources of new alternative or complementary drugs or dietary supplements. Unfortunately, there is inadequate evidence regarding safety or efficacy of many CHMs even though much money, time and energy have been spent in this field of research. Consequently, the authors believe that to get fruit from this labour, it is better to re-evaluate and test the effect of plants already reported to be efficacious towards RA than to increase the length of the intervention list. Additional work is required to define the biologically active phytoconstituents to identify better drugs for the relief of miseries of RA patients and it is believed one or more superior therapeutically effective anti-inflammatory drugs can be obtained from this source. The current regulation of herbs does not ensure that available products are safe; moreover, false and illegal marketing claims are far too common. Several changes to the regulation of these products could dramatically improve the appropriate use of herbs. Future research will be best served by the creation of national standards for the constituents of specific CHMs, greater incentives for research and the development of study designs that reduce costs and study duration.

## ACKNOWLEDGEMENTS

This research is supported jointly by National Eleven Fifth Plan Project (2006BAI04A10) and by National Science Fund for Distinguished Young Scholars (30825047).

## REFERENCES

Agarwal, S. (2010). Newer nutritional basis in the management of rheumatoid arthritis. *Electronic Physician* **2**, 33–38.

Arya, V., Kumar Gupta, V. and Kaur, R. (2011). A review on plants having anti-arthritic potential. *International Journal of Pharmaceutical Sciences Review and Research* **7**, 131–136.

Cameron, M., Gagnier, J. J. and Little, C. V. (2009). Evidence of effectiveness of herbal medicinal products in the treatment of arthritis. Part 2: Rheumatoid arthritis. *Phytotherapy Research* **23**(12), 1647–1662.

Canter, P. H., Lee, H. S. and Ernst, E. (2006). A systematic review of randomised clinical trials of Tripterygium wilfordii for rheumatoid arthritis. *Phytomedicine* **13**, 371–377.

Cheng, J. (2000). Drug therapy in Chinese traditional medicine. *Journal of Clinical Pharmacology* **40**, 445–450.

Cibere, J., Deng, Z., Lin, Y., He, Y., Wang, Z., Thorne, A., Lehman, A. J., Tsang, I. K. and Esdale, J. M. (2003). A randomized double blind placebo controlled trial of topical *Tripterygium wilfordii* in rheumatoid arthritis: Reanalyais using logistic regression. *Journal of Rheumatology* **30**, 465–467.

Efthimiou, P., Kukar, M. and Mackenzie, C. R. (2010). Complementary and alternative medicine in rheumatoid arthritis: No longer the last resort!. *HSS Journal* **6**(1), 108–111.

Ernst, E., Pittler, M. H. and Wider, B. (2007). Complementary Therapies for Pain Management, an Evidence-Based Approach. Elsevier, London.

Flaws, B. and Sionneau, P. (2005). The Treatment of Modern Western Diseases with Chinese Medicine. 2nd edn. Blue Poppy Press, Boulder.

Forster, P. J. G., Calverley, M., Hubball, S. and McConkey, B. (1979). Cheui–Fong–Tou–Geu–Wan in rheumatoid arthritis. *British Medical Journal* **2**, 308.

Funk, J. L., Frye, J. B., Oyarzo, J. N., Kuscuoglu, N., Wilson, J., McCaffrey, G., Stafford, G., Chen, G., Lantz, R. C., Jolad, S. D., Solyom, A. M., Kiela, P. R. *et al.* (2006a). Efficacy and mechanism of action of turmeric supplements in the treatment of experimental arthritis. *Arthritis and Rheumatism* **54**(11), 3452–3464.

Funk, J. L., Oyarzo, J. N., Frye, J. B., Chen, G., Lantz, R. C., Jolad, S. D., Solyom, A. M. and Timmermann, B. N. (2006b). Turmeric extracts containing curcuminoids prevent experimental rheumatoid arthritis. *Journal of Natural Products* **69**(3), 351–355.

Funk, J. L., Frye, J. B., Oyarzo, J. N. and Timmermann, B. N. (2009). Comparative effects of two gingerol-containing Zingiber officinale extracts on experimental rheumatoid arthritis. *Journal of Natural Products* **72**(3), 403–407.

Goldbach, Mansky R., Wilson, M. and Fleischmann, R. (2009). Comparison of Tripterygium wilfordii Hook F versus sulfasalazine in the treatment of rheumatoid arthritis: A randomized trial. *Annals of Internal Medicine* **151** (4), 229–240.

He, D. Y. and Dai, S. M. (2011). Anti-inflammatory and immunomodulatory effects of *Paeonia Lactiflora* Pall., a traditional Chinese herbal medicine. *Frontiers in Pharmacology* **2**(10), 1–5.

He, Y. T., Lu, A. P., Zha, Y., Yan, X., Song, Y., Zeng, S., Liu, W., Zhu, W., Su, L., Feng, X., Qian, X. and Lu, C. (2007). Correlations between symptoms as assessed in traditional Chinese medicine (TCM) and ACR20 efficacy response: A comparison study in 396 patients with rheumatoid arthritis treated with TCM or Western medicine. *Journal of Clinical Rheumatology* **13**, 317–321.

He, Y., Lu, A. P. and Lu, C. (2008). Symptom combinations assessed in traditional Chinese medicine and its predictive role in ACR20 efficacy response in rheumatoid arthritis. *The American Journal of Chinese Medicine* **36**(4), 675–683.

Hegen, M., Keith, J. C., Collins, M. and Nickerson-Nutter, C. L. (2008). Utility of animal models for identification of potential therapeutics for rheumatoid arthritis. *Annals of the Rheumatic Diseases* **67**, 1505–1515.

Herman, C. J., Allen, P., Hunt, W. C., Prasad, A. and Brady, T. J. (2004). Use of complementary therapies among primary care clinic patients with arthritis. *Preventing Chronic Disease* **1**(4), A12.

Hiramitsu, T., Yasuda, T., Ito, H., Shimizu, M., Julovi, S., Kakinuma, T., Akiyoshi, M., Yoshida, M. and Nakamura, T. (2006). Intercellular adhesion molecule-1 mediates the inhibitory effects of hyaluronan on interleukin-1 {beta}-induced matrix metalloproteinase production in rheumatoid synovial fibroblasts via down-regulation of NF-{kappa}B and p38. *Rheumatology (Oxford, England)* **45**, 824–832.

Huan, W. Y., Qiu, X. and Xu, F. L. (2008). A systematic review clinical efficacy about Xuelian treatment for rheumatoid arthritis. *China Practice Medicine* **3**, 59–60.

Izzo, A. A. and Ernst, E. (2009). Interactions between herbal medicines and prescribed drugs: An updated systematic review. *Drugs* **69**(13), 1777–1798.

Jiang, Q., Cao, W. and Tang, X. P. (2009a). Compound Tripterygium wilfordii for rheumatoid arthritis: A systematic review. *Lishizhen Medicine and Materia Medica Research* **20**, 2377–2381.

Jiang, Q., Cao, W. and Tang, X. P. (2009b). Tripterygium wilfordii extract for treating rheumatoid arthritis: Systematic review. *Zhongguo Zhong Yao Za Zhi* **34**, 2637–2643.

Kogure, T., Tatsumi, T., Sato, H., Oku, Y., Kishi, D. and Ito, T. (2010). Traditional herbal medicines (Kampo) for patients with rheumatoid arthritis receiving concomitant methotrexate: A preliminary study. *Alternative Therapies in Health and Medicine* **16**(1), 46–51.

Lesley, I. Alternative Forms of Medicine are Becoming More Popular. http://www.smudailymustang.com/?p=27201 (Last accessed Aug 1st, 2011).

Li, S. Z. (1996). Changes in blood flow in the extremities in patients with rheumatoid arthritis and treatment with integrated traditional and western medicine. *Chinese Journal of Integrated Traditional and Western Medicine* **2**(2), 115–116.

Li, H., Jia, Y. F., Pan, Y., Pan, D. J., Li, D. and Zhang, L. X. (1997). Effect of tripterine on collagen-induced arthritis in rats. *Zhongguo Yao Li Xue Bao* **18** (3), 270–273.

Li, J., Wang, Z. H., Wang, C. T., Cao, C. X., Dong, Q. F. and Jia, T. Z. (2008). Influence of Cibotium barametz and its processed samples on haemorheology index in mice with adjuvant arthritis. *Zhongguo Zhong Yao Za Zhi* **33** (17), 2170–2173.

Lipsky, P. E. and Tao, X. L. (1997). A potential new treatment for rheumatoid arthritis: Thunder god vine. *Seminars in Arthritis and Rheumatism* **26**(5), 713–723.

Little, C. V., Parsons, T. and Logan, S. (2000). Herbal therapy for treating rheumatoid arthritis. *Cochrane Database of Systematic Reviews* **4**, CD002948.

Liu, W., Zhang, L. and Xu, Z. (2006). Clinical observation on treatment of rheumatoid arthritis with *Bi Qi* capsule. *Zhongguo Zhong Xi Yi Jie He Za Zhi* **26**(2), 157–159.

Liu, W., Liu, X. Y. and Wang, Y. (2007). Effect of Chinese herbs in enhancing prednisone for treatment of refractory rheumatoid arthritis. *Zhongguo Zhong Xi Yi Jie He Za Zhi* **27**(8), 742–744.

Liu, G. L., Li, Y. C. and Shen, Y. J. (2010). Inhibitory effect of total glucosides of paeonia on the NF-κB/p65 protein expression in paws of RA rats. *Chinese journal of cellular and molecular immunology* **26**(11), 1082–1084.

Lu, A. P., Ding, X. R. and Chen, K. J. (2008). Current situation and progress in integrative medicine in China. *Chinese Journal of Integrative Medicine* **14**, 234–240.

Lu, C., Zha, Q., Chang, A., He, Y. and Lu, A. (2009). Pattern differentiation in traditional Chinese medicine can help define specific indications for biomedical therapy in the treatment of rheumatoid arthritis. *Journal of Alternative and Complementary Medicine* **15**, 1021–1025.

Ma, J., Dey, M., Yang, H., Poulev, A., Pouleva, R., Dorn, R., Lipsky, P. E., Kennelly, E. J. and Raskin, I. (2007). Anti-inflammatory and immunosuppressive compounds from Tripterygium wilfordii. *Phytochemistry* **68**(8), 1172–1178.

Manek, N. J., Crowson, C. S., Ottenberg, A. L., Curlin, F. A., Kaptchuk, T. J. and Tilburt, J. C. (2010). What rheumatologists in the United States think of complementary and alternative medicine: Results of a national survey. *BMC Complementary and Alternative Medicine* **10**, 5.

Manheimer, E., Wieland, S., Kimbrough, E., Cheng, K. and Berman, B. M. (2009). Evidence from the cochrane collaboration for traditional Chinese medicine therapies. *Journal of Alternative and Complementary Medicine* **15**, 1001–1014.

Marks, W. H. (2011). Tripterygium wilfordii Hook F versus Sulfasalazine in the treatment of rheumatoid arthritis: A well-designed clinical trial of a botanical demonstrating effectiveness. *Fitoterapia* **82**(1), 85–87.

Matta, R., Wang, X., Ge, H., Ray, W., Nelin, L. D. and Liu, Y. (2009). Triptolide induces anti-inflammatory cellular responses. *American Journal of Translational Research* **1**(3), 267–282.

McConkey, B. (2003). Adrenal corticosteroids in Chinese herbal remedies. *QJM: Monthly Journal of the Association of Physicians* **96**(1), 81–82.

Min, W. Q., Wei, Q., Li, H. Y., Zhang, Z. C., Wu, L. J., Yuan, G. H., Dou, C. R. and Shi, G. Y. (2005). A clinical study of total gluosides palony in the treatment of rheumatoid arthritis: A multicenter trial. *Chinese Journal of Rheumatology* **9**, 487–491.

Osborn, C., Baxter, G. D., Barlas, P. and Barlow, J. (2004). Complementary and alternative medicine and rheumatology nurses: A survey of current use and perceptions. *Nursing Times* **9**(2), 110–119.

Ou, Y. Q., Li, W. D., Li, X. J., Lin, Z. B. and Li, M. (2011). Sinomenine reduces invasion and migration ability in fibroblast-like synoviocytes cells co-cultured with activated human monocytic THP-1 cells by inhibiting the expression of MMP-2, MMP-9, CD147. *Rheumatology International* **31**, 1479–1485.

Qi, B. J. and Liu, C. J. (2010). Systems evaluation on traditional Chinese medicine in treating rheumatoid arthritis contrasts to methotrexate. *Modern Journal of Integrated Traditional Chinese and Western Medicine* **19**, 2214–2217.

Ramgolam, V., Ang, S. G., Lai, Y. H., Loh, C. S. and Yap, H. K. (2000). Traditional Chinese medicines as immunosuppressive agents. *Annals of the Academy of Medicine, Singapore* **29**(1), 11–16.

Scott, D. L., Wolfe, F. and Huizinga, T. W. J. (2008). Rheumatoid arthritis. *Lancet* **376**, 1094–1108.

Setty, A. R. and Sigal, L. H. (2005). Herbal medications commonly used in the practice of rheumatology: Mechanisms of action, efficacy, and side effects. *Seminars in Arthritis and Rheumatism* **34**(6), 773–784.

Shang, W., Wu, J. H. and Cai, H. (2010). Meta-analysis of total glucosides of paeony combined with methotrexate in treatment for rheumatoid arthritis. *Modern Journal of Integrated Traditional Chinese and Western Medicine* **19**, 653–656.

Shen, H. B., Bai, Y. J., Huo, Z. J., Li, W. N. and Tang, X. P. (2011). Assessment of clinical effect of therapy combining disease with syndrome on rheumatoid arthritis. *Journal of Traditional Chinese Medicine* **31**(1), 39–43.

Shi, Y., Zhang, H. and Du, X. (1999). A double blind observation for therapeutic effects of the tong luo kai bi tablets on rheumatoid arthritis. *Journal of Traditional Chinese Medicine* **19**(3), 166–172.

Sun, B., Lv, L., Lu, Z. X. and Yang, S. Y. (2008). Study on drug-guide effect of Achyranthes bidentata in Sanmiao pill in arthritic rats. *Zhongguo Zhong Yao Za Zhi* **33**(24), 2946–2949.

Sylvester, J., Liacini, A., Li, W., Dehnade, F. and Zafarullah, M. (2001). Tripterygium wilfordii Hook F extract suppresses proinflammatory cytokine-induced expression of matrix metalloproteinase genes in articular chondrocytes by inhibiting activating protein-1 and nuclear factor-kappaB activities. *Molecular Pharmacology* **59**, 1196–1205.

Tao, X. and Lipsky, P. E. (2000). The Chinese anti-inflammatory and immunosuppressive herbal remedy Tripterygium wilfordii Hook F. *Rheumatic Diseases Clinics of North America* **26**(1), 29–50.

Tao, X. L., Sun, Y., Dong, Y., Xiao, Y. L., Hu, D. W., Shi, Y. P., Zhu, Q. L., Dai, H. and Zhang, N. Z. (1989). A prospective, controlled, double-blind, crossover study of *Tripterygium wilfordii* hook F in treatment of rheumatoid arthritis. *Chinese Medical Journal* **102**, 327–332.

Tao, X., Cush, J. J., Garret, M. and Lipsky, P. E. (2001). A phase I study of ethyl acetate extract of the Chinese antirheumatic herb Tripterygium wilfordii hook F in rheumatoid arthritis. *Journal of Rheumatology* **28**(10), 2160–2167.

Tao, X., Younger, J., Fan, F. Z., Wang, B. and Lipskey, P. E. (2002). Benefit of an extract of *Tripterygium Wilfordii* Hook F in patients with rheumatoid arthritis. *Arthritis and Rheumatism* **46**, 1735–1743.

Tong, L. and Moudgil, K. D. (2007). Celastrus aculeatus Merr. suppresses the induction and progression of autoimmune arthritis by modulating immune response to heat-shock protein 65. *Arthritis Research and Therapy* **9**(4), R70.

Tsang, I. K. (2007). Establishing the efficacy of traditional Chinese medicine. *Nature Clinical Practice Rheumatology* **3**, 60–61.

Tsang, I. K., Huang, S. H. and Koehler, B. E. (2010). Traditional Chinese medicine: An opportunity for discovery. *The Journal of Rheumatology* **37**, 1989–1990.

Wang, Y., Jia, L. and Wu, C. Y. (2008). Triptolide inhibits the differentiation of Th17 cells and suppresses collagen-induced arthritis. *Scandinavian Journal of Immunology* **68**(4), 383–390.

Wiseman, N. and Ye, F. (1998). A Practical Dictionary of Chinese Medicine. Paradigm, Brookline, MA123.

Xi Bao, Y., Kwok Wong, C., Kwok Ming Li, E., Shan Tam, L., Chung Leung, P., Bing Yin, Y. and Wai Kei Lam, C. (2006). Immunomodulatory effects of lingzhi and san-miao-san supplementation on patients with rheumatoid arthritis. *Immunopharmacology and Immunotoxicology* **28**(2), 197–200.

Xu, D. S. (1996). Clinical and experimental study on RA Mixture in treatment of rheumatoid arthritis. *Chinese Journal of Integrated Traditional and Western Medicine* **2**(3), 178–183.

Xu, W. H., Xu, B. J. and He, Y. T. (2004). Chinese herbal medicines versus non-steroidal anti-inflammatory drugs for rheumatoid arthritis: A systematic review. *Chinese Journal of Basic Medicine in Traditional Chinese Medicine* **10**, 20–25.

Xu, M., Liu, L., Qi, C., Deng, B. and Cai, X. (2008). Sinomenine versus NSAIDs for the treatment of rheumatoid arthritis: A systematic review and meta-analysis. *Planta Medica* **74**(12), 1423–1429.

Yang, M. Y. (2008). Traditional Chinese Medicine for Treating Rheumatoid Arthritis: A Systematic Review. Capital Medical University.

Yang, D. S., Liu, F., Zeng, F. D. and Chen, H. (2005). Effect of sinomenine on adjuvant arthritis and its mechanisms. *Zhongguo Zhong Yao Za Zhi* **30**(17), 1361–1363.

Yu, C. (2009). The Systematic Review of the Recent Efficacy for the Treatment of RA with Chinese Herbal Alone or Chinese Herbal with Western Medicine. Chengdu University of Traditional Chinese Medicine

Yu, F., Yu, F., Li, R. and Wang, R. (2004). Inhibitory effects of the Gentiana macrophylla extract on rheumatoid arthritis of rats. *Journal of Ethnopharmacology* **95**, 77–81.

Zhang, Y. L., Ouyang, G. L. and Xiao, L. B. (2009). Research advances of mechanism of sinomenine in treating rheumatoid arthritis. *Zhong Xi Yi Jie He Xue Bao* **7**(8), 775–778.

Zhang, P., Li, J., Han, Y., Yu, X. W. and Qin, L. (2010). Traditional Chinese medicine in the treatment of rheumatoid arthritis: A general review. *Rheumatology International* **30**(6), 713–718.

Zhang, C., Jiang, M. and Lu, A. P. (2011a). Chinese herbal medicines versus disease modifying antirheumatic drugs for management of rheumatoid arthritis: A systematic review. *European Journal of Integrative Medicine* **3**, e219–e231.

Zhang, C., Jiang, M. and Lu, A. P. (2011b). Evidence based Chinese medicine for rheumatoid arthritis. *Journal of Traditional Chinese Medicine* **31**(2), 159.

Zhang, C., Jiang, M. and Lu, A. P. (2011c). A traditional Chinese medicine versus Western combination therapy in the treatment of rheumatoid arthritis: Two-stage study protocol for a randomized controlled trial. *Trials* **12**, 137.

Zheng, G., Jiang, M., He, X. J., Zhao, J., Guo, H. T., Chen, G., Zha, Q. L. and Lu, A. P. (2011). Discrete derivative: A data slicing algorithm for exploration of sharing biological networks between rheumatoid arthritis and coronary heart disease. *BioData Mining* **4**, 18.

Zhong, X., Su, N. and Chen, S. L. (2010). Meta-analysis of the efficacy and safety of total glucosides of paeony combined with immunosuppressant in the treatment of rheumatoid arthritis. *China Pharmacy* **21**, 3731–3734.

Zhou, Q. and Li, Z. G. (2003). Pharmacology and therapeutic usages of total glucosides of paeony in autoimmune diseases. *Zhongguo Xin Yao Yu Lin Chuang Za Zhi* **22**, 687–691.

Zhou, X. P., Zhou, Z. Y., Jin, M. W., Guo, L. W., Fang, T. H., Wang, H., Xu, H. Q. and Wu, M. H. (1999). Clinical and experimental study on treatment of mid-late stage rheumatoid arthritis with Shu Guan Granules. *Chinese Journal of Integrated Traditional and Western Medicine* **5**(3), 165–170.

Zhou, D., Hou, Q., Si, Q., Liu, J. and Yang, H. (2010). Concentrations of the active constituents of the Tibetan folk medicine Qinjiao (Gentiana sect. Cruciata) within and between taxonomic species across the Qinghai-Tibetan Plateau. *Chemistry and Biodiversity* **7**, 2088–2094.

Zhu, L., Wei, W., Zheng, Y. Q. and Jia, X. Y. (2005). Effects and mechanisms of total glucosides of paeony on joint damage in rat collagen-induced arthritis. *Inflammation Research* **54**(5), 211–220.

# *Ligusticum chuanxiong* and Its Decoctions: Effects and Underlying Mechanisms of Action on Stroke

CINDY L. H. YANG,[*,1] TERRY C. T. OR,[*,†,1]
JONATHAN S. H. LAU[‡] AND ALLAN S. Y. LAU[*,†,2]

**Molecular Chinese Medicine Laboratory, Li Ka Shing Faculty of Medicine, The University of Hong Kong, Pokfulam, Hong Kong Special Administrative Region, China*

*[†]Cytokine Biology Group, Department of Paediatrics and Adolescent Medicine, The University of Hong Kong, Pokfulam, Hong Kong Special Administrative Region, China*

*[‡]School of Biological Sciences, The University of Hong Kong, Pokfulam, Hong Kong Special Administrative Region, China*



[1]Contributed equally to this work.
[2]Corresponding author: E-mail: asylau@hkucc.hku.hk

Advances in Botanical Research, Vol. 62

0065-2296/12 $35.00
DOI: 10.1016/B978-0-12-394591-4.00006-4

# ABSTRACT

Stroke is not only one of the leading causes of mortality but also associated with long-term morbidity worldwide. Due to the adverse side effects and short therapeutic window of current drugs, traditional Chinese medicine (TCM) is getting more popular to treat stroke-related diseases. In this review, we will summarize recent publications of pharmacological effects of several medicinal decoctions containing *Ligusticum chuanxiong* (LCX) and its active ingredients on cerebrovascular diseases. TCM decoctions, consisting of different component herbs, have been used clinically for more than thousands of years in China. LCX, commonly found in these decoctions, is traditionally used to invigorate blood circulation and promote the flow of *Qi* according to the practice of TCM. *Qi* may be interpreted as vital energy in Chinese medicine. However, recent biomedical researches showed that LCX had anti-inflammatory, anti-oxidative and angiogenic effects. Its main ingredients including tetramethylpyrazine, Z-ligustilide and ferulic acid have been proposed to be responsible for the bioactivities of LCX. By applying modern science and technology, the efficacy of the formulae can be proven by studying their active ingredients and also determine the mechanisms of action. All these research findings will help botanical medicine to be recognized by the public.

# ABBREVIATIONS

| | |
|---|---|
| EPCs | endothelial progenitor cells |
| HIF | hypoxia-inducible factor |
| ICAM-1 | intercellular adhesion molecule-1 |
| IL-1β | interleukin-1 beta |
| iNOS | inducible nitric oxide synthase |
| LCX | *Ligusticum chuanxiong* |
| LPS | lipopolysaccharide |
| MCAO | middle cerebral artery occlusion |
| NF-κB | nuclear factor κB |
| NPCs | neural progenitor cells |
| Nrf2 | erythroid 2-related factor 2 |
| $PGE_2$ | prostaglandin E2 |
| PI3K | phosphatidylinositol 3-kinase |
| ROS | reactive oxygen species |
| TNF-α | tumour necrosis factor-alpha |

| | |
|---|---|
| VEGF | vascular endothelial growth factor |
| VSMCs | vascular smooth muscle cells |

# I. INTRODUCTION

Stroke is one of the leading causes of mortality and the primary cause of long-term disability worldwide. According to American Heart Association statistics, approximately 795,000 people experience a new or recurrent stroke every year, which gives rise to great socioeconomic burdens in United States and other countries (Lloyd-Jones, 2009). There are two major types of stroke: haemorrhagic stroke and ischaemic stroke. Ischaemic stroke, representing 87% of strokes (Lloyd-Jones, 2009), is due to restriction of blood flow to the brain. It results in insufficient supply of oxygen and nutrients including glucose to normal cellular homeostasis. This leads to a complex cascade of pathophysiological events which ultimately cause neuro-injuries in the ischaemic core within minutes of the onset of ischaemia (Barone and Feuerstein, 1999; Muir *et al.*, 2007; Rosamond *et al.*, 2007; Samson *et al.*, 2005). Many studies have been done to prevent and treat ischaemic stroke. In this review, different therapeutic approaches for treating stroke including inflammation, oxidative stress, atherosclerosis and angiogenesis will be discussed.

## A. INFLAMMATION, OXIDATIVE STRESS AND APOPTOSIS

Cerebral ischaemia causes cell necrosis in the infarcted area of the brain. These necrotic cells were found to trigger inflammatory responses (Yenari *et al.*, 2007). The process starts with the activations of immune cells including leukocytes and microglia to produce a variety of inflammatory mediators such as inducible nitric oxide synthase (iNOS) which generates nitric oxide, cytokines, reactive oxygen species (ROS) and matrix metalloproteinases (MMPs; Zheng and Yenari, 2004). Nitric oxide then combines with superoxide to form peroxynitrite which is a potent free radical. Together with other inflammatory mediators including ROS and tumour necrosis factor-alpha (TNF-$\alpha$), apoptosis is stimulated. Cytochrome *c* is released from the mitochondrial membrane. This in turn activates the downstream caspases. Ultimately, cells die from DNA fragmentation (Stenzel-Poore *et al.*, 2008). Moreover, cytokines produced by activated immune cells up-regulate the production of adhesion molecules that can facilitate adhesion of circulating leukocytes to vascular endothelia and let them infiltrate into the brain parenchyma to cause even more severe injuries (Shohami *et al.*, 1999).

Therefore, anti-inflammatory and protective of cell death against oxidative stress are important therapeutic approaches to treat ischaemic stroke-related diseases.

### B. Atherosclerosis and Angiogenesis

Besides inflammatory responses on immune cells, vascular smooth muscle cells (VSMCs) play a fundamental role in atherosclerosis (Ross and Glomset, 1973; Schwartz *et al.*, 1986). The accelerated proliferation of VSMCs has been suggested to be a characteristic feature of atherogenesis (Campbell and Campbell, 1985). In addition, neurorestorative therapies have been gaining attention recently to treat acute ischaemic strokes in the brain (Chen and Chopp, 2006). Angiogenesis is one of the important processes to restore oxygen and nutrient supply to the affected brain tissue and provide neurotrophic support to newly generated neurons (Leventhal *et al.*, 1999). As a consequence, VSMCs and endothelial cells are also important therapeutic targets on controlling stroke-related diseases.

### C. Traditional Chinese Medicine as an Alternative Medicine for Stroke

At present, few therapeutic options are available for the treatment of stroke. For instance, anti-platelet regimens including aspirin, extended-release dipyridamole and clopidogrel that block platelet aggregation to prevent recurrent cerebrovascular events (Coccheri, 2010). However, these drugs are associated with side effects including major gastrointestinal bleeding, headache, allergy and thrombotic thrombocytopenia purpura (Guthrie, 2011). The recombinant tissue-type plasminogen activator (rt-PA) therapy has lower efficacies due to its limited 3-h therapeutic window and the major side effects of haemorrhaging (Bluhmki *et al.*, 2009; Hacke *et al.*, 2004a). Other drugs were withdrawn from clinical trials in humans due to serious adverse side effects. Therefore, the development of new agents for ischaemic stroke treatment is necessary.

Traditional Chinese medicine (TCM) has been used to treat stroke for more than thousands of years in China. In helping the stroke patients, acupuncture and herbal treatments are often used, either single herbs or decoctions. The decoction formulae are not simply combined by a mixture of medicinal herbs. In fact, they are guided by TCM theory to select component herbs in the formula for treating particular diseases (Jia *et al.*, 2004). There are several prescribed formulae used for the treatment of ischaemic stroke including Buyang Huanwu decoction (BYHWD), Xuefu Zhuyu decoction (XFZYD), Taohong Siwu tang (THSWT), Xiao Xu Ming decoction (XXMD) and Shaofu Zhuyu decoction (SFZYD). The effects and mechanism of actions of these formulae are summarized in Table I.

TABLE I

*Summary of the Effects of Different Decoctions on Stroke-Related Diseases*

| Decoctions | Experimental models | Effects | References |
|---|---|---|---|
| Buyang Huanwu decoction | • Neural progenitor cells | – Promoting growth and differentiation of neural progenitor cells | Cai *et al.* (2007), Sun *et al.* (2007) |
| | • $H_2O_2$ injured Schwann 204 cells | – Rescuing the SOD activity decrease and MDH level increase | Yang *et al.* (2011) |
| | • Rats with focal cerebral ischaemic–reperfusion | – Reducing the production of inflammatory cytokine<br>– Inhibiting of caspase-1 expression | Tang *et al.* (2006) |
| | • Rats with cerebral infarction | – Reducing protein and mRNA expressions of cytokines including IL-1β and TNF-α | Yi *et al.* (2010) |
| | • HPX–XOD reaction system and Fenton reaction system | – Suppressing the superoxide and hydroxyl radicals | Lin *et al.* (2000) |
| | • Rats with transient forebrain ischaemia | – Attenuating the number of TUNEL-positive neurons<br>– Suppressing the expression of caspase-3p20 | Li *et al.* (2003) |
| | • Rats with cerebral infarction | – Promoting nerve proliferation<br>– Increasing the number of BrdU-immunoreactive cells | Tang *et al.* (2006), Tong *et al.* (2007) |
| | • Rats with cerebral ischaemia/reperfusion | – Increasing proliferation and differentiation of neural stem cells | Gao *et al.* (2009) |
| | • Gerbil with cerebral ischaemia | – Enhancing the function of hippocampal microvessels<br>– Increasing microvascular density | Wu *et al.* (2011b), Zhao *et al.* (2011) |

(*continues*)

TABLE I (*continued*)

| Decoctions | Experimental models | Effects | References |
|---|---|---|---|
| | • Rats with cerebral ischaemia | – Up-regulating VEGF mRNA and protein expression<br>– Promoting angiogenesis<br>– Enhancing the expressions of VEGF and Flk | Cai and Liu (2010), Cai *et al.* (2007) |
| Xuefu Zhuyu decoction | • Endothelial progenitor cells (EPCs) | – Inducing EPCs angiogenesis | Gao *et al.* (2010a) |
| | • Rats with thromboembolic stroke | – Inhibiting HIF-1α, iNOS, TNF-α and caspase-3 | Lee *et al.* (2011) |
| Taohong Siwu Tang | • MCAO-induced focal cerebral ischaemia in rats<br>• Glucocorticoid-induced ischemic necrosis of rabbits | – Inhibiting HIF-1α, iNOS, TNF-α and caspase-3<br>– Inhibiting of platelet activation | Wu *et al.* (2011a) |
| Xiao Xu Ming decoction | • Oxygen–glucose deprivation primary cultured neurons | – Enhancing the expression of VEGF<br>– Inhibiting apoptotic neuronal death | Qi and Chen (2009), Zhu *et al.* (2011) |
| | • Four-vessel occlusion in adult Wistar rats | – Increasing the density of neurons in the CA1 region of hippocampus of rats | Zhu *et al.* (2010) |
| Shaofu Zhuyu decoction | • $H_2O_2$-induced rat vascular smooth muscle cells (VSMCs) | – Increasing cell survival and reducing apoptosis | Liu *et al.* (2010a) |

## II. PHARMACOLOGICAL EFFECTS OF TCM DECOCTIONS ON ISCHAEMIC STROKE

### A. BUYANG HUANWU DECOCTION

BYHWD, a well-known traditional Chinese formula mentioned in "Yilin Gaicuo" (Correction on Errors in Medical Classics), is composed of *Radix Astragalis seu Hedysari*, *Radix Angelicae Sinensis*, *Radix Paeoniae Rubra*,

*Rhizoma Ligustici Chuanxiong*, *Flos Carthami*, *Semen Persicae* and *Pheretima* as recorded in the Chinese Pharmacopoeia. It is famous for its putative effects for the prevention and treatment of ischaemic cardio-cerebral vascular diseases for centuries. The therapeutic effects of BYHWD have been studied extensively (Li *et al.*, 2003; Lin *et al.*, 2000; Tang *et al.*, 2006; Yi *et al.*, 2010).

BYHWD has been demonstrated to inhibit inflammatory responses. One of the studies showed that BYHWD could reduce the production of inflammatory cytokines through the inhibition of caspase-1 in rats after focal cerebral ischaemic–reperfusion (Tang *et al.*, 2006). With BYHWD treatment, the protein and mRNA expressions of interleukin-1 beta (IL-1β) and TNF-α were reduced after cerebral infarction (Yi *et al.*, 2010). These results revealed that BYHWD protected the brain by modulating expressions of pro-inflammatory mediators (Yi *et al.*, 2010).

Evidences have accumulated to show that reactive oxygen radicals are involved in brain injuries including cerebral ischaemia and reperfusion (Sugawara and Chan, 2003). *In vitro*, BYHWD has powerful free radical scavenging activity by suppressing the production of superoxide and hydroxyl radicals generated from the hypoxanthine-xanthine oxidase reaction system and Fenton reaction system, respectively (Lin *et al.*, 2000). Further, BYHWD exhibited a protective effect through increasing superoxide dismutases activity and decreasing malondialdehyde level on $H_2O_2$ injured Schwann 204 cells (Yang *et al.*, 2011). *In vivo*, BYHWD markedly attenuated the number of TUNEL-positive neurons and suppressed the expression of caspase-3p20 to protect against ischaemic injury in the hippocampal CA1 region (Li *et al.*, 2003).

BYHWD could promote nerve proliferation in rats with ischaemic stroke as indicated by increasing the number of BrdU-immunoreactive cells in the hippocampal region of rats (Tan *et al.*, 2006; Tong *et al.*, 2007). Other studies demonstrated that BYHWD promoted growth and differentiation of neural progenitor cells (NPCs) derived from embryonic hippocampus (Cai *et al.*, 2007; Sun *et al.*, 2007). The average neurite length of NPCs grew significantly longer in culture medium containing BYHWD than that of the control serum (Cai *et al.*, 2007; Sun *et al.*, 2007). Another study demonstrated that the actions of BYHWD against cerebral ischaemia–reperfusion damage were correlated with proliferation and differentiation of neural stem cells in the hippocampal dentate gyrus (Gao *et al.*, 2009).

BYHWD promoted angiogenesis by enhancing the function of hippocampal microvessels, preventing injury, increasing microvascular density and microvascular area density (Wu *et al.*, 2011b; Zhao *et al.*, 2011). Further, it could up-regulate vascular endothelial growth factor (VEGF) mRNA and protein expression in ischaemic cerebral regions after being treated for 7–14 days (Cai and Liu, 2010). Similar study demonstrated that the therapeutic effects of BYHWD for recovery of neurological deficits were associated with the enhancement of the expressions of VEGF and its receptor (Flk) in ischaemic brains (Cai *et al.*, 2007).

### B. XUEFU ZHUYU DECOCTION

XFZYD is a well-known TCM formula for treating cardiovascular diseases. Recently, scientific studies have been demonstrated that this formula had neuroprotective effects on ischaemic diseases (Gao *et al.*, 2010a). However, it is more complicated to use as the formulation consists of 11 herbs including *Radix Angelicae Sinensis*, *Radix Bupleuri*, *Flos Carthami Tinctorii*, *Pericarpium Citri Reticulatae*, *Radix Cyathulae*, *Radix Glycyrrhizae Uralensis*, *Rhizoma Ligustici Chuanxiong*, *Radix Paeoniae Lactiflorae*, *Radix Platycodi grandiflori*, *Semen Persicae* and *Radix Rehmanniae*.

XFZYD had neuroprotective effects against ischaemic stroke by inhibiting of hypoxia-inducible factor (HIF)-1α, as well as inflammatory responses including iNOS and TNF-α in a rat model (Lee *et al.*, 2011). It can inhibit apoptosis by suppressing caspase-3 cleavage (Lee *et al.*, 2011). Further, it can induce angiogenesis via promoting proliferation of endothelial progenitor cells (EPCs; Gao *et al.*, 2010a).

### C. TAOHONG SIWU TANG

THSWT is another famous TCM formula for treating cardiovascular diseases. The formula consists of six herbs including *Radix Rehmanniae*, *Radix Paeoniae Lactiflorae*, *Radix Angelicae Sinensis*, *Rhizoma Ligustici Chuanxiong*, *Semen Persicae* and *Flos Carthami Tinctorii*.

THSWT possessed potent neuroprotection against middle cerebral artery occlusion (MCAO)-induced focal cerebral ischaemia in an *in vivo* rat model (Wu *et al.*, 2011a). The effect might be mediated by the inhibition of both HIF-1α and inflammatory responses including TNF-α and iNOS expressions (Wu *et al.*, 2011a). Moreover, it suppressed apoptosis through caspase-3 inhibition and platelet activation, resulting in a reduction in the infarct volume of the animals upon ischaemia–reperfusion brain injury (Wu *et al.*, 2011a). Another study also showed that THSWT had protective effects on glucocorticoid-induced ischaemic necrosis of femoral head of rabbits by enhancing the expression of VEGF to facilitate the generation of blood vessels and alleviate and improve ischaemia of the osseous tissue (Qi and Chen, 2009).

### D. XIAO XU MING DECOCTION

XXMD has been used clinically to treat stroke in China since Tang Dynasty (around 650 A.D.). The neuroprotective effects of XXMD were confirmed in *in vitro* and *in vivo* stroke models (Zhu and Gao, 2003; Zhu *et al.*, 2010). In an oxygen–glucose deprivation model, both XXMD extract and XXMD-containing serum inhibited apoptotic neuronal death (Zhu *et al.*, 2010). For the *in vivo* model, oral administration of XXMD can dose-dependently increase the density of neurons that survived in the CA1 region of hippocampus of adult Wistar rats. The neuroprotective effects of XXMD may be due to the inhibition of caspase-3 activity and up-regulation of anti-oncogene Bcl-2 expression in CA1 region after transient forebrain ischaemia. Further, XXMD ameliorated severe deficiencies in spatial cognitive performance induced by transient global ischaemia (Zhu *et al.*, 2010).

### E. SHAOFU ZHUYU DECOCTION

SFZYD is a TCM formula for treating blood stasis syndrome in the Qing Dynasty of China. SFZYD has been demonstrated to have protective effects on $H_2O_2$-induced oxidative injuries on rat VSMCs. One of the fractions (SF-7) of SFZYD significantly increased cell survival and reduced apoptosis of VSMCs after induced by $H_2O_2$ (Liu *et al.*, 2010a). Moreover, SF-7 could markedly increase intracellular superoxide dismutase activity and decrease the malondialdehyde level in $H_2O_2$-injured VSMCs, and suppress the generation of intracellular ROS and intracellular $Ca^{2+}$ concentration (Liu *et al.*, 2010a).

## III. PHARMACOLOGICAL EFFECTS OF *LIGUSTICUM CHUANXIONG* ON ISCHAEMIC STROKE

*Ligusticum chuanxiong* (LCX) is traditionally used to invigorate blood circulation and promote the flow of *Qi*, the meaning of vital energy in TCM. In TCM practice, it has been prescribed alone or as a component herb in some of the classical TCM formulae for the treatment of cerebrovascular diseases including ischaemic stroke, hypertension and arrhythmia (Tang and Eisenbrand, 1992). As LCX has an excellent clinical efficacy, the pharmacological effect of LCX was studied extensively as an anti-platelet (Chen and Chen, 1992; Li *et al.*, 2001; Liu, 1991), anti-coagulant (Wang *et al.*, 2008), anti-oxidant (Chao *et al.*, 2008; Cheng *et al.*, 2008a), anti-inflammatory

(Cheng *et al.*, 2008a; Fan *et al.*, 2011; Or *et al.*, 2011) and pain relief agent (Wang *et al.*, 2011b).

LCX was reported to reduce cerebral ischaemia by improving the changes of dynorphin A1-13 like immunoreactivity (ir-Dyn A1-13) levels in plasma and cerebrospinal fluid in a rabbit carotid occlusion model (Liu and Shi, 1990). The severity of brain ischaemic damage and neurologic dysfunction in LCX-treated animals was found to be lighter than that of saline-treated group (Liu and Shi, 1990). Further, LCX has better protective effects on ischaemic strokes than control drugs including papaverine, dextran and aspirin-persantin (Chen and Chen, 1992). It was demonstrated to improve brain microvascular circulation in patients with transient ischaemic attacks, through inhibiting the thrombus formation and platelet aggregation (Chen, 1992). LCX improved neurological function and reduced infarct size of the ischaemic cerebral tissues by increasing VEGF expression. Taken together, LCX could be used to protect against coronary and cerebral ischaemia (Cheng *et al.*, 2008a; Tian *et al.*, 2005).

LCX contains three major groups of compounds including phenolic acid, alkaloid and phthalide that have been proven to have pharmacological bioactivity. Among them, ferulic acid (FA; Chen *et al.*, 2010; Hsieh *et al.*, 2008a,b, 2010), tetramethylpyrazine (TMP; Fan *et al.*, 2011; Gao, 1989; Ho *et al.*, 1989; Liao *et al.*, 2004; Zhu *et al.*, 2009), Z-ligustilide (Z-lig; Du *et al.*, 2007, 2009a; Or *et al.*, 2011) and senkyunolide A (Or *et al.*, 2011) have been reported to be the main active ingredients of LCX for treating ischaemic stroke. Their structures are illustrated in Fig. 1 and the pharmacological actions of these compounds are summarized in Figs. 2 and 3 and Table II.

## IV. ACTIVE COMPOUNDS ISOLATED FROM LCX

### A. TETRAMETHYLPYRAZINE

TMP is one of the most important active compounds in LCX. It has been used for treating ischaemic stroke for decades (Lu *et al.*, 1978). TMP have been shown to exert anti-inflammatory effects in many cell types. In a recent report, TMP can suppress the production of nitric oxide in lipopolysaccharide (LPS)-stimulated N9 microglial cells by reducing the mRNA and protein expression of iNOS (Yu *et al.*, 2010b). Further investigations showed that the suppression may be due to the inhibition of phosphorylation of mitogen-activated protein kinase (MAPKs) including p38, ERK1/2 and JNK.

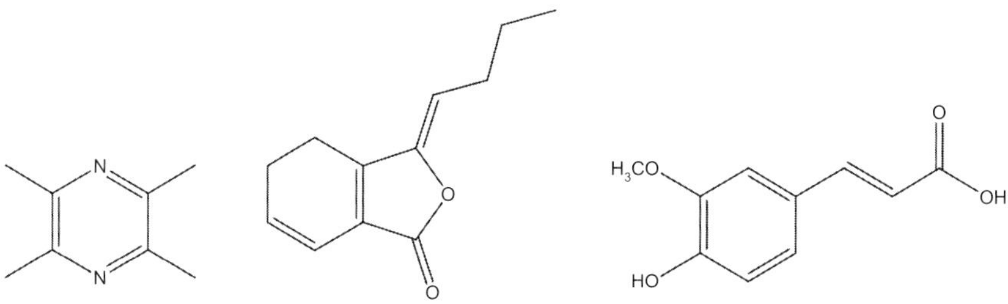

(1) (2) (3)

Fig. 1. The chemical structures of bioactive compounds of *Ligusticum chuanxiong* (LCX), which are (1) tetramethylpyrazine (TMP), (2) Z-ligustilide (Z-lig) and (3) ferulic acid (FA), respectively.

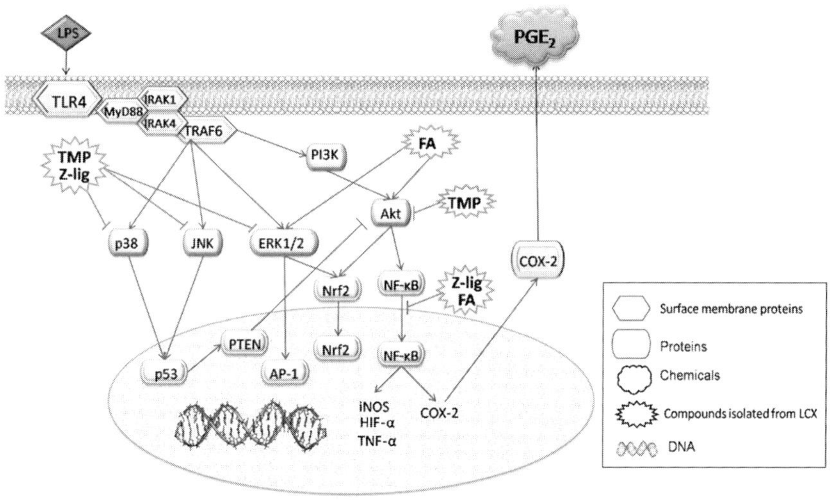


Fig. 2. Summary of the effects of tetramethylpyrazine (TMP), Z-ligustilide (Z-lig) and ferulic acid (FA) on the signalling pathways involved in inflammatory responses of the cell. (See colour plate section at the back of the book.)

Moreover, the phosphorylation of Akt but not phosphatidylinositol 3-kinase (PI3K) was decreased (Yu *et al.*, 2010b). Besides N9 microglial cell line, TMP was found to exert anti-inflammatory effects on primary rat glial cells. It can suppress the inflammatory mediators including prostaglandin E2 ($PGE_2$) production in glial cells co-stimulated with LPS and IFN-$\gamma$ (Liao *et al.*, 2004). Hence, the anti-inflammatory functions of TMP alleviate the severity of neuro-degenerative diseases including stroke. TMP also possess protective effects upon oxidative stress and excitotoxicity damage on neurons. In another study, TMP was shown to protect primary rat cerebellar granule cells

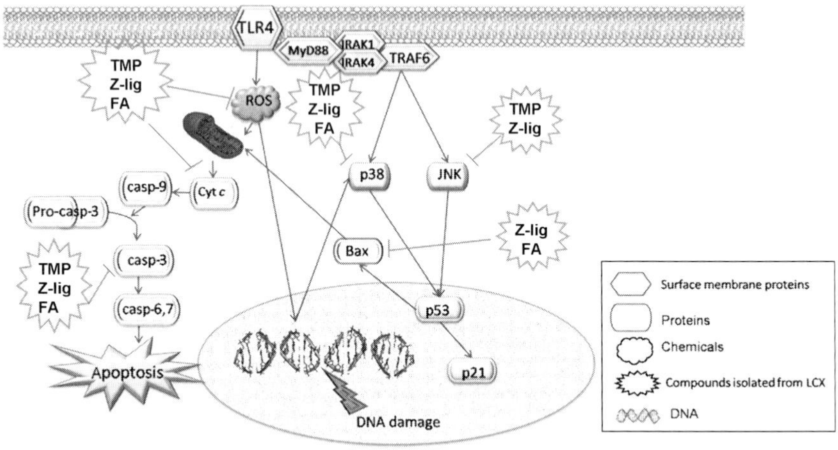


Fig. 3. Summary of the effects of tetramethylpyrazine (TMP), Z-ligustilide (Z-lig) and ferulic acid (FA) on the signalling pathways involved in oxidative stress-induced injuries. (See colour plate section at the back of the book.)

from oxidative stress challenged by $FeSO_4$. Apoptosis was reduced by inhibiting the caspase-3 activity and the gene expressions of p53 and $p21^{waf1/cip1}$ in the *in vitro* model (Zhang *et al.*, 2003). Moreover, TMP could exert protective effects on excitotoxicity induced by kainite on primary rat hippocampal neuronal cells by reduction of intracellular ROS (Shih *et al.*, 2002).

Oxidative stress in endothelial cells was found to correlate with atherosclerosis (Witztum, 1994). In a recent study, TMP could reduce the intracellular ROS formation on $H_2O_2$-induced endothelial cells by increasing activities of superoxide dismutase and glutathione peroxidase (Yu *et al.*, 2010a). Applying artificial cyclic strain on endothelial cells, oxidative stress was induced. TMP could suppress the formation of ROS by inhibiting endothelin secretion via reduced phosphorylation of ERK (Bi *et al.*, 2005). On the other hand, vasorelaxation of endothelial cells provide beneficial roles on the prevention of atherosclerosis. Moreover, the isolated rat aorta contracted when induced by norepinephrine. However, TMP could cause acute relaxation through the opening of ATP-dependent potassium channels (Rhyu *et al.*, 2010).

TMP has been studied in some ischaemic stroke models. The most commonly used is MCAO model. It was demonstrated by different groups that TMP could exert neuroprotection and reduce infarct area on rats after MCAO surgery (Hu *et al.*, 2009; Kao *et al.*, 2006; Liao *et al.*, 2004; Sheu *et al.*, 2007). TMP could suppress the expressions of TNF-$\alpha$ and HIF-1$\alpha$ in the infarct area of the brain after MCAO (Sheu *et al.*, 2007). Further, TMP

TABLE II

*Summary of the Effects of Different Bioactive Compounds of LCX on Stroke-Related Diseases*

| Tetramethylpyrazine (TMP) | *In vitro/ex vivo* | | | |
|---|---|---|---|---|
| Cells used (*in vitro*) | Inducer(s) | Functions | Mechanisms | References |
| Rat primary glial cells | LPS + IFN-γ | ↓ $PGE_2$ | Nil | Liao *et al.* (2004) |
| Murine N9 microglial cells | LPS | ↓ Nitric oxide and iNOS expression | ↓ Nuclear translocation of NF-κB<br>↓ Phosphorylation of p38 MAPK, ERK1/2, JNK and Akt<br>↓ ROS formation | Yu *et al.* (2010b) |
| Rat cerebellar granule cells | Iron (II) sulphate | ↓ Oxidative stress-induced apoptosis | ↓ Caspase-3 activity<br>↓ p53 and $p21^{waf1/cip1}$ | Zhang *et al.* (2003) |
| Primary rat hippocampus neurons | Kainate | Protect the cells from excitotoxicity | ↓ ROS formation | Shih *et al.* (2002) |
| Primary rat neural stem cells | Hypoxia | ↑ Proliferation and differentiation of NSC | ↑ The phosphorylation of ERK<br>↓ The phosphorylation of p38 | Liu *et al.* (2010c) |
| HUVEC | Cyclic strain controlled by computer | ↓ ET-1 | ↓ ROS formation<br>↓ phosphorylation of ERK | Bi *et al.* (2005) |
| HUVEC | LPS | ↓ IL-8 | ↓ Phosphorylation of ERK, p38<br>↓ Nuclear translocation of NF-κB | Yu *et al.* (2009) |
| HUVEC | Hydrogen peroxide | Protect the cells against oxidative stress damages | ↑ SOD and glutathione peroxidase<br>↓ Malondialdehyde, intracellular NO and NOS<br>↓ Cell cycle arrest in S phase | Yu *et al.* (2010a) |
| Rat pulmonary microvascular endothelial cells | Hypoxia | ↓ RPMVEC monolayer permeability<br>↓ ROS, HIF-1α and VEGF | Nil | Zhang *et al.* (2011) |

(*continues*)

TABLE II (*continued*)

| Tetramethylpyrazine (TMP) | *In vitro/ex vivo* | | | |
|---|---|---|---|---|
| Cells used (*in vitro*) | Inducer(s) | Functions | Mechanisms | References |
| Isolated rats aorta | Norepinephrine | ↑ Acute relaxation of stimulated aorta | Mediated by opening of ATP-dependent potassium channels | Rhyu *et al.* (2010) |

| Tetramethylpyrazine (TMP) | *In vivo* | | | |
|---|---|---|---|---|
| Type of animals (*in vivo*) | Disease model used (*in vivo*) | Functions | Mechanisms | References |
| Rats | MCAO | ↓ Infarction volume and brain oedemapreservation of neurons | Nil | Liao *et al.* (2004) |
| Rats | MCAO | ↓ Behavioural disturbance<br>↓ Neuronal loss and brain infarction | ↓ DNA fragmentation<br>↓ Caspase-8, caspase-9 and caspase-3 activation<br>↓ Cytochrome *c* release, Bcl-xL expression<br>↓ MCP-1 expression | Kao *et al.* (2006) |
| Rats | MCAO | ↓ Cerebral infarction volume and neurological deficit | ↓ The expression of TRX-1/TRX-2 and TRX reductase-1/TRX reductase-2 | Hu *et al.* (2009) |
| Rabbits | Subarachnoid haemorrhage | ↑ NO and eNOS expression<br>↑ Intra-endothelium calcium | Partially regulated via NO/cGMP signalling pathways | Liu *et al.* (2010b) |
| Rats | MCAO | ↓ Infarct area | ↓ HIF-1α, TNF-α<br>↓ Activated caspase-3 expression | Sheu *et al.* (2007) |

| Z-ligustilide (Z-lig) | *In vitro/ex vivo* | | | |
|---|---|---|---|---|
| Cells used (*in vitro*) | Inducer(s) | Functions | Mechanisms | References |
| Primary rat microglia | LPS | ↓ NO, TNF-α, IL-1β and MCP-1 | ↓ Nuclear translocation of NF-κB<br>↓ iNOS and COX-2 | Du *et al.* (2010) |

| | | | | |
|---|---|---|---|---|
| RAW 264.7 | LPS | ↓ NO, $PGE_2$ and TNF-α | ↓ Activation of AP-1 and NF-κB<br>↓ Degradation of IκBα<br>↓ Phosphorylation of IκB, p38 and JNK<br>↓ Intracellular ROS | Su *et al.* (2011) |
| Murine BV-2 microglia | LPS | ↓ NO and TNF-α | ↓ mRNA levels of iNOS and TNF-α | Or *et al.* (2011) |
| PC12 | Hydrogen peroxide | Attenuates $H_2O_2$-induced cell death | ↓ Intracellular ROS levels<br>↓ Bax expression<br>↓ Cleaved-caspase-3<br>↓ Cytochrome *c* release<br>↑ Bcl-2 expression | Du *et al.* (2008) |
| Primary rat vascular smooth muscle cells (VSMCs) | Nil | ↓ Proliferation and cell cycle progression | ↓ ROS formation<br>↓ The phosphorylations of p-38, JNK and ERK | Lu *et al.* (2006) |
| Rat mesenteric artery | KCl, $CaCl_2$, NA or 5-HT | ↓ Vasoconstriction | Mediated via VDCC and ROCC | Cao *et al.* (2006) |

| Z-ligustilide (Z-lig) | *In vivo* | | | |
|---|---|---|---|---|
| Type of animals (*in vivo*) | Disease model used (*in vivo*) | Functions | Mechanisms | References |
| Mice | Transient forebrain ischaemia | Neuroprotective effects with smaller infarct volume | ↓ MDA<br>↑ The activities of GSH-PX and SOD<br>↑ Bcl-2, ↓ Bax<br>↓ Caspase-3 | Du *et al.* (2006) |
| Rats | Arteriovenous shunt thrombosis | ↓ Arterial thrombus weight | Nil | Du *et al.* (2009b) |

(*continues*)

TABLE II (*continued*)

| Z-ligustilide (Z-lig) | *In vivo* | | | |
|---|---|---|---|---|
| Type of animals (*in vivo*) | Disease model used (*in vivo*) | Functions | Mechanisms | References |
| Rats | MCAO | ↓ Neurological deficit, infarct volume<br>↑ cell viability | ↑ Erythropoietin transcription via ERK pathways | Wu *et al.* (2011c) |
| Rats | Subarachnoid haemorrhage | ↓ Mortality, neurobehavioral deficits<br>↓ Brain oedema, BBB permeability<br>↓ Cerebral vasospasm | ↓ Number of apoptotic cells<br>↓ Expressions of p53<br>↓ Cleaved-caspase-3 | Wang *et al.* (2011a) |
| Rabbits | LPS-induced endotoxic shock | Neuroprotection against the endotoxic shock | ↓ TNF-α, IL-1β and NO | Zhu *et al.* (2011) |
| Rats | Amyloid β-induced neurotoxicity | ↓ Cognitive impairment and neuropathological damages | ↓ TNF-α<br>↓ NF-κB | Du *et al.* (2009a) |

| Ferulic acid (FA) | *In vitro* | | | |
|---|---|---|---|---|
| Cells used (*in vitro*) | Inducer(s) | Functions | Mechanisms | References |
| HUVEC | Gamma irradiation from 60 Co source | ↓ Oxidative stress | ↑ Translocation of Nrf2 via PI3K and ERK pathways | Gao *et al.* (2010b) |
| Primary rat cerebellar granule cells | Iron (II) Sulphate | ↓ Oxidative stress-induced apoptosis | ↓ Caspase-3 activity<br>↓ Gene expressions of p53 and p21waf1/cip1 | Zhang *et al.* (2003) |

| | | | | |
|---|---|---|---|---|
| HUVEC | Nil | Induces angiogenesis | ↑ Expressions of VEGF and PDGF<br>↑ Hypoxic-HIF-1α<br>Mediated through the MAPK and PI3K | Chen *et al.* (2010) |

| Ferulic acid (FA) | *In vivo* | | | |
|---|---|---|---|---|
| Type of animals (*in vivo*) | Disease model used (*in vivo*) | Functions | Mechanisms | References |
| Rats | MCAO | ↓ Infarct and improved neurological deficits | ↓ Superoxide radicals<br>↓ ICAM-1<br>↓ MPO<br>↓ NF-κB expression | Cheng *et al.* (2008b) |
| Rats | MCAO | Neuroprotection against the oxidative stress | ↓ Caspase-3<br>↓ ICAM-1 | Hsieh *et al.* (2008a) |
| Rats | MCAO | Provides neuroprotection against apoptosis | ↑ GABAB1<br>↓ p-38 mediated NO-induced apoptosis | Hsieh *et al.* (2010) |

was found to reduce the release of cytochrome *c* from the mitochondria as well as the expressions of caspase-3, caspase-8 and caspase-9. These led to reduced internucleosomal DNA fragmentation of the neurons, indicative of less apoptotic changes (Kao *et al.*, 2006).

### B. Z-LIGUSTILIDE

Z-lig is a phthalide component present in many medicinal plants of Umbelliferaceae family. It is also one of the major ingredients of LCX and *Angelica sinensis* with anti-asthmatic, spasmolytic and analgesic effects (Liang *et al.*, 2005; Qian *et al.*, 2006). Z-lig was found to inhibit the production of nitric oxide and TNF-α in LPS-stimulated murine microglial BV-2 cells by suppressing their mRNA transcriptions (Or *et al.*, 2011). It also exerted anti-inflammatory effects by inhibiting the translocation of nuclear factor κB (NF-κB) in primary rat microglial cells (Du *et al.*, 2010). Apart from microglial cells, Z-lig also can significantly reduce the production of nitric oxide, $PGE_2$ and TNF-α in LPS-stimulated RAW 264.7 cells (Su *et al.*, 2011), while the mRNA and protein expressions of iNOS were decreased. Further, activation of activator protein-1 and NF-κB was abrogated (Su *et al.*, 2011). Additionally, Z-lig inhibits the phosphorylation of MAPKs, including p38, ERK1/2 and JNK and the intracellular ROS level (Su *et al.*, 2011).

Z-lig could exert protective effects on neurons upon oxidative stress. In a recent study, Z-lig has been shown to rescue the cell death of PC12 cells induced by $H_2O_2$ (Du *et al.*, 2008). Its protective effect might be due to the suppression of apoptotic pathway by reducing Bax protein expressions, cytochrome *c* release into the cytoplasm, caspase-3 cleavage and also intracellular ROS formation in the neuronal cells (Du *et al.*, 2008). More recent evidence suggests that the proliferation and migration of VSMCs play important roles in the development of atherosclerosis (Zhang *et al.*, 1999). Z-lig can suppress the proliferation and cell cycle progression of VSMC through suppressing ROS production and the activation of MAPKs including ERK, JNK and p38 (Lu *et al.*, 2006). Moreover, Z-lig was found to inhibit the vasodilation and promote relaxation of isolated rat arteries upon stimulation. All the results reveal that Z-lig is beneficial to cardiovascular and cerebrovascular diseases (Cao *et al.*, 2006; Lin *et al.*, 2007).

The anti-inflammatory effects of Z-lig have been demonstrated in several *in vivo* models. In a LPS-induced endotoxic shock model, rabbits pre-treated with Z-lig had significantly lower TNF-α, IL-1β and nitric oxide levels (Zhu *et al.*, 2011). In a brain ischaemia model, Z-lig could exert neuroprotective effects by improving neurological deficit and infarct volume (Wu *et al.*, 2011c). The protection might be due to the increase in erythropoietin transcription via ERK pathways (Wu *et al.*, 2011c) as well as the decrease of the level of MDA, which in turn, increased the activities of the anti-oxidant enzyme glutathione (Du *et al.*, 2006). Moreover, Z-lig was shown to be beneficial to subarachnoid haemorrhage which is another subtype of stroke (Wang *et al.*, 2011a). The study demonstrated that Z-lig has neuroprotection and its use resulted in reduced mortality, neurobehavioral deficits, brain oedema, blood–brain barrier permeability and cerebral vasospasm (Wang *et al.*, 2011a). Further investigations demonstrated that Z-lig can reduce the number of apoptotic cells and expressions of p53 as well as caspase-3 cleavage in the injured brain (Wang *et al.*, 2011a).

## C. FERULIC ACID

FA, an important polyphenol commonly found in LCX and *A. sinensis*, possess anti-thrombotic and anti-oxidant effects in animal and human studies (Wang and Ou-Yang, 2005). As a well-known anti-oxidant, FA was shown to exert protective effects on primary rat cerebellar granule cells induced by $FeSO_4$ (Zhang *et al.*, 2003). It suppressed the oxidative stress-induced apoptosis by inhibiting the caspase-3 activity as well as p53 and $p21^{waf1/cip1}$ gene expressions (Zhang *et al.*, 2003). Moreover, FA also acted as a strong anti-oxidant in protecting endothelial cell death induced by radiation (Gao *et al.*, 2010b). The study indicated that FA can significantly increase the transcription of anti-oxidant related genes while promoted nuclear factor erythroid 2-related factor 2 (Nrf2) translocation into nuclei (Gao *et al.*, 2010b). Further, it increased the intracellular glutathione and nicotinamide adenine dinucleotide phosphate levels in radiated cells (Gao *et al.*, 2010c). The activation of Nrf2 was found to be related to PI3K and ERK signalling pathways (Gao *et al.*, 2010c). Apart from the protective effects against oxidative stress, FA could also induce angiogenesis in endothelial cells (Lin *et al.*, 2010). A recent report demonstrated that FA could promote the angiogenesis of human umbilical vein endothelial cells by increasing the expressions of VEGF, platelet-derived growth factor and HIF-1$\alpha$ (Lin *et al.*, 2010).

FA was proven to exert neuroprotective effects on rats using ischaemic stroke models. In MCAO models, FA inhibited the inflammatory responses by suppressing the expressions of iNOS and intercellular adhesion molecule-1 (ICAM-1) and inhibiting NF-κB and p38 MAPK phosphorylation in the ischaemic regions (Hsieh *et al.*, 2008a,b, 2010). Moreover, FA could protect the brain from oxidative stress-induced apoptosis. With pre-treatment of FA, the number of apoptotic cells found in the ischaemic core is reduced significantly (Hsieh *et al.*, 2008b). The effect might be due to the reduced Bax translocation, cytochrome *c* release and levels of cleaved-caspase-3 (Hsieh *et al.*, 2010).

## V. CONCLUDING REMARKS

TCM is one of the world's oldest medical systems for the prevention and treatment of ischaemic stroke for more than 3000 years in China. Unlike Western medicine, which focuses on diseases/symptoms, traditional medicine not only treats the diseases but also their underlying condition. As recorded in ancient TCM books, herbs are combined in formulae in order to enhance their properties and effects for specific diseases. By applying modern science and technology, the efficacy of the formulae can be proven by studying the active ingredients of these formulae and also determining their mechanism of actions. In the case of ischaemic stroke, several formulae have been found to have multiple pharmacological actions including anti-inflammatory, anti-oxidative, anti-apoptotic, angiogenic and neuroprotective properties. All these research findings will help TCM to be explored and search for the development of treatment modalities.

## ACKNOWLEDGEMENTS

This project was supported in part by grants from Prof. Francis S.K. Lau, Ms. Ho Yuk Ching (Ching Ping) Research Funds and Purapharm International awarded to Prof. A. Lau.

## REFERENCES

Barone, F. C. and Feuerstein, G. Z. (1999). Inflammatory mediators and stroke: New opportunities for novel therapeutics. *Journal of Cerebral Blood Flow and Metabolism* **19**, 819–834.

Bi, W. F., Yang, H. Y., Liu, J. C., Cheng, T. H., Chen, C. H., Shih, C. M., Lin, H., Wang, T. C., Lian, W. S., Chen, J. J., Chiu, H. C. and Chang, N. C. (2005). Inhibition of cyclic strain-induced endothelin-1 secretion by tetramethylpyrazine. *Clinical and Experimental Pharmacology and Physiology* **32**, 536–540.

Bluhmki, E., Chamorro, A., Davalos, A., Machnig, T., Sauce, C., Wahlgren, N., Wardlaw, J. and Hacke, W. (2009). Stroke treatment with alteplase given 3.0-4.5 h after onset of acute ischaemic stroke (ECASS III): Additional outcomes and subgroup analysis of a randomised controlled trial. *Lancet Neurology* **8**, 1095–1102.

Cai, G. X. and Liu, B. Y. (2010). Buyang Huanwu decoction increases vascular endothelial growth factor expression and promotes angiogenesis in a rat model of local cerebral ischemia. *Neural Regeneration Research* **5**, 1733–1738.

Cai, G., Liu, B., Liu, W., Tan, X., Rong, J., Chen, X., Tong, L. and Shen, J. (2007). Buyang Huanwu Decoction can improve recovery of neurological function, reduce infarction volume, stimulate neural proliferation and modulate VEGF and Flk1 expressions in transient focal cerebral ischaemic rat brains. *Journal of Ethnopharmacology* **113**, 292–299.

Campbell, G. R. and Campbell, J. H. (1985). Recent advances in molecular pathology—Smooth-muscle phenotypic changes in arterial-wall homeostasis—Implications for the pathogenesis of atherosclerosis. *Experimental and Molecular Pathology* **42**, 139–162.

Cao, Y. X., Zhang, W., He, J. Y., He, L. C. and Xu, C. B. (2006). Ligustilide induces vasodilatation via inhibiting voltage dependent calcium channel and receptor-mediated $Ca^{2+}$ influx and release. *Vascular Pharmacology* **45**, 171–176.

Chao, H. M., Lin, D. E., Chang, Y., Hsu, W. M., Lee, S. M., Lee, F. L., Chi, C. W., Pan, W. H., Liu, T. Y., Lui, W. Y., Ho, L. T., Kuo, C. D. *et al.* (2008). Ferulic acid, but not tetramethylpyrazine, significantly attenuates retinal ischemia/reperfusion-induced alterations by acting as a hydroxyl radical scavenger. *Journal of Ocular Pharmacology and Therapeutics* **24**, 461–472.

Chen, D. R. (1992). Clinical and experimental study of Ligusticum wallichii and aspirin in the treatment of transient ischemic attack. *Zhongguo Zhong Xi Yi Jie He Za Zhi* **12**, 672–674.

Chen, K. J. and Chen, K. (1992). Ischemic stroke treated with Ligusticum chuanxiong. *Chinese Medical Journal* **105**, 870–873.

Chen, J. and Chopp, M. (2006). Neurorestorative treatment of stroke: Cell and pharmacological approaches. *NeuroRx* **3**, 466–473.

Chen, Y. H., Lin, C. M., Chiu, J. H., Wu, I. H., Wang, B. W. and Pan, C. M. (2010). Ferulic acid augments angiogenesis via VEGF, PDGF and HIF-1 alpha. *The Journal of Nutritional Biochemistry* **21**, 627–633.

Cheng, C. Y., Ho, T. Y., Lee, E. J., Su, S. Y., Tang, N. Y. and Hsieh, C. L. (2008a). Ferulic acid reduces cerebral infarct through its antioxidative and anti-inflammatory effects following transient focal cerebral ischemia in rats. *The American Journal of Chinese Medicine* **36**, 1105–1119.

Cheng, C. Y., Su, S. Y., Tang, N. Y., Ho, T. Y., Chiang, S. Y. and Hsieh, C. L. (2008b). Ferulic acid provides neuroprotection against oxidative stress-related apoptosis after cerebral ischemia/reperfusion injury by inhibiting ICAM-1 mRNA expression in rats. *Brain Research* **1209**, 136–150.

Coccheri, S. (2010). Antiplatelet drugs—Do we need new options? With a reappraisal of direct thromboxane inhibitors. *Drugs* **70**, 887–908.

Du, J. R., Kuang, X., Yao, Y., Liu, Y. X., Wang, C. Y. and Qian, Z. M. (2006). Neuroprotective role of Z-ligustilide against forebrain ischemic injury in ICR mice. *Brain Research* **1102**, 145–153.

Du, J. R., Peng, H. Y., Zhang, G. Y., Kuang, X., Liu, Y. X., Qian, Z. M. and Wang, C. Y. (2007). Neuroprotective effect of Z-ligustilide against permanent focal ischemic damage in rats. *Biological and Pharmaceutical Bulletin* **30**, 309–312.

Du, J. R., Yu, Y., Wang, C. Y. and Qian, Z. M. (2008). Protection against hydrogen peroxide-induced injury by Z-ligustilide in PC12 cells. *Experimental Brain Research* **184**, 307–312.

Du, J. R., Kuang, X., Chen, Y. S., Wang, J. and Wang, Y. N. (2009a). Protective effect of Z-ligustilide against amyloid beta-induced neurotoxicity is associated with decreased pro-inflammatory markers in rat brains. *Pharmacology, Biochemistry, and Behavior* **92**, 635–641.

Du, J. R., Zhang, L., Wang, J., Yu, D. K., Chen, Y. S., He, Y. and Wang, C. Y. (2009b). Z-ligustilide extracted from Radix Angelica Sinensis decreased platelet aggregation induced by ADP ex vivo and arterio-venous shunt thrombosis in vivo in rats. *Yakugaku Zasshi* **129**, 855–859.

Du, J. R., Wang, J., Wang, Y., Kuang, X. and Wang, C. Y. (2010). Z-ligustilide attenuates lipopolysaccharide-induced proinflammatory response via inhibiting NF-kappa B pathway in primary rat microglia. *Acta Pharmacologica Sinica* **31**, 791–797.

Fan, L., Wang, K., Shi, Z., Die, J., Wang, C. and Dang, X. (2011). Tetramethylpyrazine protects spinal cord and reduces inflammation in a rat model of spinal cord ischemia-reperfusion injury. *Journal of Vascular Surgery* **54**, 192–200.

Gao, B. T. (1989). The effects of ligustrazine, aspirin and beta-histine on platelet aggregation in patients with acute ischemic stroke. *Zhonghua Shen Jing Jing Shen Ke Za Zhi* **22**, 148–151.

Gao, J. E. F. F., Lu, F. H. and Zhu, C. L. (2009). Effects of Buyang Huanwu decoction on cell proliferation and differentiation in the hippocampal dentate gyrus of aged rats following cerebral ischemia/reperfusion. *Neural Regeneration Research* **4**, 390–395.

Gao, D., Wu, L. Y., Jiao, Y. H., Chen, W. Y., Chen, Y., Kaptchuk, T. J., Lu, B., Song, J. and Chen, K. J. (2010a). The effect of Xuefu Zhuyu decoction ((sic)) on in vitro endothelial progenitor cell tube formation. *Chinese Journal of Integrative Medicine* **16**, 50–53.

Gao, Y., Ma, Z. C., Hong, Q., Wang, Y. G., Tan, H. L., Xiao, C. R., Liang, Q. D. and Zhang, B. L. (2010b). Ferulic acid protects human umbilical vein endothelial cells from radiation induced oxidative stress by phosphatidylinositol 3-kinase and extracellular signal-regulated kinase pathways. *Biological and Pharmaceutical Bulletin* **33**, 29–34.

Guthrie, R. (2011). Review and management of side effects associated with antiplatelet therapy for prevention of recurrent cerebrovascular events. *Advances in Therapy* **28**, 473–482.

Hacke, W., Donnan, G., Fieschi, C., Kaste, M., Von Kummer, R., Broderick, J. P., Brott, T., Frankel, M., Grotta, J. C., Haley, E. C., Kwiatkowski, T., Levine, S. R. *et al.* (2004). Association of outcome with early stroke treatment: Pooled analysis of ATLANTIS, ECASS, and NINDS rt-PA stroke trials. *The Lancet* **363**, 768–774.

Ho, W. K., Wen, H. L. and Lee, C. M. (1989). Tetramethylpyrazine for treatment of experimentally induced stroke in Mongolian gerbils. *Stroke* **20**, 96–99.

Hsieh, C. L., Cheng, C. Y., Ho, T. Y., Lee, E. J., Su, S. Y. and Tang, N. Y. (2008a). Ferulic acid reduces cerebral infarct through its antioxidative and anti-inflammatory effects following transient focal cerebral ischemia in rats. *The American Journal of Chinese Medicine* **36**, 1105–1119.

Hsieh, C. L., Cheng, C. Y., Su, S. Y., Tang, N. Y., Ho, T. Y. and Chiang, S. Y. (2008b). Ferulic acid provides neuroprotection against oxidative stress-related apoptosis after cerebral ischemia/reperfusion injury by inhibiting ICAM-1 rnRNA expression in rats. *Brain Research* **1209**, 136–150.

Hsieh, C. L., Cheng, C. Y., Su, S. Y., Tang, N. Y., Ho, T. Y. and Lo, W. Y. (2010). Ferulic acid inhibits nitric oxide-induced apoptosis by enhancing GABA (B1) receptor expression in transient focal cerebral ischemia in rats. *Acta Pharmacologica Sinica* **31**, 889–899.

Hu, Y. S., Jia, J., Zhang, X., Wu, Y., Wang, Q. Z., Li, N. N., Wu, C. Q., Yu, H. X. and Guo, Q. C. (2009). Protective effect of tetraethyl pyrazine against focal cerebral ischemia/reperfusion injury in rats: Therapeutic time window and its mechanism. *Thrombosis Research* **123**, 727–730.

Jia, W., Gao, W. Y., Yan, Y. Q., Wang, J., Xu, Z. H., Zheng, W. J. and Xiao, P. G. (2004). The rediscovery of ancient Chinese herbal formulas. *Phytotherapy Research* **18**, 681–686.

Kao, T. K., Ou, Y. C., Kuo, J. S., Chen, W. Y., Liao, S. L., Wu, C. W., Chen, C. J., Ling, N. N., Zhang, Y. H. and Peng, W. H. (2006). Neuroprotection by tetramethylpyrazine against ischemic brain injury in rats. *Neurochemistry International* **48**, 166–176.

Lee, J. J., Hsu, W. H., Yen, T. L., Chang, N. C., Luo, Y. J., Hsiao, G. and Sheu, J. R. (2011). Traditional Chinese medicine, Xue-Fu-Zhu-Yu decoction, potentiates tissue plasminogen activator against thromboembolic stroke in rats. *Journal of Ethnopharmacology* **134**, 824–830.

Leventhal, C., Rafii, S., Rafii, D., Shahar, A. and Goldman, S. A. (1999). Endothelial trophic support of neuronal production and recruitment from the adult mammalian subependyma. *Molecular and Cellular Neuroscience* **13**, 450–464.

Li, M., Handa, S., Ikeda, Y. and Goto, S. (2001). Specific inhibiting characteristics of tetramethylpyrazine, one of the active ingredients of the Chinese herbal medicine 'chuanxiong', on platelet thrombus formation under high shear rates. *Thrombosis Research* **104**, 15–28.

Li, X. M., Bai, X. C., Qin, L. N., Huang, H., Xiao, Z. J. and Gao, T. M. (2003). Neuroprotective effects of Buyang Huanwu Decoction on neuronal injury in hippocampus after transient forebrain ischemia in rats. *Neuroscience Letters* **346**, 29–32.

Liang, M. J., He, L. C. and Yang, G. D. (2005). Screening, analysis and in vitro vasodilatation of effective components from Ligusticum chuanxiong. *Life Science* **78**, 128–133.

Liao, S. L., Kao, T. K., Chen, W. Y., Lin, Y. S., Chen, S. Y., Raung, S. L., Wu, C. W., Lu, H. C. and Chen, C. J. (2004). Tetramethylpyrazine reduces ischemic brain injury in rats. *Neuroscience Letters* **372**, 40–45.

Lin, C. C., Yen, F. L., Hsu, F. F. and Lin, J. M. (2000). Anti-hypercholesterolaemia, antioxidant activity and free radical scavenger effects of traditional Chinese medicine prescriptions used for stroke. *Journal of Pharmacy and Pharmacology* **52**, 1387–1393.

Lin, G., Chan, S. S. K. and Cheng, T. Y. (2007). Relaxation effects of ligustilide and senkyunolide A, two main constituents of Ligusticum chuanxiong, in rat isolated aorta. *Journal of Ethnopharmacology* **111**, 677–680.

Lin, C. M., Chiu, J. H., Wu, I. H., Wang, B. W., Pan, C. M. and Chen, Y. H. (2010). Ferulic acid augments angiogenesis via VEGF, PDGF and HIF-1 alpha. *The Journal of Nutritional Biochemistry* **21**, 627–633.

Liu, Z. (1991). Effects of Ligusticum wallichii on the plasma levels of beta-thromboglobulin, platelet factor 4, thromboxane B2 and 6-keto-PGF1 alpha in patients with acute cerebral infarction. *Zhong Xi Yi Jie He Za Zhi* **11**, 711–713.

Liu, Z. and Shi, Y. (1990). Effects of Ligusticum wallichii on the plasma and CSF levels of dynorphin A1-13 in rabbits under acute experimental cerebral ischemia. *Zhong Xi Yi Jie He Za Zhi* **10**, 160–161.

Liu, L., Duan, J. A., Tang, Y. P., Ma, H. Y., Su, S. L., Guo, J. M. and Hua, Y. Q. (2010a). The protective effects of the active fraction of shaofu zhuyu decoction on hydrogen peroxide-induced oxidative injury in vascular smooth muscle cells. *Molecules* **15**, 5066–5078.

Liu, X. Z., Shao, Z. K., Li, J. W., Zhao, Z. H., Gao, C. and Sun, Z. (2010b). Effects of tetramethylpyrazine on nitric oxide/cGMP signaling after cerebral vasospasm in rabbits. *Brain Research* **1361**, 67–75.

Liu, Y., Tian, Y. M., Chen, X. L., Zhang, H. X., Shi, Q. D., Zhang, J. E. F. F. and Yang, P. B. (2010c). Tetramethylpyrazine promotes proliferation and differentiation of neural stem cells from rat brain in hypoxic condition via mitogen-activated protein kinases pathway in vitro. *Neuroscience Letters* **474**, 26–31.

Lloyd-Jones, D. (2009). Heart disease and stroke statistics—2009 update: A report from the American Heart Association Statistics Committee and Stroke Statistics Subcommittee. *Circulation* **119**, 21–118.

Lu, C. S., Yin, H. M., Hung, Y. H., Pao, C. and Sung, C. F. (1978). Radix-salviae-miltiorrhizae and Rhizoma-ligustici-wallichii in coronary heart disease. *Chinese Medical Journal* **4**, 43–46.

Lu, Q., Qiu, T. Q. and Yang, H. (2006). Ligustilide inhibits vascular smooth muscle cells proliferation. *European Journal of Pharmacology* **542**, 136–140.

Muir, K. W., Tyrrell, P., Sattar, N. and Warburton, E. (2007). Inflammation and ischaemic stroke. *Current Opinion in Neurology* **20**, 334–342.

Or, T. C. T., Yang, C. L. H., Law, A. H. Y., Li, J. C. B. and Lau, A. S. Y. (2011). Isolation and identification of anti-inflammatory constituents from Ligusticum chuanxiong and their underlying mechanisms of action on microglia. *Neuropharmacology* **60**, 823–831.

Qi, Z. X. and Chen, L. (2009). Effect of Chinese drugs for promoting blood circulation and eliminating blood stasis on vascular endothelial growth factor expression in rabbits with glucocorticoid-induced ischemic necrosis of femoral head. *Journal of Traditional Chinese Medicine* **29**, 137–140.

Qian, Z. M., Du, J. R., Bai, B., Kuang, X., Yu, Y., Wang, C. Y., Ke, Y., Xu, Y. J. and Tzang, A. H. C. (2006). Ligustilide inhibits spontaneous and agonists- or K+ depolarization-induced contraction of rat uterus. *Journal of Ethnopharmacology* **108**, 54–58.

Rhyu, M. R., Kim, E. Y. and Kim, J. H. (2010). Endothelium-independent vasorelaxation by Ligusticum wallichii in isolated rat aorta: Comparison of a butanolic fraction and tetramethylpyrazine, the main active component of Ligusticum wallichii. *Biological and Pharmaceutical Bulletin* **33**, 1360–1363.

Rosamond, W., Flegal, K., Friday, G., Furie, K., Go, A., Greenlund, K., Haase, N., Ho, M., Howard, V., Kissela, B., Kittner, S., Lloyd-Jones, D. *et al.* (2007). Heart disease and stroke statistics—2007 update: A report from the American Heart Association Statistics Committee and Stroke Statistics Subcommittee. *Circulation* **115**, 69–171.

Ross, R. and Glomset, J. A. (1973). Atherosclerosis and the arterial smooth muscle cell: Proliferation of smooth muscle is a key event in the genesis of the lesions of atherosclerosis. *Science* **180**, 1332–1339.

Samson, Y., Lapergue, B. and Hosseini, H. (2005). Inflammation and ischaemic stroke: Current status and future perspectives. *Revista de Neurologia* **161**, 1177–1182.

Schwartz, S. M., Campbell, G. R. and Campbell, J. H. (1986). Replication of smooth muscle cells in vascular disease. *Circulation Research* **58**, 427–444.

Sheu, J. R., Chang, Y., Hsiao, G., Chen, S. H., Chen, Y. C., Lin, J. H., Lin, K. H. and Chou, D. S. (2007). Tetramethylpyrazine suppresses HIF-1 alpha, TNF-alpha, and activated caspase-3 expression in middle cerebral artery occlusion-induced brain ischemia in rats. *Acta Pharmacologica Sinica* **28**, 327–333.

Shih, Y. H., Wu, S. L., Chiou, W. F., Ku, H. H., Ko, T. L. and Fu, Y. S. (2002). Protective effects of tetramethylpyrazine on kainate-induced excitotoxicity in hippocampal culture. *Neuroreport* **13**, 515–519.

Shohami, E., Ginis, I. and Hallenbeck, J. M. (1999). Dual role of tumor necrosis factor alpha in brain injury. *Cytokine & Growth Factor Reviews* **10**, 119–130.

Stenzel-Poore, M. P., Doyle, K. P. and Simon, R. P. (2008). Mechanisms of ischemic brain damage. *Neuropharmacology* **55**, 310–318.

Su, Y. W., Chiou, W. F., Chao, S. H., Lee, M. H., Chen, C. C. and Tsai, Y. C. (2011). Ligustilide prevents LPS-induced iNOS expression in RAW 264.7 macrophages by preventing ROS production and down-regulating the MAPK, NF-[kappa]B and AP-1 signaling pathways. *International Immunopharmacology* **11**, 1166–1172.

Sugawara, T. and Chan, P. H. (2003). Reactive oxygen radicals and pathogenesis of neuronal death after cerebral ischemia. *Antioxidants & Redox Signaling* **5**, 597–607.

Sun, J., Bi, Y., Guo, L., Qi, X., Zhang, J., Li, G., Tian, G., Ren, F. and Li, Z. (2007). Buyang Huanwu Decoction promotes growth and differentiation of neural progenitor cells: Using a serum pharmacological method. *Journal of Ethnopharmacology* **113**, 199–203.

Tan, X. H., Qu, H. D., Peng, K., Chen, Y. Y., Tong, L., Shen, J. G. and Zhu, C. W. (2006). Effects of Buyanghuanwu decoction on nerve proliferation in rats with sequelae of ischemic stroke. *Nan Fang Yi Ke Da Xue Xue Bao* **26**, 189–192.

Tang, W. and Eisenbrand, G. (1992). Ligusticum chuanxiong Hort. Springer-Verlag.

Tang, Y. H., Li, H. and Chen, B. Y. (2006). Effect of active fraction of buying huanwu decoction on caspase expression in rats after focal cerebral ischemic reperfusion. *Zhongguo Zhong Xi Yi Jie He Za Zhi* **26**, 533–537.

Tian, J. W., Fu, F. H., Jiang, W. A. L. L., Wang, C. Y., Sun, F. and Zhang, I. P. (2005). Protective effect of logistical chuanxiong phthalides on focai cerebral ischemia in rats and its related mechanism of action. *Zhongguo Zhong Yao Za Zhi* **30**, 466–468.

Tong, L., Tan, X. H. and Shen, J. G. (2007). Comparative study of Buyang Huanwu Decoction and the different combinations of its ingredients on neurogenesis following ischemic stroke in rats. *Zhongguo Zhong Xi Yi Jie He Za Zhi* **27**, 519–522.

Wang, B. H. and Ou-Yang, J. P. (2005). Pharmacological actions of sodium ferulate in cardiovascular system. *Cardiovascular Drug Reviews* **23**, 161–172.

Wang, S., Gao, Z., Chen, X. M., Lian, X. J., Zhu, H. S., Zheng, J. and Sun, L. Z. (2008). The anticoagulant ability of ferulic acid and its applications for improving the blood compatibility of silk fibroin. *Biomedical Materials* **3**, 044106.

Wang, S. L., Chen, D., Tang, J. P., Khatibi, N. H., Zhu, M., Li, Y. B., Wang, C. Y., Jiang, R. and Tu, L. (2011a). Treatment with Z-ligustilide, a component of angelica sinensis, reduces brain injury after a subarachnoid hemorrhage in rats. *Journal of Pharmacology and Experimental Therapeutics* **337**, 663–672.

Wang, Y. H., Liang, S., Xu, D. S., Lin, X., He, C. Y., Feng, Y. and Hong, Y. L. (2011b). Effect and mechanism of senkyunolide I as an anti-migraine compound from Ligusticum chuanxiong. *Journal of Pharmacy and Pharmacology* **63**, 261–266.

Witztum, J. L. (1994). The oxidation hypothesis of atherosclerosis. *The Lancet* **344**, 793–795.

Wu, C. J., Chen, J. T., Yen, T. L., Jayakumar, T., Chou, D. S., Hsiao, G. and Sheu, J. R. (2011a). Neuroprotection by the traditional Chinese medicine, Tao-Hong-Si-Wu-Tang, against middle cerebral artery occlusion-induced cerebral ischemia in rats. *Evidence-based Complementary and Alternative Medicine* **2011**, 803015.

Wu, X. G., Li, Y. X., Liu, H. X., Yin, Y. H., Zhao, S. M. and Guo, Y. Y. (2011b). Microvessel changes in the gerbil hippocampus after cerebral ischemia and reperfusion by Buyang Huanwu decoction pretreatment. *Neural Regeneration Research* **6**, 656–660.

Wu, X. M., Qian, Z. M., Zhu, L., Du, F., Yung, W. H., Gong, Q. and Ke, Y. (2011c). Neuroprotective effect of ligustilide against ischemia-reperfusion injury via up-regulation of erythropoietin and down-regulation of RTP801. *British Journal of Pharmacology* **164**, 332–343.

Yang, S., Gao, Q., Xing, S., Feng, X., Peng, L., Dong, H., Bao, L., Zhang, J., Hu, Y., Li, G., Song, T., Li, Z. *et al.* (2011). Neuroprotective effects of Buyang Huanwu decoction against hydrogen peroxide induced oxidative injury in Schwann cells. *Journal of Ethnopharmacology* **137**, 1095–1101.

Yenari, M. A., Wang, Q. and Tang, X. N. (2007). The inflammatory response in stroke. *Journal of Neuroimmunology* **184**, 53–68.

Yi, J., Huang, X., Yu, Y., Cai, G. X. and Liu, B. Y. (2010). Effect of Buyang Huanwu decoction on interleukin-1beta and tumor necrosis factor-alpha expression in rats after cerebral infarction. *Zhongguo Wei Zhong Bing Ji Jiu Yi Xue* **22**, 599–601.

Yu, C., Li, X. Y., He, J. L., Liu, H. T. and Li, W. M. (2009). Tetramethylpyrazine suppresses interleukin-8 expression in LPS-stimulated human umbilical vein endothelial cell by blocking ERK, p38 and nuclear factor-kappa B signaling pathways. *Journal of Ethnopharmacology* **125**, 83–89.

Yu, C., Li, W. M., Liu, H. T., Li, X. Y., Wu, J. Y., Xu, G., Teng, Y. Z. and Ding, S. T. (2010a). The effect of tetramethylpyrazine on hydrogen peroxide-induced oxidative damage in human umbilical vein endothelial cells. *Basic & Clinical Pharmacology* **106**, 45–52.

Yu, C., Liu, H. T., Du, Y. G., He, J. L., Chen, W. J., Li, W. M., Yang, Z. and Wang, Y. X. (2010b). Tetramethylpyrazine inhibits production of nitric oxide and inducible nitric oxide synthase in lipopolysaccharide-induced N9 microglial cells through blockade of MAPK and PI3K/Akt signaling pathways, and suppression of intracellular reactive oxygen species. *Journal of Ethnopharmacology* **129**, 335–343.

Zhang, X. J., Tenner, T. E. and Lombardini, J. B. (1999). Inhibition of rat vascular smooth muscle cell proliferation by taurine and taurine analogues. *Biochemical Pharmacology* **57**, 1331–1339.

Zhang, Z. H., Wei, T. T., Hou, J. W., Li, G. S., Yu, S. Z. and Xin, W. J. (2003). Iron-induced oxidative damage and apoptosis in cerebellar granule cells: Attenuation by tetramethylpyrazine and ferulic acid. *European Journal of Pharmacology* **467**, 41–47.

Zhang, L., Deng, M. and Zhou, S. (2011). Tetramethylpyrazine inhibits hypoxia-induced pulmonary vascular leakage in rats via the ROS-HIF-VEGF pathway. *Pharmacology* **87**, 265–273.

Zhao, S. M., Wu, X. G., Li, Y. X., Liu, H. X., Yin, Y. H. and Guo, Y. Y. (2011). Microvessel changes in the gerbil hippocampus after cerebral ischemia and reperfusion by Buyang Huanwu decoction pretreatment. *Neural Regeneration Research* **6**, 656–660.

Zheng, Z. and Yenari, M. A. (2004). Post-ischemic inflammation: Molecular mechanisms and therapeutic implications. *Neurological Research* **26**, 884–892.

Zhu, X. H. and Gao, T. M. (2003). Effects of Xiaoxuming Decoction on neuronal injury in hippocampus after transient forebrain ischemia in rats. *Journal of Neurochemistry* **87**, 135.

Zhu, X. L., Xiong, L. Z., Wang, Q., Liu, Z. G., Ma, X., Zhu, Z. H., Hu, S., Gong, G. and Chen, S. Y. (2009). Therapeutic time window and mechanism of tetramethylpyrazine on transient focal cerebral ischemia/reperfusion injury in rats. *Neuroscience Letters* **449**, 24–27.

Zhu, X. H., Li, S. J., Hu, H. H., Sun, L. R., Das, M. and Gao, T. M. (2010). Neuroprotective effects of xiao-xu-ming decoction against ischemic neuronal injury in vivo and in vitro. *Journal of Ethnopharmacology* **127**, 38–46.

Zhu, H. B., Shao, M., Qu, K., Liu, K., Zhang, Y. Q., Zhang, L., Lian, Z. Q., Chen, T. T., Liu, J. E. F. F., Wu, A. L. and Tang, Y. (2011). Effects of ligustilide on lipopolysaccharide-induced endotoxic shock in rabbits. *Planta Medica* **77**, 809–816.

# Potential of 'Omics' Technologies for Implementation in Research on Phytotherapeutical Toxicology

THOMAS EFFERTH[*,1] AND HENRY JOHANNES GRETEN[†,‡]

*Department of Pharmaceutical Biology, Institute of Pharmacy and Biochemistry, Johannes Gutenberg University, Mainz, Germany
†Heidelberg School of Chinese Medicine, Heidelberg, Germany
‡Biomedical Sciences Institute Abel Salazar, University of Porto, Porto, Portugal



[1]Corresponding author: E-mail: efferth@uni-mainz.de

Advances in Botanical Research, Vol. 62

0065-2296/12 $35.00
DOI: 10.1016/B978-0-12-394591-4.00007-6

## ABSTRACT

High toxicity is the most common reason why new agents drop out of drug development in the pharmaceutical industry. There is hope that toxicogenomics facilitates the early detection of toxic effects and their molecular mechanisms of action during preclinical studies to remove potentially toxic substances from the development. Herbal remedies consist of mixtures of different herbs, which represent a considerable source of heterogeneity and toxicity. They may be caused by botanical misidentification, contamination with pesticides, heavy metals, organic solvents, microbials and radioactivity. Intentional faked herbal products may contain chemical drugs or hormones. Approaches to apply 'omics'-based technologies (genomics, transcriptomics, proteomics, metabolomics) to medicinal plant research may help to detect toxicity related to phytotherapy. Toxic reactions in the liver are common to many natural xenobiotic compounds, since orally ingested herbal drugs are first metabolized in the liver by phase I and II enzymes and are then excreted by phase III transporters. Current toxicogenomics approaches need to be adapted to herbal toxicology research in terms of appropriate *in vitro* and *in vivo* test models, experimental reproducibility, the development of predictive models, suitable software tools for data analysis and databases, the identification and validation of biomarkers, as well as quantitative structure–activity relationship modelling. Although toxicogenomics of medicinal plants and herbal products is still in its infancy, omics technologies represent attractive methods for the prediction of desirable and undesirable effects of phytotherapy.


## I. INTRODUCTION

The classical paradigm of molecular biology was that a gene codes is transcribed to mRNA, which is translated to a protein. The past years showed that the situation is much more complicated. Transcription from DNA to RNA is influenced by DNA polymorphisms, DNA methylation and histone deacetylation. DNA can be translated to non-coding micro-RNA, which hybridizes to mRNA and suppression of translation. One gene can code for several mRNA species by alternative splicing. Post-translational modifications (phosphorylation, glycosylation, methylation, acetylation, farnesylation, etc.), may alter the function of proteins, which result in different metabolic products. The comprising measurement of all DNA alterations, mRNA expressions, protein expressions and modifications and metabolic products in a cell or tissue is subject of the 'omics' technologies (Fig. 1). DNA microarrays serve as suitable high throughput tool for simultaneous analysis of multiple genes. Major practical applicability of DNA microarrays remains in DNA mutation and polymorphism analysis. Chavan *et al.* (2006) highlighted applications of DNA microarrays in pharmacodynamics, pharmacogenomics, toxicogenomics (TGx) and quality control of herbal drugs and extracts.

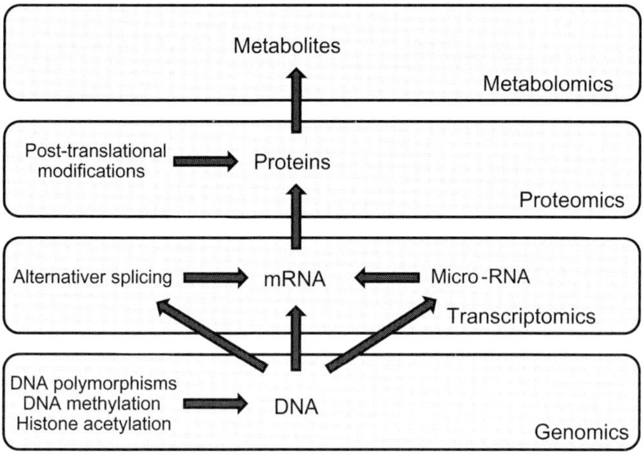


Fig. 1. Cellular targets for 'omics' technologies.

The entire expressed mRNA molecules (transcriptome) are subject of transcriptomics. Proteomics concerns the expression profiles of proteins (proteome). The results obtained from proteomic methods cannot be directly equated with those of genomics, because proteins are often modified, for example by alternative splicing of RNA, post-translational modifications protein folding and others. The changes are crucial to different functional states of proteins. There are other 'omics' technologies such as metabolomics, which examined endogenous metabolic intermediates and end products and degradation products from the fat, carbohydrate or protein metabolism.

In a tight sense (*sensu strictu*), genomics is restricted to DNA. In a broader sense (*sensu latu*), genomics stands for all 'omics' technologies, since the entire diversity of cellular molecules has its origin in the DNA code.

The goal of toxicogenomics, -proteomics and -metabolomics is to measure mRNA, protein and metabolite expression profiles at a genome-wide level: (1) to uncover mechanisms of xenobiotic effects and cellular networks, (2) to develop new methods of drug screening and toxicity and (3) to identify new biomarkers for clinical toxicity studies. Microarrays have become a key technology in toxicology.

High toxicity is the most common reason why new agents drop out of drug development. This is often found relatively late in development after completion of the preclinical phase of clinical trials. At worst, unacceptable side effects even show up only when the drug is approved and on the market. Examples from the past are thalidomide or Baycoll. The hope is that TGx facilitates the early detection of toxic effects and their molecular mechanisms of action during preclinical studies to remove potentially toxic substances from the development. This assumption is based on the hypothesis that TGx can detect low toxicity of compounds, which is below the detection limit of conventional methods. TGx was rapidly embraced by the pharmaceutical industry as useful tool to identify safer drugs in a quicker, more cost-effective manner (Yang *et al.*, 2004). Studies have already demonstrated the benefits of applying gene expression profiling towards drug safety evaluation, both for identifying mechanisms underlying toxicity, as well as for providing a means to identify safety liabilities early in the drug discovery process.

Herbal remedies consist of mixtures of different herbs, which represent a considerable source of heterogeneity and toxicity. These biological variables and inconsistencies can contribute to sometimes unpredictable effects of herbal mixtures and hinders the identification both of pharmacological active and toxic components. Whereas variations in cultivation and processing can lead to herbal products with more or less activity, toxic reactions associated with herbal products may be a considerable problem. They may be caused by botanical misidentification, contamination with pesticides, heavy metals, organic solvents, microbials and radioactivity. Intentional faked herbal products may contain chemical drugs or hormones. The full dimension of this problem has recently been discussed (Efferth and Kaina, 2011). Apart from the contaminations and adulterations, toxic reactions may also occur by interactions of herbs with western drugs (Efferth and Kaina, 2011; Kober *et al.*, 2008).

There has been an increasing number of reports on the adverse reactions associated with herbal consumption. For many of these adverse reactions, the underlying biochemical mechanisms are unknown, but bioactivation of herbal compounds to generate reactive intermediates have been implicated. Chen *et al.* (2011) reviewed the metabolic activation of herbal compounds, molecular targets and the toxicity relevance. Some herbal compounds are converted to toxic or even carcinogenic metabolites by phase I [e.g. cytochrome P450s (CYPs)] and less frequently by phase II enzymes, for example aristolochic acids in *Aristolochia* spp. are metabolized to highly reactive cyclic nitrenium ions reacting with macromolecules DNA and activation of H-ras and myc oncogenes and ultimately leading to carcinogenesis. Teucrin A and teuchamaedryn A, two diterpenoids found in germander (*Teucrium chamaedrys*) which are used for slimming herbal supplements are converted by CYP3A4 to reactive epoxides and cause severe hepatotoxicity. Some naturally occurring alkenylbenzenes (e.g. safrole, methyleugenol and estragole) and flavonoids (e.g. quercetin) can undergo bioactivation in the liver resulting in reactive intermediates capable of forming DNA adducts. The hepatotoxicity of kava is possibly due to intracellular glutathione depletion and/or quinone formation. Moreover, several herbal compounds including capsaicin from chili peppers, dially sulfone in garlic, methysticin and dihydromethysticin in kava, oleuropein in olive oil and resveratrol found in grape seeds are mechanism-based (suicide) inhibitors of various CYPs. System-biological 'omics' approaches may provide more insights into mechanistic aspects of herb-induced toxicities, and contribute to bridging the relationships between herbal bioactivation, protein/DNA adduct formation and the toxicological consequences.

# II. 'OMICS'-BASED TOXICOLOGY

## A. TRANSCRIPTOMICS

Expression profiles of mRNA can be measured by using cDNA arrays. cDNA samples with a length of 200–400 base pairs can be immobilized on a solid support and hybridized with mRNA of the test sample. Fragments of 200 bp exhibit stable hybridization properties independent of SNPs and GT-varying content. Cross-hybridization with fragments of 400 bp can occur by repetitive elements and non-specific interactions. A length of 200–400 bp is sufficient to distinguish individual genes of large gene families with a high degree of homology. In addition to cDNA microarrays, oligonucleotide arrays can be used. The samples have a length of 50–70 bases. High-throughput analyses of oligonucleotide arrays are preferred for accurate quantification of mRNA expression compared with cDNA arrays. Bioinformatic methods such as hierarchical cluster analysis or self-organizing maps are used for the evaluation.

## B. PROTEOMICS

The proteome is estimated to be 30–10 times larger than the transcriptome. This variability is caused by covalent modifications, cell–cell interactions, protein–protein and protein–ligand interactions. The proteome is dynamic, since such modifications regulate protein functions. The genome is largely static in nature. It is possible to pursue two directions proteomic analysis:

- The quantitative determination of total protein expression of a tissue or a cell (expression proteomics). Typical setting are (1) the investigation of signalling pathways under physiological and pathological conditions, (2) the influence of drugs on protein expression and (3) new disease markers and similar issues.
- The subcellular localization of proteins and protein–protein interactions. Systematic identification of protein complexes allows the mapping of cellular proteins of different functional units. This research field is called cell-map proteomics.

Two-dimensional gel electrophoresis can separate several thousand protein spots from each other and allows the determination of molecular weight and isoelectric point (pI) and the separation of isoforms. The liquid is suitable for the separation of complex protein mixtures as standard (e.g. enzymatic digestion). The two-dimensional liquid chromatography has a high separation capacity, is technically complicated. Other methods for proteome analysis represent the multidimensional chromatography and capillary electrophoresis.

Proteins, which are separated with one of these methods have to be identified in the next step, for example by matrix-assisted laser desorption/ionization (MALDI), electrospray ionization (ESI) and high-performance liquid chromatography. MALDI and ESI can be coupled with mass analysis methods: time of flight (TOF), ion traps, quadrupole, etc. MALDI-TOF is a widely used methodology for the rapid and accurate determination of proteins, which were separated by two-dimensional gel electrophoresis. The measurement and data-based identification of proteins takes automatically place allowing the evaluation of several thousand spectra per day.

*ESI*: The formation of ions is produced by spraying an aqueous organic or aqueous solution with high voltage from a tip. This results in small droplets with positive or negative charge. The spray is formed under pressure and is sucked through an opening in the mass analysis system, which is under vacuum. This produces ions which are focused to an ion beam.

*HPLC*: The test sample is placed in a liquid (mobile) phase and pressed at high pressure through a column (stationary phase). The molecules diffuse at different rates depending on their size through the column.

Tandem mass spectrometry represents the coupling of two analytical devices known as tandem mass spectrometry (MS/MS). For example, ESI-MS/MS is used for sequencing of peptides and to detect covalent protein modifications. These include deamination, phosphorylation, *N*-acetylation, glycosylation, hydroxylation, methoxylation, oxidation, nitration, glutathionylation, ubiquitination, ADP-ribosylation, palmoylation, carboylation, formylation, myristoylation, etc.

Changes in gene expression after toxin exposure is a result of altered binding of transcription factors to DNA. Approximately, 2000 transcription factors are responsible for gene expression in the genome. Transcription factors bind to promoter or enhancer sequences. This allows the binding of RNA polymerase to the transcription start site of genes to initiate transcription with other proteins in multi-protein complexes. Thus, these factors have access to DNA and alter the chromatin structure. The connection between DNA and histones are locally loosened. Complex networks of signalling pathways and crosstalk mechanisms are necessary to accomplish the fine regulation of these processes. With the chromatin immunoprecipitation technique, such processes are investigated in living cells. This method is used to identify target genes for transcription factors. These proteins are covalently bound to their DNA-binding sites (cross-linking). Then, specific protein–DNA complexes are isolated by immunoprecipitation. After removal of cross-links and purification of DNA-binding sequences by PCR, cloning and sequencing takes place.

## C. METABOLOMICS

A central question for the prediction of toxic effects is how gene expression is related to certain pathophysiological conditions. One method that goes beyond the measurement of gene and protein expression patterns is metabolomics. This refers to the holistic measurement of complex time-dependent concentration distribution and activity of metabolites in biological systems, the flux of endogenous metabolites in cells, tissues and bio-fluids (blood, urine, saliva). These include not only small molecules but also carbohydrates, peptides and lipids. The determination of metabolic profiles has two main applications:

- The definition of normal and pathological conditions by determining metabolic metabolites in bio-fluids (blood, urine, saliva) and tissue extracts.
- The identification of drug metabolites as a basis for uncovering the molecular mechanisms of action and toxic reactions of medicines.

A detection method is nuclear magnetic resonance (NMR) spectroscopy. Traditionally, NMR is used for structural analysis of molecules in chemistry. In the 1990s, NMR methods were developed for the analysis of bio-fluids and tissue extracts. Often liquid chromatography with NMR and mass spectrometry are combined (LC-NMR–MS). Large databases with metabolic models serve as benchmark to assess the toxicity of substances. An example is the COMET (Consortium for Metabonomics Toxicity) database, which contains more than 100,000 NMR spectra.

## D. BIOINFORMATICS

The application of genomics, proteomics and metabonomics generate vast amounts of data. Therefore, computer-based methods are necessary for data analysis. Bioinformatics provides mathematical models to answer complex questions about cellular molecular processes, such as

- the alteration of gene expression patterns under exposure to toxic substances,
- the functional annotation of genes,
- the reconstruction of pathways and systems biology networks,
- the prediction of effects and side effects.

Different strategies have been developed to achieve this goal: since the basic functions of genes are often similar across species boundaries, cross-species comparisons are useful to set up hypotheses about gene function. This approach is referred to as co-evolutionary profile analysis. Important model organisms for this purpose are, for example, *Drosophila melanogaster*, *Caenorhabditis elegans* and *Saccharomyces cerevisiae*. Genetic information on different species is deposited in the NCBI genome database (http://www.ncbi.nlm.nih.gov/COG/), for example, on groups of orthologues genes. A long-term goal is to identify functional pathways to derive networks and biological systems which are relevant for responds to toxins.

The amount of data stored in different databases allows the exploration of new patterns of relationships, which these data were not originally collected for. This technique is known as data mining, which provides tools for the *in silico* simulation of real situations to

- create testable hypotheses and to examine networks,
- identify new targets for drug development,
- identify new compounds and predict their effect,
- predict toxicity and risk assessments carried out.

In toxicology, predictive *in silico* disease models shall be modelled on an individual basis for each patient allowing statements on prevention, diagnosis, treatment efficacy and side effects.

## III. EXPERIMENTAL MODELS

Phytotherapy is frequently discussed with some reluctance among physicians and scientists from academic medicine. The therapeutic activity is doubted to exceed placebo effects, but repeated case reports in the literature about poisoning by herbal drugs give the impression that medicinal herbs may exert considerable toxicity and especially hepatotoxicity.

Toxic reactions in the liver are common to many natural and synthetic xenobiotic compounds, since orally ingested drugs are first metabolized in the liver by phase I and II enzymes and are then excreted by phase III transporters. Hepatotoxicity is one of the main reasons, why new candidate drugs fail in the drug development process. It is also the major single reason, why approved medications are withdrawn from the market. The Committee for Medicinal Products for Human Use of EMEA developed guidelines for the pharmaceutical industry for the early detection of hepatotoxicity in non-clinical studies (EMEA, 2006, 2008). This guideline points out the usefulness of new technologies such as TGx to study hepatotoxic modes of actions. Further, the EMEA committee recommends the use of metabolically competent hepatocyte cultures.

The problem of current TGx approaches is that they still need maturing and are expensive, slow, highly variable, often qualitative and do not easily yield information useful to decide whether to progress a drug candidate or not. Most crucially, the reliable prediction of toxicity earlier in the discovery/development process is still an issue (Wills, 2007). Therefore, methodological research is necessary to develop experimental settings which are able to develop reliable test models for industrial routine applications. Issues are the most appropriate *in vitro* and *in vivo* test models, experimental reproducibility, the development of predictive models, suitable software tools for data analysis and databases, the identification and validation of biomarkers, as well as quantitative structure–activity relationship (QSAR) modelling.

## A. *IN VITRO* AND *IN VIVO* TEST MODELS

In the past, hepatotoxicity has been tested by experiments in rats or mice—which reveal considerable differences in the metabolism of xenobiotic compounds in comparison to human beings. To overcome this problem and to reduce animal experiments, human or rat hepatocytes have been used. The advantage of the HepG2 cell line is that it is a permanently growing cell line, which allows direct comparability between test results overtime. However, HepG2 cells lost substantial functions of normal liver cells, including a decrease of expression of cytochrome P450 monooxigenase (CYP) phase I enzymes (Hewitt and Hewitt, 2004). Primary hepatocytes isolated from liver tissue and cultured *in vitro* also show a substantial loss of their metabolic capacity (Binda *et al.*, 2003). This loss can partially be prevented by sandwich cultures, where hepatocytes are embedded between two layers of gelled collagen (Farkas and Tannenbaum, 2005). The sandwich configuration mimics the extracellular matrix of the liver resulting in matrix-induced polarization of the hepatocytes (Tuschl *et al.*, 2009). Recently, liver slices were used. Liver slices have the advantage that hepatocytes remain in their natural environment and that other liver cells, such as Kupffer, stellate and endothelial cells are also present, which may also contribute to hepatotoxic effects (Graaf *et al.*, 2007). However, liver biopsies are frequently being taken from diseased patients rather than from healthy volunteers. Biopsies from patients with inflammatory liver diseases or other diseases show altered metabolic reactions. Even normal liver tissue from patients suffering from hepatocellular carcinoma may not fully correspond to the metabolic liver capacity of healthy subjects. In addition, the inter-individual variation has to be considered which bear the danger of irreproducible results. Recently, the HepaRG cell line has been generated, which is a promising alternative (Guillouzo *et al.*, 2007). HepaRG cells polarize in cell culture and show a stable expression of phase I, II and III enzymes and transporters resulting in a high metabolic capacity (Jossé *et al.*, 2008; Kanebratt and Andersson, 2008). Taken together, we conclude that the HepaRG cell line may represent a suitable model to analyse the toxic effects of medicinal plants and herbal products by toxicogenomic methods.

In addition to hepatotoxicity, other organ toxicities also have to be considered as potential hazards of phytotherapy, that is nephrotoxicity, cardiotoxicity, haematotoxicity, immunotoxicity, gastrointestinal toxicity, mutagenicity, embryotoxicity and genotoxicity. Also, allergies and various skin reactions may take place. Many cell models have been described as test systems to investigate these organ toxicities. Further, animal models are still indispensable in toxicology. In addition to rat, mice and other rodents, more sophisticated animal models have been established in recent years, for example, transgenic and knockout mice, humanized mice as well as knockout mutants of *Escherichia coli*, *S. cerevisiae*, *C. elegans*, *D. melanogaster*, zebrafish (*Danio rerio*) and others. Toxicogenomic techniques applied to these animal models are attractive tools for toxicology studies with medicinal plants. Cell lines and animal models are only one side of the coin, since the situation in human subjects may be different than in *in vitro* and *in vivo* models. In the past few years, many groups attempted to developed toxicogenomic methods, which also have the potential for early detection of toxic effects in human beings.

Gou and Gu (2011) demonstrated the potential application of a new transcriptional effect level index (TELI) to convert information-rich toxicogenomic data into integrated and quantitative endpoints. A library of transcriptional fusions of green fluorescent protein (GFP) that includes different promoters for 91 stress-related genes in *E. coli* K12, MG1655 was employed to evaluate the gene expression alteration induced by exposure to four nanomaterials. The TELI values exhibit a characteristic ‘sigmoid’-shaped toxicity dose–response curve. Further, TELI values determined for genes in each stress response category (e.g. oxidative stress, DNA repair) indicated the mode of action-related comparative transcriptional level toxicity among compounds and revealed information of toxic response pathways such as different DNA damage and repair mechanisms among the nanomaterials.

Rooney *et al.* (2008) produced 99 molecular bar-coded *E. coli* gene-deletion mutants specific to DNA repair and damage signalling pathways, and each bar-coded mutant can be tracked in pooled format using bar-code specific microarrays. Our design adapted bar-codes developed for the *S. cerevisiae* gene-deletion project, which allowed us to utilize an available microarray product for pooled gene-exposure studies. Microarray-based screens were used for *en masse* identification of individual mutants sensitive to methyl methanesulphonate (MMS). As expected, gene-deletion mutants specific to direct, base excision and recombinational DNA repair pathways were identified as MMS sensitive in our pooled assay, thus validating our resource. Further, when comparing microarray to traditional plate-based screens both overlapping and distinct results were obtained, which is a novel technical finding, with discrepancies between the two approaches explained by differences in output measurements (DNA content vs. cell mass). The authors discovered that Deltatag and DeltadinG cells showed a filamentation phenotype after MMS exposure, thus accounting for the discrepancy.

Prediction of developmental toxicity *in vitro* could be based on short-time toxicogenomic endpoints in embryo-derived cell lines. Microarray studies in P19 mouse embryo carcinoma cells and mouse embryos have indicated that valproic acid (VPA), an inducer of neural tube defects, deregulated the expression of many genes, including those critically involved in neural tube development. Jergil *et al.* (2011) exposed undifferentiated R1 mouse embryonic stem cells to VPA and VPA analogues and used CodeLink whole-genome expression microarrays to define VPA-responsive genes correlating with teratogenicity. Compared with the non-teratogenic analogue 2-ethyl-4-methylpentanoic acid, VPA and the teratogenic VPA analogue (*S*)-2-pentyl-4-pentynoic acid deregulated a much larger number of genes. Five genes (of ~2500 array probes correlating with the separation) were sufficient to effectively separate teratogens from non-teratogens. A large fraction of the target genes correlating with teratogenicity are functionally related to embryonic development and morphogenesis, including neural tube formation and closure. These marker genes showed a positive predictive value for the teratogens butyrate and trichostatin A, which like VPA and (*S*)-2-pentyl-4-pentynoic acid are known histone deacetylase (HDAC) inhibitors. This indicates that HDAC inhibition may be a major mechanism by which VPA induces gene deregulation and possibly teratogenicity.

Elferink *et al.* (2011) applied precision-cut human liver slices as *in vitro* method for the prediction of human specific toxicity by TGx. The liver slices contain all cell types of the liver in their natural architecture. This is important since drug-induced toxicity often is a multicellular process. Toxicogenomic analysis of rat liver slices was highly predictive for rat *in vivo* toxicity. Observed changes in gene expression were associated with cytoskeleton remodelling, extracellular matrix and cell adhesion, but only minor changes were observed for toxicity-related genes of absorption, distribution, metabolism and excretion (ADME). Principal component analysis showed that changes in gene expression were not associated with age, sex or source of the human livers. The results indicated that precision-cut human liver slices are relatively stable during 24 h of incubation and represent a valuable model for human *in vitro* hepatotoxicity testing despite the human inter-individual variability.

Molecular cancer epidemiology is used to identify human environmental cancer risks by applying a range of effect biomarkers, which tend to be non-specific and do not generate insights into underlying modes of action. Toxicogenomic technologies may improve on this by providing the opportunity to identify molecular biomarkers consisting of altered gene expression profiles. Van Leeuwen *et al.* (2008) investigated associations of gene expression with blood and urinary measures of biomarkers of exposure, early phenotypic effects and tumour markers were investigated. Individual gene expression was measured by means of quantitative polymerase chain reaction in peripheral blood cells of 398 individuals. Importantly, Van Leeuwen *et al.* (2008) observed multiple significant correlations of particular gene expressions with blood and urinary measures of various environmental carcinogens. The authors concluded that gene expression profiling appears promising as a tool for biological monitoring in relation to environmental exposures in humans.

## B. EXPERIMENTAL REPRODUCIBILITY

Scott *et al.* (2011) designed a prospective collaborative study to determine the level of inter- and intra-laboratory reproducibility between three independent laboratories. The genotoxic carcinogen benzo[a]pyrene (B[a]P) and the human hepatoma cell line HepG2 were used to generate three comparable toxicogenomic datasets. High levels of technical reproducibility were demonstrated using a widely employed gene expression microarray platform. While differences at the global transcriptome level were observed between the test centres, a common subset of 400 B[a]P-responsive genes was identified at all test centres which included many genes previously reported in the literature as B[a]P responsive.

Pooling of RNA samples is generally applied to obtain samples that represent the average signal of biological replicates of a single treatment. A direct comparison was made to separately hybridized RNA samples. Pronk *et al.* (2011) observed that pooling RNA from compounds of a class substantially increased power to detect significantly regulated genes between classes because variability between pooled samples was much lower. Biomarkers selected from the pooled and separate design performed equally in classification of compounds to their class and relevant processes were found enriched in both designs. Consequently, pooling of RNA of different compound treated samples can be applied to determine class specific biomarkers and processes at much reduced cost and with limited loss of accuracy.

## C. PREDICTIVE MODELS

Uehara *et al.* (2011) developed a robust gene-based prediction model for early assessment of potential hepatocarcinogenicity of chemicals in rats by using the TGx database, TG-GATEs (Genomics-Assisted Toxicity Evaluation System developed by the Toxicogenomics Project in Japan). The positive training set consisted of high- or middle-dose groups that received six different non-genotoxic hepatocarcinogens during a 28-day period. The negative training set consisted of high- or middle-dose groups of 54 non-carcinogens. The best classifier yielded prediction accuracies for hepatocarcinogenicity of 99% sensitivity and 97% specificity in the training dataset. Pathway analysis of feature genes revealed that the mitogen-activated protein kinase p38- and phosphatidylinositol-3-kinase-centred interactome and the v-myc myelocytomatosis viral oncogene homologue-centred interactome were the two most significant networks, indicating that these genes might be common in early stages of carcinogenesis for both genotoxic and non-genotoxic carcinogens in the rat liver. Further characteristics associated with early response to liver injury were up-regulated genes involved in apoptosis/necrosis, inflammation, oxidative stress and metabolic enzymes (Wang *et al.*, 2011).

Hrach *et al.* (2011) treated primary rat hepatocyte maintained as sandwich cultures were treated with 13 well-known model compounds. Changes in the global gene expression profile were measured and subsequently used for the establishment of a predictive classification model. A subset of 724 genes was capable of discriminating compounds with a misclassification rate of 7.5%. The classification of compounds according to transient changes in global gene expression allowed the correct prediction independently from the knowledge of their underlying toxic mechanisms. These results demonstrate the possibility of *in vitro* gene expression models to contribute to candidate selection early in drug discovery by improving the predictivity of toxicological studies and thereby reducing animal usage in toxicology.

## D. SOFTWARE TOOLS FOR DATA ANALYSIS AND DATABASES

At the FDA's National Center for Toxicological Research (NCTR), Fang *et al.* (2009) developed a public microarray data management and analysis software called ArrayTrack, that is also used in the routine review of genomic data submitted to the FDA. ArrayTrack stores a full range of information related to DNA microarrays and clinical and non-clinical studies as well as the digested data derived from proteomics and metabonomics experiments. In addition, ArrayTrack provides a rich collection of functional information about genes, proteins and pathways drawn from various public biological databases for facilitating data interpretation. Many data analysis and visualization tools are available with ArrayTrack for individual platform data analysis, multiple omics data integration and integrated analysis of omics data with study data. Importantly, gene expression data, functional information and analysis methods are fully integrated so that the data analysis and interpretation process is simplified and enhanced. Using ArrayTrack, users can select an analysis method from the ArrayTrack tool box, apply the method to selected microarray data, and the analysis of results can be directly linked to individual gene, pathway and Gene Ontology analysis. ArrayTrack is publicly available online (http://www.fda.gov/nctr/science/centers/toxicoinformatics/ArrayTrack/index.htm) and the prospective user can also request a local installation version by contacting the authors.

The U.S. National Toxicology Program (NTP) is an interagency programme, whose mission is to evaluate agents of public health concern by developing and applying the tools of modern toxicology and molecular biology. The NTP has published more than 500 reports of the findings and conclusions from its toxicology and carcinogenesis studies (Chhabra *et al.*, 2003). TGx, an emerging scientific field that examines the expression of thousands of genes simultaneously in response to chemical exposure, holds promise for future application to better understand the underlying mechanisms of chemical toxicity. The programme participates in the preparation of national and international toxicity testing guidelines and the findings from NTP studies are widely used for risk assessments by international organizations and federal agencies. The NTP maintains databases that contain toxicity, and health and safety information on a large number of chemicals. These databases are available from the NTP website (http://ntp-server.niehs.nih.gov) and are accessed over 100,000 times a month from around the world.

The Toxicogenomics Project is a 5-year collaborative project by the Japanese government and pharmaceutical companies in 2002. Its aim is to construct a large-scale toxicology database of 150 compounds orally administered to rats. The test consists of conventional toxicology together with the microarray-based gene expression in liver. Takashima *et al.* (2006) examined whether methylcellulose or corn oil employed as vehicle affect the analysis of gene expression. The authors found that corn oil alone affected the food consumption and biochemical parameters mainly related to lipid metabolism, and this accompanied typical changes in the gene expression. Most of the genes modulated by corn oil were related to cholesterol or fatty acid metabolism (e.g. CYP7A1, CYP8B1, 3-hydroxy-3-methylglutaryl-coenzyme A reductase, squalene epoxidase, angiopoietin-like protein 4, fatty acid synthase, fatty acid binding proteins), suggesting that the response was physiologic to the oil intake. Many of the lipid-related genes showed circadian rhythm within a day, but the expression pattern of general clock genes (e.g. period 2, arylhydrocarbon nuclear receptor translocator-like, D site albumin promoter binding protein) were unaffected by corn oil, suggesting that the effects are specific for lipid metabolism.

The Comparative Toxicogenomics Database (CTD) is another public resource that promotes understanding about the effects of environmental chemicals on human health. Currently, CTD describes over 184,000 molecular interactions for more than 5100 chemicals and 16,300 genes/proteins. Davis *et al.* (2009) have leveraged this dataset of chemical–gene relationships to compute similarity indices following the statistical method of the Jaccard index. These scores were used to produce lists of comparable genes ('GeneComps') or chemicals ('ChemComps') based on shared toxicogenomic profiles. These metrics provide a novel way to view and classify genes and chemicals and will help advance testable hypotheses about environmental chemical–gene disease networks. This database is freely available at http://ctd.mdibl.org/.

## E. BIOMARKERS

Current approaches to biomarker discovery through gene expression profiling usually utilize a single or few strains of rodents, limiting the ability to detect biomarkers that may represent the wide range of toxicity responses typically observed in genetically heterogeneous human populations. To enhance the utility of animal models to detect response biomarkers for genetically diverse populations, Harrill *et al.* (2009) used a laboratory mouse strain diversity panel. Specifically, mice from 36 inbred strains derived from *Mus mus musculus*, *Mus mus castaneous* and *Mus mus domesticus* origins were treated with a model hepatotoxic agent, acetaminophen. Gene expression profiling was performed on liver tissue collected at 24 h after dosing. The authors identified 26 population-wide biomarkers of response to acetaminophen hepatotoxicity in which the changes in gene expression were significant across treatment and liver necrosis score but not significant for individual mouse strains. Importantly, most of these biomarker genes are part of the intracellular signalling involved in hepatocyte death and include genes previously associated with acetaminophen-induced hepatotoxicity, such as cyclin-dependent kinase inhibitor 1A (p21) and interleukin 6 signal transducer (Il6st), and genes not previously associated with acetaminophen, such as oncostatin M receptor (Osmr) and MLX interacting protein like (Mlxipl). This multistrain approach may facilitate the understanding genotype-independent toxicity responses and facilitate identification of novel targets of therapeutic intervention.

### F. QUANTITATIVE STRUCTURE–ACTIVITY RELATIONSHIP MODELLING

QSAR modelling and TGx are typically used independently as predictive tools in toxicology. Low *et al.* (2011) evaluated the power of several statistical models for predicting drug hepatotoxicity in rats using different descriptors of drug molecules, namely, their chemical descriptors and TGx profiles. The records were taken from the Toxicogenomics Project rat liver microarray database containing information on 127 drugs (http://toxico.nibio.go.jp/datalist.html).

*In silico* approaches based on QSAR are rapid and inexpensive and thus have been investigated for such purposes. A more expensive approach based on short-term animal studies with TGx represents another attractive option for this application. Liu *et al.* (2011) developed predictive models for rodent liver carcinogenicity using gene expression data generated from short-term animal models at different time points and QSAR. The study was focused on the prediction of non-genotoxic carcinogenicity since the genotoxic chemicals can be inexpensively removed from further development using various *in vitro* assays individually or in combination. Liu *et al.* (2011) identified 62 chemicals whose hepatocarcinogenic potential was available from the National Center for Toxicological Research liver cancer database (NCTRlcdb). The authors found that the TGx models outperformed QSAR in every aspect of modelling. For example, the TGx models' predictive accuracy was much higher for an independent validation set than that of a QSAR model.

## IV. HERBAL HEPATOTOXICITY AND TOXICOGENOMICS

The liver is the principal organ for the metabolism of herbs. Therefore, hepatotoxicity is the most frequent toxicity observed with the consumption of herbal products. Natural products are frequently activated by phase I enzymes. This results in electrophilic metabolites and free radicals as byproducts. These can be toxic, because they bind covalently to proteins, lipids and DNA and cause oxidative stress. Hepatotoxicity may be caused by damaging certain organelles (mitochondria, endoplasmic reticulum, cell nucleus). At the molecular level, disruption of signal transduction and transcription factors, and consequently alterations of gene expression profiles may take place. Further, glutathione or ATP may be depleted. This can lead to induction of apoptosis or necrosis. On the other hand, oxidative stress and covalent binding to nucleophilic molecules (e.g. DNA) may activate both the innate and the adaptive immune system and induce defence responses in liver tissue. Both factors contribute to pathogenesis in the liver. The severity of hepatotoxicity depends on several factors: exposure duration and concentration of toxic substances, environmental and genetic factors that influence the drug metabolism and transport, as well as repair and regeneration processes.

Acute liver toxicity may manifest as acute hepatitis, cholestasis or as a mixed appearance end in the development of hepatocellular carcinoma. Jaundice is a typical symptom of severe hepatitis. Cholestasis develops from injury or inflammation of the hepatic ducts. Even if it may not be life-threatening in all cases, it may lead to destruction of the hepatic ducts and in rare cases to liver cirrhosis.

Hepatotoxicity is at least in some cases, predictable, dose dependent and reproducible in an animal model. Toxic reactions of this type are referred to as direct toxic damage. In most cases, however, hepatotoxicity is very unpredictable, not dose dependent and not reproducible in an animal model (idiosyncratic damage). Here, reactions of the immune system may be responsible. Many food additives and medicinal herbs exert hepatotoxic effects. This is particularly noteworthy, as herbal products are often misunderstood by the public as 'natural medicine' with no side effects.

TGx of medicinal plants and herbal products is still in its infancy. An illustrative example how TGx facilitates the detection of hepatotoxicity not visible in animal experiments was given by Anderson *et al.* (2009). Butterbur extracts (*Petasites hybridus*) are recommended for the prevention of migraine, but pharmacovigilance reports may be suggestive of rare hepatobiliary toxicity. To identify mechanisms of hepatotoxicity, Anderson *et al.* (2009) investigated liver function *in vitro* at $>$170-fold of therapeutic $C_{max}$ levels. Only with extracts rich in petasin (37% petasin) and at high and well above therapeutic doses, liver toxicity was observed. A toxicogenomic approach applied to hepatocyte cultures enabled hypothesis generation and was highly suggestive for extracts high in petasin content to impair bile acid transport and lipid and protein metabolism. As a matter of fact, neither chronic rat *in vivo* nor rat *in vitro* studies predicted reliably hepatotoxicity, therefore reemphasizing the utility of human-based *in vitro* investigations for the development of safe medicinal products. Finally, TGx enabled the characterization of a novel butterbur extract with no signals for hepatotoxicity.

## V. CONCLUSIONS AND PERSPECTIVES

Omics technologies represent attractive methods for the prediction of desirable and undesirable drug effects. Testing of readily accessible bio-fluids or tissue samples should result in the identification of meaningful biomarkers. Biomarkers are biochemical and molecular biological parameters. Clinical parameters such as neuroimaging (magnetic resonance imaging and positron emission tomography) are not counted among the biomarkers. Biomarkers may be effectively used to detect intoxications caused by medicinal plants and herbal products at an early stage.

With the help of pharmacogenetic biomarkers, therapeutic interventions can be adapted for each individual patient. This is the goal of individualized therapy. According to the results of pharmacogenetic screening prior to therapy effective drugs are chosen and used with individual optimal dose. At the same time, undesirable side effects less effective drugs are avoided. Before and during therapy, biomarkers can be used to monitor the effectiveness and toxicity of herbal medications. After phytotherapy, biomarkers are used for monitoring the sustainability of the therapeutic success and the absence of toxic side effects.

## REFERENCES

Anderson, N., Meier, T. and Borlak, J. (2009). Toxicogenomics applied to cultures of human hepatocytes enabled an identification of novel petasites hybridus extracts for the treatment of migraine with improved hepatobiliary safety. *Toxicological Sciences* **112**(2), 507–520.

Binda, D., Lasserre-Bigot, D., Bonet, A., Thomassin, M., Come, M. P., Guinchard, C., Bars, R., Jacqueson, A. and Richert, L. (2003). Time course of cytochromes P450 decline during rat hepatocyte isolation and culture: Effect of L-NAME. *Toxicology In Vitro* **17**(1), 59–67.

Chavan, P., Joshi, K. and Patwardhan, B. (2006). DNA microarrays in herbal drug research. *Evidence Based Complementary and Alternative Medicine* **3**(4), 447–457.

Chen, X. W., Serag, E. S., Sneed, K. B. and Zhou, S. F. (2011). Herbal bioactivation, molecular targets and the toxicity relevance. *Chemico-Biological Interactions* **192**(3), 161–176.

Chhabra, R. S., Bucher, J. R., Wolfe, M. and Portier, C. (2003). Toxicity characterization of environmental chemicals by the US National Toxicology Program: An overview. *International Journal of Hygiene and Environmental Health* **206**(4–5), 437–445.

Davis, A. P., Murphy, C. G., Saraceni-Richards, C. A., Rosenstein, M. C., Wiegers, T. C., Hampton, T. H. and Mattingly, C. J. (2009). GeneComps and ChemComps: A new CTD metric to identify genes and chemicals with shared toxicogenomic profiles. *Bioinformation* **4**(4), 173–174.

Efferth, T. and Kaina, B. (2011). Toxicities by herbal medicines with emphasis to traditional chinese medicine. *Current Drug Metabolism* **12**, 989–996.

Elferink, M. G., Olinga, P., van Leeuwen, E. M., Bauerschmidt, S., Polman, J., Schoonen, W. G., Heisterkamp, S. H. and Groothuis, G. M. (2011). Gene expression analysis of precision-cut human liver slices indicates stable expression of ADME-Tox related genes. *Toxicology and Applied Pharmacology* **253**(1), 57–69.

European Medicines Agency, Committee for Medicinal Products for Human Use (CHMP) Draft (2006). Guideline on detection of early signals of drug-induced hepatotoxicity in non-clinical studies. http://www.rsihata.com/updateguidance/emea/2006/15011506en.pdf. (Accessed October 01, 2011).

European Medicines Agency, Committee for Medicinal Products for Human Use (CHMP) Draft (2008). Non-clinical guideline on drug-induced hepatotoxicity. http://www.ema.europa.eu/docs/en_GB/document_library/Scientific_guideline/2009/09/WC500003355.pdf. (Accessed October 01, 2011).

Fang, H., Harris, S. C., Su, Z., Chen, M., Qian, F., Shi, L., Perkins, R. and Tong, W. (2009). ArrayTrack: An FDA and public genomic tool. *Methods in Molecular Biology* **563**, 379–398.

Farkas, D. and Tannenbaum, S. R. (2005). Characterization of chemically induced hepatotoxicity in collagen sandwiches of rat hepatocytes. *Toxicological Sciences* **85**(2), 927–934.

Gou, N. and Gu, A. Z. (2011). A new transcriptional effect level index (TELI) for toxicogenomics-based toxicity assessment. *Environmental Science and Technology* **45**(12), 5410–5417.

Graaf, I. A., Groothuis, G. M. and Olinga, P. (2007). Precision-cut tissue slices as a tool to predict metabolism of novel drugs. *Expert Opinion on Drug Metabolism & Toxicology* **3**(6), 879–898.

Guillouzo, A., Corlu, A., Aninat, C., Glaise, D., Morel, F. and Guguen-Guillouzo, C. (2007). The human hepatoma HepaRG cells: A highly differentiated model for studies of liver metabolism and toxicity of xenobiotics. *Chemico-Biological Interactions* **168**(1), 66–73.

Harrill, A. H., Ross, P. K., Gatti, D. M., Threadgill, D. W. and Rusyn, I. (2009). Population-based discovery of toxicogenomics biomarkers for hepatotoxicity using a laboratory strain diversity panel. *Toxicological Sciences* **110**(1), 235–243.

Hewitt, N. J. and Hewitt, P. (2004). Phase I and II enzyme characterization of two sources of HepG2 cell lines. *Xenobiotica* **34**(3), 243–256.

Hrach, J., Mueller, S. O. and Hewitt, P. (2011). Development of an in vitro liver toxicity prediction model based on longer term primary rat hepatocyte culture. *Toxicology Letters* **206**(2), 189–196.

Jergil, M., Forsberg, M., Salter, H., Stockling, K., Gustafson, A. L., Dencker, L. and Stigson, M. (2011). Short-time gene expression response to valproic acid and valproic acid analogs in mouse embryonic stem cells. *Toxicological Sciences* **121**(2), 328–342.

Jossé, R., Aninat, C., Glaise, D., Dumont, J., Fessard, V., Morel, F., Poul, J. M., Guguen-Guillouzo, C. and Guillouzo, A. (2008). Long-term functional stability of human HepaRG hepatocytes and use for chronic toxicity and genotoxicity studies. *Drug Metabolism and Disposition* **36**(6), 1111–1118.

Kanebratt, K. P. and Andersson, T. B. (2008). Evaluation of HepaRG cells as an in vitro model for human drug metabolism studies. *Drug Metabolism and Disposition* **36**(7), 1444–1452.

Kober, M., Pohl, K. and Efferth, T. (2008). Molecular mechanisms underlying St. John's wort drug interactions. *Current Drug Metabolism* **9**(10), 1027–1037.

Liu, Z., Kelly, R., Fang, H., Ding, D. and Tong, W. (2011). Comparative analysis of predictive models for nongenotoxic hepatocarcinogenicity using both toxicogenomics and quantitative structure-activity relationships. *Chemical Research in Toxicology* **24**(7), 1062–1070.

Low, Y., Uehara, T., Minowa, Y., Yamada, H., Ohno, Y., Urushidani, T., Sedykh, A., Muratov, E., Kuz'min, V., Fourches, D., Zhu, H., Rusyn, I. *et al.* (2011). Predicting drug-induced hepatotoxicity using QSAR and toxicogenomics approaches. *Chemical Research in Toxicology* **24**(8), 1251–1262.

Pronk, T. E., van der Veen, J. W., Ezendam, J., Van Loveren, H. and Pennings, J. L. (2011). Effects of pooling RNA from samples treated with different compounds for determining class specific biomarkers and processes in toxicogenomics. *Toxicology In Vitro* **25**, 1841–1847.

Rooney, J. P., Patil, A., Zappala, M. R., Conklin, D. S., Cunningham, R. P. and Begley, T. J. (2008). A molecular bar-coded DNA repair resource for pooled toxicogenomic screens. *DNA Repair (Amsterdam)* **7**(11), 1855–1868.

Scott, D. J., Devonshire, A. S., Adeleye, Y. A., Schutte, M. E., Rodrigues, M. R., Wilkes, T. M., Sacco, M. G., Gribaldo, L., Fabbri, M., Coecke, S., Whelan, M., Skinner, N. *et al.* (2011). Inter- and intra-laboratory study to determine the reproducibility of toxicogenomics datasets. *Toxicology* **290**, 50–58.

Takashima, K., Mizukawa, Y., Morishita, K., Okuyama, M., Kasahara, T., Toritsuka, N., Miyagishima, T., Nagao, T. and Urushidani, T. (2006). Effect of the difference in vehicles on gene expression in the rat liver—Analysis of the control data in the Toxicogenomics Project Database. *Life Sciences* **78**(24), 2787–2796.

Tuschl, G., Hrach, J., Walter, Y., Hewitt, P. G. and Mueller, S. O. (2009). Serum-free collagen sandwich cultures of adult rat hepatocytes maintain liver-like properties long term: A valuable model for in vitro toxicity and drug-drug interaction studies. *Chemico-Biological Interactions* **181**(1), 124–137.

Uehara, T., Minowa, Y., Morikawa, Y., Kondo, C., Maruyama, T., Kato, I., Nakatsu, N., Igarashi, Y., Ono, A., Hayashi, H., Mitsumori, K., Yamada, H. *et al.* (2011). Prediction model of potential hepatocarcinogenicity of rat hepatocarcinogens using a large-scale toxicogenomics database. *Toxicology and Applied Pharmacology* **255**(3), 297–306.

van Leeuwen, D. M., Gottschalk, R. W., Schoeters, G., van Larebeke, N. A., Nelen, V., Baeyens, W. F., Kleinjans, J. C. and van Delft, J. H. (2008). Transcriptome analysis in peripheral blood of humans exposed to environmental carcinogens: A promising new biomarker in environmental health studies. *Environmental Health Perspectives* **116**(11), 1519–1525.

Wang, T., Papoutsi, M., Wiesmann, M., Decristofaro, M., Keselica, M. C., Skuba, E., Spaet, R., Markovits, J., Wolf, A., Moulin, P., Pognan, F., Vancutsem, P. *et al.* (2011). Investigation of correlation among safety biomarkers in serum, histopathological examination, and toxicogenomics. *International Journal of Toxicology* **30**(3), 300–312.

Wills, Q. (2007). SimuGen Ltd: Reliable, early prediction of drug toxicity with toxicogenomics, human cell culture and computational models. *Pharmacogenomics* **8**(8), 1081–1084.

Yang, Y., Blomme, E. A. and Waring, J. F. (2004). Toxicogenomics in drug discovery: From preclinical studies to clinical trials. *Chemico-Biological Interactions* **150**(1), 71–85.

# Herbal Medicine: Toxicity and Recent Trends in Assessing Their Potential Toxic Effects

CHIT SHING JACKSON WOO, JONATHAN SEE HAN LAU AND HANI EL-NEZAMI[1]

*School of Biological Sciences, Faculty of Science, The University of Hong Kong, Hong Kong SAR, China*



## ABSTRACT

Herbal medicine has been used for thousands of years. It is estimated that 80% of world population rely on traditional herbal medicine for primary health care. In recent years, herbal remedies have been considered as dietary supplement for disease prevention and as alternative/complementary medicine. A wide variety of herbal medicines are readily available in the market all over the world. With the rising utilisation of herbal products, safety and efficacy of herbal medicine have

[1]Corresponding author: E-mail: elnezami@hku.hk

Advances in Botanical Research, Vol. 62 0065-2296/12 $35.00

DOI: 10.1016/B978-0-12-394591-4.00009-X

become a public health concern. Adverse health effects associated with herbal products could be attributed to both inherent toxic effects of herbal medicine and toxicities induced by adulterants/contaminants. Increasing evidence, regarding side effects of herbal medicine, has highlighted the demand and necessity of toxicological studies for herbal products. Toxicology constitutes an essential role in the development of herbal medicines. With the advancements of analytical techniques and molecular technology, coupling with the conventional test systems, the '-omic-' technology makes a significant contribution to the predictive and preclinical toxicology of herbal medicine. In this chapter, side effects related to herbal medicine and its adulterants/contaminants are summarised. The recent application of '-omic-' technology for toxicological evaluation of herbal products is also illustrated.

# I. INTRODUCTION

Herbal medicine is also known as phytochemicals or botanical medicine. According to World Health Organization (WHO), herbal medicine includes 'herbs, herbal materials, herbal preparations and finished herbal products, that contain active ingredient parts of plants or other plant materials or combinations thereof' (Robinson and Zhang, 2011). Herbal medicine is generally considered as an integral part of dietary supplement. There is a growing interest in herbal medicine due to its long history of application and general belief that herbs are natural and intrinsically safe. According to WHO, approximately 4 billion of people, 80% of the world population, rely on traditional herbal medicine for their primary health care (Akerele, 1993). In recent years, utilisation of herbal remedies as a dietary supplement for disease prevention or as alternative/complementary medicine (CAM) for disease treatment has become increasingly popular. A wide variety of herbal medicine/products are readily available in the market all over the world. It is estimated that a significant percentage of population in developed countries such as Canada (70%), France (49%), Australia (48%), USA (42%) and Belgium (31%) have used CAM at least once for health care (WHO, 2002). The increasing growth in the use of herbal medicine indirectly indicates the dissatisfaction with the conventional medicine in developed countries (Chan, 2003). This chapter attempt to review the current knowledge in relation to the safety of herbal medicine. Additionally, an overview on the application of various '-omics-' technologies to assess the safety of herbal medicine is presented.

# II. RECENT CONCERNS WITH REGARD TO THE SAFETY OF HERBAL MEDICINES

Traditionally, herbal medicine was discovered through trial and error. The accumulated experience across generations forms the basis of today's traditional medicine. Although a long history of traditional use may represent the evidence of safety and efficacy, this kind of experience is not reliable. It may not provide the information regarding reactions that are subtle and with a prolonged latency period (De Smet, 2004). There are numerous studies revealing the association between herbal medicine and adverse effects (De Smet, 2004; Huxtable, 1990). Still, the mechanisms of vast majority of herbal medicine are largely unknown and have not been evaluated in randomised clinical trials. Assessment of the safety and efficacy of herbal medicine has been a growing concern nowadays. Toxicities related to herbal medicine could be attributed to two major factors, intrinsic (direct) and extrinsic (indirect) (Drew and Myers, 1997). Herbal toxicities that are directly in relation to the presence of active chemical constituents in the herbs such as ephedrine-type alkaloids in Ma Huang are regarded as the intrinsic effects. Toxicities related to the extrinsic factors including contamination, adulteration and misidentification of herbal products are generally associated with the presence of foreign toxic substances instead of the herbs themselves.

### A. INTRINSIC ADVERSE EFFECTS

Adverse effects of herbal medicine could be classified into four categories as in orthodox medicine, according to De Smet (1995) (Table I). *Type A* (acute/augmented) reactions include overdose reactions and interactions with pharmaceuticals. It is mainly related to the inherent pharmacological properties of herbal products. From the toxicological perspective, whether the substance is a remedy or poison, it all depends on the dose level. Although herbal medicine has been demonstrated to be an effective remedy for thousands of years, inappropriate consumption/over dosage of herbal medicine could result in adverse drug effects (ADRs). Various organs and systems could be affected such as liver, kidney, digestive system, nervous system and cardiovascular system (Zeng and Jiang, 2010). *Type B* (bizarre/idiosyncratic) reactions are the most common adverse reactions induced by herbal products (Smolinske, 2005). Reactions can range from dermatitis to anaphylactic shock, which is the most severe type of allergic reaction. *Type C* (chronic/cumulative) reactions are chronic effects resulting from long-term therapy, which are well-known and anticipative. *Type D* (delayed) reactions are not commonly reported because of the lack of systemic evaluation for herbal medicine. More delayed effects of herbal medicine are expected to become apparent in the future. Table II shows the adverse effects related to commonly used herbs.

In addition, herbal remedies are always prescribed as a mixture of medicinal plants and have been widely used as the complementary medicine together with the conventional medicine. Interaction between herbs, drugs and foods could occur when they are concurrently present in the body

TABLE I
*Types of Adverse Drug Reaction*[a]

| Drug reactions | Characteristics | Example |
| --- | --- | --- |
| Type A | Pharmacologically predictable<br>Dose dependent<br>Common | Hepatotoxicity induced by Kava–kava (Escher *et al.*, 2001; Humberston *et al.*, 2003; Stickel *et al.*, 2003). |
| Type B | Idiosyncratic and Allergic<br>Pharmacologically unpredictable<br>Dose/time independent<br>Rare<br>Environmental and genetic dependent | Acute asthma, hives and anaphylaxis caused by Echinacea (Mullins and Heddle, 2002). |
| Type C | Developed due to long-term use<br>Well described<br>Anticipative | Hypokalemic paralysis due to long-term licorice ingestion (Lin *et al.*, 2003) |
| Type D | Delayed effects (e.g. mutagenicity, carcinogenicity and teratogenicity) | Carcinogenesis associated to aristolochic acids in *Aristolochia* species (De Smet, 1997) |

[a]Modified from De Smet (1995) and Edwards and Aronson (2000).

TABLE II
*Adverse Effects Related to Commonly Used Herbs*[a]

| Herb | Adverse effects |
| --- | --- |
| Echinacea | Acute asthma, Anaphylaxis |
| Ginkgo biloba | Gastrointestinal symptoms, headache, nausea, vomiting |
| Ginseng | Diarrhoea, euphoria, headache, hypertension, hypotension, insomnia, mastalgia, nausea, vaginal bleeding |
| Kava | Hepatitis, reversible yellowish discolouring of skin; nails and hair, visual disturbances, dizziness, stupor, gastrointestional discomfort, extrapyramidal effects (rare) |
| Liquorice root | Hypokalemia, hypertension, arrhythmias, edema |
| Ma Huang | Hallucination, paranoia |
| Saw palmentto | Constipation, decreased libido, diarrhoea, headache, hypertension, nausea, urine retention |
| St. John's wort | Nausea, allergic reactions |

[a]Information compiled from Edzard (1998) and Ernst (2002a).

(Table III). These interactions can occur in both pharmacokinetic and pharmacodynamic manner. Pharmacokinetic interaction may alter absorption, distribution, metabolism or excretion of drug, hence affecting the therapeutic properties. Pharmacodynamic interaction influences the molecular target that mediates different physiological responses. The presence of various constituents may modulate both pharmacological and toxicological effects.

### B. EXTRINSIC ADVERSE EFFECTS

Apart from the intrinsic herbal toxicity, herbs-related toxic effects may be due to the presence of toxic substances other than the herbals themselves. Contamination, adulteration and misidentification of medicinal plants are the common possibilities in relation to the adverse effects. Due to the poor

TABLE III

*Summary of Herb–Drug Interactions with Commonly Used Herbs*[a]

| Herb | Drug | Interaction |
|---|---|---|
| Garlic | Chlorpropamide | Hypoglycaemia |
| | Paracetamol | Changes in pharmacokinetic variables |
| | Warfarin | Increase international normalised ratio (INR) |
| Ginkgo biloba | Aspirin | Spontaneous hyphaema |
| | Thiazide diuretic | Increased blood pressure |
| | Trazodone | Coma |
| | Warfarin | Intracerebral haemorrhage |
| Ginseng | Pheneizine | Insomnia, headache, tremulousness, mania |
| | Warfarin | Decrease INR |
| Kava | Aplrazolam | Lethargic and disoriented state |
| | Levodopa | Increase in the duration and number of 'off' periods |
| St. John's wort | Amitriptyline | Decrease amitriptyline and its metabolite nortriptyline plasma concentrations |
| | Cyclosporin | Decrease cyclosporine plasma concentration |
| | Dextromethorphan | Increase metabolism of dextromethorphan |
| | Digoxin | Decrease digoxin plasma concentration |
| | Indinavir | Decrease indinavir plasma concentration |
| | Loperamide | Brief episode of acute delinium |
| | Nefazodone | Nausea, vomiting, headache |
| | Oral contraceptive | Alter menstrual bleeding |
| | Paroxetine | Nausea, weakness, fatigue, groggy, lethargic state |
| | Phenprocoumon | Decrease anticoagulant effect |
| | Sertaline | Dizziness, nausea, vomiting, headache, epigastric pain, anxiety, confusion |
| | Theophylline | Decrease theophylline plasma concentration |
| | Warfarin | Reduce INR |

[a]Information extracted from Izzo and Ernst (2001).

quality control of the production and preparation of herbal products, the occurrence of heavy metals and microbial toxins has been commonly reported worldwide.

*1. Contamination*

Heavy metal contamination has continued to be a global concern because of their cumulative behaviour and potential health hazards on humans (Rao *et al.*, 2011). Heavy metal contamination of herbal products is a serious problem in Asia (Bateman *et al.*, 1998). Among the heavy metals, lead (Pb), mercury (Hg), cadmium (Cd) and arsenic (As) are commonly found in herbal medicine (Chan, 2003) and of special public health concern due to their potential toxic effects at very low concentrations (Das *et al.*, 1997; Zahir *et al.*, 2005). Medicinal plants are subjected to heavy metal contamination during growth, development and processing (Shad *et al.*, 2008). Contamination of agricultural soil, irrigation water and air resulting from pollution, fertilisers and pesticides contribute to the uptake and accumulation of heavy metals in medicinal plants (Ajasa *et al.*, 2004). Ingestion of heavy metals results in the accumulation of heavy metals in different organs and may interfere with the normal functions, thus eliciting serious health effects. Therefore, WHO recommends that the presence of heavy metals have to be checked from medicinal plants that form the raw materials for the final product. The maximum permissible limits of toxic metals such as Pb, As and Cd are also regulated (WHO, 1998).

Fungal/microbial contamination has been a global concern for decades. In spite of the extensive research on fungal contamination in foods, there is also increasing evidence with regard to the occurrence of microbial toxins in herbal products (Bugno *et al.*, 2006; Gautam *et al.*, 2009, Sewram *et al.*, 2006). Along the production and distribution process, medicinal plants are subjected to both field and post-harvest contaminations by various toxic fungi and bacteria. Mycotoxins, one type of microbial toxins, are secondary metabolites produced by fungi in particular *Aspergillus*, *Penicillium* and *Fusarium* species. Mycotoxins have been demonstrated to induce diverse toxic effects including hepatotoxicity, nephrotoxicity, immunotoxicity, neurotoxicity, carcinogenicity and teratogenicity. Because of their detrimental impact on human health, mycotoxin exposure and their health implications have been increasingly recognised especially over the past three decades and become a subject of international importance (FAO, 1991). Among numerous of mycotoxins, those causing health hazards to human are always of the great concern. Aflatoxins, ochratoxin A, fumonisins, deoxynivalenol and zearalenone have been detected in medicinal plants worldwide (Gray *et al.*, 2004; Martins *et al.*, 2001; Omurtag and Yazicioglu, 2004;

Sewram *et al.*, 2006; Tassaneeyakul *et al.*, 2004; Yang *et al.*, 2005; Yue *et al.*, 2010). The occurrence of mycotoxins in different herbal products indicates that mycotoxin contamination in herbal medicine is in a widespread manner. The presence of mycotoxins in herbal medicines may lead to drug interaction, influence the therapeutic effect and potentiate the adverse effects.

2. *Adulteration*

Pharmaceutical adulteration is another safety concern of herbal medicine. Adulteration refers to the presence of chemical substances that are not labelled or prescribed as part of the intended use in herbal medicine (Huang *et al.*, 1997). A number of herbal remedies such as the Black Pearl Pills for arthritis (Gertner *et al.*, 1995) have been reported to contain unlabeled pharmaceutical ingredients (Table IV). In Taiwan, an average of 23.7% of traditional Chinese medicine (TCM) was found to contain drugs with more than half of the adulterated TCM with at least two adulterants (Huang *et al.*, 1997). Because adulterants are not indicated, it may raise the possibility of overdose and interactions, hence, serious outcomes (Ergil *et al.*, 2002). For instance, an epidemic hepatotoxicity associated with the adulteration of *N*-nitroso-fenfluramine in Chinese weight loss herbal products was reported in Japan (Kanda *et al.*, 2003).

Adulteration also accounts for the presence of heavy metals in herbal medicines. Pb, As and Hg are some commonly used active ingredients in

TABLE IV

*Some Adulterants Found in Herbal Medicines*[a]

| Heavy metals | Drugs | |
|---|---|---|
| Aluminium | Acetaminophen | Ibuprofen |
| Arsenic | Aminopyrine | Indomethacin |
| Cadimum | Aspirin | Ketoprofen |
| Copper | Betamethasone | Mefenamic acid |
| Lead | Caffeine | Papaverine |
| Mercury | Chlordiazepoxide | Paracetamol |
| Thallium | Chlormezanone | Phenacetin |
| Tin | Chloroxazone | Penylbutazone |
| Zinc | Corticosteroids | Piroxicam |
| | Diazepam | Prednisolone |
| | Diclofenac | Salicylamide |
| | Ephedrine | Theophylline |
| | Ethoxybenzamide | Thiazide diuretics |
| | Hydrochlorothiazide | |

[a]Information compiled from Bensoussan and Myers (1996), Huang *et al.* (1997) and Ernst and Thompson Coon (2001).

some traditional medicines at high level (Chan *et al.*, 1993; Ernst, 2002b). However, improper preparation of those herbal remedies containing heavy metals may result in intoxication. A number of reports regarding heavy metal intoxications associated with consumption of herbal medicine have been published. For instance, Pb poisoning in babies occurred after consumption of herbal mixtures containing Po Ying Tan (Chan, 1994).

### 3. *Misidentification*

Medicinal plants and herbal remedies are easily misidentified because of similar appearance, confused nomenclature and complexity of processed products (Zhao and Li, 2007). There is always a risk of misidentification and mislabelling. One of the key episodes of misidentification is the advertent substitution of plantain (*Plantago major*) with Woolly foxglove (*Digitalis lanata*) in the herbal products due to the similarity of leaf appearance. It resulted in hospitalisation of a 23-year-old woman for complete heart block after the intake of an herbal cleansing product Chomper (Slifman *et al.*, 1998).

## III. INCREASING DEMAND OF TOXICOLOGICAL EVALUATION OF HERBAL MEDICINES

Although the use of herbal medicines is increasingly popular nowadays, the majority of herbal products have been marketed without reliable scientific evidence and any mandatory safety and toxicity evaluation in most countries (Bandaranayake, 2006). The number of reports of ADRs related to herbal medicine from China have been increased from 173,000 in 2005 to 547,000 by 2007 after the State Food and Drug Administration improved regulation of ADR monitoring in 2004 (Shaw, 2010). Due to the increasing evidence regarding side effects of herbal medicine, countries have published guidance documents controlling the herbal products. There is an increasing demand of toxicological evaluation for herbal medicines.

*In vivo* studies are considered as the gold standard in toxicology testing. However, ethical issues have arisen, especially in the case of chronic and subacute toxicity testing. For carcinogenicity testing, the course of the assay extends over the lifetime of the animal, which is 2 years in the case of rodents. Associated with animal testing are increasing costs for upkeep. The International Conference on Harmonization (ICH) has made a commitment for reduced animal use in genotoxicity studies by utilising specific *in vitro* models rather than the whole organism (ICH, 2008). For instance, the use of cell models for $IC_{50}$ (inhibition concentration of 50%) determination, genotoxicity assays and identification of mechanistic pathways.

Herbal medicine differs from Western medicine in that a concoction usually contains multiple active components. For example, Yin Chen Hao Tang (YCHT) is a TCM formula which has been shown to contain 45 compounds (Wang *et al.*, 2008b). The number of active components has led to discussion of organ toxicity and interactions between active components and contaminants. Due to the multitude of active components in herbal medicine, toxicity testing can be difficult as an isolated compound may have low toxic effects but could be potentiated by another compound within the same concoction. Also, the therapeutic dose in herbal medicine may be close to toxic levels. One example is nux vomica, which contains strychnine (Behpour *et al.*, 2011). Five to ten milligramme of nux vomica can result in an adverse reaction in adults, while 30 mg can result in death (Liu, 1998). As such, identifying acute toxicity, lethal dose and $IC_{50}$ is important, organ toxicity has also been associated with herbal medicine and is one of the largest issues preventing its widespread use as a complementary medicine in Western nations. Organ toxicity is a wide reaching term with regard to damage resulting from tissue specific cellular death caused by chemical interactions or by products from another series of interactions. Multiple organ damage has been observed after usage of herbal medicine and is the common result of chronic or subchronic exposure to toxic agent (Jha and Rathi, 2008). A number of advanced analytical techniques and biological experimental models have been used in toxicological studies and safety evaluation. In addition to the conventional toxicological studies such as the *in vitro* and *in vivo* test systems, a newly evolved systems biological approach has been suggested for predictive and preclinical toxicology in herbal medicinal research. The '-omic-' technologies like genomics, proteomics and metabonomics are high-throughput platforms, which act as powerful tools for toxicological studies of herbal medicine. Toxicogenomics is an integration of toxicology and molecular science to study the changes of mRNA expression, protein expression and metabolites profiling in response to toxic agent exposure (Oberemm *et al.*, 2005; Youns *et al.*, 2010) (Table V).

## A. TOXICOTRANSCRIPTOMICS

There has been a rising trend in the application of DNA microarrays for toxicological study of herbal medicine, which falls under the umbrella term of toxicotranscriptomics. In toxicology, genomics is mainly referring to transcriptomics for the depiction of gene transcriptional level in response to toxic agent exposure (Ulrich-Merzenich *et al.*, 2007). When an organism is

TABLE V

*Toxicogenomics Technologies and Limitations*[a]

| Name | Definition | Limitations |
|---|---|---|
| Toxicotranscriptomics | Study of a toxin's effects on gene transcription | Limited to gene transcription and no evidence of physiological difference |
| Toxicoproteomics | Study of a toxin's effect on translational expression level and activity | Toxic effects may be localised or have no observable effect |
| Toxicometabonomics | Identification of metabolites and biomarkers after interactions with toxins | Biomarker presence may not indicate toxic effect |
| Systems biology | Integration of toxicotranscriptomics, toxicoproteomics and toxicometabolomics for rapid identification for new compounds | Predictive ability but requires experimental confirmation |

[a]Information compiled from Cutler (2003), Youns *et al.* (2010) and Zhang *et al.* (2010).

exposed to xenobiotics, alteration of cellular gene expression occurs due to the modulation of specific signalling pathways. Gene transcriptions that are positively or negatively affected by treatment can be identified (Barrett and Kawasaki, 2003). The DNA microarray, also known as biochip or gene chip, is a common tool in genomics study. Oligonucleotide-based array and cDNA array are the two major methodologies in transcriptomics study (Oberemm *et al.*, 2005). Because the DNA microarray technology allows parallel monitoring of gene expression for tens of thousands of genes in an equivalent amount of time, this revolutionary platform provides the most comprehensive profile of gene expressions and allows pattern comparison of gene expression in dose and time context (Hamadeh *et al.*, 2001; Wetmore and Merrick, 2004). In a study on *Echinacea purpurea* extract, up-regulation of NFκB, CCL2, CCL5 and IL-8 was observed at 4 h, while IFN-α and LILRB3 mRNA were decreased indicating immune modulatory effects of this extract using the microarray analysis (Wang *et al.*, 2008a).

Although the microarray technology has been shown to be a powerful tool, there are certain limitations. Due to several factors such as cross hybridisation and sequence-specific binding anomalies, DNA microarray is only semiquantitative (Hamadeh *et al.*, 2002). Also, it is not very sensitive towards individual changes in gene transcription and therefore is best used as a screening protocol (Fabian *et al.*, 2011). To complement the downside of microarray, quantitative polymerase chain reaction is also involved in toxicotranscriptomics studies by quantifying the levels of gene transcription related to heat shock, apoptosis, inflammation and acute toxicity and other related genes (Fabian *et al.*, 2011). The expression of organ specific or non-organ specific markers can be used as an indicator of cell damage (Fabian *et al.*, 2011). By comparing the gene expression profile of different tissues, these markers can indicate whether a compound has more than one target organ. With the information on DNA transcription currently available, it is possible to identify the mechanism of action based on the amount of knowledge on signalling pathways. For instance, toxicotranscriptomics analysis of a natural herbal toxin aristolochic acid (AAI) has provided information regarding molecular mechanism underlying pathogenesis of AAI-associated nephropathy and expression profiling of AAI-induced carcinogenicity (Gao *et al.*, 2006). Anderson *et al.* (2009) also applied the toxicogenomic approach for the investigation of hepatotoxicity induced by *Petasites hybridus* extract.

However, the genomics approach does not provide any details on mRNA stability and false positives due to expression of genes simply resulting from stress or changes in the environment could also occur. Another major downfall of toxicotranscriptomics is that changes in gene transcription may not reflect the corresponding protein levels or activities due to events such as phosphorylation or cleavage (Anderson and Seilhamer, 1997). Protein expression level has to be analysed by proteomics approach.

## B. TOXICOPROTEOMICS

Toxicokinetics, toxicometabonomics and toxicoproteomics provide additional and holistic information in systems toxicology to fill the gap between genomics and physical onset of symptoms resulting from exposure to toxins (Adourian *et al.*, 2008). Protein synthesis is the downstream process of gene expression. Changes in gene transcriptional level may not reflect corresponding protein levels or activity due to the involvement of series complex transcriptional and translational events such as RNA splicing, transcriptional silencing, phosphorylation, glycosylation and others. For instance, despite any changes in mRNA levels in response to exposure to a concoction, intracellular protein levels may not change, but there may be increased or decreased activation of key proteins instead, such as the mitogen activated protein kinase family. This is one of the advantages over toxicotranscriptomics as the addition or removal of compounds within a concoction allows for the study of protein activity and the role each compound has (Cho, 2007). This plays into looking at the greater picture of a concoction's mechanistic effect. Toxicoproteomics can also be used to reveal further downstream effects of mRNA changes, such as the expression of cellular surface proteins. Changes in cell surface markers can affect the level of response a cell displays to exogenous factors, and possibly one of the reasons why herbal medicine is able to benefit users suffering from chronic disease and to boost overall health. This complicated downstream process explains the infeasibility of protein characterisation solely based on gene expression analysis and highlights the importance of proteomics technology.

Traditional two-dimensional polyacrylamide gel electrophoresis (2D-PAGE) coupled with mass spectrometry (MS) for protein identification is still the key methodology in proteomics studies (Rabilloud, 2002). Due to limited application for large series of samples, low amounts of sample material and basic- and low-abundance proteins, a number of alternative methodologies have been developed such as matrix-assisted laser desorption and ionisation (MALDI) or surface-enhance laser desorption and ionisation (SELDI) associating with time of flight (TOF) or MS based on the relative mass of a protein and isoelectric point for enhancement of the traditional method (Oberemm *et al.*, 2005; Wang *et al.*, 2011a; Williams *et al.*, 2003). In toxicological study with regard to herbal medicine, the proteomic approach was used to identify protein alterations in response to toxic substance exposure. It helps to reveal or characterise the mode of action and disease process by monitoring proteins involved in the corresponding process (Oberemm *et al.*, 2005). One example is the determination of several apoptosis-related proteins alteration in HeLa cells for the investigation of cytotoxic mechanisms induced by tubeimoside-1 from Tu Bei Mu (Xu *et al.*, 2009). With the employment of 2D-PAGE and MALDI–TOF-MS analysis, dioscin, an extract from the root of *Polygonatum zanlanscianense* pamp, was shown to exhibit cytotoxicity towards human myeloblast leukaemia HL-60 cells (Wang *et al.*, 2006). The authors also suggested that mitochondria were the major cellular target of cytotoxicity involving multi-apoptosis pathways. These examples successfully illustrated the involvement of proteomics approach in toxicological studies of herbal medicine. There also has been ongoing development of protein microarrays (antibody arrays) (Gatzidou *et al.*, 2007). This protein microarray is the enzyme linked immunosorbent assay (Cutler, 2003). Due to its high protein specificity, these assays allow detailed investigation into the mechanism of action for toxic compounds.

The genomics and proteomics approaches only provide characterisation of transcriptional and translational alterations. Neither of these approaches could provide information on the dynamic metabolic status of a living system (Hamadeh *et al.*, 2002). Metabonomics approach has been included in toxicogenomics together with the other two approaches to reveal full profile of toxic substances.

### C. TOXICOMETABONOMICS

Pathological and physiological alterations caused by toxic agents could lead to changes in relative concentrations of endogenous biochemicals (Hamadeh *et al.*, 2002). Metabonomics refers to the measurement of metabolites present within tissue or fluid, which is related to physiological changes towards medication, environment or diseases in animals or humans (Ulrich-Merzenich *et al.*, 2007). It can be used as a real-time monitoring of physiological changes based on the body fluid composition differences in response to the exposure of toxic substances (Weiss and Kim, 2011). Nuclear magnetic resonance spectroscopy ($^{1}$H NMR) is the most widely applied analytical technology for metabolite identification and structural characterisation (Trock, 2011). For instance, hepatotoxicity of Huang-yao-zi was determined in a metabonomics study using $^{1}$H NMR for the quantification of biomarkers like creatine and dimethylglycine (Liu *et al.*, 2010). In addition to NMR, MS coupled with modern analytical platforms such as UPLC–MS and LC–MS have been increasingly applied. In a recent study of a herbal product, eight endogenous metabolites relating to nephrotoxicity induced by Morning Glory Seed have been characterised with UPLC–MS (Ma *et al.*, 2010). As toxins are metabolised and react to form specific biomarkers, toxicometabonomics approach allows rapid elucidation of toxic constituent serving as an early detection system of toxic response. DNA or protein adduct formation resulting from reaction between the toxic substances and biomolecules are the common biomarker of toxic effects. By determining the amount of adducts, it is possible to reveal the site of toxicity and severity of response. There are a number of metabonomic-based methods have been used in herbal toxicity investigation. For instance, coupling with animal study, perturbation of metabolic profile was noticed from biofluids analysis of rats treated with the main toxic substances in Fu Zi, aconitum alkaloid, by $^{1}$H NMR and GC/TOF-MS (Sun *et al.*, 2009). Pathological outcomes of nephrotoxicity induced by AAI were also revealed by Ni *et al.* (2007) using a combined GC–MS and LC–MS metabolic profiling approach.

Because of the broad-spectrum metabolic profiling generated by metabonomics, it is now recognised as a top–down approach integrating with other '-omic-' studies (Lindon *et al.*, 2007). However, although a number of information can be obtained from metabonomic studies such as target organ identification, pattern recognition and elucidation of mechanism, current chromatographic driven metabonomic approaches also encounter many difficulties such as unequivocal identification of metabolite candidates due to failure in full spectrum and quantitative analysis (Liu *et al.*, 2011). In addition, structure of biomarker can be changed during the isolation and purification processes leading to false positive results when comparing metabolite profiles.

## D. SYSTEMS BIOLOGY

Systems biology involves the integration of genomics, proteomics and metabonomics into one large database for comparison. As toxin exposure is not limited to one area of study, systems biology allows prediction of symptoms and organismal effects (Wetmore and Merrick, 2004). This is done by developing a library of data on the interactions of a compound at gene, protein and metabolite levels and comparing a compound's resulting profile to the database. Due to the specificity of DNA arrays, or MS/MS profiles, mechanistic prediction of a compound's interaction is easier by comparing the profile to the known information (Wetmore and Merrick, 2004). The development of MetaCore as a pathway analysis system applicable to microarray, proteomics and metabolomics will play a key role in the future of systems biology (Wang *et al.*, 2009). This has resulted from the emerging field of bioinformatics, as the creation of a large database accessible by researchers worldwide is now possible.

Systems biology is still an emerging field and has not been well defined due to its immense potential, far-reaching grasp and integrative properties. By identifying biomarkers via metabonomics, there can be a shift towards identifying the mechanistic pathway of action of herbal medicine concoctions. Wang *et al.* (2009) treated rats with YCHT, then identified one of the biomarkers as ceramide, thereby providing an area of future research on a mechanistic level. Another example of integration is the use of metabonomics and chromatographic fingerprinting to identify the bioactivity of Chenpi dependent upon storage time (Wang *et al.*, 2011b). Chemical fingerprinting can play a role, as different compounds may result in the same metabolite being produced. Another aspect of chemical fingerprinting is that different herbs may contain the same compounds, albeit in different amounts. Chemical analysis of *Oldenlandia diffusa* and *O. corymbosa* indicated that *O. diffusa* had higher levels of E-6-*O*-*p*-coumaroylscandoside methyl ester and *E*-6-*O*-*p*-coumaroylscandoside ester-10-methyl ether (Liang *et al.*, 2007). From this, chemical fingerprinting can help to identify if different herbs will have the same mechanism of action due to the presence of the same compound. Also, by using herbal medicine profiling and systems biology, there is a high probability of identifying a compound or a concoction which can have a similar mechanism of action as known pharmaceutical or displaying antagonistic properties (Li *et al.*, 2009). Systems biology can also reduce time spent investigating some compounds as toxic effects may be predicted early on via systems biology.

However, there is the possibility of false positives when using systems biology or a similar '-omic-' profile to another compound with a different effect upon cells or organisms. Thus, systems biology should be utilised as a guidance system for future experiments rather than as a be all and end all. Another benefit of using systems biology to identify mechanistic pathways is that it can act as a bridge between herbal medicine and Western medicine.

## IV. CONCLUSIONS

With the growing market of herbal medicine, the safety of herbal remedies have arisen public concern due to the lack of proper pharmacological and toxicological data. Increasing evidence showing the adverse effects related to herbal medicine further highlights the demand and necessity in toxicological evaluation. Recent advancement of technologies and molecular biology allow the analysis of complex reactions in response to multitude of herbal products at molecular level. This sophisticated '-omic-' technology is in a unique position to contribute to the prediction and preclinical toxicity testing of herbal medicine in future.

## REFERENCES

Adourian, A., Jennings, E., Balasubramanian, R., Hines, W. M., Damian, D., Plasterer, T. N., Clish, C. B., Stroobant, P., McBurney, R., Verheij, E. R., Bobeldijk, I., van der Greef, J. *et al.* (2008). Correlation network analysis for data integration and biomarker selection. *Molecular BioSystems* **4**, 249–259.

Ajasa, A. M. O., Bello, M. O., Ibrahim, A. O., Ogunwande, I. A. and Olawore, N. O. (2004). Heavy trace metals and macronutrients status in herbal plants of Nigeria. *Food Chemistry* **85**, 67–71.

Akerele, O. (1993). Nature's medicinal bounty: Don't throw it away. *World Health Forum* **14**, 390–395.

Anderson, L. and Seilhamer, J. (1997). A comparison of selected mRNA and protein abundances in human liver. *Electrophoresis* **18**, 533–537.

Anderson, N., Meier, T. and Borlak, J. (2009). Toxicogenomics applied to cultures of human hepatocytes enabled an identification of novel petasites hybridus extracts for the treatment of migraine with improved hepatobiliary safety. *Toxicological Sciences* **112**, 507–520.

Bandaranayake, W. M. (2006). Quality control, screening, toxicity, and regulation of herbal drugs. *In* Modern Phytomedicine, (I. Ahmad, F. Aqil and M. Owais, eds.), pp. 25–57. Wiley-VCH Verlag GmbH & Co. KGaA, Weinheim, Germany.

Barrett, J. C. and Kawasaki, E. S. (2003). Microarrays: The use of oligonucleotides and cDNA for the analysis of gene expression. *Drug Discovery Today* **8**, 134–141.

Bateman, J., Chapman, R. D. and Simpson, D. (1998). Possible toxicity of herbal remedies. *Scottish Medical Journal* **43**, 7–15.

Behpour, M., Ghoreishi, S. M., Khayatkashani, M. and Motaghedifard, M. (2011). A New Method for the Simultaneous Analysis of Strychnine and Brucine in

Strychnos nux-vomica Unprocessed and Processed Seeds Using a Carbon-paste Electrode Modified with Multi-walled Carbon Nanotubes. Phytochemical Analysis Advance online publication.
Bensoussan, A. and Myers, S. P. (1996). Towards a Safer Choice—The Practice of Traditional Chinese Medicine in Australia. Faculty of Health, University of Western Sydney Macarthur, Campbelltown, NSW.
Bugno, A., Almodovar, A. A. B., Pereira, T. C., Pinto, T.d.J. A. and Sabino, M. (2006). Occurrence of toxigenic fungi in herbal drugs. *Brazilian Journal of Microbiology* **37**, 47–51.
Chan, T. Y. (1994). The prevalence use and harmful potential of some Chinese herbal medicines in babies and children. *Veterinary and Human Toxicology* **36**, 238–240.
Chan, K. (2003). Some aspects of toxic contaminants in herbal medicines. *Chemosphere* **52**, 1361–1371.
Chan, T. Y. K., Chan, J. C. N., Tomlinson, B. and Critchley, J. (1993). Chinese herbal medicines revisited: A Hong Kong perspective. *Lancet* **342**, 1532–1534.
Cho, W. C. (2007). Application of proteomics in Chinese medicine research. *The American Journal of Chinese Medicine* **35**, 911–922.
Cutler, P. (2003). Protein arrays: The current state-of-the-art. *Proteomics* **3**, 3–18.
Das, P., Samantaray, S. and Rout, G. R. (1997). Studies on cadmium toxicity in plants: A review. *Environmental Pollution* **98**, 29–36.
De Smet, P. A. (1995). Health risks of herbal remedies. *Drug Safety: An International Journal of Medical Toxicology and Drug Experience* **13**, 81–93.
De Smet, P. A. (1997). Adverse effects of herbal remedies. *Adverse Drug Reaction Bulletin* **183**, 695–698.
De Smet, P. A. G. M. (2004). Health risks of herbal remedies: An update[ast]. *Clinical Pharmacology and Therapeutics* **76**, 1–17.
Drew, A. K. and Myers, S. P. (1997). Safety issues in herbal medicine: Implications for the health professions. *The Medical journal of Australia* **166**, 538–541.
Edwards, I. R. and Aronson, J. K. (2000). Adverse drug reactions: Definitions, diagnosis, and management. *Lancet* **356**, 1255–1259.
Edzard, E. M. D. (1998). Harmless Herbs? A Review of the Recent Literature. *The American Journal of Medicine* **104**, 170–178.
Ergil, K. V., Kramer, E. J. and Ng, A. T. (2002). Chinese herbal medicines. *The Western Journal of Medicine* **176**, 275–279.
Ernst, E. (2002a). The risk-benefit profile of commonly used herbal therapies: Ginkgo, St. John's Wort, Ginseng, Echinacea, Saw Palmetto, and Kava. *Annals of Internal Medicine* **136**, 42–53.
Ernst, E. (2002b). Toxic heavy metals and undeclared drugs in Asian herbal medicines. *Trends in Pharmacological Sciences* **23**, 136–139.
Ernst, E. and Thompson Coon, J. (2001). Heavy metals in traditional Chinese medicines: A systematic review. *Clinical Pharmacology and Therapeutics* **70**, 497–504.
Escher, M., Desmeules, J., Giostra, E. and Mentha, G. (2001). Hepatitis associated with Kava, a herbal remedy for anxiety. *British Medical Journal* **322**, 1097.
Fabian, G., Farago, N., Feher, L. Z., Nagy, L. I., Kulin, S., Kitajka, K., Bito, T., Tubak, V., Katona, R. L., Tiszlavicz, L. and Puskas, L. G. (2011). High-density real-time PCR-based in vivo toxicogenomic screen to predict organ-specific toxicity. *International Journal of Molecular Sciences* **12**, 6116–6134.
FAO (1991). Review: Food for the Future. *Food, Nutrition and Agriculture* **1**, 72.
Gao, S., Dickman, K., Dong, H., Miller, F., Zavadil, J. and Grollman, A. P. (2006). Toxicogenomic Analysis of Aristolochic Acid-Induced Epithelial Injury and Carcinogenesis. AACR Meeting Abstracts449-a.

Gatzidou, E. T., Zira, A. N. and Theocharis, S. E. (2007). Toxicogenomics: A pivotal piece in the puzzle of toxicological research. *Journal of Applied Toxicology* **27**, 302–309.

Gautam, A. K., Sharma, S. and Bhadauria, R. (2009). Detection of toxigenic fungi and mycotoxins in medicinally important powdered herbal drugs. *The Internet Journal of Microbiology* **7**.

Gertner, E., Marshall, P. S., Filandrinos, D., Potek, A. S. and Smith, T. M. (1995). Complications resulting from the use of Chinese herbal medications containing undeclared prescription drugs. *Arthritis and Rheumatism* **38**, 614–617.

Gray, S. L., Lackey, B. R., Tate, P. L., Riley, M. B. and Camper, N. D. (2004). Mycotoxins in root extracts of american and asian Ginseng Bind Estrogen receptors α and β. *Experimental Biology and Medicine* **229**, 560–568.

Hamadeh, H. K., Bushel, P., Paules, R. and Afshari, C. A. (2001). Discovery in toxicology: Mediation by gene expression array technology. *Journal of Biochemical and Molecular Toxicology* **15**, 231–242.

Hamadeh, H. K., Amin, R. P., Paules, R. S. and Afshari, C. A. (2002). An overview of toxicogenomics. *Current Issues in Molecular Biology* **4**, 45–56.

Huang, W., Wen, K. and Hsiao, M. (1997). Adulteration by synthetic therapeutic substances of traditional Chinese medicines in Taiwan. *Journal of Clinical Pharmacology* **37**, 344–350.

Humberston, C. L., Akhtar, J. and Krenzelok, E. P. (2003). Acute hepatitis induced by kava kava. *Journal of Toxicology, Clinical Toxicology* **41**, 109–113.

Huxtable, R. J. (1990). The harmful potential of herbal and other plant products. *Drug Safety: An International Journal of Medical Toxicology and Drug Experience* **5**(Suppl. 1), 126–136.

ICH (2008). ICH harmonized tripartite guideline: Guidance on genotoxicity testing and data interpretation for pharmaceuticals intended for human use, S2 (R1) Step 2 version 5.3. International Conference on Harmonisation of technical requirements for registration of pharmaceuticals for human use.

Izzo, A. A. and Ernst, E. (2001). Interactions between herbal medicines and prescribed drugs: A systematic review. *Drugs* **61**, 2163–2175.

Jha, V. and Rathi, M. (2008). Natural medicines causing acute kidney injury. *Seminars in Nephrology* **28**, 416–428.

Kanda, T., Yokosuka, O., Tada, M., Kurihara, T., Yoshida, S., Suzuki, Y., Nagao, K. and Saisho, H. (2003). N-nitroso-fenfluramine hepatotoxicity resembling chronic hepatitis. *Journal of Gastroenterology and Hepatology* **18**, 999–1000.

Li, P., Yang, L. P. and Gong, Y. W. (2009). Application of systems biology technology in research of traditional Chinese medicine. *Journal of Traditional Chinese Medicine* **29**, 153–157.

Liang, Z., Jiang, Z., Ho, H. and Zhao, Z. (2007). Comparative analysis of oldenlandia diffusa and its substitutes by high performance liquid chromatographic fingerprint and mass spectrometric analysis. *Planta Medica* **73**, 1502–1508.

Lin, S. H., Yang, S. S., Chau, T. and Halperin, M. L. (2003). An unusual cause of hypokalemic paralysis: Chronic licorice ingestion. *The American Journal of the Medical Sciences* **325**, 153–156.

Lindon, J. C., Holmes, E. and Nicholson, J. K. (2007). Metabonomics in pharmaceutical R & D. *FEBS Journal* **274**, 1140–1151.

Liu, H. (1998). A report of 2 cases of Nux Vomica poisoning. *GuangMing Journal of Traditional Chinese Medicine* **13**, 52.

Liu, Y., Huang, R., Liu, L., Peng, J., Xiao, B., Yang, J., Miao, Z. and Huang, H. (2010). Metabonomics study of urine from Sprague,ÄìDawley rats exposed to Huang-yao-zi using 1H NMR spectroscopy. *Journal of Pharmaceutical and Biomedical Analysis* **52**, 136–141.

Liu, P., Liu, S., Tian, D. and Wang, P. (2011). The applications and obstacles of metabonomics in traditional chinese medicine. *Evidence-Based Complementary and Alternative Medicine* **2012**, 945824.

Ma, C., Bi, K., Zhang, M., Su, D., Fan, X., Ji, W., Wang, C. and Chen, X. (2010). Metabonomic study of biochemical changes in the urine of morning glory seed treated rat. *Journal of Pharmaceutical and Biomedical Analysis* **53**, 559–566.

Martins, M. L., Martins, H. M. and Bernardo, F. (2001). Fumonisins B1 and B2 in black tea and medicinal plants. *Journal of Food Protection* **64**, 1268–1270.

Mullins, R. J. and Heddle, R. (2002). Adverse reactions associated with echinacea: The Australian experience. *Annals of Allergy, Asthma & Immunology: Official Publication of the American College of Allergy, Asthma, & Immunology* **88**, 42–51.

Ni, Y., Su, M., Qiu, Y., Chen, M., Liu, Y., Zhao, A. and Jia, W. (2007). Metabolic profiling using combined GC-MS and LC-MS provides a systems understanding of aristolochic acid-induced nephrotoxicity in rat. *Federation of European Biochemical Societies* **581**, 707–711.

Oberemm, A., Onyon, L. and Gundert-Remy, U. (2005). How can toxicogenomics inform risk assessment? *Toxicology and Applied Pharmacology* **207**, 592–598.

Omurtag, G. Z. and Yazicioglu, D. (2004). Determination of fumonisins B1 and B2 in herbal tea and medicinal plants in Turkey by high-performance liquid chromatography. *Journal of Food Protection* **67**, 1782–1786.

Rabilloud, T. (2002). Two-dimensional gel electrophoresis in proteomics: Old, old fashioned, but it still climbs up the mountains. *Proteomics* **2**, 3–10.

Rao, M. M., Kumarmeena, A. and Galib (2011). Detection of toxic heavy metals and pesticide residue in herbal plants which are commonly used in the herbal formulations. *Environmental Monitoring and Assessment* **181**, 267–271.

Robinson, M. M. and Zhang, X. (2011). Traditional Medicines: Global Situation. Issues and Challenges, The World Medicines Situation2011.

Sewram, V., Shephard, G. S., van der Merwe, L. and Jacobs, T. V. (2006). Mycotoxin contamination of dietary and medicinal wild plants in the Eastern Cape Province of South Africa. *Journal of Agricultural and Food Chemistry* **54**, 5688–5693.

Shad, K. A., Lajbar, k., Iqbal, H., Khan, B. M. and Akhtar, N. (2008). Profile of heavy metals in selected medicinal plants. *Pakistan Journal of Weed Science Research* **14**, 101–110.

Shaw, D. (2010). Toxicological risks of Chinese herbs. *Planta Medica* **76**, 2012–2018.

Slifman, N. R., Obermeyer, W. R., Aloi, B. K., Musser, S. M., Correll, W. A., Cichowicz, S. M., Betz, J. M. and Love, L. A. (1998). Contamination of botanical dietary supplements by digitalis lanata. *The New England Journal of Medicine* **339**, 806–811.

Smolinske, S. C. (2005). Herbal product contamination and toxicity. *Journal of Pharmacy Practice* **18**, 188–208.

Stickel, F., Baumüller, H.-M., Seitz, K., Vasilakis, D., Seitz, G., Seitz, H. K. and Schuppan, D. (2003). Hepatitis induced by Kava (Piper methysticum rhizoma). *Journal of Hepatology* **39**, 62–67.

Sun, B., Li, L., Wu, S., Zhang, Q., Li, H., Chen, H., Li, F., Dong, F. and Yan, X. (2009). Metabolomic analysis of biofluids from rats treated with Aconitum alkaloids using nuclear magnetic resonance and gas chromatography/time-of-flight mass spectrometry. *Analytical Biochemistry* **395**, 125–133.

Tassaneeyakul, W., Razzazi-Fazeli, E., Porasuphatana, S. and Bohm, J. (2004). Contamination of aflatoxins in herbal medicinal products in Thailand. *Mycopathologia* **158**, 239–244.

Trock, B. J. (2011). Application of metabolomics to prostate cancer. *Urologic Oncology* **29**, 572–581.

Ulrich-Merzenich, G., Zeitler, H., Jobst, D., Panek, D., Vetter, H. and Wagner, H. (2007). Application of the "-Omic-" technologies in phytomedicine. *Phytomedicine: International Journal of Phytotherapy and Phytopharmacology* **14**, 70–82.

Wang, Y., Cheung, Y. H., Yang, Z., Chiu, J.-F., Che, C.-M. and He, Q.-Y. (2006). Proteomic approach to study the cytotoxicity of dioscin (saponin). *Proteomics* **6**, 2422–2432.

Wang, C.-Y., Staniforth, V., Chiao, M.-T., Hou, C.-C., Wu, H.-M., Yeh, K.-C., Chen, C.-H., Hwang, P.-I., Wen, T.-N., Shyur, L.-F. and Yang, N.-S. (2008a). Genomics and proteomics of immune modulatory effects of a butanol fraction of echinacea purpurea in human dendritic cells. *BioMed Central Genomics* **9**, 1–20.

Wang, X., Sun, W., Sun, H., Lv, H., Wu, Z., Wang, P., Liu, L. and Cao, H. (2008b). Analysis of the constituents in the rat plasma after oral administration of Yin Chen Hao Tang by UPLC/Q-TOF-MS/MS. *Journal of Pharmaceutical and Biomedical Analysis* **46**, 477–490.

Wang, J., van der Heijden, R., Spruit, S., Hankermeier, T., Chan, K., van der Greef, J., Xu, G. and Wang, M. (2009). Quality and safety of Chinese herbal medicines guided by a systems biology perspective. *Journal of Ethnopharmacology* **126**, 31–41.

Wang, D.-Z., Dong, H.-P., Cheng, L., Xie, Z.-X., Lin, L. and Hong, H.-S. (2011a). Identification and Characterization of Cell Wall Proteins of a Toxic Dinoflagellate Alexandrium catenella using 2-D DIGE and MALDI TOF-TOF Mass Spectrometry. Evidence-Based Complementary and Alternative Medicine Advance online publication.

Wang, X., Sun, H., Zhang, A., Sun, W., Wang, P. and Wang, Z. (2011b). Potential role of metabolomics apporoaches in the area of traditional Chinese medicine: As pillars of the bridge between Chinese and Western medicine. *Journal of Pharmaceutical and Biomedical Analysis* **55**, 859–868.

Weiss, R. H. and Kim, K. (2011). Metabolomics in the Study of Kidney Diseases. Nature Reviews Nephrology Advance online publication.

Wetmore, B. A. and Merrick, B. A. (2004). Toxicoproteomics: Proteomics applied to toxicology and pathology. *Toxicologic Pathology* **32**, 619–642.

WHO (1998). Quality Control Methods for Medicinal Plant Materials. World Health Organization, Geneva, Switzerland.

WHO (2002). WHO Traditional Medicine Strategy 2002–2005. World Health Organization, Geneva, Switzerland.

Williams, E. A., Coxhead, J. M. and Mathers, J. C. (2003). Anti-cancer effects of butyrate: Use of micro-array technology to investigate mechanisms. *Proceedings of the Nutrition Society* **62**, 107–115.

Xu, Y., Chiu, J.-F., He, Q.-Y. and Chen, F. (2009). Tubeimoside-1 exerts cytotoxicity in HeLa cells through mitochondrial dysfunction and endoplasmic reticulum stress pathways. *Journal of Proteome Research* **8**, 1585–1593.

Yang, M.-H., Chen, J.-M. and Zhang, X.-H. (2005). Immunoaffinity column clean-up and liquid chromatography with post-column derivatization for analysis of aflatoxins in traditional Chinese medicine. *Chromatographia* **62**, 499–504.

Youns, M., Hoheisel, J. D. and Efferth, T. (2010). Toxicogenomics for the prediction of toxicity related to herbs from traditional Chinese medicine. *Planta Medica* **76**, 2019–2025.

Yue, Y.-T., Zhang, X.-F., Pan, J., Ou-Yang, Z., Wu, J. and Yang, M.-H. (2010). Determination of deoxynivalenol in medicinal herbs and related products by GC–ECD and confirmation by GC–MS. *Chromatographia* **71**, 533–538.

Zahir, F., Rizwi, S. J., Haq, S. K. and Khan, R. H. (2005). Low dose mercury toxicity and human health. *Environmental Toxicology and Pharmacology* **20**, 351–360.

Zeng, Z. P. and Jiang, J. G. (2010). Analysis of the adverse reactions induced by natural product-derived drugs. *British Journal of Pharmacology* **159**, 1374–1391.

Zhang, A., Sun, H., Wang, Z., Sun, W., Wang, P. and Wang, X. (2010). Metabolomics: Towards understanding traditional Chinese medicine. *Planta Medica* **76**, 2026–2035.

Zhao, Z. Z. and Li, Y. S. (2007). Easily Confused Chinese Medicines in Hong Kong. Chinese Medicine Merchants Association Ltd, Hong Kong.

# Development of Drugs from Plants: Regulation and Evaluation

MICHAEL RIEDER[*,†,‡,1] AND JOHN R. BEND[*,†,§]

*Department of Paediatrics, Division of Clinical Pharmacology, Schulich School of Medicine & Dentistry, University of Western Ontario, Children's Hospital, London, ON, Canada

†Department of Physiology & Pharmacology, Schulich School of Medicine & Dentistry, University of Western Ontario, London, ON, Canada

‡Department of Medicine, Schulich School of Medicine & Dentistry, University of Western Ontario, London, ON, Canada

§Department of Pathology, Schulich School of Medicine & Dentistry, University of Western Ontario, London, ON, Canada



[1]Corresponding author: E-mail: mrieder@uwo.ca

Advances in Botanical Research, Vol. 62

0065-2296/12 $35.00
DOI: 10.1016/B978-0-12-394591-4.00010-6

## ABSTRACT

The drug development process is a carefully regulated, very expensive, long and complex process designed to bring safe and effective drugs to the market. The process begins with preclinical studies that identify and evaluate promising compounds that are then studied in a series of carefully regulated steps to bring the products from the laboratory to the clinic. This process is closely regulated, even after approval is given for the drug to be marketed. The classical drug development process can be applied to drugs derived from plants, with the acknowledgement that there are both special opportunities and special challenges in using the drug regulatory process to bring drugs derived from plants to market. The drug development and regulatory process is facing a number of challenges, many of which are not unique to drugs derived from plants but rather reflect the changing nature of biomedical science and drug regulation. Some of these changes may in fact facilitate the development of plant-derived therapeutic agents, and recent successes in this area point to the considerable promise for development of drugs from plants.

## I. PLANTS AND HUMAN HEALTH

Plants serve a wide range of essential uses, including providing food, clothing and key elements of shelter. There is also a long tradition of medicinal use of plants, dating well before the dawn of organized medical care. Since antiquity, plants, parts of plants (flowers, roots, leaves and seeds) and preparations made from plants have also been used for the prevention and treatment of disease. This has on rare occasion translated into the development of specific targeted therapeutic products; for example, the use of digoxin for cardiac disease derived from the observation made by William Withering that folk remedies derived from the *Digitalis purpurea*—the Foxglove plant—seemed to improve the condition of patients with dropsy. In this case, Withering—an English physician and botanist—was asked to investigate why a folk healer in Shropshire was able to improve the condition of patients with dropsy—oedema usually related to congestive heart failure—who had not had relief using the best medical practices of the day. Withering used his botanical background to evaluate the approximately 20 herbs used in the mixture and concluded that only one—Foxglove—was likely to be biologically active. He subsequently demonstrated this in a series of what we would now call clinical trials to evaluate the most effective and safe dose. In 1785, he published his work in his landmark paper, *An Account of the Foxglove and Some of Its Medical Uses: With Practical Remarks on Dropsy and Other Diseases.* This paper is remarkable in its candour, in that Withering acknowledged that he had over-treated many patients to the point of toxicity. The honest acknowledgement of the potential hazards of developing drugs from plants forms an informed foundation upon which to build our understanding of how to develop drugs from plants and why the many steps involved are necessary to ensure that these drugs are both effective and safe.

Over the past several decades, increasing interest in folk or traditional medicine has highlighted the potential for the development of drugs from plants. This has also accompanied the 'flattening of the world' described by Thomas Friedman, in that globalization has opened new markets—and new opportunities—worldwide, including the desire by traditional medical practitioners to move more closely to the developed world medical mainstream and for patients to have traditional medications more readily available.

## II. THE HISTORY OF SPECIFIC THERAPY

While organized systems of medical care date back to Hammurabic Mesopotamia and Five Kingdoms China, specific therapy—in terms of safe and effective drugs—is much more recent. Drug therapy for most of history has been conducted using preparations of mixtures of compounds that would now be considered traditional or folk medicine, with no requirement for proof of efficacy or safety for compounds or mixtures used to treat patients. This changed radically in 1935, when—shortly after Fleming described the inhibition of *Streptococcal* growth by the products of *Penicillium*—Dömagk demonstrated the antimicrobial activity of Prontosil$^{TM}$, which is dependent upon inhibition of bacterial folic acid synthesis by the sulphanilamide component of this agent to inhibit bacterial growth which then provided the compound the specific ability to combat *Streptococcal* infection (Weinshilboum, 1987). The ability to provide effective antimicrobial therapy with a single pure chemical that could be readily synthesized was a major advance. A useful perspective is that in 1934 there were less than 10 drugs that could be described as effective and safe. The provision of safe, effective and relatively inexpensive drugs provided clinicians with an unprecedented enhancement in therapeutics in an era when effective drugs were few and infectious disease-related deaths were very common. This also led to the 'Therapeutic Revolution', the explosive increase in new synthetic drugs and therapeutic approaches that occurred after World War II (Weinshilboum, 1987). Nevertheless, therapeutic tragedies and adverse drug reactions forced a change in the process by which drugs were evaluated and approved in the United States, including the formation of the Food and Drug Administration (FDA; Bates *et al.*, 1995; Choonara and Rieder, 2002). The first of these episodes was the Elixir of Sulfanilamide Tragedy, in which more than 100 children died of renal toxicity when the antibiotic sulfanilamide was dissolved in the solvent ethylene glycol—a potent nephrotoxin—and marketed without any evaluation of safety. This tragedy—the proximate reason for the development of the American Food and Drug Adminstration—highlighted the need for new therapeutic products to be not only effective but also safe. These changes were quickly adopted, in more or less similar forms, by most other developed countries and formed the basis for the current system of drug development, evaluation for efficacy and safety and regulation (http://www.fda.gov/Drugs/default.htm).

## III. DRUG DEVELOPMENT AND THE DRUG APPROVAL PROCESS

The process by which drugs are developed is expensive, highly regulated and oriented to the goal of bringing effective new drugs that are safe to the market in a timely manner (Lipsky & Sharp, 2001). The current cost of bringing a new drug from concept to the market can take as long as 10 years and has been estimated to be as high as 1.8 billion American dollars, and this has created pressure to be innovative and use new approaches to streamline drug development (Paul *et al.*, 2010).

In the developed world, while there remain some differences between countries in the drug approval and regulatory process, these are now relatively minor as a result of the International Congress on Harmonization, which has had a global impact. A long series of meetings has resulted in the widespread adoption of policies and practices such as Good Manufacturing Practice and Good Clinical Practice that establish clear and comprehensive guidelines on how to develop, manufacture and evaluate novel therapeutic entities.

There are two different scenarios for approval of therapeutic agents obtained from plants, affecting the route by which regulatory approval is sought. Whereas some plant-derived drugs—such as quinine—have always been regulated via the conventional drug approval process, other herbal medications have become available using regulatory mechanisms that have labelled them as dietary products, including nutraceuticals. Approval by this route is typically associated with a less rigorous requirement to demonstrate efficacy and safety. While this is of advantage in facilitating market access, this typically does not provide the robust safety or efficacy data needed to guide evidence-based practice. This factor impacts on the willingness of many health care providers—notably younger providers, who have been trained in evidence-based systems—to use these treatments. This also impacts on the ability of patients who use these treatments to have them provided by private or public insurers, as reimbursement frequently is dependent on the status of

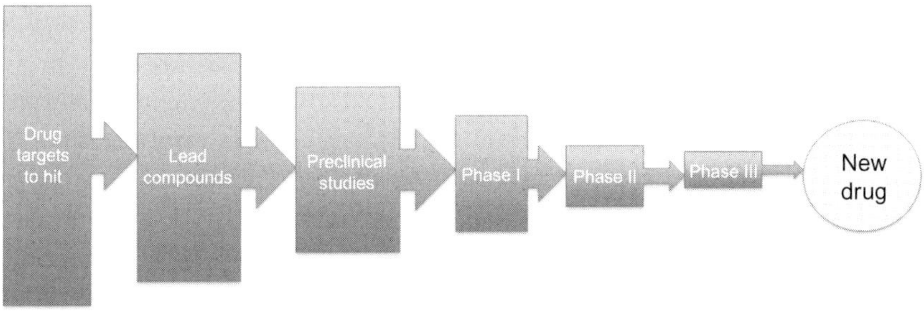


Fig. 1. An overview of the drug development process. The size of the bars represents the loss of compounds over the process. Phase IV (post-marketing surveillance) studies occur following approval of the drug. (For colour version of this figure, the reader is referred to the online version of this chapter.)

the treatment with respect to the drug regulatory authority. Thus, although it is costly, lengthy and onerous, there are clear benefits to having drugs derived from plants that undergo approval using the conventional drug approval process. In addition, the considerable interest in traditional medicine by the public—as demonstrated by hundreds of millions of dollars of sales—suggests that such products would have ready acceptance by the general public.

There are generally considered to be three major stages in the drug approval process: the first major stage is preclinical, the second is clinical evaluation and the third is post-approval surveillance (Fig. 1). During each stage and phase of drug development, the candidate therapeutic compounds are evaluated rigorously. Given the time and cost involved in drug development, go or no-go decisions are crucial in selecting which compounds are likely to result in a new therapeutic entity versus those for which there is a fatal flaw. There is a significant rate of attrition at each of the stages in pre-approval drug development, and even after approval some drugs are removed from the market. The stages of drug developments will be considered in turn.

## IV. PRECLINICAL STUDIES

The development of drugs begins with understanding the condition or disease of interest (Paul *et al.*, 2010). Drug development does not occur in a vacuum—the goal of the process is to bring a safe and effective drug to the clinic, and thus it is crucial for the goal to be clear at the start of the therapy development process. The admonitions of von Clausewitz as to selection and maintenance of objectives are nowhere more relevant than in the drug development process. Thus, groups who are developing novel therapeutic agents must have a complete and comprehensive understanding, insofar as this is possible, about the disease or health condition of interest. As an example, if the goal was to develop a new drug to treat asthma, interventions could target airway inflammation, mucous production or bronchoconstriction. When selecting where intervention might occur, a key consideration is how to intervene. Thus, in addition to identifying a potential intervention, it is important to also consider potential therapeutic targets that modulate these processes (Swinney and Anthony, 2011). At this point, it should be noted that there have been, simplistically put, two different strategic approaches to new drug development (Swinney and Anthony, 2011). The first—which historically has been the most common—has been phenotypic screening, in which screening batteries have been developed to assess the impact of potential candidate molecules on the phenotype of interest, which often has occurred despite imperfect knowledge of the precise mechanisms driving the disorder in question. A more recent approach has been to use precise molecular or receptor targets based on a mechanistic understanding of the disease of interest. Despite the obvious attractive nature of the latter, it has been suggested that the increasing use of this approach has in fact been associated with a lower success rate for new therapeutic agents than was phenotypic screening (Fiordellis *et al.*, 2006). Returning to our example of asthma, potential targets to intervene in processes such as bronchoconstriction might include ß receptors or targets involving airway inflammation might include targets such as platelet-activating factor. The decision as to which strategy to pursue is an important one, and if the disease or health condition of interest is of complex aetiology—especially if it is known or likely that there several mechanistic possibilities that produce the condition of interest—the phenotypic screening approach may be preferable.

Understanding the nature of the potential target(s) then allows the identification of candidate molecules. To return to the classical example cited earlier in the development of digoxin, one of the key steps was that Withering—who was a trained botanist in addition to being a physician—recognized that, of the 20 plants that were used in the folk remedy he studied, only one—*D. purpurea*—was likely to be biologically relevant. This was of key importance, in that it then permitted Withering to focus his considerable intelligence and energy on determining the dose most likely to be effective while at the same time least likely to produce harm using a single plant. Had he not been able to focus his studies, it is unlikely that studies using all 20 plants would have succeeded in bringing digoxin forward as a new therapeutic entity.

The traditional approach has been to evaluate large libraries of compounds using an appropriate screening technique—often an *in vitro* assay—to detect compounds of interest, typically by impacting on one of the already identified relevant drug target(s) identified in the initial considerations outlined above. While this approach has historically been effective, there is an extremely high attrition rate which, when coupled with the costs of clinical evaluation, presents a major problem in the timely and cost-effective delivery of new therapeutic entities (Garnier, 2008). Thus, there is a constant search for innovation in the identification and characterization of new target molecules. As an example, screens are increasingly guided by targeted searches based on structure–function relationships, which can be undertaken in conjunction with *in silico* modelling to determine which compounds might best impact the identified therapeutic target to best produce the desired response (Chin *et al.*, 2004; Mishra, 2011). In the case of drugs derived from plants that have been used medicinally, this can often be guided to some extent by the traditional uses of products prepared from these plants. In this case, the number of potential compounds may be smaller and the search for promising candidate molecules may be more focused. Conversely, using approaches that have traditionally used combinations of plants may be more complex as there is a much wider range of combinations possible. Many remedies used in traditional Chinese medicine, for example, are prepared using a four-herb model. The potential that this is necessary is clear from classical pharmacology—as an example, the efficacy of co-trimoxazole is based on the synergy provided by two compounds which inhibit different portions of the bacterial folic acid synthetic pathway. A more recent example is the success that highly active antiretroviral therapy enjoys with respect to controlling HIV replication—control achieved by targeting multiple steps in viral pathogenesis and replication. While this may appear daunting, in fact this can and has been accomplished, a recent example being the work conducted by Dr. YC Cheng and his team at Yale in developing adjunctive therapy for the treatment of cancer based on a four-herb preparation used in traditional Chinese medicine (Lam *et al.*, 2010).

An additional step that is needed, however, is the purification of relevant compounds and their complete chemical characterization. It is also very important to have a reliable and adequate supply of the compounds of interest for the next series of studies required. This is of special relevance in the case of drugs derived from plants, in that there can be considerable batch to batch variability when the starting ingredients of compounds are grown rather than synthesized chemically or created with recombinant technology. This also has clear and crucial relevance in the totality of the drug development strategy, in that ultimately there must be sufficient and reproducible quantities of compound prepared according to Good Manufacturing Practices if the drug is to be commercially relevant. The importance of ensuring a robust and reliable supply of raw material cannot be overemphasized, notably given the variation that can occur in plant products based on differences in species and in agricultural practice. Recent advances in technology such as DNA fingerprinting and the development of repositories for plants used in traditional medicine offer considerable promise in addressing this issue.

In addition, while other studies are being conducted throughout the drug development process, there is an ongoing need for formulation research, the ultimate goal being to produce a dosage form that is readily usable in the clinical environment. As the vast majority of drugs are given orally, the drug in most cases must be able to survive the transition through the gastrointestinal tract in order to be delivered in therapeutically relevant concentrations to the small intestine, where the majority of drug absorption occurs. The drug delivery system—capsule or tablet—must be sufficiently small to be swallowed. The potential impact of excipients—the non-drug components of the capsule or tablet, which for orally administered drugs will necessarily be present in much larger quantities than the drug—must also be considered. In addition, the formulation must be stable through a range of temperatures and other environmental conditions.

Special attention must be given to drugs that are intended to be administered in a controlled or delayed release format. This is of relevance for useful drugs that have a relatively short half-life. In many cases, rather than reject a potentially therapeutically compound—especially when there is a potentially significant benefit or the compound has been shown to be very active—a useful strategy may be to develop a dosage form that releases the medication slowly over time rather than as a single bolus in the gastrointestinal tract, as is the case with conventional release formulations. The classical approach has been to use a matrix formulation derived from a relatively insoluble substance such as wax, which dissolves slowly and thus releases the drug over time. There have been a number of innovations over the past two decades in the creation of novel approaches to slow or controlled release, including the use of osmotic pumps to slowly release medication from capsules with precision-drilled drug egress points. The latter approach has been used with great success in the therapy of children with attention deficit disorder.

A further consideration is that a liquid formulation may be necessary for drugs for which children are a target population. As well, there are reasonable numbers of adults—especially among the rapidly growing numbers of seniors in most of the world—who find taking medication in solid form difficult. This can be due to neurological problems or difficulty with swallowing. There also are a number of healthy normal adults who for a variety of reasons find it very difficult to swallow solid dosage forms of medication.

When developing a liquid dosage form, issues such as temperature stability and palatability are even more important than with tablet or capsule formulations (Rieder, 2010). It is increasingly clear that palatability is in fact a major influence on adherence with therapy. Strategies to address palatability issues must also evaluate the potential impact of these strategies on the stability of the drug formulations developed. Another approach is to use alternate drug delivery systems. There is a large and emerging science in novel delivery systems to address this issue, including the use of microparticles, melt formulations and gels. Many of these products offer promise not only for children but also for adults who have difficulties in taking current solid dosage forms.

When a series of candidate molecules have been identified, further screening is needed to decide which compounds should move forward to further studies. The biological activity of the compounds needs to be assessed both *in vitro* and *in vivo* in light of therapeutically relevant concentrations. Using our earlier example, a compound being developed for therapy of asthma that exerts potent anti-inflammatory effects at concentrations from 1 to 5 M is not a likely candidate for further development as a drug, as the administration of a drug which requires such high concentrations in the blood or lung to have an effect is likely to have significant undesired effects when administered to experimental animals or patients. The potential impact of the novel therapeutic agent on other or related targets—especially if they are not relevant to the disease being treated—must also be assessed. It may be possible to attenuate undesired effects with chemical modification of the original molecule. Examples are the derivatives of oleic acid, a compound derived from traditional Chinese medicine, which are being developed to treat renal disease in the setting of diabetes. In this case, a series of chemical modifications have been made to the parent compound to enhance desired biological activities. When suitable candidate molecule(s) with suitable potency have been identified, a series of additional investigations must be undertaken.

Detailed studies are needed to determine the fate of the compound(s) of interest in biological systems. Some of these studies can be done *in vitro* using cultured cells or a variety of cell-free models. These studies can include modelling of drug metabolism, for example, using microsomal or recombinant *in vitro* drug metabolizing systems. However, ultimately it is necessary to conduct pharmacokinetic studies of the drug in animal models. In addition to studying the fate of the compound itself—typically using the conventional rubric of absorption, distribution, metabolism and excretion—additional consideration is normally given to drug metabolites, notably if they are active or if they could impact on other metabolic processes. Again returning to our example of drugs for asthma, a potentially useful compound may not be considered further if, as an example, it was a potent irreversible inhibitor of CYP3A4, the isozyme responsible for the metabolic elimination of approximately 50% of the drugs in common use for humans. Similarly, if a drug was primarily cleared via metabolism mediated by CYP2D6, this could be a major issue in moving forward. CYP2D6 is polymorphic, with at least 80 different single-nucleotide polymorphisms described to date. Phenotypically CYP2D6 is considered to exist in at least three forms. The most common form is the extensive metabolizer phenotype, the frequency of which is approximately 90% in most Northern European populations. The poor metabolizer phenotype has markedly less enzyme activity, related to the homozygous presence of non-functional alleles, present in approximately 9% of Northern European populations. The ultrarapid metabolizer phenotype, in contrast, shows markedly increased enzyme activity related to the presence of extra copies of functional alleles, present in roughly 1% of Northern European populations. This is especially relevant as the distribution of the ultrarapid metabolizer phenotype varies widely in different regions, with a 10% frequency in the Mediterranean littoral and a frequency as high as 20% in the Middle East and in the Horn of Africa (Sistonen *et al.*, 2007). This has had significant implications for drug safety for drugs such as codeine, a prodrug metabolized by CYP2D6 to morphine which accounts for much of its therapeutic efficacy. Finally, if a drug is metabolized to a toxic metabolite, this may preclude further drug development, notably as reactive drug metabolites have been implicated in a number of serious drug toxicities (Rieder, 2009).

In addition to studies of drug disposition and clearance, an extensive series of Safety Pharmacology—the umbrella term used to describe toxicity and adverse drug reaction studies—must be conducted (Parasuraman, 2011). The core battery for Safety Pharmacology studies includes evaluation of drug effects on the cardiovascular, central nervous, reproductive and respiratory systems plus supplemental studies as relevant with respect to the genitourinary system, the gastrointestinal system, the autonomic nervous system and other relevant organ systems (Table I). These studies include *in vitro* toxicology studies which are used as a screen and which can be adapted to evaluate issues such as the impact of the compound(s) of interest on relevant pathways of drug metabolism (Table II). In addition to studies of formulation and metabolism, *in vitro* studies are typically also used to evaluate potential genotoxicity as an economical prelude to lifetime carcinogenicity studies. As well, further mechanistic studies may be undertaken, notably if novel toxicities are identified. These studies can often be done reasonably quickly. However, while these studies are both important and

TABLE I
*Safety Batteries in Drug Development*

| Core safety battery | Supplemental safety battery |
|---|---|
| Central nervous system | Gastrointestinal system |
| Cardiovascular system | Autonomic nervous system |
| Respiratory system | Renal/urinary system |
| | Other organ systems |

TABLE II
*Typical* In Vitro *Studies for Safety Pharmacology*

| Types of study |
|---|
| Basic chemistry |
| Formulation studies |
| Genotoxicity studies |
| Metabolism studies |
| Mechanism studies |

provide key insights, the use of any *in vitro* system is limited in that *in vivo* biology can be very difficult to extrapolate from *in vitro* work. As well, *in vitro* work can provide false conclusions as to both safety and efficacy that can only be proved or refuted in living systems. Thus, it is crucial at this point in drug development to move *in vivo.*

The core of preclinical Safety Pharmacology studies consists of a series of *in vivo* investigations in experimental animals (Table III). A series of studies are conducted, with work such as long-term studies and reproductive studies often conducted in parallel. The short-term studies are designed to provide key data on pharmacokinetics and drug metabolism that can be used to guide dosing for the next series of studies. That being said, it is important to remember that the metabolic pathways of relevance in rodents—a typical species used for short-term studies—may not be germane to the metabolic pathways used by humans. Long term or repeated dose studies are designed to help understand drug kinetics and safety over time. The caveat expressed above with respect to application of this work to humans continues to apply. In addition, it must be acknowledged that, even with the use of two species, the populations being studied are very homogenous. One laboratory rat or rabbit is very likely to be much more similar to the other than are any two patients. As well, the extremes of age—the very young and the very old—are typically not studied, as given the duration of the studies and the life spans of most species used, this work is essentially always conducted in adult animals.

TABLE III
*Animal Studies for Safety Pharmacology*

| Type of Study | Characteristics |
|---|---|
| Short-term studies | – Pharmacokinetic studies<br>– Conducted in two species (adult animals)<br>– Toxicity evaluated after a single dose<br>– Usually evaluated via two routes of administration |
| Long-term studies | – Repeated dose studies<br>– Conducted in two species (adult animals); often rat, non-rat (e.g. rabbit)<br>– Use placebo and at least three doses<br>– Duration 4–52 weeks |
| Reproductive studies | – Usually done using rats or rabbits<br>– Dose range used, often with very high doses<br>– Fertility and embryofoetal development studied<br>– Post-natal development can also be studied |
| Carcinogenicity studies | – Typically done using rodents<br>– Start age 6 weeks, duration 104 weeks |
| Sensitivity studies | – Conducted with compounds likely to produce allergic responses<br>– Typically use the guinea pig model |

Carcinogenicity studies are designed to detect pro-carcinogens, and many human pro-carcinogens can in fact be detected in animal studies. However, these studies are known to have both a false-negative and false-positive rate. The goal of reproductive studies is to detect human teratogens, and all human teratogens have produced effects in animal models. That being said, the number of drugs that are human teratogens is very small, and the detection of a teratogenic signal in an animal model requires both expertise and a certain amount of luck. As an example, while all human teratogens have been shown to be teratogenic in animal models, some well-known human teratogens do not produce teratogenicity in certain animal models. As well, teratogenic effects often occur in a specific exposure window during pregnancy, and thus timing of drug dosing becomes crucial.

Sensitivity studies are conducted for compounds that are likely allergens and are intended to identify those compounds likely to elicit an allergic response during clinical trials. However, the most commonly used model—the guinea pig—is far from ideal. Guinea pigs are extraordinarily sensitive to allergens, and there is a very poor correlation between allergic events that are common in the guinea pig model versus those that are very common in humans. There are several alternate systems—such as local lymph node response—that are being evaluated. In addition, adverse immune reactions that are directed against a drug metabolite may not be seen if the parent drug alone is used for sensitivity testing, notably if production of a reactive metabolite occurs via metabolic activation systems that differ in the guinea pig from those in humans.

In order to better correlate with results in humans, there has been work undertaken to evaluate potential therapeutic compounds using transgenic cells or animals. These are cells or animals that are genetically modified to express human proteins. While these systems can be extremely useful for selected studies, there are often unknowns—notably with transgenic animals—as to what else may or may not have been expressed and, in the case of complex systems, if all the signalling or activation pathways are expressed and functionally related as would be the case in humans. There is considerable ongoing research to better understand these systems, as they offer tremendous potential for better being able to extrapolate animal work to the human condition.

One key consideration that is very difficult to evaluate in any of the animal studies described above is the issue of confounders. Succinctly put, experimental animals do not smoke, work with chemicals or take other drugs—legal or recreational. Patients are tremendously complex, and it is not possible to adequately evaluate the multiple factors that may alter—sometimes in a very significant way—the efficacy and safety of drugs in patients.

The goal of preclinical safety pharmacology studies is to identify those compounds that are likely to cause toxicity or adverse drug reactions when given to humans. It should be emphasized that these studies identify compounds likely to cause adverse events, and that absolute risk reduction (i.e. finding a compound that will never produce an adverse effect in a patient) is neither realistic nor attainable. That being said, human risk assessment from animal studies is very conservative, such that the standard approach to assign a human NOEL (no observable effects level) or NOAEL (no observable adverse effects level) is to take the NOEL found in animal studies and divide by a safety factory, typically 10 to account for species differences (i.e. the human NOEL is one-tenth the experimental animal NOEL). The completion of the entire core Safety Battery—including the repeated dose and long-term toxicity animal studies—can take from 2 to 4 years and costs more than $2 million.

While these studies are being conducted, data is collected and analysed on a regular basis. A number—indeed, often the majority—of compounds will not 'make the grade' and proceed to clinical testing, and a crucial part of the drug development process is having systems and expertise to make the hard decisions as to which compounds are likely to be winners (in the sense of producing a safe and effective therapeutic agent) and which are not (Goodwin *et al.*, 2011).

## V. CLINICAL STUDIES

When the preclinical studies have been completed and analysed, the next step is to seek regulatory approval for testing of the compound(s) of interest in people. To obtain this approval, the complete collected data must be submitted to a national drug regulatory agency and reviewed. Experts at the agency—for example, FDA in the United States—will review the data and direct a series of questions arising from the evaluation of the submitted data to the group seeking approval (typically a pharmaceutical company). Only after all questions and concerns have been addressed to the satisfaction of the reviewers at the drug regulatory agency will issue approval to designate the compound(s) as an Investigational New Drug (IND) be provided. This appellation means precisely what is implied, that is, that this is a new drug being investigated for safety and efficacy in patients based on extensive testing of the pure compound for efficacy and safety in experimental animals.

Having an assigned IND then permits investigators to proceed to test the approved compound(s) in patients. There are four stages of assessment of new drugs in humans, designated Phases I–IV. Of these, Phases I–III are conducted prior to approval of the drug for general use, while Phase IV involves post-marketing surveillance studies (Orloff *et al.*, 2009). The premarketing studies have three primary objectives, these being to determine if the pharmacological effects seen in preclinical studies translate into pharmacological effects in patients, to determine if the drug is sufficiently safe for clinical use and to determine if the drug is likely to be useful therapeutically.

Phase I studies or First in Man studies occur when the drug is first given to people. Phase I studies are rigorous and carefully scrutinized investigations that require both special expertise and dedicated facilities. For these studies, the drug is administered—often in a range of doses—to healthy volunteers who are then closely monitored over a given period of time, during which they are fed standardized meals and have blood and urine taken on a regular basis. Phase I studies are non-blinded and non-therapeutic; they are often conducted in healthy volunteers—frequently young adult men—and are designed to evaluate drug safety as well as to provide key data on the pharmacokinetics of the drug in humans. The exception to this involves chemotherapeutic drugs, where Phase I studies are often done in patients with cancer, as the inherent toxicity of most drugs developed for chemotherapy precludes testing in healthy volunteers. It should be noted that most Phase I studies involve relatively small numbers of subjects, rarely more than 100 and often less. Interestingly, Phase I studies share a common issue with preclinical animal studies, in that the potential impact of confounders are essentially not studied during Phase I studies as the participants are typically selected to be as free of confounding variables as possible.

The data from Phase I studies are then analysed carefully to provide insights into the pharmacokinetics of the drug in humans as well as to detect predictable toxicities. At this point, hard decisions must be made as to which compounds can proceed to Phase II testing (Goodwin *et al.*, 2011). There is a reasonably high attrition rate of compounds during Phase I studies, with as many as 50–75% of the chemicals tested being rejected due to unfavourable pharmacokinetics or toxicity profiles. Using our earlier example, if a very promising anti-inflammatory drug under developed for treatment of asthma had to be taken six times a day due to a very short half-life in humans, this would probably be a fatal flaw that would preclude further development of this specific compound unless the drug formulation could be suitably modified to permit the same efficacy while allowing for a longer dosing interval. This points to the need, noted earlier, for ongoing formulation research during the drug development process.

Phase II studies represent the first set of evaluations to investigate the potential therapeutic efficacy of the compound(s) of interest. These studies are conducted with patients who have the disease for which the drug is being developed to treat. These studies are typically fairly small—often being conducted in 200 or less patients—and use patients who are often defined by very rigid inclusion and exclusion criteria. It is at this point that the potential impact of confounders—such as other drugs—begins to be assessed. Patients in Phase II studies are very closely monitored. Clinical trial design is often single blinded and may use either concurrent controls or may compare patients to their baseline status. Drug dosing is often done by dose escalation, as the goal of Phase II studies is to better understand how the drug works in patients so that dose regimens can be developed for the more definitive Phase III studies to follow. In addition to providing that key insights on dosing, Phase II trials can also serve to evaluate useful therapeutic end points and help to determine which patient populations might obtain the greatest benefit from the drug under study.

The data derived from Phase II studies are again closely scrutinized to make the hard decision as to which drugs should proceed to Phase III trials. This is a critical decision because Phase III trials are typically much larger—and much more costly—than Phase I and II studies. The attrition rate of compounds in Phase II studies averages 50%.

Phase III represents the final stage of pre-marketing evaluation, and often the largest investment. Phase III studies are the definitive series of investigations that compare the drug being evaluated with the usual or 'gold standard' therapy for the disease or condition of interest. These studies typically are conducted in settings similar to what is expected to be usual use of the drug in clinical practice and use robust clinical trial strategies including double-blind randomization. Thus, in Phase III studies, the potential effect(s) of confounders can be assessed and evaluated. The use of placebo in a Phase III study can only be justified if there is no conventional therapy. Given the large number of active therapeutic agents available, most Phase III studies are active drug controlled. The only exceptions for this are conditions for which there is no known therapy, which is increasingly less common. A cross-over design can be used for drugs intended for chronic disorders, in which subjects are randomized to either the 'gold standard' drug or the drug under study and after a suitable period of evaluation are switched to the other arm of the study. As these are active drug-controlled studies—often using drugs as active controls that are very effective—sample size can be very large, and it is expected for Phase III studies to require thousands of patients. In order to achieve timely completion of these studies, this means that Phase III studies are also often multi-centre, which increases cost and complexity.

The cost, complexity and time requirement for Phase III studies have presented an urgent need for innovation. A number of approaches are being studied to reduce the time and cost of these studies. One approach is to reduce the variability in the population being studied. As an example, if a disorder is known to have variable severity or phenotypes that occur on a genetic basis, or if the therapeutic response occurs mechanistically via pathways under well-identified genetic control, one approach to develop a more homogenous study population is to screen subjects prior to enrolment, such that a more genetically—and presumably phenotypically—homogenous population is used for the Phase III study. This has obvious attractive features in that the intervention being assessed may be more effective or safer—or perhaps both—in a pre-selected population, which could sharply reduce sample size and the time and cost of the study, especially as the cost of genetic testing is a fraction of what it has been and that this cost is likely to continue to decline. However, this creates a problem that will be manifested after drug approval, when the drug enters general use. While it is possible that the labelling for the drug recommends genetic testing prior to use, there are few regulations currently in place in most jurisdictions—beyond limiting reimbursement—that will mandate this. Thus, it is possible—and indeed highly likely—that the drug will be used among populations and patient groups very different from the subjects for whom the drug was approved. A second approach is to challenge the conventional dogma of the randomized blinded clinical trial with the use of novel clinical trial designs (Orloff *et al.*, 2009). These include strategies such as play the winner or other designs that use the results of the study conducted to date to inform how future subjects are randomized and treated (Goodwin *et al.*, 2011). Other approaches include extensive pre-study modelling and carefully designed Proof of Concept studies, which again offer the potential for finding differences with much smaller sample sizes; this approach has been used with success in finding novel therapies for rare genetic disorders (Lachmann *et al.*, 2009).

At the conclusion of Phase III studies, the data should support the case that the drug being developed is safe and efficacious. The data—and this means all the data, including the preclinical studies—can now be submitted to the drug regulatory agency as a New Drug Submission seeking regulatory approval for marketing of the drug as an approved therapeutic entity. The submission process involves submitting many volumes of data, which are then carefully reviewed. The process involves considerable dialogue between the company seeking approval and the drug regulatory agency, which only when successfully completed, results in the granting of a Drug Identification Number (DIN) and the approval to market the drug in the respective jurisdiction. This can be a lengthy process, and it would not be unusual for the approval process to take 1–2 years after the submission of the complete data set.

## VI. POST-MARKETING STUDIES

The drug can now be marketed in the regions in which it has been approved for the approved indications, that is, to treat or prevent the disease(s) for which regulatory approval has been granted. At this point, Phase IV studies begin. Part of the granting of regulatory approval is the requirement that the company that markets the drug will conduct post-marketing surveillance. That is to say, the company is mandated to evaluate the safety of the drug once it enters general use. That being said, there are few agencies that have mechanisms to ensure compliance, and the enthusiasm for and efficacy of post-marketing surveillance varies widely.

The reasons that post-marketing surveillance is needed are several. First, while the studies done in Phases I—III will detect common toxicities and adverse drug reactions, less common adverse drug reactions may only appear when the drug enters general use. An example is rofecoxib (Vioxx$^{TM}$), a selective inhibitor of cyclooxygenase-2 developed for the therapy of arthritis as a drug that would be safer with respect to the risk of gastrointestinal bleeding than conventional non-steroidal anti-inflammatory drugs. One of the pivotal Phase III studies was the VIGOR trial, a well constructed and expertly designed trial involving more than 8000 patients in 22 countries to assess the safety and efficacy of rofecoxib versus naproxen among patients with rheumatoid arthritis (Bombardier *et al.*, 2000). This study clearly demonstrated fewer adverse gastrointestinal adverse events among the patients randomized to receive rofecoxib than in the naproxen arm. However, there were more adverse reactions characterized as thrombotic event producing myocardial ischaemia in the rofecoxib arm than in the naproxen arm, the difference in myocardial infarction rates in the naproxen group versus rofecoxib being a difference between 0.1% and 0.4%. This was initially ascribed to a protective effect in the naproxen group. However, when rofecoxib entered the market, it rapidly became one of the most frequently prescribed drugs, and it was appreciated that there was indeed a population of patients for whom selective inhibition of cyclooxygenase-2 led to thrombosis on the basis of unopposed cyclooxygenase-1 effects. When this was confirmed as part of the safety analysis of the APPROVe trial—a study of the use of rofecoxib in prevention of intestinal polyps—rofecoxib was withdrawn from the market (Weir *et al.*, 2003). It has been estimated that as many as 80,000 patients worldwide may have had adverse cardiac events related to rofecoxib use prior to the withdrawal of the drug from the market (Juni *et al.*, 2004).

A second reason that post-marketing surveillance is important is indication creep. While the drug development process is conventionally viewed in a linear fashion (Fig. 1), in fact the life cycle of a drug may be more accurately viewed as circular (Fig. 2). As noted above, drugs enter the market for a specific indication, this being the foundation for the preclinical and clinical studies that supported the drug approval. However, once drugs enter the market, clinicians can prescribe them for a number of indications. As experience with the drug accumulates, the drug may be used for new indications—sometimes quite different from the indications for which the drug was approved. Indeed, for some populations—such as children and pregnant women—such 'off-label' or 'off-indication' prescribing is the norm, not the exception (Choonara, 2007; Rieder, 2010). The use of the drug for off-label indications can lead to unexpected adverse drug reactions, as the disease being treated and concurrent conditions—such as the use of other medications—are often quite different from the original condition evaluated for the original drug approval. As an example, lamotrigine was approved for marketing as an anticonvulsant. However, it has subsequently been discovered that lamotrigene is effective for a number of other conditions, including mood disorders. This has now become one of the most common indications for which lamotrigene is prescribed. As lamotrigene has been increasingly used in younger patients, it has also been found that age—specifically being less than age 13—is a major independent risk factor for the development of

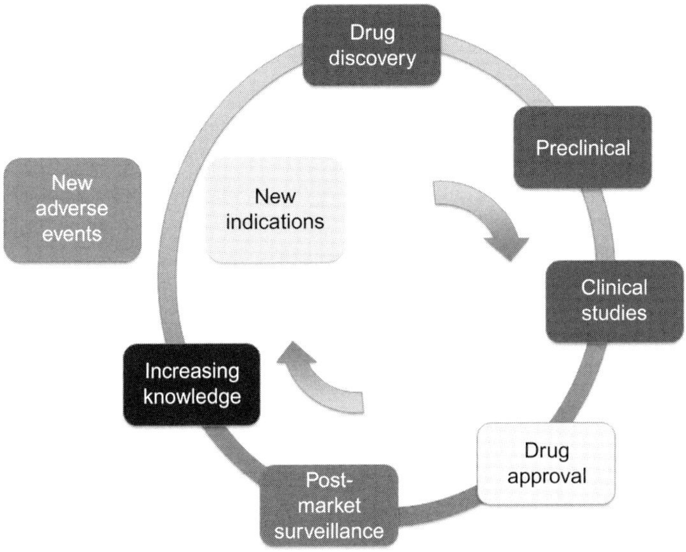


Fig. 2. The life cycle of a drug. After drugs that enter the market through the drug approval process, there is a dynamic balance of new knowledge leading to new indications—and new uses for the drug—and new toxicities—and new adverse drug reactions. Thus, the process of drug development, evaluation, approval and use—which conventionally is thought of as linear—is in fact more appropriately viewed as circular (concept adapted from Dr. David Lee, Health Canada). (For colour version of this figure, the reader is referred to the online version of this chapter.)

serious drug-related rash. Even when the same adverse events profile occurs, the rate can be markedly different in different age groups or genders. As an example, the rate of serum sickness like reactions to the cephalosporin antibiotic cefaclor is 1% among toddlers, which is 100-fold more common than among adults (Kearns *et al.*, 1998). Similarly, the rate of serious hepatic injury due to valproic acid is 1:5000 in adults and 1:500 among toddlers. There differences may be related to the significant differences in metabolic activation capacity on a per-kilogram basis for toddlers compared to adults. As noted above, many of the clinical studies done pre-marketing are conducted largely among adult men, and there is often a lack of data with respect to effects in women—and typically a total paucity of effects during pregnancy, as pregnancy is often an explicit exclusion criteria for pre-marketing clinical studies. Given that women appear to have a higher risk for adverse drug reactions than men, the relative lack of women in preclinical studies is a crucial information gap that can only be filled in, at the present time, by post-marketing surveillance.

In addition to indication creep and age and gender issues, there are population considerations. Many drugs are tested in fairly homogenous populations, and as genetic testing is increasingly used to guide subject selection for clinical trials, this is likely to be increasingly the case. However, in the real world, drugs are used in patients who have a wide anthropological and genetic background, and unexpected toxicity based on population-specific genetics has produced therapeutic tragedies when toxicity that did not occur in the initially studied population occurs. An example is post-operative death from codeine toxicity, to which patients from the Mediterranean littoral and the Horn of Africa are markedly more susceptible than are the Northern European populations for which codeine was originally evaluated related, in great part due to the polymorphisms in codeine metabolism described above (Madadi *et al.*, 2010). This is highly germane with respect to drugs derived from plants, as many traditional herbal medicines have been used—and often used for centuries—in fairly uniform populations. Bringing drugs derived from these sources to the global market considerably widens the diversity of patients being treated.

As well, the issue of effectiveness becomes important following marketing of the drug. The pre-marketing clinical studies are conducted with rigour and care. Once the drug enters the market, drug use in the clinical setting is typically less well monitored, and there are numerous examples of drugs that have been proven to be efficacious in clinical trials that have been less effective in clinical practice than might be anticipated.

## VII. CHALLENGES

The drug development process is tightly linked to the drug approval and regulatory process, and there is considerable pressure on all elements of the system. To begin with the drug regulatory system, there is increasing pressure for transparency and enhanced drug safety. Society, generally, is much more risk-averse than has been the case historically, and this has translated into increased public and political pressure on drug regulatory agencies to ensure that both new and old drugs are safe and effective. While it is entirely reasonable to expect that the drug approval process evaluates drug safety, it is entirely unreasonable to expect that drugs will be free of adverse events. Simply put, potent drugs are likely to have potent adverse drug reactions. The minimal rate of adverse drug reactions for most drugs is approximately 5%, and while there is considerable work underway to better characterize the rate and risks for adverse drug reactions, it is likely in the short term that this will not decrease. What is reasonable is that the rate of and risk factors for common and important adverse drug reactions should be appreciated and understood and that this information should be part of the information shared with patients when therapy is being prescribed and dispensed.

At the same time that there is pressure for increased drug safety, there is also pressure for faster drug approval. As noted above, the cost and time required for drug development has become a major concern for industry. The current time taken to bring a drug to market—typically 7–10 years—presents a challenge to industry but also to patients. Patient advocacy groups, notably for diseases for which current therapies are unavailable or suboptimal, have appropriately called for more timely approval of therapies.

This raises an additional major challenge, which may be the most significant one, which is the productivity gap (Booth and Zemmel, 2004). Simply put, there is a wide and increasing gap between the amount of resources being put into drug development and the number of compounds emerging as new therapeutic entities (Paul *et al.*, 2010). The number of new molecular entities being approved has declined over the past decade, and while the number of truly novel first-in-class has remained constant (at about five per year), the number of drugs likely to generate revenue versus the amount invested in new drug development has fallen (Paul *et al.*, 2010). This is the result of a series of separate developments that have come together to produce fundamental challenges to the way in which drugs have been developed (Paul *et al.*, 2010). One issue is noted above, that being drug safety. There is an increasing concern by the public—which has been accompanied by rising demands on drug regulators—for drug safety. Society, in general, is much more risk-aversive than has been the case in the previous century, which at the extreme includes advocating for the unachievable goal of zero risk. While this impacts all industrial sectors, this is a particular challenge for the pharmaceutical industry, given that as noted above most commonly used drugs have adverse drug reaction rates of 5%, with rates increasing in the case of more complex patients and more complex therapeutic regimens. Safety concerns have resulted in drugs being withdrawn from the market—sometimes appropriately and sometimes not—and in increased cost for safety pharmacology studies.

The cost of basic drug discovery is also a factor. As biomedical science has developed, the cost of new technology and new approaches to target identification and new compound evaluation has increased. In the case of screening strategies using targeted approaches, as noted above, this has not always resulted in an increased stream of new therapeutic entities (Swinney and Anthony, 2011).

The cost of clinical trails is a major concern. It could be argued that the pharmaceutical industry is a victim of its own success. The Therapeutic

Revolution has resulted in a large number of potent and effective drugs, and the 'blockbuster' discoveries of drugs that would be used by large numbers of patients globally are less and less common. In the case of some disease classes for which there are numerous safe and effective drugs, this has two implications. First, finding a clinically relevant difference will become harder and harder, given that Phase III trials will now have to use an active control. As well, these trials will necessarily be both larger and more complex, adding substantially to the cost of drug development. Finally, if a new therapeutic entity is developed, it will have to compete in an increasingly cost- and effectiveness-driven environment with other drugs, including many drugs whose patents have expired and are available as generics. The decision by many innovative pharmaceutical companies to shift their research direction to new therapeutic areas is in no small part due to concerns about the potential market placement of new therapeutic entities.

The current situation is unsustainable and as a consequence both industry and drug regulators are working to develop innovative solutions. In addition to streamlining and enhancing the drug development process, consideration is being given to changing the drug regulatory process, in part in response to the need for more timely drug approval and in part in recognition of the actual life cycle of a drug (Fig. 2). These changes may include earlier licensure of promising compounds with the caveat that this earlier licensure is accompanied by rigorous and careful evaluation of the effectiveness and safety of the drug in 'real world' conditions. It will be important to follow the changes that industry and drug regulatory agencies make in response to the looming crisis in drug development.

How does this apply to drugs derived from plants? Given that many of these drugs will be used as adjunctive therapy, many of these considerations apply, notably with respect to drugs used for complex diseases for which there are therapies currently available. Investigators or groups seeking to develop drugs from plants would be well advised to given careful thought as to how innovation can be applied to address the time and cost of drug development.

## VIII. CONCLUSIONS

Plants have been and are likely to continue to be an important source of new therapeutic agents, including semi-synthetic derivatives designed to improve the pharmacokinetic profile of the compound being developed. The process by which drugs are developed, approved and enter the market is complex, expensive and daunting, but provides important assurances to safety and efficacy that are essential for these agents to enter mainstream clinical practice. The recent work by Dr. YC Cheng's team at Yale in demonstrating that a four-herb mixture derived from traditional Chinese medicine is effective in reducing gastrointestinal side effects of chemotherapy is an example of how medicines derived from plants can begin the move to the medical mainstream (Lam *et al.*, 2010). This and other recent examples of success in developing new drug entities from herbal medicines suggest that there is great potential for further development of drugs from plants. There are very real challenges including time and financial pressures on the current drug development process, and innovation and reform are urgently needed to ensure that there is a regular flow of new therapeutic products. Some of these reforms may make the process of developing drugs from plants both more swift and straightforward (Orloff *et al.*, 2009; Swinney and Anthony, 2011).

## REFERENCES

Bates, D. W., Cullen, D. J., Laird, N. *et al.* (1995). Incidence of adverse drug events and potential adverse drug events. Implications for prevention. *Journal of the American Medical Association* **274**, 29–34.

Bombardier, C., Laine, L., Reicin, A., Shapiro, D., Burgos-Vargas, R., Davis, B., Day, R., Ferraz, M. B., Hawkey, C. J., Hochberg, M. C., Kvien, T. K., Schnitzer, T. J. *et al.* (2000). Comparison of upper gastrointestinal toxicity of rofecoxib and naproxen in patients with rheumatoid arthritis. VIGOR Study Group. *The New England Journal of Medicine* **343**, 1520–1528.

Booth, B. and Zemmel, R. (2004). Prospects for productivity. *Nature Reviews Drug Discovery* **3**, 451–456.

Chin, D. N., Chuqaui, C. E. and Singh, J. (2004). Integration of virtual screening into the drug discovery process. *Mini Reviews in Medicinal Chemistry* **4**, 1053–1065.

Choonara, I. (2007). Regulation of drugs for children in Europe. *British Medical Journal* **335**, 1221–1222.

Choonara, I. and Rieder, M. J. (2002). Drug toxicity and adverse drug reactions in children—A brief historical review. *Paediatric and Perinatal Drug Therapy* **5**, 12–18.

Fiordellis, C. S., Manolis, A. S., Paris, H. and Karabinis, A. (2006). Rethinking target discovery in polygenic diseases. *Current Topics in Medicinal Chemistry* **6**, 1791–1798.

Garnier, J. (2008). Rebuilding the R & D engine in big pharma. *Harvard Business Review* **86**, 68–76.

Goodwin, R., Giaccone, G., Calvert, H., Lobbezoo, M. and Eisenhauer, E. A. (2011). Targeted agents: How to select the winners in preclinical and early clinical studies. *European Journal of Cancer* **48**, 170–178.

Juni, P., Nartey, L., Reichenbach, S., Sterchi, R., Dieppe, P. A. and Egger, M. (2004). Risk of cardiovascular events and rofecoxib: Cumulative meta-analysis. *Lancet* **364**, 2021–2029.

Kearns, G. L., Wheeler, J. G., Rieder, M. J. and Reid, J. (1998). Serum sickness-like reaction to cefaclor: Lack of in vitro cross-reactivity with loracarbef. *Clinical Pharmacology and Therapeutics* **63**, 686–693.

Lachmann, H. J. *et al.* (2009). Use of canakinumab in the cyropyrin-associated periodic syndrome. *The New England Journal of Medicine* **360**, 2416–2425.

Lam, W., Busson, S., Guan, F., Jiang, Z., Zhang, W., Gullen, E. A., Liu, S. H. and Cheng, Y. C. (2010). The four-herb Chinese medicine PHY906 reduces chemotherapy-induced gastrointestinal injury. *Science Translational Medicine* **18**, 45ra59.

Lipsky, M. S. and Sharp, L. K. (2001). From idea to market: The drug approval process. *Journal of the American Board of Family Medicine* **14**, 362–367.

Madadi, P., Hildebrandt, D., Gong, I. Y., Schwarz, U. I., Ciszkowski, C., Ross, C. J., Sistonen, J., Carleton, B. C., Hayden, M. R., Lauwers, A. E. and Koren, G. (2010). Fatal hydrocodone overdose in a child: Pharmacogenomics and drug interactions. *Pediatrics* **126**, e986–e989.

Mishra, N. K. (2011). Computational modeling of P450s for toxicity prediction. *Expert Opinion on Drug Metabolism and Toxicology* **7**, 1211–1231.

Orloff, J., Douglas, F., Pinheiro, J., Levison, S., Branson, M., Chaturvedi, P., Ette, E., Gallo, P., Hirsch, G., Mehta, C., Patel, N., Sabir, S. *et al.* (2009). The future of drug development: Advancing clinical trial design. *Nature Reviews Drug Discovery* **8**, 949–957.

Parasuraman, S. (2011). Toxicological screening. *Journal of Pharmacology and Pharmacotherapeutics* **2**, 74–79.

Paul, S. M., Mytelka, D. S., Dunwiddie, C. T., Persinger, C. C., Munos, B. H., Lindborg, S. R. and Schacht, A. L. (2010). How to improve R&D productivity: The pharmaceutical industry's grand challenge. *Nature Reviews Drug Discovery* **9**, 203–214.

Rieder, M. J. (2009). Immune mediation of hypersensitivity adverse drug reactions: Implications for therapy. *Expert Opinion on Drug Safety* **8**, 331–343.

Rieder, M. J. (2010). If children ruled the pharmaceutical industry: The need for pediatric formulations. *Drug News & Perspectives* **23**(7), 458–464.

Sistonen, J., Sajantila, A., Lao, O., Corander, J., Barbjuani, G. and Fuselli, S. (2007). CYP2D6 worldwide genetic variation shows high frequency of altered activity variants and no continental structure. *Pharmacogenetics and Genomics* **17**, 95–101.

Swinney, D. C. and Anthony, J. (2011). How were new medicines discovered? *Nature Reviews Drug Discovery* **10**, 507–519.

Weinshilboum, R. M. (1987). The therapeutic revolution. *Clinical Pharmacology and Therapeutics* **42**, 481–484.

Weir, M. R., Rhoda, S., Sperling, R. S., Reicin, A. and Gertz, B. J. (2003). Selective COX-2 inhibition and cardiovascular effects: A review of the rofecoxib development program. *American Heart Journal* **146**, 591–604.

# AUTHOR INDEX

Note: Page numbers followed by "*f*" indicate figures, and "*t*" indicate tables.

**G**



**H**

**I**



**J**

**K**



**L**

# SUBJECT INDEX

Note: Page numbers followed by "*f*" indicate figures, and "*t*" indicate tables.

## S



## T

**U**



**V**



**W**



**X**



**Z**

Paul Pui-Hay But *et al.*, Fig. 1. Morphological characters shared by *Stemona*. (A) connective extension, (B) sterile apical appendage of anther, (C) anther, (D) filament, (E) articulate pedicel in *S. japonica*, and (F) seeds with white elaiosome.

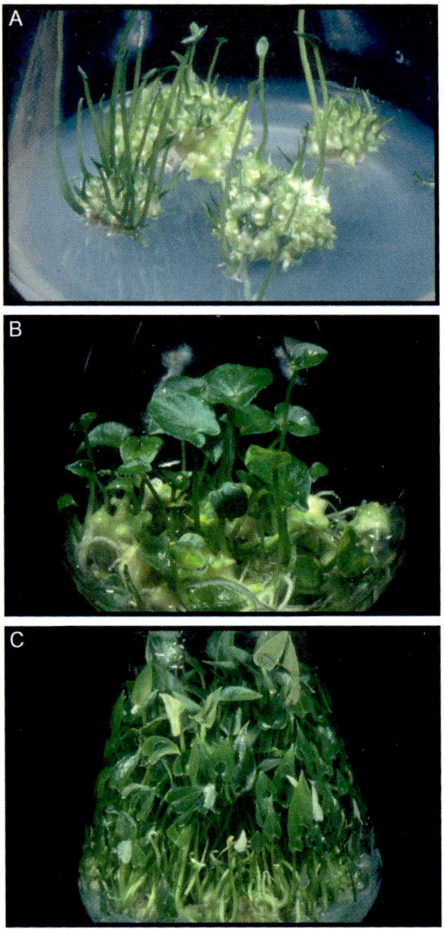


Sushim Kumar Gupta *et al.*, Fig. 1. (A) Multiple shoots induction in protocorm-like bodies in *P. ternata* (2 weeks after culture in MS basal agar medium). (B) Shoot regeneration in protocorm-like bodies in *Pinellia ternata* (4 weeks after culture in MS basal liquid medium). (C) Shoot elongation in *P. ternata* (4 weeks after culture in MS basal, agar medium).

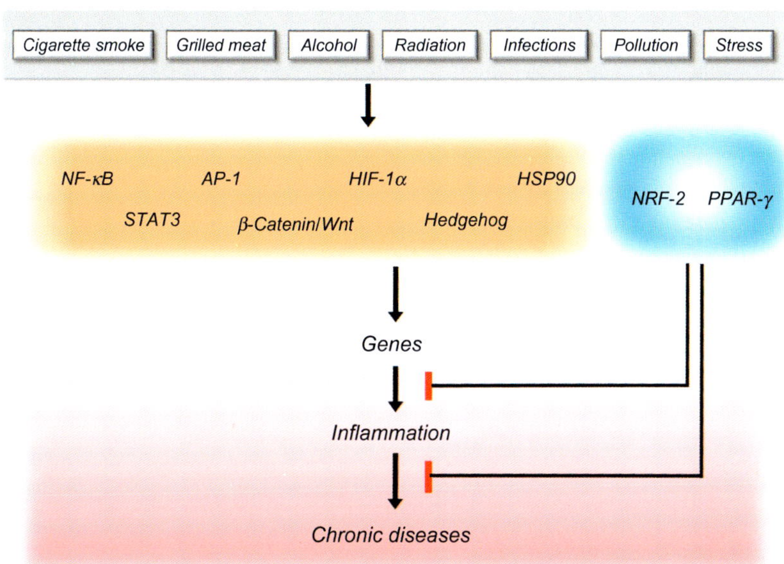


Bokyung Sung *et al.*, Fig. 1. List of transcription factors known to modulate chronic diseases.

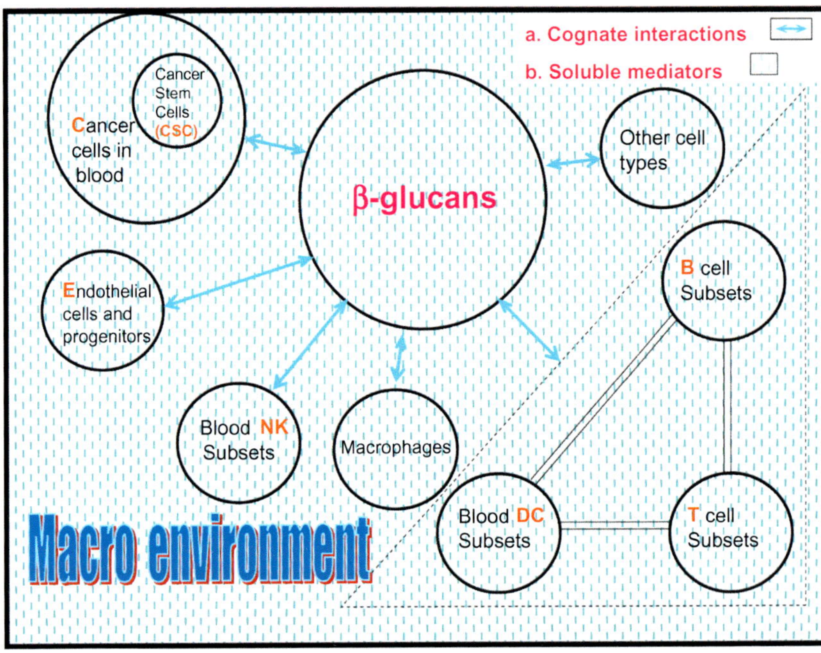


Daniel Man-yuen Sze and Godfrey Chi-Fung Chan, Fig. 1. The effects of β-glucans on the macroenvironment of cancer. A schematic diagram representing the cancer cell macroenvironment. β-glucans have been shown to exhibit multi-directional immunomodulatory activities with many different cells types. These activities can be summarized as either cognate interactions that involve many different membrane surface molecules or secreting cytokines and chemokines as soluble mediators in blood and cellular fluids.

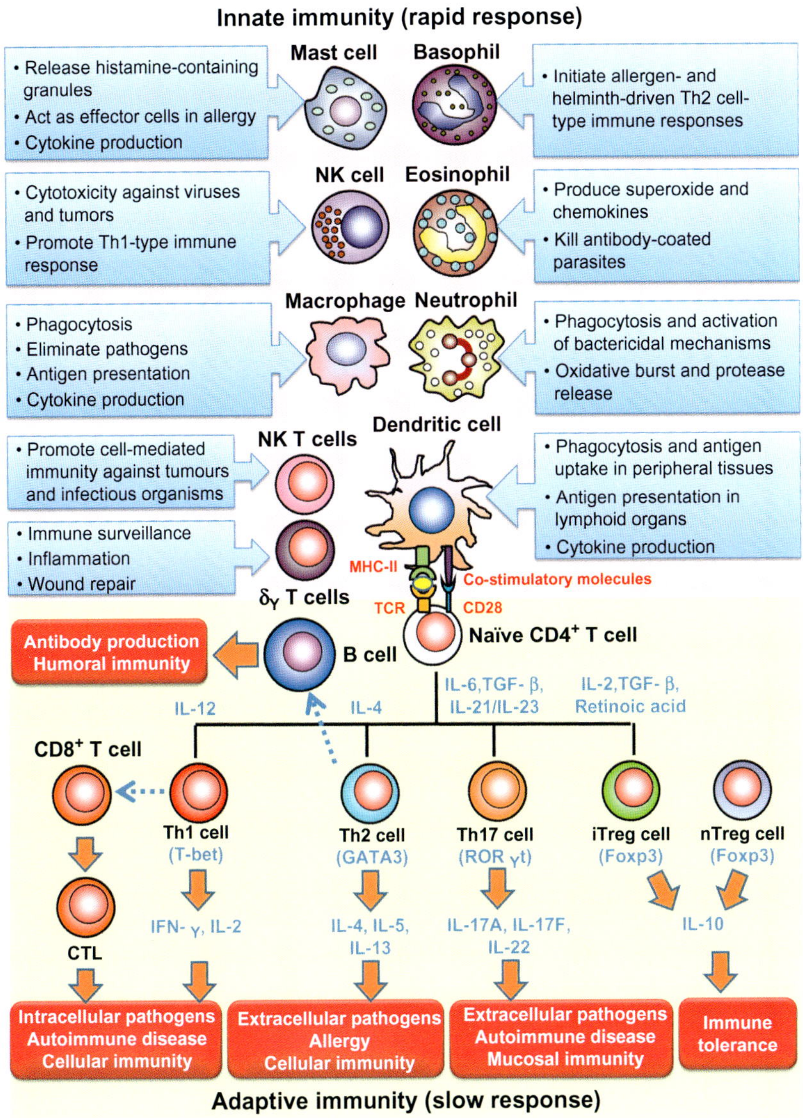


Chih-Chun Wen *et al.*, Fig. 2. Characteristics and functions of various innate and adaptive immune cells in the immune system. The immune system can be divided into innate immunity and adaptive immunity. The innate immune system involves the participation of dendritic cells, macrophages, mast cells, granulocytes (neutrophils, eosinophils and basophils), NK cells, NKT cells, δγ T cells and others. The key functions of each cell type are described in the blue grid. The adaptive immune system involves $CD4^+$ T cells, $CD8^+$ T cells, B cells and others. $CD4^+$ T cells can differentiate into Th1, Th2, Th17 and inducible Treg (iTreg) cells under different microenvironments specialized by interactive cytokines and chemokines, and distinct activation of specific transcription factors. $CD8^+$ T cells are responsible for confirming cytotoxicity against virus-infected cells or tumour cells. Treg cells are generally grouped into two classes, iTreg cells and natural Treg (nTreg) cells. They can regulate specific immune responses, especially immune tolerance, to maintain immune homeostasis. Abbreviations: Th1, T-helper type 1; Th2, T-helper type 2; Th17, T-helper type 17; Treg, T regulatory cells; NK, natural killer cells; NK T cells, natural killer T cells; MHC-II, major histocompatibility complex class II; TCR, T cell receptor; IFN-γ, interferon-gamma; TGF-β, transforming growth factor-beta; IL-2, interleukin-2; IL-4, interleukin-4; IL-6, interleukin-6; IL-12, interleukin-12; IL-17, interleukin-17; IL-21, interleukin-21; IL-23, interleukin-23.

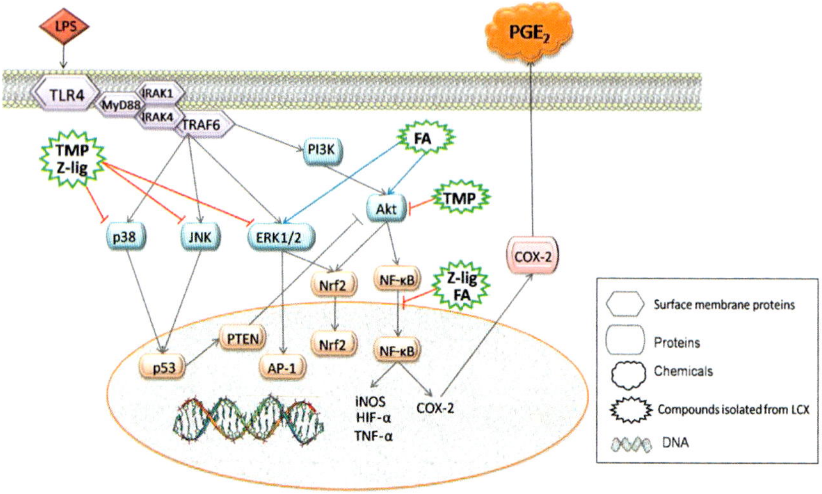


Cindy L.H. Yang *et al.*, Fig. 2. Summary of the effects of tetramethylpyrazine (TMP), Z-ligustilide (Z-lig) and ferulic acid (FA) on the signalling pathways involved in inflammatory responses of the cell.

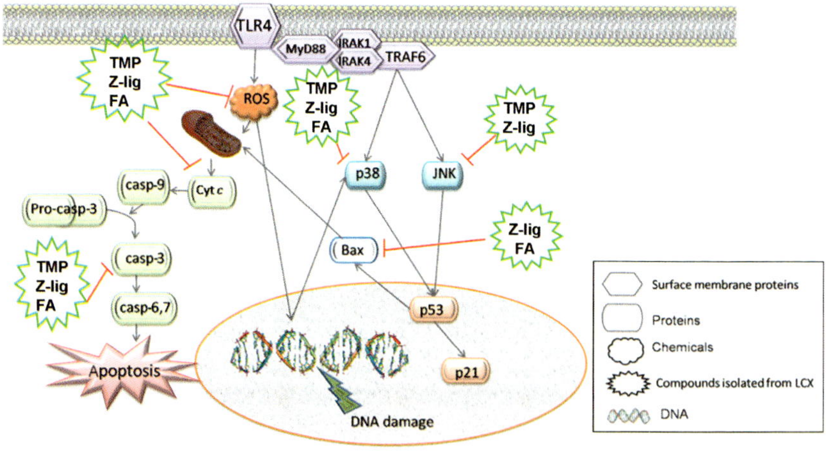


Cindy L.H. Yang *et al.*, Fig. 3. Summary of the effects of tetramethylpyrazine (TMP), Z-ligustilide (Z-lig) and ferulic acid (FA) on the signalling pathways involved in oxidative stress-induced injuries.